IUTAM Symposium on Vibration Control of Nonlinear Mechanisms and Structures

SOLID MECHANICS AND ITS APPLICATIONS
Volume 130

Series Editor: G.M.L. GLADWELL
Department of Civil Engineering
University of Waterloo
Waterloo, Ontario, Canada N2L 3GI

Aims and Scope of the Series

The fundamental questions arising in mechanics are: *Why?*, *How?*, and *How much?* The aim of this series is to provide lucid accounts written by authoritative researchers giving vision and insight in answering these questions on the subject of mechanics as it relates to solids.

The scope of the series covers the entire spectrum of solid mechanics. Thus it includes the foundation of mechanics; variational formulations; computational mechanics; statics, kinematics and dynamics of rigid and elastic bodies: vibrations of solids and structures; dynamical systems and chaos; the theories of elasticity, plasticity and viscoelasticity; composite materials; rods, beams, shells and membranes; structural control and stability; soils, rocks and geomechanics; fracture; tribology; experimental mechanics; biomechanics and machine design.

The median level of presentation is the first year graduate student. Some texts are monographs defining the current state of the field; others are accessible to final year undergraduates; but essentially the emphasis is on readability and clarity.

For a list of related mechanics titles, see final pages.

IUTAM Symposium on

Vibration Control of Nonlinear Mechanisms and Structures

Proceedings of the IUTAM Symposium held in Munich, Germany, 18–22 July 2005

Edited by

H. ULBRICH

Technical University Munich, Garching, Germany

and

W. GÜNTHNER

Technical University Munich, Garching, Germany

Springer

A C.I.P. Catalogue record for this book is available from the Library of Congress.

ISBN-10 1-4020-4160-8 (HB)
ISBN-13 978-1-4020-4160-0 (HB)
ISBN-10 1-4020-4161-6 (e-book)
ISBN-13 978-1-4020-4161-7 (e-book)

Published by Springer,
P.O. Box 17, 3300 AA Dordrecht, The Netherlands.

www.springeronline.com

Printed on acid-free paper

Printed in the Netherlands.

TABLE OF CONTENTS

DEDICATION

Contributions of the IUTAM Symposium on Vibration Control of Nonlinear Mechanisms and Structures are dedicated to Professor Friedrich Pfeiffer on the occasion of his seventieth birthday, which he celebrated on 22 February 2005.

Friedrich Pfeiffer originated from Wiesbaden, where he was awarded his university-entrance diploma from the Realgymnasium in 1955 as the best of his class. Afterwards, he studied mechanical engineering at the Technische Hochschule Darmstadt from 1955 to 1961, supported by a scholarship from the "Studienstiftung des deutschen Volkes". From 1961 to 1965, he was a research assistant at the TH Darmstadt Institute for Aeronautics, under the supervision of Professor Günther Bock. He received his Dr.-Ing. summa cum laude in 1965 based on his doctoral thesis "Abwindkorrekturen für Flügel beliebiger Pfeilung in offenen und geschlossenen kreisrunden Windkanälen mit Bodenplatte".

This was the foundation of his storybook career in industry. In the year 1966 he started as a development engineer at the department of aerospace of Bölkow GmbH in Ottobrunn, where he soon became project manager, then department manager, head of the department, and finally a member of the company management. There he was responsible for research and development, supervising about 1000 employees and with annual sales of approximately 100 million Euro.

During his 16 years at the company, Professor Pfeiffer was always at the top of the ongoing current research in the field of the dynamics of rockets and satellites. When he was appointed to a professorship at the newly founded "Bundeswehrhochschule Hamburg" in 1973 he declined the offer. But almost 10 years later he was eventually attracted to university life and became a full professor for mechanics at the "Technische Universität München – Lehrstuhl B für Mechanik", where he was the successor of Professor Magnus.

Professor Pfeiffer has educated many thousands of students in the basic principles of mechanics and in the field of multi-body dynamics of robots and walking robots. He was very popular among students, who respected him as an excellent teacher. And as an outstanding researcher, it was easy for him to inspire his research and teaching assistants, of which more than 80 did their Ph.D. under his supervision.

More than 200 papers in international journals and five books demonstrate Professor Pfeiffer's exceptional talents in research and teaching, especially his remarkable ability to put complex theory into engineering practice. He focused his scientific research mainly on dynamics and control of large mechanical systems with friction and clearance but also always kept an eye on the mechanical fundamentals and worked on different problems from industry. The name Pfeiffer is primarily linked to groundbreaking work about non-smooth dynamics of multi-body systems with unilateral constraints and biological oriented walking robots, for which he was awarded the "Körber Preis". Professor Pfeiffer is editor and co-editor of many internationally renowned journals in the area of non-linear dynamics and robotics. For many years he was active on different committees, among those the board of "Studienstiftung des deutschen Volkes", member of the senate of the DFG (German Research Foundation), dean of the Faculty of Mechanical Engineering, a member of the senate of the "Technische Universität München" and, since 2000, president of the Society of Applied Mathematics and Mechanics ("Gesellschaft für Angewandte Mathematik und Mechanik, GAMM"). He has received numerous honours of which only a few are listed here: honorary doctorate (Dr. h. c. and Dr.-Ing. E. h.) from the universities of Moscow and Dresden, "Bundesverdienstkreuz am Bande", and the appointment as an IEEE fellow and ASME fellow.

Students, colleagues, and fellow scientists wish him first of all health, happiness, and vitality. We hope that he continues being active and giving advice in the years to come.

Munich, July 2005 Heinz Ulbrich

PREFACE

During the last decades, the growth of micro-electronics has reduced the cost of computing power to a level acceptable to industry and has made possible sophisticated control strategies suitable for many applications. Vibration control is applied to all kinds of engineering systems to obtain the desired dynamic behavior, improved accuracy and increased reliability during operation. In this context, one can think of applications related to the control of structures' vibration isolation, control of vehicle dynamics, noise control, control of machines and mechanisms and control of fluid-structure-interaction. One could continue with this list for a long time.

Research in the field of vibration control is extremely comprehensive. Problems that are typical for vibration control of nonlinear mechanisms and structures arise in the fields of modeling systems in such a way that the model is suitable for control design, to choose appropriate actuator and sensor locations and to select the actuators and sensors.

The objective of the Symposium was to present and discuss methods that contribute to the solution of such problems and to demonstrate the state of the art in the field shown by typical examples. The intention was to evaluate the limits of performance that can be achieved by controlling the dynamics, and to point out gaps in present research and give links for areas of future research. Mainly, it brought together leading experts from quite different areas presenting their points of view.

The book evolved from the International Symposium on Vibration Control of Nonlinear Mechanisms and Structures, held in Munich, Germany, from 18 to 22 July 2005. This Symposium was initiated by the International Union of Theoretical and Applied Mechanics. The main topics of the Symposium were:

- Control of machines and mechanisms
- Control of vehicles
- Active control of noise and vibration in structures
- Active vibration isolation of mechanisms
- Control of nonsmooth dynamics
- Actuators and sensors
- Vibration control of fluid-structure interaction

- Numerical methods in real-time-control

A Scientific Committee was appointed by the Bureau of IUTAM with the following members:

- Heinz Ulbrich, Munich, Germany (Chairman)
- Sunil K. Agrawal , Delaware, USA
- Steven R. Bishop, London, UK
- Felix L. Chernousko, Moscow, Russia
- Mikio Horie, Yokohama, Japan
- Ali H. Nayfeh, Blacksburg, USA
- Werner Schiehlen, Stuttgart, Germany
- Gábor Stépán, Budapest, Hungary

This committee selected the participants to be invited and the papers to be presented at the Symposium. As a result of this procedure, more than 60 active scientific participants from 14 different countries followed the invitation. There were three key lectures selected to give an overview of the different fields to be covered by the Symposium followed by 31 papers. All the papers presented at the Symposium are included in this book.

Since many of the presentations are related to more than one of these topics, the papers in this book are arranged in alphabetical order with respect to the family name of the first author, starting at the beginning with the three key papers. The papers cover a wide range of engineering applications of the vibration control of nonlinear mechanisms and structures. The presentations and discussions during the Symposium will certainly stimulate further theoretical and experimental investigations in the related research fields.

The editors wish to thank both the participants of this IUTAM Symposium and the authors of the papers for their valuable contributions to the important field of vibration control of nonlinear mechanisms and structures. Special thanks are given to the invited lecturers, all the lecturers and the sessions chairmen for making this Symposium a success.

The organizer gratefully acknowledges the financial support and/or effective help in the preparation of the Symposium

- Deutsche Forschungsgemeinschaft (German Research Foundation)
- International Union of Theoretical and Applied Mechanics (IUTAM)
- BMW Research Group
- Bayerisches Staatsministerium für Wissenschaft und Kunst (Bavarian Ministry of Science and Arts)
- Technical University of Munich (TUM)

The main contribution to the success of the Symposium was the great help and excellent work of the staff of the Institute of Applied Mechanics of the TUM and the Local Organizing Committee.

Special thanks are given to Dipl.-Ing. Sandor Riebe and to Dipl.-Ing. Wolfgang Günthner, they did an excellent job, as well as to Ms. Rita Schneider and Ms. Manuela Müller-Philipp for their great administrative help.

In addition, many thanks are due to Springer (formerly Kluwer Academic Publishers) and Karada Publishing Services for their efficient cooperation.

Munich, July 2005 Heinz Ulbrich

WELCOME ADDRESS

Magnifizenz Professor Schilling, Chairman Professor Ulbrich,
Dear Colleagues from all over the world,
Ladies and Gentlemen,

It is my honor and pleasure to welcome all of you on behalf of the International Union of Theoretical and Applied Mechanics, and its President Professor Ben Freund from Brown University. Let me use this Opening Ceremony for a short look on the past and present activities of IUTAM.

Organized meetings between scientists in the field of mechanics were initiated 83 years ago, namely in 1922, when Professor Theodore von Kármán and Professor Tullio Levi-Civita organized the world's first conference in hydro- and aero-mechanics. Two years later, in 1924, the First International Congress was held in Delft, the Netherlands, encompassing all fields of mechanics, that means analytical, solid and fluid mechanics, including their applications. From then on, with the exception of the year 1942, International Congresses in Mechanics have been held every four years.

In particular, when the mechanics community reassembled in Paris for the Sixth Congress in 1946, out of the Congress series an international union was formed, and as a result IUTAM was created and statutes were adopted. After one year, in 1947, the Union was admitted to ICSU, the International Council for Science. This council coordinates activities among various other scientific unions to form a tie between them and the United Nations Educational, Scientific and Cultural Organization, well known as UNESCO.

Today, IUTAM forms the international umbrella organization of about 50 national Adhering Organizations of mechanics from nations all over the world. Furthermore, a large number of international scientific organizations of general or more specialized branches of mechanics are connected with IUTAM as Affiliated Organizations. As a few examples, let me mention: the European Mechanics Society (EUROMECH), the International Association of Computational Mechanics (IACM), the International Association for Vehicle System Dynamics (IAVSD), and the International Commission of Acoustics (ICA).

Within IUTAM the only division used so far is related to solid and fluid mechanics as indicated by our two Symposia Panels. But more recently nine

Working Parties with up to five members each have been established by the General Assembly of IUTAM devoted to specific areas of mechanics. These areas are:

- Non-Newtonian Fluid Mechanics and Rheology,
- Dynamical Systems and Mechatronics,
- Mechanics of Materials,
- Materials Processing,
- Biomechanics,
- Nano- and Micro-Scale Phenomena in Mechanics,
- Geophysical and Environmental Mechanics,
- Education in Mechanics and Capacity Building.

The terms of reference of the Working Parties include to make recommendations to the General Assembly regarding timely subjects for IUTAM Symposia, to maintain contact with the relevant Affiliated Organizations and sister International Unions, to identify important growth areas of the field, and to assist the Bureau and the General Assembly in discussions on position statements. Professor Friedrich Pfeiffer whom I am greeting too, is Chairman of the Working Party on Dynamical Systems and Mechatronics.

IUTAM carries out an exceptionally important task of scientific cooperation in mechanics on the international scene. Each national Adhering Organization of IUTAM, like the German Committee for Mechanics (DEKOMECH), is represented by a number of scientists in IUTAM's General Assembly. In particular, the German delegates with IUTAM are

Professor Ulrich Gabbert, Otto von Guericke University of Magdeburg;
Professor Christian Miehe, University of Stuttgart;
Professor Wolfgang Schröder, Rheinisch-Westfälische Technische Universität Aachen (RWTH Aachen);
Professor André Thess, Ilmenau University of Technology.

Mechanics is a very well developed science in Germany represented at most universities and some national laboratories. Since 1949 there has been held more than 280 IUTAM symposia worldwide. Out of them 31 symposia where organized in Germany with two symposia in Munich.

In 1977, just 28 years ago, the first IUTAM Symposium was held at the Technical University of Munich under the chairmanship of Professor Kurt Magnus. This first Munich Symposium was devoted to the Dynamics of Multibody Systems with the result that a new branch of mechanics called "Multibody Dynamics" was created.

Twenty-one years later in 1998 a second IUTAM Symposium took place here in Munich. Professor Friedrich Pfeiffer was chairman of the Symposium on Unilateral Multibody Contacts which was very successful again. Since that time contact problems are considered as an attractive research field in multibody dynamics, too.

As I mentioned before, IUTAM organizes not only symposia but also international congresses all over the world. Last year the 21st International Congress of Theoretical and Applied Mechanics was held in Warsaw, Poland. With 1515 participants the Warsaw Congress was a major event in mechanics also described as the Olympics of Mechanics. The Twenty-second International Congress of Theoretical and Applied Mechanics will be held in Adelaide, Australia, from 24th to 30st August 2008, which means in three years from now. Announcements of this forthcoming congress will be widely distributed and published in many scientific journals. The German members elected to the standing Congress Committee of IUTAM are Professor Edwin Kreuzer, Technical University Hamburg-Harburg, and Professor André Thess, Ilmenau University of Technology.

The present Symposium is exceptionally interesting because it deals with new developments in mechanics. The Symposium covers important approaches:

- Control of machines and mechanisms,
- Active control of vibration in structures,
- Control of multibody dynamics,
- Sensors and observers,
- Applications to cranes and robots.

IUTAM found that the proposal of Professor Ulbrich for such a symposium was not only very timely, but also well justified in the outstanding research carried out in this field at the Technical University of Munich. Thus, the proposal for the Symposium was readily accepted and granted by the General Assembly of IUTAM. There is no doubt that IUTAM considers vibration control of nonlinear systems as an important field of mechanics.

Finally, I would like to mention that to sponsor a scientific meeting is one thing, but to organize one is another. A heavy burden is placed on the shoulders of the Chairman and his associates who are in charge of the scientific program and the practical local arrangements. All who have tried this before know very well how much work has to be done in organizing a meeting like this one.

Thus, we should be thankful, not only to the International Scientific Committee, but also to the Chairman, Professor Heinz Ulbrich, and his associates who assisted him in carrying the heavy load and responsibility.

It is up to you now, Ladies and Gentlemen, to harvest the fruits of the Organizer's work. Contribute your share to make this IUTAM Symposium a meeting that will be long remembered as a very successful one!

On behalf of IUTAM, I greet you all and wish you great success!

Werner Schiehlen
Past President of IUTAM, University of Stuttgart, Germany

1 K. Krüger
2 L. Ginzinger
3 W. Günthner
4 S. Okada
5 M. Blajer
6 Y. Okada
7 K. Blajer
8 K. Schlacher
9 V. Babitsky
10 W. Blajer
11 B. Lohmann
12 M. Friedrich
13 E. Babitsky
14 P. Wagner
15 A. Kugi
16 M. Förg
17 R. Schneider
18 H. Gattringer
19 T. Mizuno
20 P. Hagedorn
21 S. Bremer
22 T. Hardeman
23 K. Nonami
24 K. Seto
25 H. Bremer
26 C. Schiehlen
27 R. Zander
28 D. Söffker
29 H. Ulbrich
30 D. Bestle
31 T. Thümmel
32 R. Ulbrich
33 U. Spaelter
34 N. Sobolev
35 I. Krajcin
36 G. Hagedorn
37 F. Pfeiffer
38 R. Pfeiffer
39 S. Nayfeh
40 M Sahinkaya
41 W. Schiehlen
42 A. Nayfeh
43 J. Dijk
44 T. Sammet
45 A. Ratering
46 P. Kiriazov
47 M. Müller-Philipp
48 F. Barrot
49 N. Guse
50 A. Ulbrich
51 H. Bleuler
52 Y. Li
53 G. Stepan
54 C. Wen

Not on the photo:
M. Bachmayer
H. Baier
B. Beadle
A. Bockstedte
G. Brandenburg
M. Buhl
M. Buss
S. Elliott
R. Freymann
M. Herrnberg
J. Höfeld
M. Horie
K. Kostadinov
K. Kühnlenz
M. Neubauer
M. Otter
J. Paschedag
C. Paulitsch
S. Riebe
F. Schiller
R. Skelton
M. Sobotka
U. Stöbener
D. Wisselmann
K. Wong
D. Wollherr
C. Zauner

ACTIVE CONTROL OF STRUCTURAL VIBRATION

Stephen J. Elliott
Institute of Sound and Vibration Research, University of Southampton, Highfield, Southampton SO17 1BJ, UK

Abstract: In its most general form, active control can be used to arbitrarily modify the vibration response of a structure, although in engineering applications the objective is normally to attenuate the vibration. Many actuators and sensors are required to control the vibrations of a large structure, and conventionally these are connected using a single centralised controller, designed using a detailed model of the structure. Very selective control is possible using this approach, which can, for example, modify some modes with little effect on others. However such centralised controllers may not be stable if the response of the structure changes or individual transducers fail, and can become very complicated as the structure becomes large. An alternative approach is to use multiple local control loops in a decentralised arrangement. With careful choice of the actuators and sensors, these loops are simple to design and can be guaranteed stable even if the structural response of the system changes or some transducers fail. Each actuator, sensor and controller can be constructed as a self-contained module, and although the resulting vibration control is not selective such a decentralised system can perform almost as well as a centralised system in reducing the global response of the structure.

Decentralised control for vibration isolation with multiple active mounts will be discussed in this paper as an example of an engineering structure, and this will be contrasted with the way that decentralised control loops enhance the vibration of the inner ear in the active hearing mechanism.

Key words: active control isolation, decentralised control, active hearing.

1. INTRODUCTION

The control of sound and vibration in systems with many degrees of freedom requires multiple actuators and multiple sensors. Conventionally, all the actuators are driven by a single, centralised controller, which is also supplied with signals from all the sensors. Such centralised controllers are often designed using a model of the system under control, such as a modal model, and have the advantage that the controller can be designed so that individual

H. Ulbrich and W. Günthner (eds), IUTAM Symposium on Vibration Control of Nonlinear Mechanisms and Structures, 1–19.

modes can be influenced to different extents [1–3]. The type of actuators and sensors used and their positioning are chosen to best influence or observe these modes. There are two potential disadvantages of centralised controllers First, their performance, and sometimes their stability, can be threatened if the system changes, or if a transducer fails, so that the assumed model is no longer an accurate one. Second, a great deal of wiring is required on large systems to connect all the actuators and sensors to the single controller, and the complexity of the controller rises sharply as the size of the system increases.

In a fully decentralised system, however, a number of independent controllers are used to drive individual actuators from individual sensors. Care must be taken in the selection of the types of sensors and actuators used, however, to ensure their stability. Such a control strategy has the advantages of reduced complexity and wiring and, if all the controllers are the same, it also has the advantage of modularity, so that increasingly complex systems can be controlled with a larger number of identical simple units, and scalability, so the complexity of the system scales only linearly with the number of modes. Because such units only act locally, however, individual global properties of the system, such as mode amplitudes, cannot be selectively controlled.

In this paper decentralised control will be discussed in both the design of controllers for the reduction of vibration in engineering structures, and also in the active mechanism that appears to operate in the inner ear to enhance the vibration of the basilar membrane.

The centralised and decentralised control of vibration in engineering structures is discussed first, using collocated pairs of dual actuators and sensors. Ideal duality in the actuator and sensor pair can rarely be achieved in practical active vibration control systems. The concept of a passive controller, however, which relies on such an idealisation of dual actuator and sensor pairs, can be used, to motivate the combined selection of the control strategy and the details of the actuator and sensor design. The practical example described is the active isolation of a piece of delicate equipment from base vibration. Here conventional passive mounts provide good isolation at high frequencies, but can amplify the base vibration at low frequency resonances. By applying an active force in parallel with the active mount, controlled by the absolute velocity above the mount, "skyhook" damping can be realised to attenuate these low frequency resonances, without affecting the high frequency performance of the passive mounts. The reaction force acting on the base structure, which is generally flexible, spoils the duality of the force acting on the equipment and the sensed velocity. It is fortunate, however, that although this reaction force can be significant, it does not unduly affect the passivity of the system under control. Thus local control loops closed round individual mounts can be shown to be unconditionally stable, even in the event of failures.

This decentralised approach to vibration reduction is then contrasted with the active mechanism in the inner ear, which also appears to be achieved with local feedback loops. Although the details of these control loops are not well understood, they are clearly nonlinear in their action and this nonlinearity gives rise to a number of important auditory effects. Although the mechanics of the inner ear are distributed and nonlinear, and their details are rather complicated, the study of nonlinearities in the ear using simple engineering models appears to hold promise.

2. CENTRALISED AND DECENTRALISED CONTROL OF VIBRATION

The conventional approach to controlling the dynamics of large structures, with many degrees of freedom, is to work in terms of the structure's modes. The dynamic behaviour is approximated by the sum of a finite number of modal contributions, actuators and sensors are chosen so that they can observe and control these modes, and a feedback controller is designed to modify their response. Often the objective is not to actively change the shapes of the modes, but to independently modify the natural frequencies and damping ratios of a number of modes of interest, which has been termed independent modal-space control, IMSC, by Meirovitch [1].

In a large, well defined, structure it may not be difficult to numerically model the first 10 or so modes, using finite elements for example, or to experimentally identify a similar number of modes using modal analysis. It becomes increasingly difficult, however, to distinguish between the different modal contributions beyond this number, particularly in built-up structures with a high modal overlap. Under these conditions the modal approach to control becomes inappropriate. If we are considering the control of large structures, particularly at the higher frequencies for which sound radiation is important, then a reliable modal model of the structure is generally not available and alternative approaches to active control have to be found.

If local, fixed-gain, control loops are closed around dual actuator/sensor pairs [4], the control system is an example of direct output feedback control [1] and has some very attractive stability properties [5]. Since each local feedback loop can only absorb power from the structure, this is an example of a dissipative controller [6, 7], and no combinations of positive gains can destabilise the system. Each force actuator driven by a collocated velocity sensor in fact synthesises an entirely passive damper. In practice vibration velocity is often estimated by integrating the output of an accelerometer, and so the velocity is measured relative to the inertial reference frame. The integrator singularity at dc is avoided since rigid-body modes are not being controlled. The actively-synthesised dampers thus operate with respect to this reference frame and are called skyhook or inertial dampers [8].

Ideal local velocity control thus has very attractive robustness properties, since it is unconditionally stable, whatever the response of the structure to which it is attached, and in spite of any failures that may occur in other control loops. The design of such local loops to achieve a specific performance objective in controlling the structure is less obvious than it was for modal control, however. Clearly the actuator/sensor pairs should be positioned so that they efficiently couple into the modes to be controlled. The individual gains could be optimised to minimise a specific quadratic cost function, using state space methods for example. In general, however, decentralised control is not selective since it does not allow the independent specification of the modal responses that is the attractive feature of modal control. If, for example, a large number of force actuator/velocity sensor pairs with equal feedback gains are uniformly distributed over a structure then the damping of an equal number of the structural modes is increased to approximately the same extent [9].

Apart from the completely decentralised controller described above, there are a number of other strategies for reducing the complexity of fully-coupled controllers in systems with many actuators and sensors, for example based on clustering or hierarchical control [10–12].

3. ACTIVE ISOLATION OF VIBRATION

One important application of active vibration control is to the isolation of sensitive equipment from the vibrations of a base structure. Compliant passive mounts are often used for vibration isolation, and their performance can be quantified in terms of the transmissibility of the complete system, which is the ratio of the velocity of the equipment to be isolated to the velocity of the base structure, as a function of frequency. The transmissibility of a compliant passive mount, modelled as a spring and a damper, when isolating a simple equipment structure, modelled as a lumped mass, from base vibrations are shown in Figure 1. The transmissibility for this single degree of freedom system can be written as

$$\mathrm{T} = \frac{v_e}{v_b} = \frac{1 + 2j\zeta_{\text{pass}}\Omega}{1 - \Omega^2 + 2j\zeta_{\text{pass}}\Omega}, \tag{1}$$

where v_e and v_b are the complex velocities of the equipment and base at a given excitation frequency. $\Omega = \omega/\sqrt{k/m}$ is the normalised excitation frequency where k and m are the stiffness of the mount and the mass of the equipment and the passive damping ratio is $\zeta_{\text{pass}} = c/2\sqrt{km}$ where c is the passive damping in the mount.

Figure 1 shows the calculated transmissibility for two values of mount damping, corresponding to the damping ratios $\zeta_{\text{pass}} = 0.1$ and $\zeta_{\text{pass}} = 0.6$.

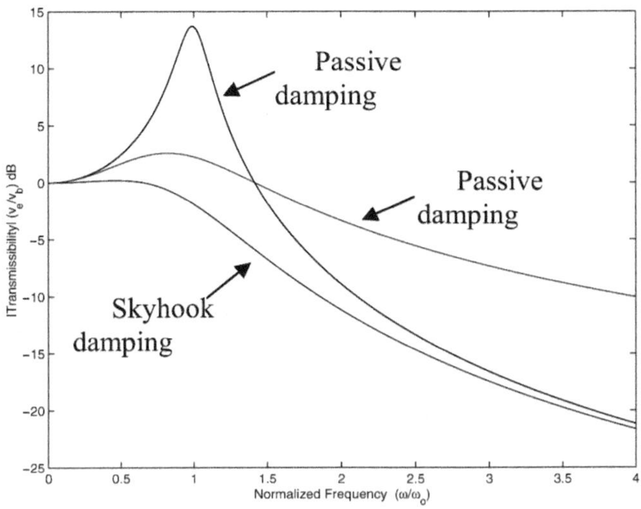

Figure 1. Transmissibility of a single degree of freedom system for the isolation of base vibration with passive damping, having damping ratio of $\zeta_{\text{pass}} = 0.1$, $\zeta_{\text{pass}} = 0.6$ and skyhook damping with damping ratio $\zeta_{\text{act}} = 0.6$.

If the mount is lightly damped the transmissibility is low at higher frequencies, and so the isolating performance is good, but the transmissibility is greater than 0 dB, i.e. the vibration is amplified, at excitation frequencies close to the natural frequency of the equipment mass on the isolator stiffness. If the passive damping is increased, the amplification at resonance is reduced, but significant degradation of high frequency performance results from the increased impedance of the damper at high frequencies. This illustrates a classic trade-off in the design of passive isolation systems [13].

One method of actively controlling the resonance without degrading high frequency performance is to use a feedback system to generate a force on the equipment that is proportional to the absolute velocity of the equipment as illustrated in Figure 2. This generates skyhook damping [8] and if the base structure has a low mobility so that the reaction force acting on the base structure in Figure 2 does not affect the equipment velocity, the action of the feedback loop is to synthesise the additional skyhook damper, as also shown in Figure 2. The transmissibility then becomes

$$\mathrm{T} = \frac{1 + 2j\zeta_{\text{pass}}\Omega}{1 - \Omega^2 + 2j(\zeta_{\text{pass}} + \zeta_{\text{act}})\Omega}, \tag{2}$$

where the active damping ratio is $\zeta_{\text{act}} = h/2\sqrt{km}$ where h is the gain in the feedback loop. The transmissibility with $\zeta_{\text{act}} = 0.6$ is also shown in Figure 1 and it can be seen that the performance is significantly better than the passive

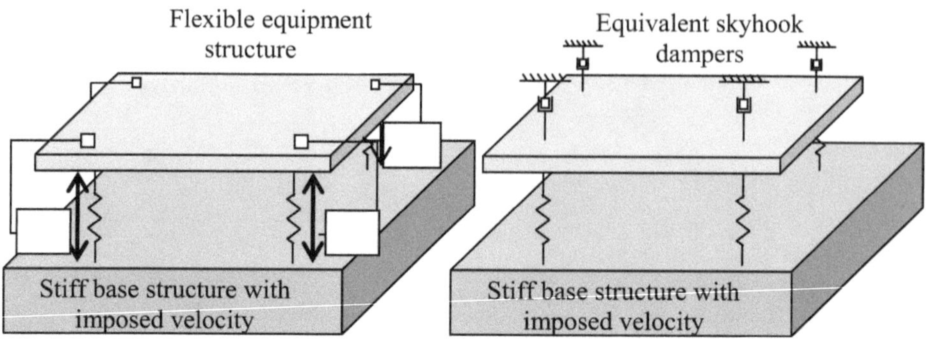

Figure 2. A multichannel active sky damper in which the velocity at each mount is fed back only to a force actuator in parallel with the passive mount (left), the equivalent mechanical system (centre) and the block diagram (right).

system with the same damping ratio at all frequencies, and is as good as the lightly damped passive system at high frequencies.

Apart from the clear performance advantages, the control loop shown in Figure 2 is also very robust [14]. This can be illustrated by considering a single active feedback loop in which $G(j\omega)$ is the plant response from secondary force input to velocity sensor output. If the mobility of the base structure is low then $G(j\omega)$ is equal to the input, or point, mobility of the mounted equipment structure, so that $v = G(j\omega)f$. The mechanical power supplied to equipment structure by this force is

$$W = \tfrac{1}{2}\mathrm{Re}[f^*v] = \tfrac{1}{2}|f|^2\mathrm{Re}[G(j\omega)] \tag{3}$$

and since W must be positive then so must $\mathrm{Re}[G(j\omega)]$, and the plant is said to be passive. The stability properties of the feedback loop can be described in terms of the system's Nyquist plot. Since $H(j\omega) = h > 0$ and the real part of $G(j\omega)$ is also positive, the Nyquist plot for the single degree of freedom active isolation feedback loop is entirely on the right hand side of the imaginary axis. However complicated the dynamics of the equipment are, its input mobility must be passive and so the Nyquist plot will always stay on the right hand side of the imaginary axis and is thus well away from the instability point $(-1, 0)$ so that the control strategy is inherently robust. Serrand [15] and Elliott et al. [14, 16] have shown that even if the mobility of the base structure does become significant, then the stability of the feedback loop is not generally threatened. In practice it is the low frequency phase shifts in the transducer conditioning amplifiers that limit the maximum feedback gains that can be used before instability, although with careful design, reductions in transmissibility of 30 dB at resonance can readily be achieved.

We now consider the stability and performance of a multichannel vibration isolation system, in which each active mount is modelled as above and is controlled by an independent feedback control loop.

One has to be careful in analysing the stability of this multichannel feedback system. The feedback loops can no longer be regarded as independent since they are coupled via the overall dynamics of the equipment structure, which is potentially flexible. The stability of a multichannel feedback control system can again be determined using a generalisation of the Nyquist plot, in which the locus of the real and the imaginary parts of eigenvalues of the matrix $\mathbf{G}(j\omega)\mathbf{H}(j\omega)$ are plotted as ω varies from $-\infty$ to ∞ [17]. In this case the controller is decentralised and each loop is assumed to have a constant gain, so that $\mathbf{H}(j\omega) = h\mathbf{I}$. Also the vector of velocities at the mounting points is related to the vector of actuator forces by $\mathbf{v} = \mathbf{G}(j\omega)\mathbf{f}$, where $\mathbf{G}(j\omega)$ is the fully populated plant response. If the actuators and sensors are collocated and the base structure is solid, $\mathbf{G}(j\omega)$ is the input mobility matrix for the mounted equipment structure, which is symmetric due to reciprocity. The total mechanical power supplied by all the actuator forces can then be written as

$$W = \tfrac{1}{2}\mathrm{Re}[\mathbf{v}^H\mathbf{f}] = \tfrac{1}{2}\mathrm{Re}[\mathbf{f}^H\mathbf{G}(j\omega)\mathbf{f}] \tag{4}$$

and since W must be positive, then $\mathbf{G}(j\omega)$ must be passive. In this case passivity implies that both

$$\mathrm{Eig}[\mathrm{Re}\,\mathbf{G}(j\omega)] \geq 0 \quad \text{and} \quad \mathrm{Re}[\mathrm{Eig}\,\mathbf{G}(j\omega)] \geq 0. \tag{5}$$

The second condition is less well known than the first and can be demonstrated by writing the eigendecomposition of the normal, symmetric but complex matrix $\mathbf{G}$ as $\mathbf{Q}^H\mathbf{\Lambda}\mathbf{Q}$ where $\mathbf{Q}$ and $\mathbf{\Lambda}$ are generally complex. The power supplied to the system is thus proportional to

$$\mathrm{Re}[\mathbf{f}^H\mathbf{G}\mathbf{f}] = \mathrm{Re}[\mathbf{q}^H\mathbf{\Lambda}\mathbf{q}] = \mathrm{Re}\left[\sum_i \lambda_i |q_i|^2\right], \tag{6}$$

where $\mathbf{q} = \mathbf{Q}\mathbf{f}$, and since the power must be positive for all inputs, so must the real parts of the eigenvalues of $\mathbf{G}$. This condition ensures that each of the loops in the generalised Nyquist diagram behave in a similar way to those of the single channel system and are strictly on the right hand side of the imaginary axis, ensuring that the control system is very robust. In particular, the stability is ensured for any degree of flexibility in the equipment structure, no matter how strongly the actuators and sensors are coupled. Exactly the same argument can be used to show that a control system with only three independently-acting active mounts is robustly stable, and so the stability of the system is not compromised by failure of individual mounts.

Figure 3. Experimental arrangement of 4 active mounts isolating the equipment structure from the vibration of the buss structure.

Figure 3 shows an experimental arrangement in which a simple equipment structure is supported on a vibrating plate by four active mounts [18]. The active force at each mount is provided by a shaker that acts through the mount, as illustrated in Figure 4, and is driven from the integrated output of an accelerometer mounted on the equipment close to the mount. Each active mount is thus of exactly the same design and contains actuator, sensor and independently-acting controller. A number of such identical modules could thus be used to isolate more complicated pieces of equipment with no increase in the complexity of the system.

Figure 5 shows the magnitude of the individual frequency responses of the elements of the 4×4 plant matrix $\mathbf{G}(j\omega)$, all on the same scales, as measured on the experimental system. The diagonal terms correspond to input mobilities and have alternating poles and zeros [3], whereas the off-diagonal terms correspond to transfer mobilities. The measured matrix is reasonably symmetric at each frequency as one would expect from reciprocity. Figure 6 shows the generalised Nyquist plot for this control system, as calculated from the measured plant frequency responses. It should be noted that the calculation of a set of eigenvalues that vary smoothly with frequency from a matrix of experimental data is not a trivial matter, as discussed by Huang [19]. Even though the plant matrix in Figure 5 has strong cross coupling, its eigenvalues still have almost entirely positive real parts, as one would expect from equation (5). The

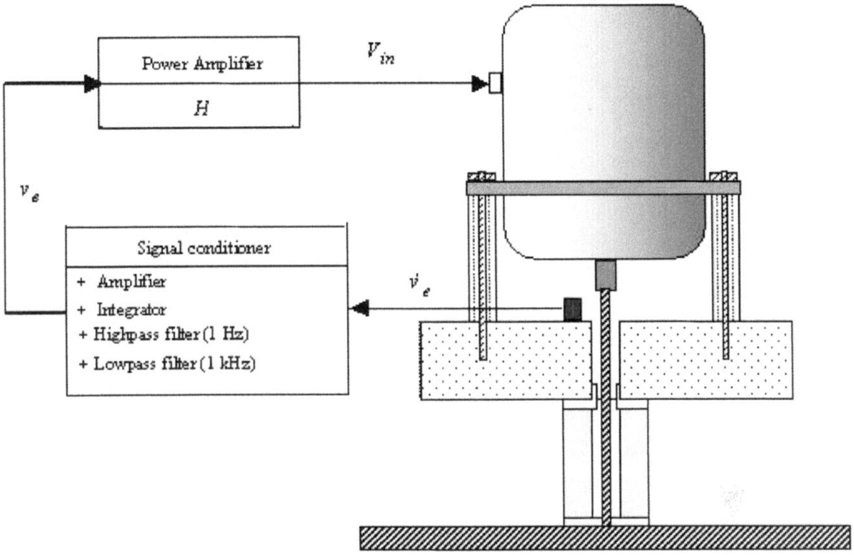

Figure 4. Details of a single control loop, in which the integrated output of the accelerometer is fed back to the shaker to generate skyhook damping.

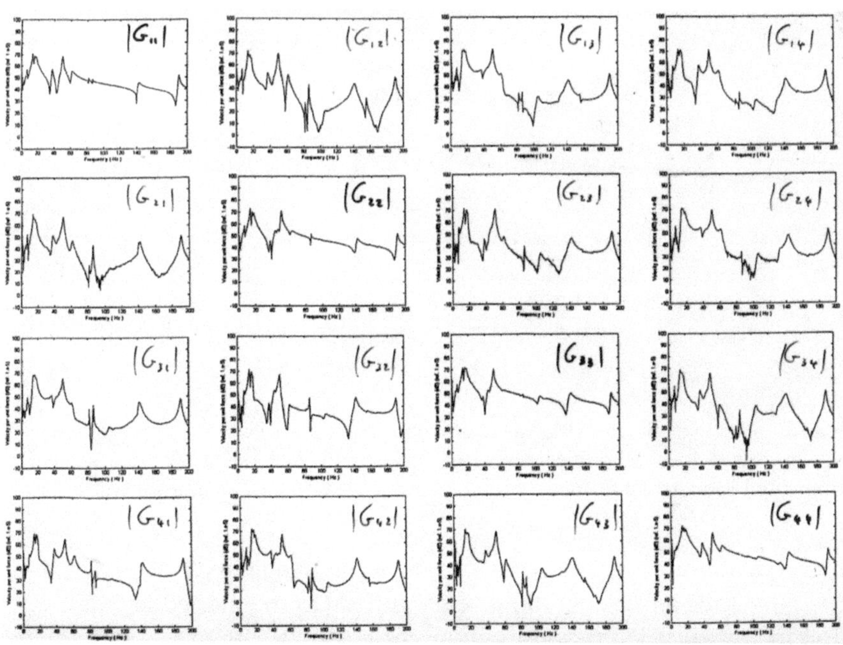

Figure 5. The magnitudes of the elements of the 4 × 4 plant matrix from shaker inputs to accelerometer outputs measured on the experimental active isolation system.

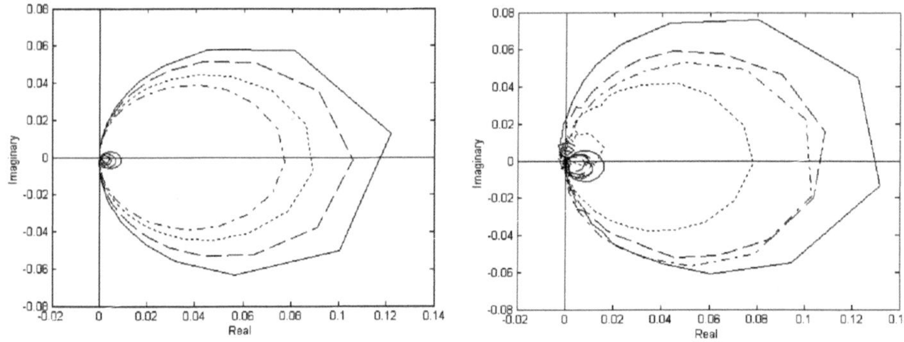

Figure 6. Generalised Nyquist plots of the 4 × 4 plant matrix for the isolation system; from simulation (left) and from experiments (right).

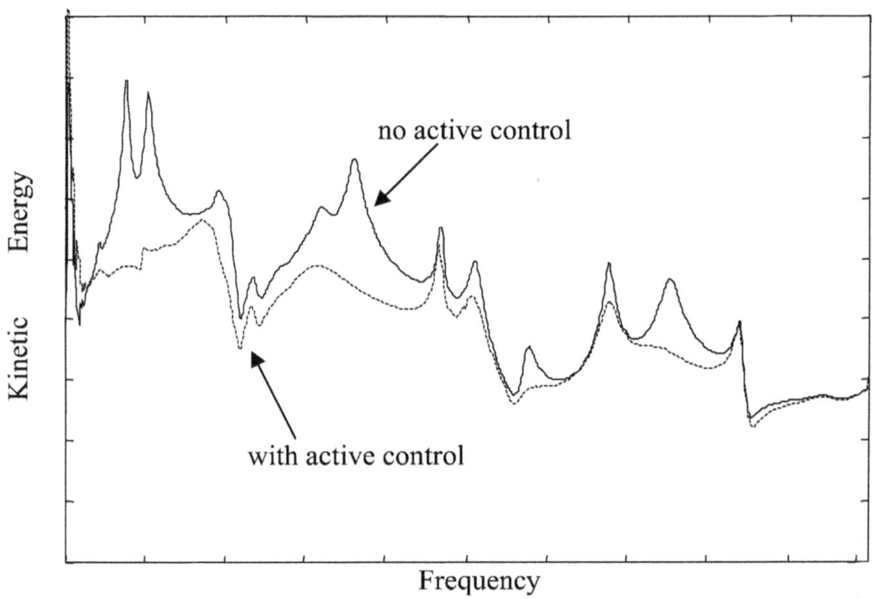

Figure 7. Total kinetic energy of the equipment structure, calculated from the four velocity signals measured at the mounting points of the experimental active isolation system before control (solid) and after 4 channels of local feedback control are implemented (dashed).

slight incursion of the plots onto the left of the imaginary axis is due to phase shifts in the transducer conditioning amplifiers, as mentioned above.

The measured velocities at the four corners of the equipment structure can be used to calculate the amplitudes of its rigid body modes and hence its total kinetic energy. Figure 7 shows the kinetic energy in the experimental system when driven by the randomly-excited base structure alone (no active control), and with all of the four feedback control systems connected (with active control). The peaks at about 20 Hz are due to the heave and pitch resonances of the equipment structure on its mounts, and these have been attenuated by more than 20 dB by the active mounts. The peaks in the kinetic energy above 20 Hz are due to resonances of the plate that forms the base structure in this case. Even though the feedback system was not originally designed to control these resonances, they make a strong contribution to the plant responses in Figure 5 and so the loop gain is large enough to achieve reductions of up to 15 dB in the equipment's kinetic energy at these frequencies. If the kinetic energy is measured over a wider frequency range, it is also seen that the action of the feedback control system does not significantly increase the vibration of the equipment at higher frequencies, as predicted by the simple models at the start of this section. The local feedback loops can thus be designed to control the low frequency transmissibility, while lightly-damped passive mounts can be used to provide high frequency isolation.

4. LOCAL CONTROL IN THE INNER EAR

The high sensitivity and exquisite frequency selectivity of our hearing has long suggested that there is an active amplification process taking place in the inner ear. These active processes are not well understood, but it is interesting to compare some of the current models for these active processes with the types of active vibration control system discussed above, in order to compare these two examples of distributed control. The snail-shaped cochlea in the inner ear contains two fluid chambers, separated by the basilar membrane, which are excited at the end by the bones of the middle ear [20]. The basilar membrane is much stiffer near the entrance to the cochlea than it is at the other end, which results in a distribution of natural or characteristic frequencies along its length. Figure 8 shows a diagrammatic representation of the ear, in which the spiral structure of the inner ear has been straightened out and the mass and stiffness of the individual parts of the limp basilar membrane are made explicit. The lower part of this figure shows the approximate distribution of characteristic frequencies, along the length of the cochlea in the human ear.

The coupling between the inertia of the fluid and the dynamics of the basilar membrane can, with various simplifying assumptions, be analysed to give a wave equation for the propagation of disturbances within the cochlea, see, for example, de Boer [21];

$$\frac{\partial^2 p(x,\omega)}{\partial x^2} - \frac{2j\omega\rho/h}{Z(x)} p(x,\omega) = -j\omega\rho q(x,\omega), \tag{7}$$

where $p(x,\omega)$ is the complex pressure difference between the two chambers, which are filled with fluid of density ρ, x is the position along the cochlea, h is the height of the chamber and $q(x,\omega)$ is a complex source term. $Z(x,\omega)$

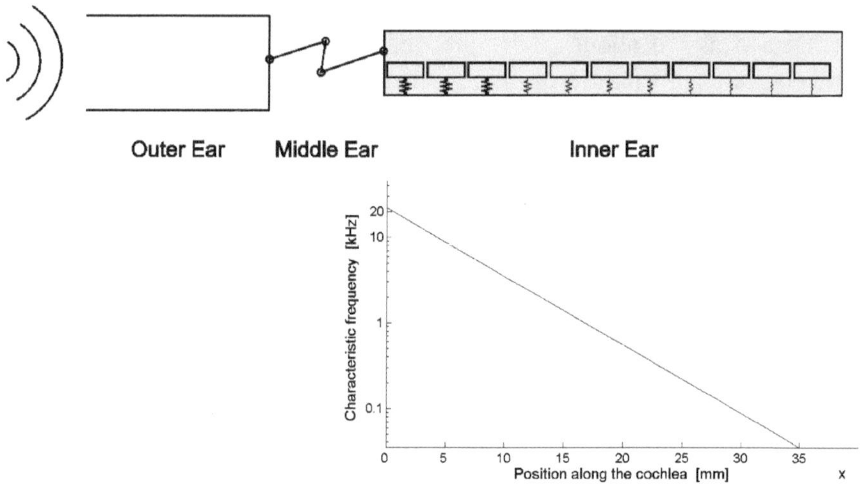

Figure 8. Idealised representation of the outer, middle and inner ear, showing the basilar membrane as a series of mass-spring-damper systems distributed down the cochlea, together with the distribution of the natural or characteristic frequencies of these single degree of freedom systems.

is the mechanical impedance of the basilar membrane at position x and frequency ω. The basilar membrane is fairly limp and this impedance is generally taken to be the point impedance of an isolated lumped parameter system. To a first approximation the dynamics at any one point on the cochlea can thus be approximated by a single degree of freedom, mass-spring-damper system, with a fixed mass, but a stiffness which varies in such a way as to give the distribution of characteristic frequencies discussed above. For a given excitation frequency, ω, this wave equation has different types of solution for different positions along the basilar membrane compared with the characteristic place, x_o, for which the characteristic frequency is equal to ω_o.

If $x < x_o$, so that $\omega < \omega_o$, $Z(x, \omega) \approx k(x)/j\omega$, since the basilar membrane is stiffness-controlled in this region below resonance. The homogeneous wave equation then becomes

$$\frac{\partial^2 p(x, \omega)}{\partial x^2} + \frac{2\omega^2 \rho}{hk(x)} p(x, \omega) = 0 \tag{8}$$

which has a *propagating* wave solution

$$p(x, \omega) = p_o e^{-j\omega x / c(x)}, \tag{9}$$

where $c(x)$ is the wavespeed given by $(hk(x)/2\rho)^{1/2}$ which falls as x increases, becoming zero at the characteristic place, x_o.

If $x > x_o$, so that $\omega > \omega_o$, $Z(x, \omega) \approx j\omega m(x)$, since the basilar membrane is mass-controlled in this region above resonance. The homogeneous wave equation then becomes

$$\frac{\partial^2 p(x, \omega)}{\partial x^2} + \frac{2\rho}{hm(x)} p(x, \omega) = 0 \tag{10}$$

which has an *evanescent* solution

$$p(x, \omega) = p_o e^{-x/\ell(x)}, \tag{11}$$

where $\ell(x)$ is decay length given by $(hm(x)/2\rho)^{1/2}$, although since $m(x)$ is almost constant, $\ell(x)$ is almost independent of x and has a value of about 1 mm in the human cochlea.

The overall effect is that at a given frequency a wave propagates along the cochlea, with a wavelength that gets smaller and an amplitude that gets larger as the wavespeed decreases with increasing x, until the wave stalls at the characteristic place for this frequency and the pressure then rapidly decays away. This gives rise to the characteristic travelling wave envelope widely used to describe the action of the cochlea as a frequency discriminator [20, 21]. Another way of looking at the dynamic response of the cochlea is to consider the frequency response at one particular position. Figure 9 shows the magnitude of the basilar membrane velocity at a position, x_o, to sinusoidal excitation at various frequencies, simulated using a numerical approximation to the wave equation in equation (8) [22]. The lower line corresponds to the "passive" model of the cochlea mechanics outlined above. This gives a heavily damped resonance, which would provide much coarser frequency discrimination than is known to exist at low levels in the healthy cochlea. In order to obtain tuning curves that agree with modern measurements of the frequency response of physiologically healthy basilar membranes, some kind of active hearing mechanism must be introduced.

There is still considerable controversy about the exact form of this active mechanism, but one model that explains many features found in measurements is that put forward by Neely and Kim [23]. This model not only accounts for the vertical motion of the basilar membrane but also the motion of the tectorial membrane, which sits just above the basilar membrane in the organ of Corti, as shown in Figure 10. The shearing between the basilar and tectorial membranes activates the 3,500 inner hair cells, which are believed to be the main sensory structure in the ear, but is also influenced by the 12,000 outer hair cells, which are believed to act both as local sensors and fast-acting actuators [20]. The presence of these "motile" cells in the ear, whose length changes with applied potential, is essential to the active hearing mechanism. The overall dynamics of the active cochlea are represented in [23] by the two degree of freedom model also shown in Figure 10, in which the lower mass and spring

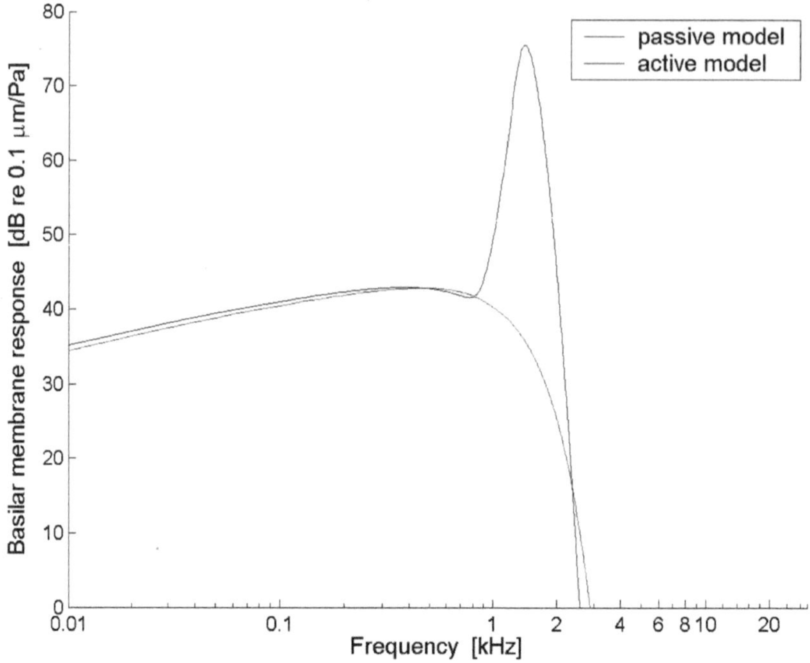

Figure 9. Simulated frequency response of the basilar membrane velocity at the passive mid-point of the cochlea, for the passive, single degree of freedom, model of the basilar membrane dynamics (red) and the active two-degree of freedom model (blue).

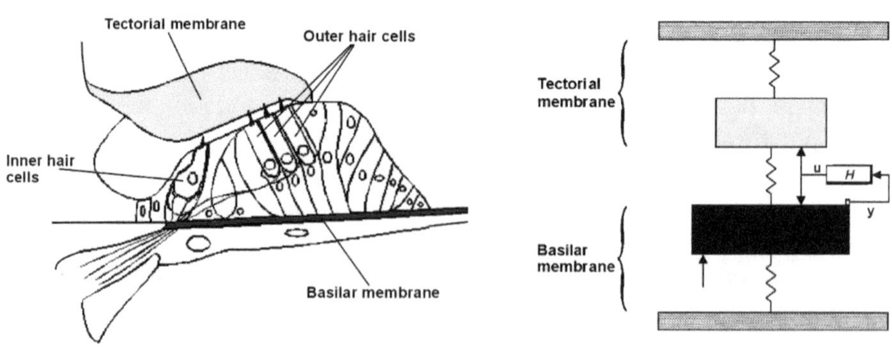

Figure 10. The detailed mechanical structure of the organ of Corti (left) and the simplified model of the dynamics, including active feedback loop (right).

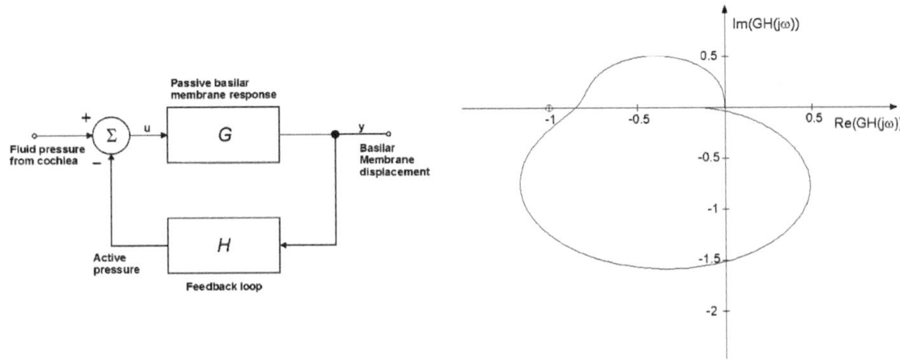

Figure 11. Block diagram of the active hearing mechanism as a feedback loop, and the resulting Nyquist plot for low-level excitation.

are associated with the basilar membrane and the upper mass and spring model the movement of the tectorial membrane. The overall response of this active system, including the feedback loop driving the outer hair cells from the basilar membrane motion, is also shown in Figure 9, and has a much sharper resonance than the passive model.

The Nyquist plot of the open loop frequency response, for the mechanical constants given by Neely and Kim [13], gives a series of plots like those in Figure 11 for different positions on the basilar membrane. The Nyquist plots pass close to the $(-1, 0)$ point, particularly at the high frequency end of the cochlea, and this provides the sharper peak in the frequency response seen for this active model in Figure 9, which is much more like that measured experimentally at low excitation levels.

Some care must be exercised when using these models, however, since even though the dynamics of individual parts of the basilar membrane are stable when considered in isolation, there is no guarantee that the overall model of the different parts of the cochlea, coupled via the fluid chambers, will also be stable. In particular, it may not be obvious using only a frequency-domain analysis for the coupled set of closed-loop systems whether the system is stable or not. A more complete model of cochlea mechanics is more complicated still, since it is heavily nonlinear. The extremely sharp tuning curves described above are only observed for very low level excitation of the cochlea [20]. As the sound pressure level increases, the tuning curves become broader and the system is more stable. One way in which this nonlinearity could be incorporated into the active model illustrated in Figure 9 would be to make the gain of the feedback loop level-dependent. This could be achieved by incorporating the saturating nonlinear response of the outer hair cells into the feedback loop [24, 25].

The stability of such a dispersive, active, nonlinear system is difficult to analyse theoretically. There is evidence, however, that in very quiet environments the active hearing mechanisms in the ears of many people are not entirely stable. This evidence comes from measurements of spontaneous otoacoustic emissions, obtained by placing a small microphone in a person's ear canal and recording the output for some time with very low levels of external noise [20]. In about 60% of women and 40% of men, the spectrum of the sound recorded has a series of between 3 and 8 clear peaks, typically between 1 kHz and 5 kHz. The frequency of these peaks is fairly stable over time but differs from one person to another. These spontaneous otoacoustic emissions seem to be clear evidence that the active hearing process in many people's ears is unstable, unless excited at a sufficiently high level. Their origin in the hair cells is also supported by studies which show that they disappear when the physiological condition of the cochlea is damaged. Although we do not have a complete understanding of the active processes within the ear, it seems clear that this is an example of a distributed biological control system, whose gain is somehow regulated to be just at the point of instability under quiet conditions, when the maximum sensitivity is required.

There are many other examples of nonlinear effects in the auditory system and by observing the motion of the basilar membrane it can be shown that some of these are directly due to the dynamics of the inner ear, rather than any subsequent neural processing. Although some of these effects have a clear engineering description, such as aural harmonics and combination tones [26], others require more careful consideration. The masking of one sound by another, for example, is thought to be due to the nonlinear interaction between the masking sound and the original sound at specific places along the basilar membrane, causing suppression of the original unmasked response [27].

The nonlinearity in the active feedback loop causes high sound pressures to be amplified less than lower sound pressures and thus compresses the dynamic range of the sound, in a similar manner to an automatic gain control. This compression can be clearly seen by plotting the measured level of the basilar membrane vibration against the measured level of the applied pressure, which typically yields a line with a slope of 0.2 to 0.5 dB/dB [20]. The compression acts to help reduce the dynamic range over which we can hear, typically 120 dB, down to the 40 dB dynamic range of the inner hair cells, which act as sensors within the inner ear. This compression is often taken for granted in the audiology literature and is only noticed when people have hearing loss caused by damage to the active mechanisms, in which case they can only comfortably hear over a much reduced dynamic range of pressure. This is termed loudness recruitment [21].

So although there is a difference in emphasis and in terminology between the audiological and engineering descriptions of nonlinearity, as listed in

Table 1. The description of various effects found in the hearing literature and the engineering description of the probably cause.

Description in Hearing	Engineering Description
Spontaneous emissions	Limit cycle oscillations
Aural harmonics	Harmonic distortion
Combination tones	Intermodulation distortion
Masking	Suppression
Lack of recruitment	Compression

Table 1, there is considerable scope for engineering analysis to contribute to the understanding of nonlinear effects caused by the active hearing process.

5. CONCLUSIONS

Centralised control systems can perform well, but their stability is potentially sensitive to transducer failures, they require a good deal of wiring and their complexity increases rapidly with the size of the system. In a fully decentralised control system only local feedback loops are implemented, and if the actuator and sensor are both dual and collocated, constant gain feedback controllers are guaranteed to be stable, even if some of the transducers fail. The actuator, sensor and controller for each loop could also be implemented as a single module and only larger numbers of these identical modules be needed to control larger systems.

Practical arrangements are generally more complicated than the ideal case of collocated dual actuators and sensors, since the secondary force requires another structure to react off. Some effort must be devoted to ensuring that the system is not destabilised by the secondary force reacting off the base structure in the case of isolation. Nevertheless, impressive reductions in vibration can be achieved by such decentralised systems. It is not possible to selectively control different modes on a structure using decentralised control, however.

As a final demonstration of the prevalence of a decentralised approach, the active process in the inner ear is considered as a distributed control system. Although we only have an imperfect knowledge of the exact mechanisms of the active hearing process that makes our ears so acute, it is clear that the outer hair cells play an important role. These cells can respond mechanically to an electrochemical excitation with a bandwidth of tens of kilohertz, and are thought to generate nonlinear feedback loops acting on the basilar membrane, tuned closed to instability to amplify its response. At very low levels of excitation these loops appear to be automatically tuned very close to the point of instability, so that some of them are not stable and have limit cycles that give rise to spontaneous otoacoustic emissions in most peoples' ears. Very little is known about this automatic tuning process and it may be that by studying this process more will be learnt about how to tune distributed active control systems.

REFERENCES

[1] Meirovitch, L., *Dynamics and Control of Structures*, Wiley, 1990.

[2] Fuller, C.R., Elliott, S.J. and Nelson, P.A., *Active Control of Vibration*, Academic Press, 1996.

[3] Preumont, A., *Vibration Control of Active Structures, An Introduction*, 2nd edition, Kluwer Academic Publishers, 2002.

[4] Sun, J.Q., Some observations on physical duality and collocation of structural control sensors and actuators, *J. Sound Vib.* **194**, 1996, 765–770.

[5] Balas, M.J., Direct velocity feedback of large space structures, *J. Guidance and Control* **2**, 1979, 252–253.

[6] Benhabib, R.J., Iwens, R.P. and Jackson, R.L., Stability of large space structure control systems using positivity concepts, *J. Guidance and Control* **4**, 1981, 487–493.

[7] Gawronski, W.K., *Dynamics and Control of Structures, A Modal Approach*, Springer Verlag, 1998.

[8] Karnopp, D., Active and semi-active vibration isolation, *ASME J. Mechanical Design* **117**, 1995, 177–185.

[9] Elliott, S.J., Distributed control of sound and vibration, in *Proc. ACTIVE04*, 2004.

[10] Frampton, K.D., Decentralized vibration control in a launch vehicle payload fairing, in *Proc. of the ASME Symposium on Active Noise Control in Transportation Systems*, Vol. 29, 2002, pp. 155–160.

[11] Chou, K.C., Guthart, G.S. and Flamm, D.S., A multiscale approach to the control of MIMO systems, in *Proc. IEEE Conf. on Decision and Control*, 1995, pp. 3706–3711.

[12] Aubrun, J.N., Theory of the control of structures by low-authority control, *J. Guidance and Control* **3**, 1980, 444–451.

[13] Ungar, E.E., *Vibration Isolation*, in *Noise and Vibration Control Engineering*, L. Beranck and I.L. Ver (eds), Wiley, Chichester, 1992, Chapter 11.

[14] Elliott, S.J., Serrand, M. and Gardonio, P., Feedback stability limits for active isolation systems with reactive and inertial actuators, *J. Vibration and Acoustics* **123**, 2001, 250–261.

[15] Serrand, M. and Elliott, S.J., Multichannel feedback control for the isolation of base-excited vibration, *J. Sound Vib.* **234**(4), 2000, 681–704.

[16] Elliott, S.J., Benassi, L., Brennan, M.J., Gardonio, P. and Huang, X., Mobility analysis of active isolation systems, *J. Sound Vib.* **271**, 2004, 297–321.

[17] Skogestad, S. and Postlethwaite, I., *Multivariable Feedback Control*, Wiley, New York, 1996.

[18] Kim, S-M., Elliott, S.J. and Brennan, M.J., Decentralized control for multichannel active vibration isolation, *IEEE Trans. on Control Systems Technology* **9**(1), 2001, 93–100.

[19] Huang, X., Elliott, S.J. and Brennan, M.J., Active isolation of a flexible structure from base vibration, *J. Sound Vib.* **263**, 2003, 357–376.

[20] Pickles, J.O., *An Introduction to the Physiology of Hearing*, 2nd edition, Academic Press, London, 1988.

[21] de Boer, E., Auditory physics. Physical principles in hearing theory III, Physics Report, *Review Section of Physics Letters* **203**(3), 1991, 125–231.

[22] Le Henaff, B., Elliott, S.J. and Maury, C., Modelling wave propagation in the cochlea, ISVR Technical Memorandum 925, University of Southampton, 2003.

[23] Neely, S.T. and Kim, D.O., A model for active element in cochlear biomechanics, *J. Acoust. Soc. Am.* **79**(5), 1986.

[24] Yates, G.K., Basilar membrane nonlinearity and its influence on auditory nerve rate-intensity functions, *Hearing Research* **50**, 1990, 145–162.

[25] Elliott, S.J. and Harte, J.M., Models for compressive nonlinearities in the cochlea, ISVR Technical Memorandum 913, University of Southampton, 2003.

[26] Gelfand, S.A., *Hearing*, 4th edition, Marcel Dekker, 2004.

[27] Moore, B.C.J., *An Introduction to the Psychology of Hearing*, 5th edition, Academic Press, 2003.

CONTROL OF SHIP-MOUNTED CRANES

Ali H. Nayfeh, Ziyad N. Masoud, Nader A. Nayfeh and Eihab Abdel-Rahman
Department of Engineering Science and Mechanics, Virginia Polytechnic Institute and State University, Blacksburg, VA 24061, USA
anayfeh@vt.edu

Abstract: We developed an integrated systems solution, which enables the transfer of cargo from one pitching, rolling, yawing, surging, heaving, and swaying ship to another pitching, rolling, yawing, surging, heaving, and swaying ship adjacent to it without damage to the ships or the material, personnel, or equipment. The total solution has been demonstrated through computer simulations, which account for the interaction of the two ships and the waves. Also, the total solution has been demonstrated in the Computer Automated Virtual Environment (CAVE) at Virginia Tech. As a part of the total systems solution, we also developed a process to ensure safe and expeditious ship approach, connection of the ships, minimization of the absolute as well as relative motions of the two ships, dynamic handling of the moored-ship assembly, and separation of the ships in an open ocean environment and in sea conditions up to sea state five, as defined by the Pierson–Moscowitz scale.

The simulation is based on a dynamic model of the ships and their interactions with each other, waves, winch/mooring system, inflatable fenders, cranes, and hoisted cargo. The model was used to produce a design that minimizes the absolute and relative motions of moored ships and prevents damaging metal-on-metal contact between them. Then, the motions were used to tune the crane control system, which was in turn used to provide the specifications of the cranes needed to safely and reliably transfer materials between the two ships. The crane control system consists of a pendulation control component, a motion prediction component, and a soft-landing component. The latter will protect against impact between the hoisted cargo and both of the ship decks throughout the pick-up, maneuvering, and deposit steps (the transfer maneuver).

Key words: time-delay control, sea basing, pendulation control, container cranes, ship-mounted cranes.

1. INTRODUCTION

The United States Navy is aiming for a transformational sea basing system, which would permit sustaining logistic operations with significantly reduced reliance on land bases. This would be particularly useful in areas where suit-

H. Ulbrich and W. Günthner (eds), IUTAM Symposium on Vibration Control of Nonlinear Mechanisms and Structures, 21–35.

able secure shore or port facilities are not available, or where over-all costs for shore bases would be uneconomically high. This requires the development of a high capacity, high reliability at-sea capability to transfer fuel, cargo, vehicle, and personnel in rough seas while underway from commercial container ships to large sea basing ships and then to smaller ships. The wave-induced motion of the crane ship can produce large pendulations of the cargo being hoisted and cause the operations to be suspended. To solve this problem, the Office of Naval Research chose three teams to develop concepts to enable the transfer. One of these teams is lead by Lockheed-Martin and comprises Virginia Tech, MacGregor Cranes, D&K Engineering, Sea Systems, and Synthesis Partners.

Our team has developed an integrated systems solution, which enables the transfer of cargo from one pitching, rolling, yawing, surging, heaving, and swaying ship to another pitching, rolling, yawing, surging, heaving, and swaying ship adjacent to it without damage to the ships or the material, personnel, or equipment. This particular proposed transfer process is a skin-to-skin replenishment in a more rigorous sea environment than sea state 3. The integrated system includes a process to ensure safe and expeditious ship approach, connection of ships, minimization of the absolute and relative motions between the ships, dynamic handling of the moored-ship assembly, separation of the ships in an open ocean environment and in sea states up to and including sea state 5, and safe cargo transfer.

2. SHIP STABILIZATION

As part of the integrated solution, the Navy ship (called MPF(F)) will deploy active anti-roll tanks and fenders and mooring lines to secure the two ships together, minimize their absolute and relative motions, and prevent damaging metal-on-metal contact. To accomplish this, we have developed an active anti-roll/mooring system as follows. The smaller containership is placed in the lee of the larger ship, the MPF(F), as shown in Figure 1. This shelters the smaller and livelier ship from incident waves. The heading of the MPF(F) is restricted to keep incident waves within a range of either off head seas on the bow or off following seas on the stern. The speeds of both ships are restricted to a range within 3–8 knots. We found that this configuration minimizes the roll motions of both ships and the relative yaw between them. Active anti-roll tanks are installed on the MPF(F) to control its heave, pitch, roll, and yaw, thereby allowing it to act as an absorber to the motions of the containership. The MPF(F) will deploy 4–8 inflatable fenders along the parallel sections (amidships) of the two ships. The fenders are placed fore and aft of the center of gravity of the MPF(F). The fenders are used as a passive shock absorber to provide a soft ride for the containership over the MPF(F) and prevent skin-to-skin contact between them. The modularity of the inflatable fenders allows for a balance

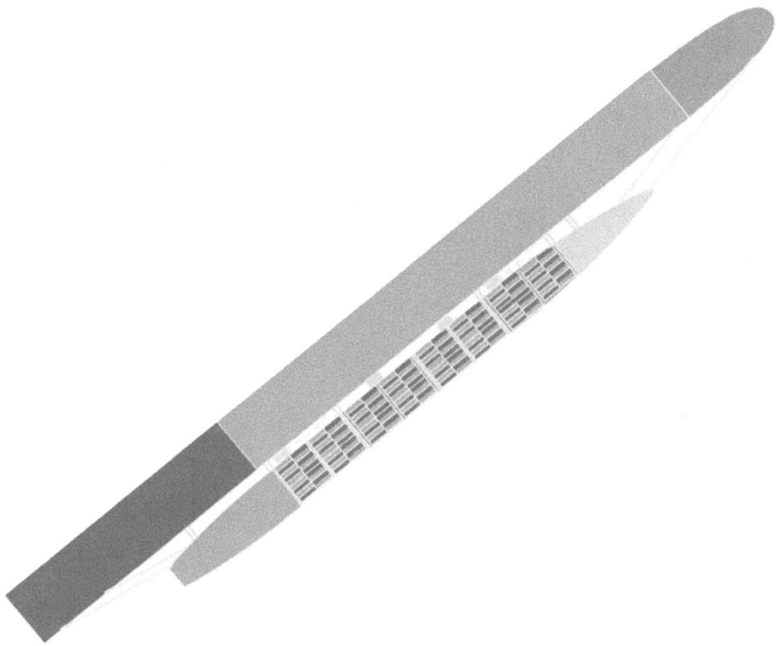

Figure 1. Configuration of moored-ship assembly.

between absorbing the energy of the relative ship motions and maintaining a minimum standoff distance between them. Active mooring winches and lines are installed on and deployed by the MPF(F). Fifteen to twenty-five winches are placed along one side of the MPF(F) and act on the containership through bow lines, stern lines, spring lines, and breast lines. Some of the breast lines are attached above the deck of the containership and dubbed “High Breast Lines” and others are deployed at the level of the containership deck and dubbed “Low Breast Lines.” The breast and spring lines are distributed fore and aft of the centers of gravity of the two ships.

We simulated the motions of two notional ships (the Bob Hope and the Argonaut) connected and stabilized using a preliminary design of the active mooring system (4 fenders and 22 mooring lines). The simulations were performed using the Multiple-Body version of the Large Amplitude Motions Program (LAMP) of SAIC. All mooring lines were rated to carry a maximum load of 200 KN. The lines were designed to carry active, continuously varying loads defined by a closed-loop, feedback control strategy for each line. The control strategy uses the line length and the states of both ships to define the load for each winch. We used a conservative estimate of the damping effect of the anti-roll tanks on the motion of the Bob Hope. We succeeded in stabilizing the relative and absolute motions of both ships in sea state 4 for all wave headings

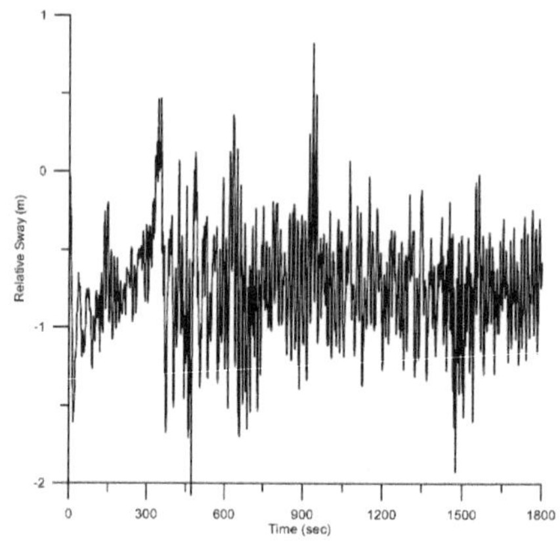

(a) relative sway

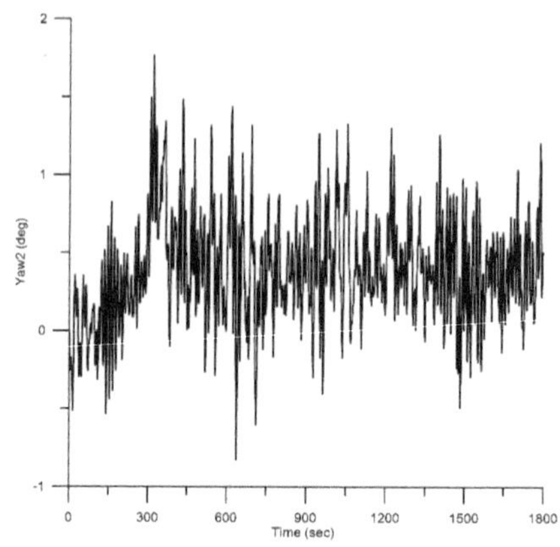

(b) relative yaw

Figure 2. Relative (a) sway and (b) yaw between the Bob Hope and the Argonaut in SS4.

in a range of 15°–25° off head and following seas and for speeds ranging from 3–8 knots. We show in Figures 2–5 the responses of the ships over a period of 30 minutes in sea state 4 with the incoming wave being incident on the Bob Hope stern at 20° off following seas.

These simulations and others show that the system maintains a skin-to-skin configuration with the relative sway being less than 1.5 m, the minimum ship separation being more than 5 m, and the relative yaw being less than 1°. Moreover, the system controls the heaves of the Bob Hope and Argonaut to less than 0.6 m and 1.0 m, pitch angles to less than 0.25° and 1.5°, and roll angles to less than 0.5° and 1.3°, respectively. Consequently, this system enables the crane-control system to transfer cargo softly and safely between these two ships in sea state 4.

3. CRANE-CONTROL SYSTEM

We have developed a system for transferring safely and reliably cargo from a surging, heaving, swaying, pitching, rolling, and yawing vessel to another surging, heaving, swaying, pitching, rolling, and yawing vessel in rough seas. The system consists of four components: sensors, a pendulation controller, a soft pick-up/landing controller, and a motion predictor. A sensor system measures and monitors the absolute motion and orientation of the MPF(F) ship and the relative motion between the crane tip and target location. This information is fed into the controller and used to correct the operator commanded crane motions. The safe pick-up/landing component depends on the motion predictor. Combination of these two components protects against impact between the hoisted cargo and the decks of both ships throughout the pick-up and landing

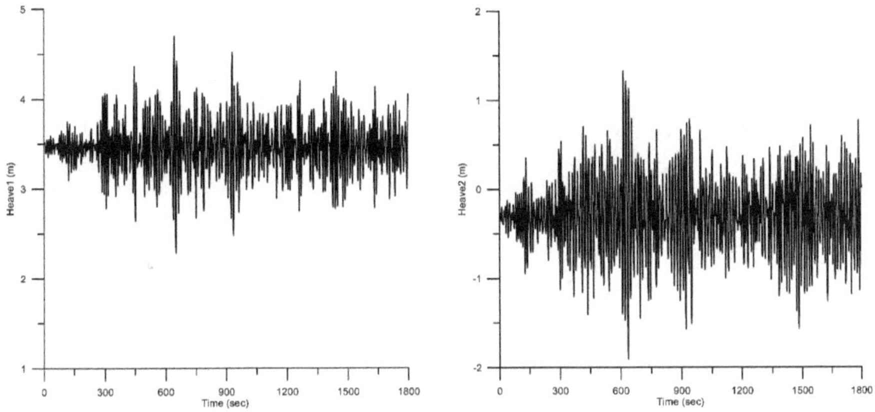

(a) Bob Hope heave (b) Argonaut heave

Figure 3. Heave of the (a) Bob Hope and (b) Argonaut in SS4.

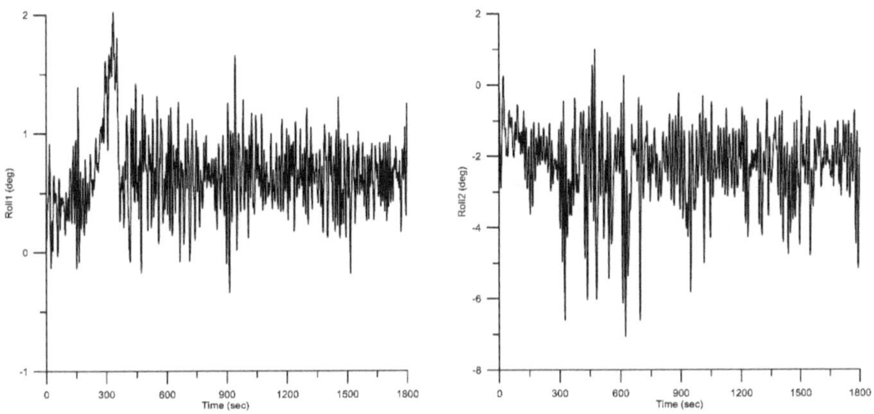

(a) Bob Hope roll (b) Argonaut roll

Figure 4. Roll of the (a) Bob Hope and (b) Argonaut in SS4.

stages. The pendulation controller prevents cargo pendulations throughout the transfer maneuver, ensuring both safety and fast turnover. The four components are integrated into a single system, which is designed so that the crane operator can manually maneuver the cargo using standard input devices. The integrated system is transparent to the operator, thereby maximizing safety and speed while eliminating the need for special operator training. A schematic of the integrated system is shown in Figure 6. In this paper, we concentrate on the pendulation controller.

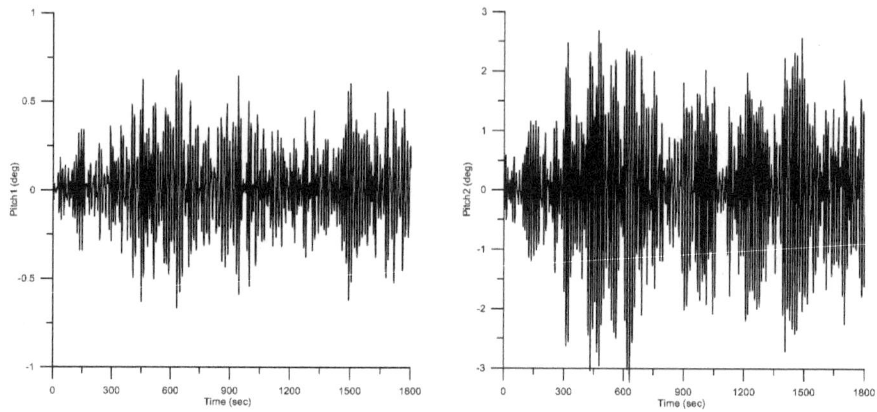

(a) Bob Hope pitch

(b) Argonaut pitch

Figure 5. Pitch of the (a) Bob Hope and (b) Argonaut in SS4.

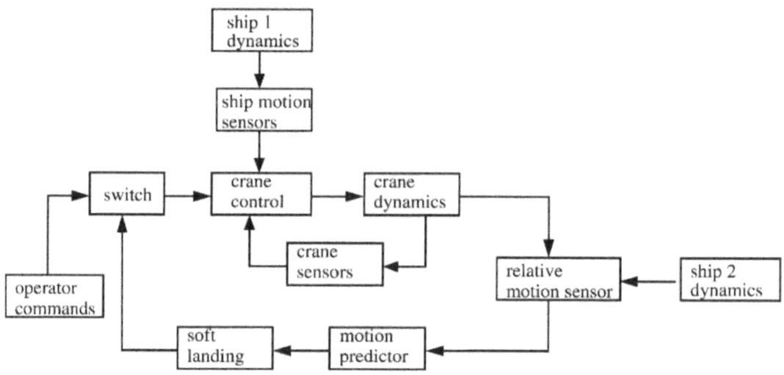

Figure 6. A diagram of the crane-control system.

4. PENDULATION CONTROLLER

The pendulation controller is based on a time-delayed position-feedback strategy, which has been applied to four of the most commonly used cranes: ship-mounted boom cranes, tower cranes, telescopic cranes, and quay-side container cranes. The controller is more advanced than the current technology. Instead of compensating for the time delays inherent in the crane motors, the controller makes use of them. The controller is insensitive to the weight of the payload and has a low sensitivity to the length of the hoist line, which is one of the most limiting parameters in the current crane control theory.

We discuss only feedback controllers because feedforward controllers demand predefined maneuvers. Moreover, optimal control techniques and input shaping techniques are limited by the fact that they are extremely sensitive to variations in the parameter values about the nominal values and to changes in the initial conditions and external disturbances and that they require "highly accurate values of the system parameters" to achieve satisfactory system response (Zinober and Fuller, 1973; Virkkunen and Marttinen, 1988; Yoon et al., 1995). While a good design can minimize the controller's sensitivity to changes in the payload mass, it is much harder to alleviate the controller's sensitivity to changes in the cable length. In fact, Singhose et al. (1997) showed that input shaping techniques are sensitive to the pendulation natural frequency. As a result, they suffer significant degradation in crane maneuvers that involve hoisting. Furthermore, input shaping and optimal control techniques require a predetermined endpoint of the transport process. This makes them less practical because most crane operations are coordinated visually by the crane operator.

Linear controllers and static feedback linearization control techniques have very poor performance and usually fail due to the highly nonlinear nature of the payload oscillations (d'Andrea-Novel and Levine, 1989). Moreover, these control strategies are of limited use when excitations are introduced through the base of the crane, as in ship-mounted cranes. Imazeki et al. (1998) used a 35-ton active mass-damper system to control the pendulations in one plane of a barge-mounted crane. The system reduced the payload pendulations by 67%. Patel et al. (1987), McCormick and Witz (1993), and Witz (1995) presented a three-dimensional nonlinear model of a ship-mounted boom crane. Based on computer simulations, they found that "The hook load does not significantly affect the roll motions of the vessel." Thus it is safe to decouple the motions of the ship from that of the crane and ship.

Henry et al. (2001) and Masoud et al. (2003, 2003a, 2003b) developed a strategy by which cargo pendulations of a crane payload are significantly suppressed by forcing the suspension point of the payload hoisting cable to track inertial reference coordinates, which consist of a percentage of the delayed motion of the payload in the inertial horizontal plane relative to the suspension point superimposed on the operator commanded motion. For boom cranes, the in and out-of the plane of boom and crane tower (in-plane and out-of-plane) are controlled by simply actuating the luff and slew angles of the boom. These degrees of freedom already exist in ship-mounted cranes, and therefore modification to the hardware of current cranes would be limited to the addition of a few sensors and electronics to execute the control algorithm. The control strategy is based on time-delayed position feedback of the payload cable angles. This control algorithm is superimposed transparently on the input of the crane operator, which eliminates any special training requirements for crane operators and furnishes smoother and faster transport operations. The stability of the in-plane cargo pendulations using delayed position feedback controller was in-

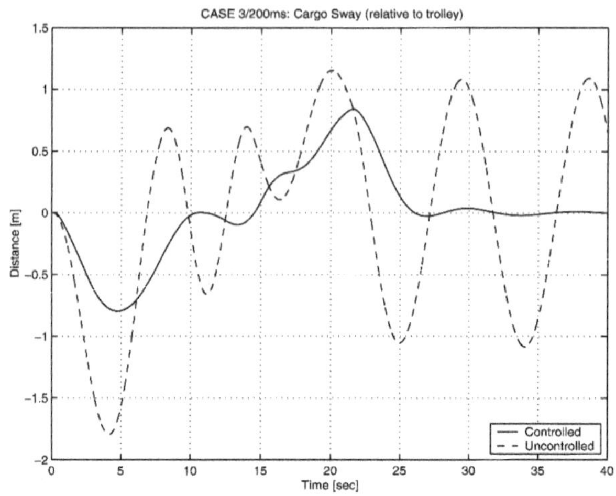

Figure 7. Payload sway of a 65-ton container crane.

vestigated by Henry et al. (2001) and was proven robust under significantly large base excitations and large initial disturbance conditions.

The effectiveness of the control strategy has been demonstrated using numerical simulations of computer models of ship-mounted (the T-ACS crane), rotary, and container cranes. Furthermore, the control strategy has been applied to, and tested on, experimental scaled models of rotary cranes operating in both rotary and gantry modes of operation and the T-ACS crane excited by the motion of a platform with three degrees of freedom, which correspond to the heave, pitch, and roll of a ship. In this paper, we summarize the performance of the controller for quay-side and ship-mounted cranes.

4.1 Quay-Side Container Cranes

The performance of the controller has been demonstrated using a fully nonlinear computer model of a 65-ton container crane with its actual hoisting assembly, including a 15-ton spreader bar. The operation requirement is to move the container from a ship at port to a waiting truck 50 m away in 21.5 seconds. At the beginning of the transfer maneuver, the container is raised from 35 m below the trolley to 20 m and then is lowered 15 m to a waiting truck. The container pendulation has to settle to less than 50 mm within 5 seconds after the end of the transfer. The simulations in Figure 7 show that the container settles to less than 50 mm in 4 seconds after the end of the traversing motion although the lowering maneuver was concluded at 28.5 seconds.

In the summer of 2002, the performance of the controller was demonstrated using IHI's 1/10-scale model of a 65-ton container crane, Figure 8, at their

Figure 8. 1/10 scale model of a 65-ton container crane.

Yokohama Research Facility, Japan. The crane model has all essential degrees of freedom of the actual crane. These degrees of freedom are actuated using driving mechanisms comparable to those used on the actual crane. The model is equipped with the sensors necessary to measure the states of the system. The model has a 4.8 m track, a 15-kg spreader bar, and a 65-kg payload. The payload consists of four 250 by 250 mm steel slabs, mounted to the bottom of the spreader bar. The trolley is driven via a cable-pulley system. Four cables are used as a hoisting mechanism. The distance separating the front and rear hoisting cables at the trolley is 282 mm. The four cables are then attached to the spreader bar. The distance between the front and rear hoisting cables on the spreader bar is 141 mm. A heavy set of cables supported on one side of the spreader bar, shown in Figure 8, is used to transmit signals from and to a number of sensors and actuators on the spreader bar. These sensors and actuators are used for experiments conducted by IHI. DC brushless rotary motors are used to drive the trolley and the hoisting mechanism. Each motor is equipped with an optical encoder to track its motion. Amplifiers with internal PI velocity tracking controllers are used to drive the crane motors. An optical encoder attached to one of the hoisting cables was used to measure the payload sway.

To demonstrate the robustness of the controller, we simulated several cases covering a variety of payload handling maneuvers. We then tested the same cases on the experimental scaled model. In addition to these cases, the controller performance was tested experimentally in the manual mode of operation using direct input from the crane joysticks. Only controlled experimental tests were conducted due the hazardous conditions, which would result from un-

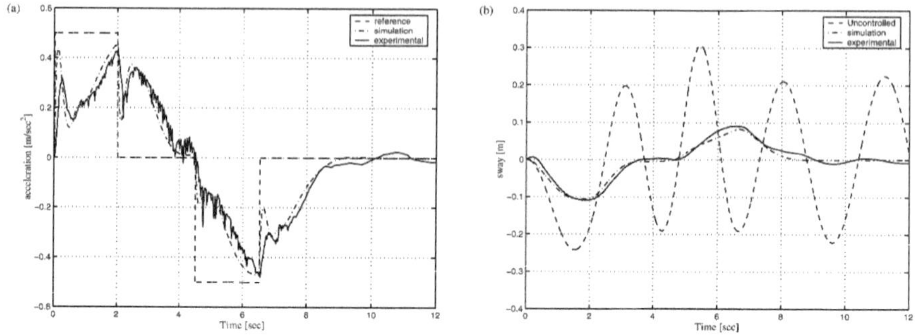

Figure 9. (a) Reference and controlled trolley accelerations (b) controlled and uncontrolled payload sway.

controlled tests, as predicted by the numerical simulations. The commanded trolley-acceleration profile for all of the tested cases started with a 2-second acceleration phase with a magnitude of 0.5 m/s^2. The trolley then coasted for 2.5 seconds at a speed of 1 m/s. Finally, a deceleration of 0.5 m/s^2 for 2 seconds brought the trolley to a stop at the target position in 6.5 seconds.

In a sample case, the center of gravity of the payload was initially placed at 3.5 m below the trolley. The payload was then raised 1.5 m in the first 4.5 seconds of the transfer maneuver with a maximum acceleration of 0.4 m/s^2. The controlled experimental acceleration significantly matched the simulated controlled acceleration. The simulation and experimental results show that within 1.7 seconds of the end of the commanded maneuver, the amplitude of the payload sway of the controlled system was reduced to less than 5 mm, as shown in Figure 9.

The controller performance was also demonstrated in the manual mode. Again, the controller significantly reduced the pendulations throughout the maneuvers to the extent that they could not be observed by the naked eye. Hence, we installed a laser on the load and pointed it at the wall behind the apparatus; even so, the observed pendulations were very small, they were less than 5 mm. Twenty-five people drove the controlled crane and no one was able to destabilize the controller.

Recently, we built a 1/10-scaled model of the 65-ton container crane at Virginia Tech.

4.2 Ship-Mounted Cranes

The performance of the controller has been also demonstrated using a fully nonlinear three-dimensional computer model of the US Navy T-ACS ship. The center of gravity of the hoisted cargo is 27.1 m below the boom tip, making the natural frequency of the payload pendulation 0.096 Hz. A linear damping factor of 0.002 was used. An inextensible massless rigid link was used to

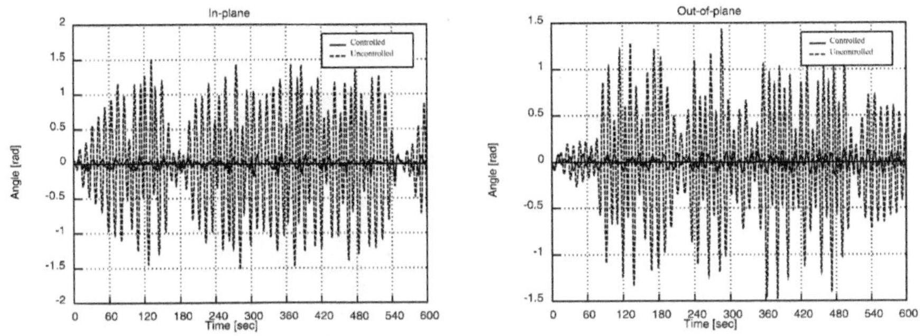

(a) in-plane angles (b) out-of-plane angles

Figure 10. Computer simulations: (a) in-plane and (b) out-of-plane angles of the payload cable as functions of time. The ship was excited sinusoidally in roll and pitch at the natural frequency of the payload pendulation and sinusoidally in heave at twice the natural frequency of the payload pendulation and the crane was performing a slewing action through 90° and back every 40 seconds.

model the hoisting cable. The payload is excited via primary resonance and principal parametric resonance by setting the ship's pitch and roll frequencies equal to the pendulation frequency and the heave frequency equal to twice the pendulation frequency.

Three sets of simulations were performed. In the first set, the crane was oriented so that the boom was extended over the side of the ship perpendicular to its axis. The boom luff angle was set equal to 45°. In the uncontrolled simulation, the in-plane and out-of-plane pendulation angles grew rapidly to approximately 70° and 65°, respectively . On the other hand, in the controlled case, these angles remained within 1.5° and 1°, respectively.

At the beginning of the second set of simulations, the crane was oriented so that the boom was extended over the side of the ship perpendicular to its axis. The boom luff angle was set equal to 45°. The controller was turned off, and the crane operator executed a sinusoidal slewing action through 90° and back in every 40 seconds. The same simulation was then repeated with the controller turned on. The results of the controlled and uncontrolled in-plane and out-of-plane angles of the hoisting cable are shown in Figure 10. Clearly, the controller is very effective in drastically reducing the pendulations.

To demonstrate the robustness of the controller, the crane was oriented so that the boom was extended over the side of the ship and normal to the ship axis. The boom luff angle was set equal to 45°. The in-plane angle was given an initial 60°. The crane was subjected to the same roll, pitch, and heave excitations as in the two previous simulations. While the uncontrolled response grew to approximately 100°, the controlled response dropped rapidly and remained within 2°.

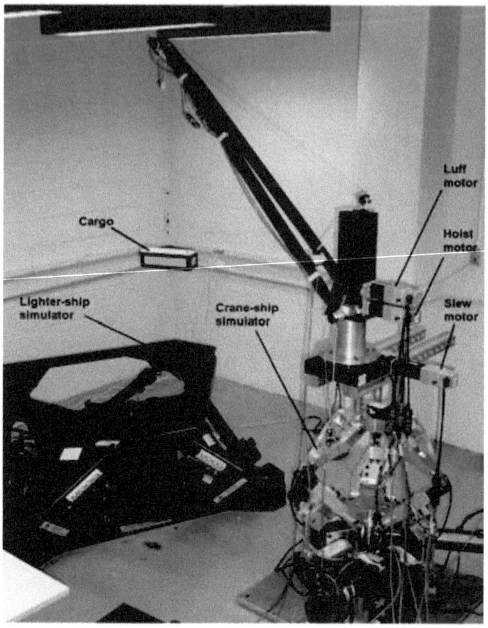

Figure 11. A 1/24 scale model of the T-ACS crane mounted on a three-degrees-of-freedom platform.

To validate the theory and the computer simulations, we built an experimental model. A 1/24-scale model of a T-ACS crane was mounted on the moving platform of a Carpal wrist mechanism, as shown in Figure 11. A rotary platform with a 1:45 gear ratio was used to give the crane its slewing degree of freedom. The platform used is capable of producing arbitrary independent roll, pitch, and heave motions. A desktop computer supplies the rolling, pitching, and heaving commands to the platform motors. The platform was driven to simulate the motion of the crane ship. A 1/24-scale model of an 8 ft by 8 ft by 20 ft container weighing 20 tons was used as a payload. The center of gravity of the payload was located 1 m below the boom tip. This length yields a pendulation frequency of 0.498 Hz.

Experiments were carried out for the worst-case scenario of sinusoidal motions at the critical frequencies. Throughout these experiments, the platform and the crane model were excited sinusoidally by 2° in roll and 1° in pitch at the pendulation frequency (0.498 Hz) and by 1.27 cm in heave at twice the pendulation frequency. Two sets of experiments were conducted. In the first set, the crane model was extended over the side of the modeled ship and perpendicular to its axis. The boom luff angle was set equal to 45°. In the uncontrolled case, as shown in Figure 12, the excitation caused the amplitude of the pendulation angles to grow rapidly, and the experiment had to be stopped at 10 seconds when the in-plane pendulation angle reached approximately 70°. The same

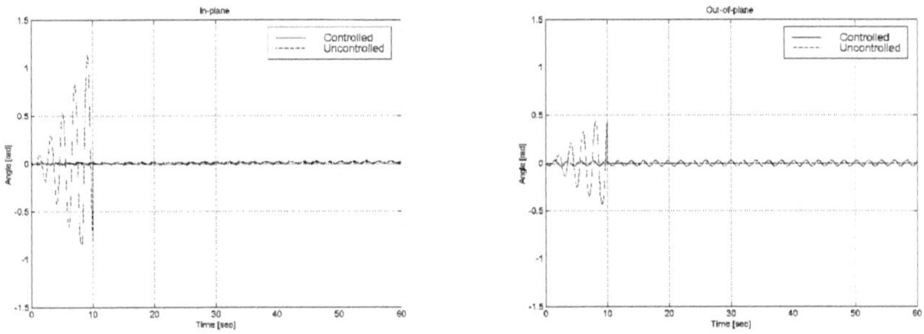

Figure 12. Experimental results: in-plane and out-of-plane oscillation angles of the payload.

experiment was then repeated with the controller turned on. The maximum amplitudes of the in-plane and out-of-plane pendulation angles remained less than 1.5° and 2°, respectively.

In the second set, the crane model was initially extended over the side of the modeled ship and perpendicular to its axis. The boom luff angle was set equal to the crane model and was extended over the side of the modeled ship, perpendicular to its axis. The boom luff angle was set equal to 45°. The crane operator performed a sinusoidal slewing action from 0° to 90° every 8 seconds. In the uncontrolled case, the excitation together with the slewing action caused the amplitude of the pendulation angles to grow rapidly, and the experiment had to be stopped at 10 seconds when the in-plane pendulation angle reached approximately 70°. The same experiment was then repeated with the controller turned on. The maximum amplitude of the in-plane and out-of-plane pendulation angles remained less than 2°.

In the third set, the experiment started with the controller turned off. After a few seconds, and when the in-plane pendulation angle of the payload had increased to over 20°, the controller was turned on. After the controller was turned on, the pendulation angles of the payload dropped to less than 1° in 10 seconds, as shown in Figure 13.

Figure 14 shows the maximum crane rate of a MacGreggor boom crane and the requirements of actuation in the luff direction for the same crane. Three sea states were simulated for the Bob Hope Argonaut in sea state 5 with 20° following off stern waves, sea state 4 with 15° waves off head seas, and sea state 5 with 15° waves off head seas. In each case, the maximum requirements of the controller were far less than the available power.

5. CONCLUDING REMARKS

We have designed and implemented a controller based on delayed position feedback that is applicable to most commercial and military cranes: ship-

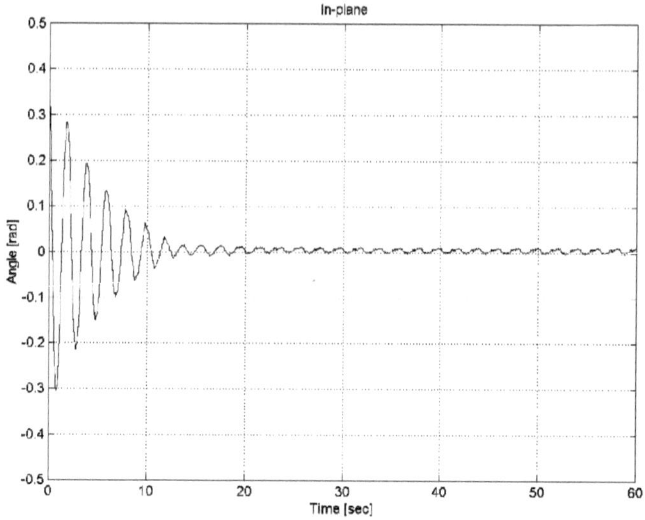

Figure 13. Experimental results: in-plane payload oscillation angle with 20° initial disturbance.

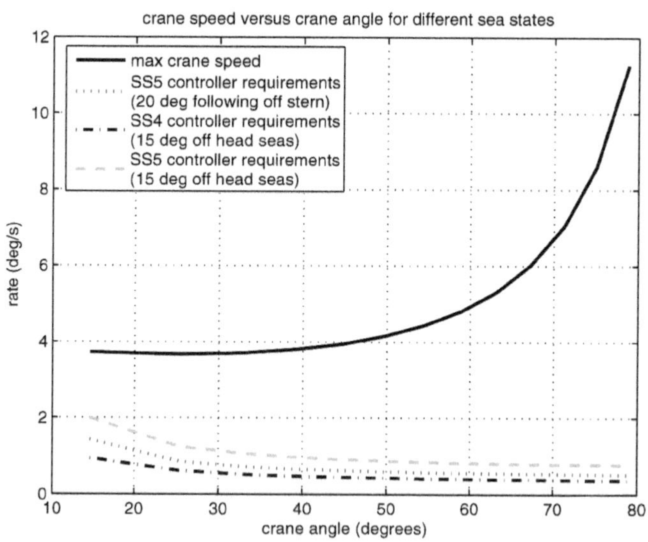

Figure 14. A comparison of the maximum crane rate with the maximum requirements of the controller for different sea states.

mounted, rotary, and container cranes. Significant reductions in the pendulation angles can be achieved with relatively small control inputs. The system can be operated in both the automated and manual modes. Neither the trajectory nor the end point of the transport maneuver need be predefined.

REFERENCES

d'Andrea-Novel, B. and Levine, J. (1989). Modeling and Nonlinear Control of an Overhead Crane. In *Proceedings of the International Symposium MTNS-89*, Vol. 2, pp. 523–529.

Henry, R., Masoud, Z., Nayfeh, A. and Mook, D. (2001). Cargo Pendulation Reduction on Ship-Mounted Cranes via Boom-Luff Angle Actuation. *Journal of Vibration and Control*, 7:1253–1264.

Imazeki, M., Mutaguchi, M., Iwasaki, I. and Tanida, K. (1998). Active Mass Damper for Stabilizing the Load Suspended on a Floating Crane. *IHI Engineering Review*, 31(2):61–69.

Masoud, Z.N. and Nayfeh, A.H. (2003). Sway Reduction on Container Cranes using Delayed Feedback Controller. *Nonlinear Dynamics*, 34(3–4):347–358.

Masoud, Z.N., Nayfeh, A.H. and Al-Mousa, A. (2003a). Delayed Position-Feedback Controller for the Reduction of Payload Pendulations of Rotary Cranes. *Journal of Vibration and Control*, 9(1):257–277, 2003a.

Masoud, Z.N., Nayfeh, A.H. and Mook, D.T. (2003b). Cargo Pendulation Reduction of Ship-Mounted Cranes. *Nonlinear Dynamics*, 35(3):299–311.

McCormick, F. and Witz, J. (1993). An Investigation into the Parametric Excitation of Suspended Loads During Crane Vessel Operations. *Underwater Technology, V*, 19:30–39.

Patel, M., Brown, D. and Witz, J. (1987). Operability Analysis for a Monohull Crane Vessel. *Transactions of the Royal Institute of Naval Architects*, 129:103–113.

Singhose, W.E., Porter, L.J. and Seering, W.P. (1997). Input Shaped Control of a Planar Crane with Hoisting. In *Proceedings of the American Control Conference*, Albuquerque, NM, pp. 97–100.

Virkunnen, J. and Martinnen, A. (1988). Computer Control of a Loading Bridge. In *Proceedings of the IEE Control Conference: Control '88*, Oxford, UK, pp. 484–488.

Witz, J. (1995). Parametric Excitation of Crane Loads in Moderate Sea States. *Ocean Engineering*, 22:411–420.

Yoon, J.S., Park, B.S., Lee, J.S. and Park, H.S. (1995). Various Control Schemes for Implementation of the Anti-Swing Crane. In *Proceedings of the ANS 6th Topical Meeting on Robotics and Remote Systems*, Monterey, CA, pp. 472–479.

Zinober, A.S. and Fuller, A.T. (1973). The Sensitivity of Normally Time-Optimal Control of System to Parameter Variations. *International Journal of Control*, 17:673–703.

MASTER-SLAVE BILATERAL CONTROL OF HYDRAULIC HAND FOR HUMANITARIAN DEMINING

Kenzo Nonami
Department of Electronics and Mechanical Engineering, Chiba University, Japan

Abstract: At present, the Sensing and Access Control R&D for Humanitarian Mine Action project has been carrying out by the Japan Science and Technology Agency (JST). Chiba University's group join to this project. Our effort is to develop a small vehicle for mine detection and clearance. Specifically we are concerned with the development of a 6-degree-of-freedom robot arm for exposing buried anti-personnel mines, with plans of field-testing in Afghanistan in 2005. This paper describes the current state of development in the anti-personnel mine exposure and clearance system, focusing on the robot arm and the master-slave bilateral control hand system.

Key words: humanitarian demining, mine action, master-slave bilateral control hand system, mine hunter vehicle.

1. INTRODUCTION

Since 1999, our group has been involved in the research and development [1] of landmine detection and clearance technology with the aim of maximizing safety and increasing the overall speed in humanitarian demining as shown in Figure 1.

Currently Chiba University's group is participating in the Japan Science and Technology Agency's (JST) Sensing and Access Control R&D for Humanitarian Mine Action project [2] with our effort to develop a small vehicle named Mine Hunter Vehicle with two arms, one is the SCARA arm having mine detection sensor, the other one is the multiple link arm for mine clearance as shown in Figure 2. Specifically we are concerned with the development of a 6-degree-of-freedom robot arm for exposing buried anti-personnel mines, with plans of field-testing in Afghanistan sometime around the middle of 2005. We must therefore produce results in a relatively short period of time, and so we have divided the project into several tasks to be tackled.

H. Ulbrich and W. Günthner (eds), IUTAM Symposium on Vibration Control of Nonlinear Mechanisms and Structures, 37–53.

Figure 1. Mine detection robot COMET-III.

Figure 2. Mine hunter vehicle of Chiba University Group's robot in JST project.

Other groups participating in the JST project are Hirose's group from Tokyo Institute of Technology and Fukuda's group from Nagoya University including Tadano Ltd except advanced sensor R&D groups [3, 4]. The three groups have the same ultimate goal, but are taking different approaches. Hirose's group has put a lot of effort into its Mine hand device and a buggy vehicle with metal detector. It is low cost and structurally simple. Fukuda's group is developing a huge crane based on existing technology for GPR based mine detection. Our team, in cooperation with Fuji Heavy Industries and Sato's group from Tohoku University, and Arai's group from University of Electro-Communications has proposed a two-ton class small remote controlled vehicle-based system. The system is still in development, and we believe that when it is complete it will be effective in a practical setting.

This paper describes the current state of development in the anti-personnel mine exposure and clearance system, focusing on the robot arm. In particular, we have achieved the reasonable performance of a 6-degree-of-freedom robot with multi-function tool by means of nonlinear control based on "LOOK AT TABLE" scheme and also the master-slave hand.

2. PROJECT OUTLINE

The tasks prescribed to our group concerning the JST project are as follows.

(1) Anti-personnel mine detection.

(2) Anti-personnel mine exposure.

The task of landmine detection is being undertaken by Sato's group at Tohoku University, who are developing an array-style ground penetrating radar (SAR-GPR) [3, 5]. Fuji Heavy Industries and a member of our group are responsible for the development and control of an arm that the sensor will be mounted on. The rest of our group is focused on the task of anti-personnel mines exposure. After a mine is exposed, its disposal will be carried out in a usual manner, for example the mine may be exploded by placing a charge next to it. The disposal of anti- personnel mines lies outside the requirements laid out for our project, however we aim to complete a system capable of removing them.

3. HYDRAULIC ROBOT ARM AND CONTROL SYSTEM

3.1 The Hydraulic Robot Arm

The controlled object, the multifunction robot arm (Figure 3), has six joints and an end effector consisting of a gripper, electromagnet, drill, and air jet (Figures 4 and 5). The sixth joint is a tool changer that allows for switching between end effector tools as needed by the application. The arm is to be mounted on a mine detection vehicle that is being manufactured by Fuji Heavy Industries. The arm's primary specifications are listed in Table 1.

Hydraulic cylinders with proportional electromagnetic valves actuate the six joints. The speeds at which the cylinders extend and retract are proportional to the voltages applied to the valves. The gripper on the end effector is also hydraulically driven, but it employs an ON/OFF valve. The electromagnet, drill, and air jet are all activated by switch.

The arm's link configuration and joint ranges are shown in Figure 6. Each joint, with the exception of the base joint (Joint 1), has an absolute value encoder to measure the joint angle and eliminate noise. Joint 1 currently uses a potentiometer to measure the angle, but it will also be switched to an encoder

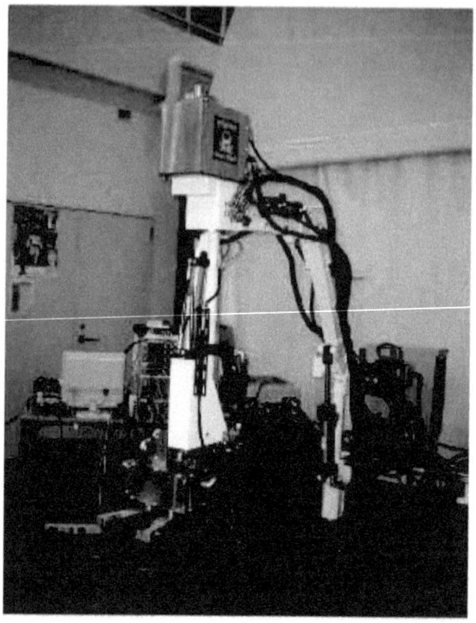

Figure 3. Multifunction robot arm.

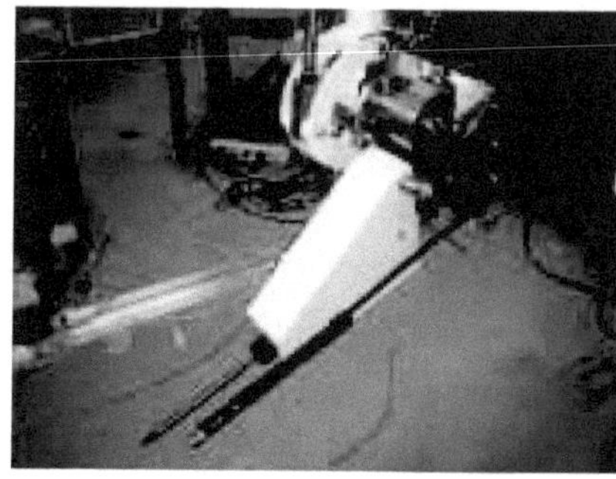

Figure 4. Drill and air jet nozzle.

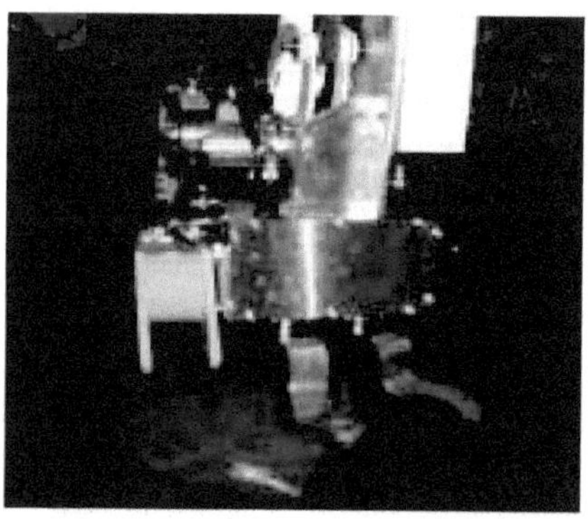

Figure 5. Gripper and electromagnet.

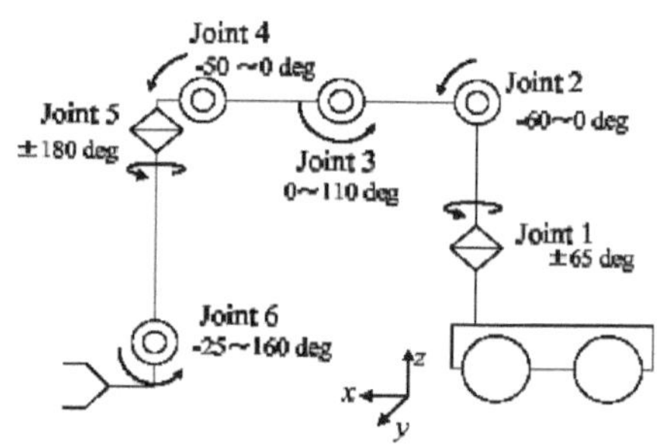

Figure 6. Configuration of multifunction robot arm.

in the near future. The main problem encountered with potentiometers is one of noise, and the coping method currently being used is to calculate a moving average of data over the last five samples. Strictly speaking this method produces a delay, but considering the desired rotational speed of the arm, this delay is not perceived as a problem.

Table 1. Specifications of the multifunction robot arm.

Length/height/width	2112 mm × 1536 mm × 307 mm
Degrees of freedom	6
Drive system	Hydraulic pressure (13.7 MPa, 10 L/min)
Position sensors	5 Absolute value encoders, 1 potentiometer
Range of motion	2-meter radius, 130-degree fan-shaped region
Mass	200 kg
Onboard tools	Drill, high-pressure air jet, electromagnet, gripper
2-finger gripper strength	10 kgf
Tool changer range of motion	180°
Position encoder precision	Horizontal direction: ±10 mm Vertical direction: ±10 mm
High-pressure air jet	Tank capacity: 24 m^3, Pressure: 10 Mpa, 1 tank lasts 200 times
Control methods	Vision-based, Joystick, etc.

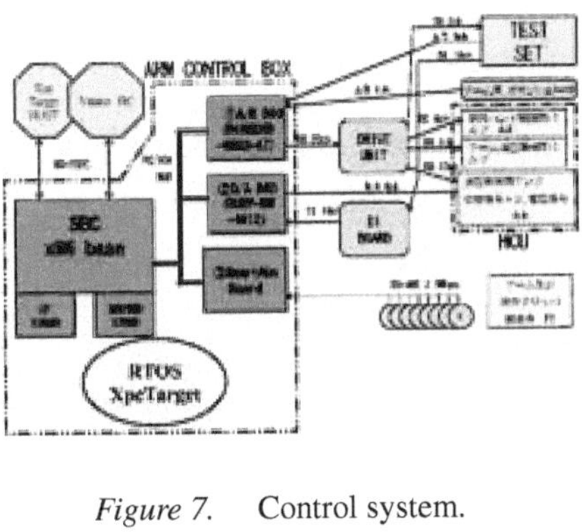

Figure 7. Control system.

Figure 8. PC/104.

3.2 Control System

The control system shown in Figures 7 and 8 is currently being developed for the MATLAB/Simulink xPC Target operating system. xPC Target is a real-time operating system for x86 processors that allows rapid prototyping using Simulink models. The target PC is a PC/104 system from Diamond Systems. The controller is developed in Simulink on a host PC and uploaded to the target PC for execution. Communication between the host PC and target PC is via RS232, and while the controller is running, the operator can manage the target

PC and issue commands from the host PC. We have found xPC Target to be an extremely useful tool for prototyping, but in the future we will have to move to an embedded system. Realtime Workshop converts a program composed in Simulink to C code. These tools along with Stateflow are very effective for developing large-scale systems in a short time. In the future we are considering using this C code and developing an embedded system.

4. IMPLEMENTATION TASKS

The tasks currently being worked on are described here. They will be combined with the aim of constructing the anti-personnel mine exposure system described in Section 6. These tasks are the core of our research and are currently under development.

4.1 Environment Recognition with Stereovision

Mine detectors can be used to determine the position of buried anti-personnel mines, but they cannot identify obstacles on the terrain surface that may hinder the robot arm. Stereovision can be used to generate a 3D map of the terrain and detect obstacles. The 3D terrain map and sensor data can be combined into a virtual reality scene to assist the operator in target selection. Terrain data and obstacle position data will also be transmitted to the arm?s controller and used in trajectory generation.

4.2 Reference Generation with Online Inverse Kinematics

When the positions of obstacles and buried mines are known, the various joints of the robot arm must be controlled to achieve target positions. In practice the positions of obstacles and mines will always be different, so arm positioning cannot be solved offline in advance. Therefore this system will always use reference generation with online inverse kinematics to solve the positioning problem.

4.3 Online Trajectory Generation

Vibration and jerky behavior in the arm is minimized by generating a smooth trajectory reference like Fig.9 between the current position and the reference position. The trajectory is calculated according to Eq. (1)

$$y = \frac{x^p - x^n}{2} \cos\left\{\frac{2\pi}{T}(t - t_n)\right\} + \frac{x^p + x^n}{2}. \tag{1}$$

Here, x_p is the initial reference position, x_n is the next reference position, T is the period in seconds, and t_n is the predefined time at which x_n is achieved. y is an intermediate reference point along the smooth trajectory. And t is the present time, t should be satisfied the following condition as

$$t_n \leq t \leq t_n + \frac{T}{2}. \tag{2}$$

4.4 Proportional Control with Input Compensation Using Table-Lookup

A traditional analytical approach to controller design was difficult due to time constraints and strong nonlinearities inherent in the system. As an alternative, a tuning-based algorithm was developed that worked around the nonlinear elements and yet proved to have reasonable performance. Section 5 discusses the algorithm.

5. CONTROL TECHNIQUE

The hydraulic manipulator has various nonlinearities such as valve dead zones, hydraulic cylinder stick-slip, and inertial changes depending on the posture of the arm [6, 7]. These nonlinearities make theoretical modeling and modeling from system identification very difficult, and development of a model-based control system would be very time-consuming. Due to limited time constraints within the project, we had to sidestep some common practices for improving control performance.

5.1 Creating the Input Compensation Table

The two primary factors contributing to nonlinear behavior in the control subject are the hydraulic elements and changes in arm posture. To determine the real system response with these nonlinear elements, we took experimental measurements of the cylinder velocities and joint angles for a range of input voltages to the valves shown in Figure 10. The measurements were used to construct reference tables for the voltage-velocity-angle relationship of every valve. Examples of voltage-velocity-angle relations are given in Figure 11. The plots show the measurement results for joint 2.

Sample data for the tables was collected and averaged for 20 sampling cycles, each lasting 1 minute. Averaging over a large number of sample sets increases the reliability of the data.

While the data was being collected, the hydraulic cylinders exhibited some small intermittent movements causing some data points to be unreliable. The voltage-velocity data sets were approximated by first and second order curves to smooth over the irregularities, as shown in Figure 12. The tables then allow an input voltage to be calculated for a desired rotational velocity. It is not necessary to change the compensation table, if the tool on end-effector

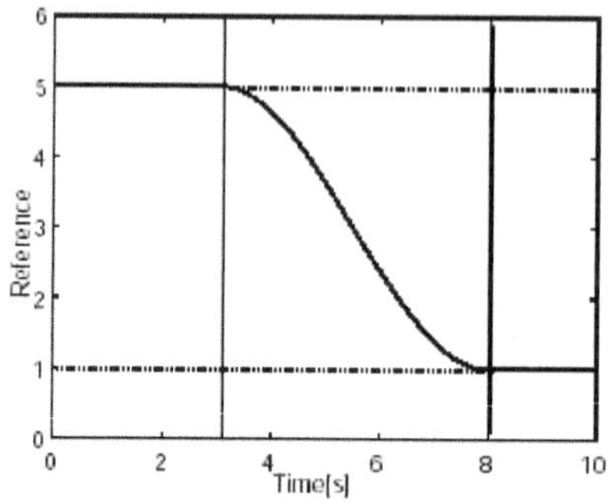

Figure 9. Online trajectory planning.

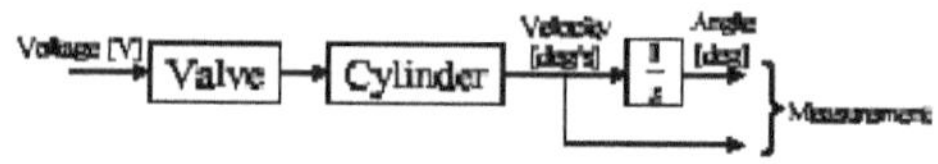

Figure 10. Measurement of velocity-voltage-angle relations.

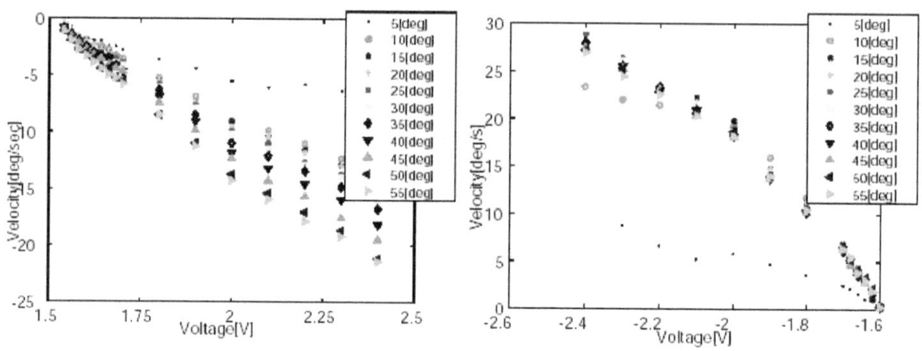

Figure 11. Velocity-voltage-angle relations.

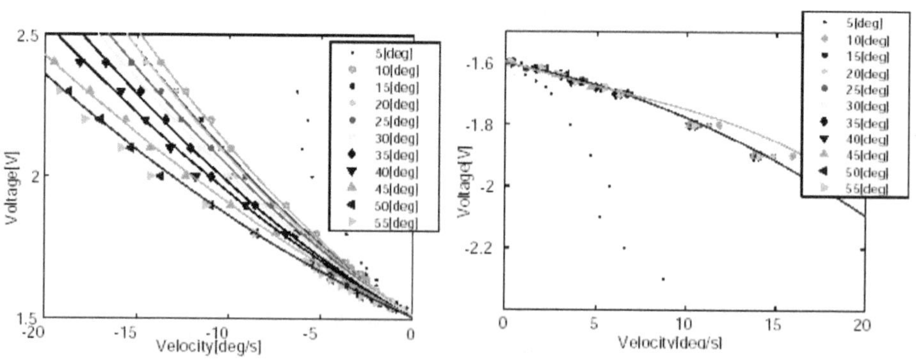

Figure 12. Reference table curves for link 2.

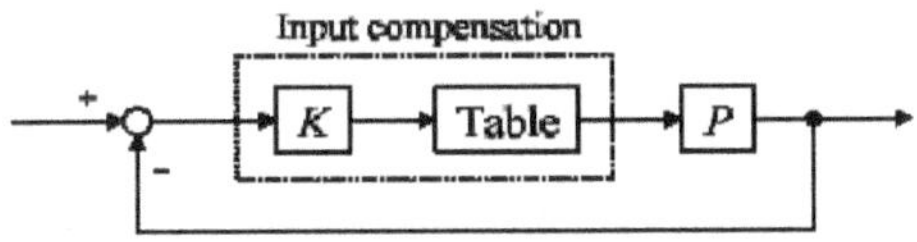

Figure 13. Block diagram of proportional control using input compensation.

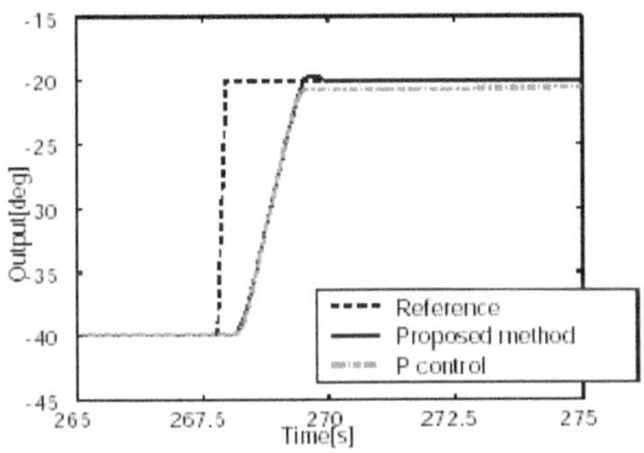

Figure 14. Rising step response.

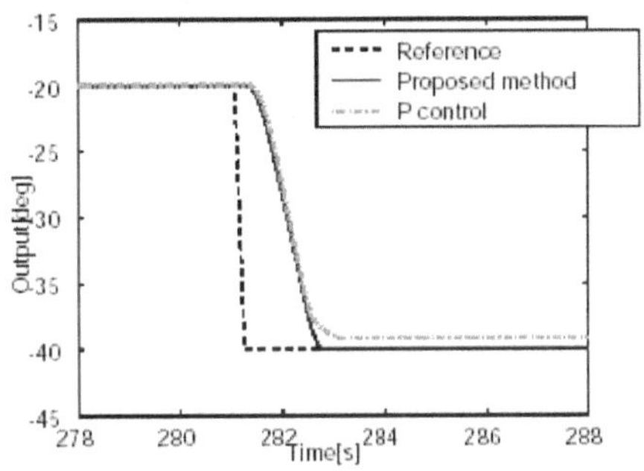

Figure 15. Falling step response.

is changed. Because the total dynamics does not change comparing with the changes in arm posture.

5.2 Experimentation

Experimentation was carried out to compare the performance two control strategies – a proportional controller with dead zone compensation and a new controller with the input compensation tables added. Figure 13 shows a block diagram of the proposed new controller. Tuning was introduced for the dead zone compensation controller while observing the arm's response to prevent abrupt motion.

5.2.1 Performance Comparison for a Single Joint. The step response and sine wave response of the system with both the proportional controller and input-compensated proportional controller are displayed in Figures 14–16. As before, these are the results for link 2. The figures show the reduction of dead zone effects and the lag of the arm's motion behind the reference signal. The proposed controller was able to compensate for dead zone effects without abrupt motion in the system response.

5.2.2 Trajectory Following. In practice several joints will move at the same time, and not just one, so the input-compensated proportional controller was tested for the case of multiple joints moving. Specifically, a trajectory was

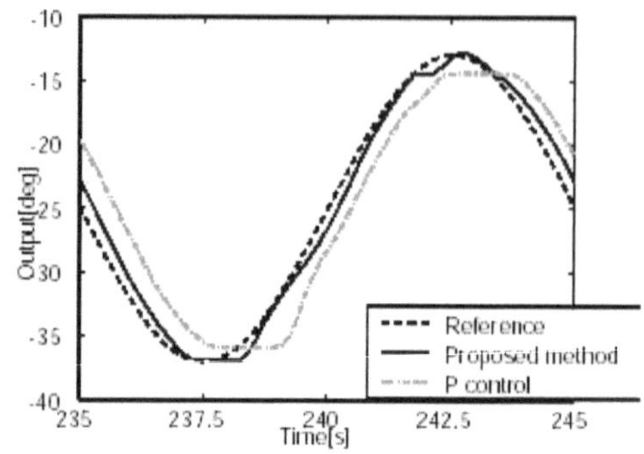

Figure 16. Sine wave response.

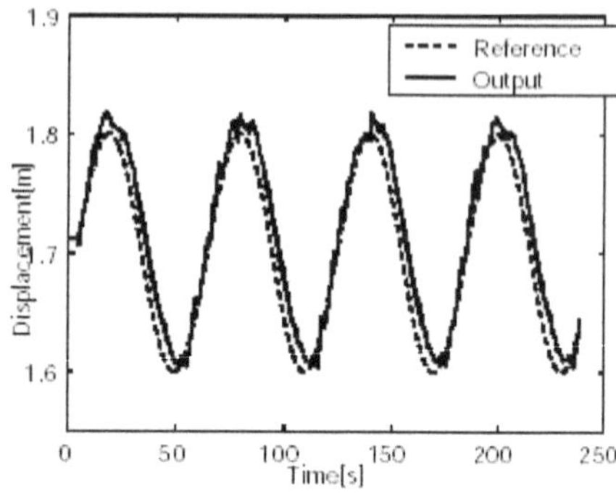

Figure 17. Response along x-axis.

planned to move the arm's end effector forward 20 cm along the x-axis at a constant height of 20 cm by actuating joints 2 through 4. Figures 16 and 17 show the results. It can be seen that compliance with the reference deteriorates when multiple joints are actuated, compared to single joint motion. This is likely due to a drop in hydraulic pressure to the cylinders because there is only one pump. From a practical standpoint it is unrealistic for each joint to have its own pump. Trajectory following has become an essential task because joystick operation is planned for the robot arm. Improving the performance has become a primary focus.

Because of the poor trajectory following performance, when a reference point is supplied, the present algorithm moves the joints one at a time in sequence, rather than all at once, and forces the control input to zero when the error reduces to within an acceptable level.

6. THE ANTI-PERSONNEL MINE EXPOSURE AND CLEARANCE SYSTEM

The tasks described in the previous sections are being joined together to construct part of a comprehensive anti-personnel mine exposure and clearance system to meet a set of functional requirements laid out by JST. The requirements are listed as follows:

(1) Removal of rocks and other obstacles from the mine detection area.

(2) Removal of spent ammunition and other small metal objects from the mine detection area.

(3) Identification of the position of buried mines using SAR-GPR.

(4) Breaking up of bedrock or loosening of hard soil surrounding buried mines.

(5) Clearing of rocks and soil from the surfaces of buried mines, thereby exposing them.

The goal is to meet these functional requirements, though the means of implementing them are not explicitly specified. Our research is directed toward achieving all of these tasks except the third and we have decided on a suitable approach for each. The gripper will be used for grasping objects in Task 1. The electromagnet will be used to collect metal objects in Task 2. The drill will be used to loosen soil in Task 4. Finally, the high-pressure air jet will be used to clear away soil in Task 5.

The prototype of air spade have been deployed to Afghanistan and Cambodia [7]. Its objective is to provide a safe standoff capability to rapidly excavate anti-personnel mines and anti-tank mines, doing so at a faster rate than any other current method. The air spade is simple to operate and transport. It is not intended for use in heavily vegetated areas. This air spade is operated by deminer. Our prototype is completely different from the prototype of air spade because our air jet nozzle positioning and posture is a full autonomous or a tele-operated control based on stereovision technology.

For electromagnet action, the end effector follows the orbital trajectory shown in Figures 18 and 19. For other actions, data from stereovision is used. The operating procedure we are following to expose anti-personnel mine is outlined below.

(1) The terrain and obstacles are detected using stereovision.

(2) The gripper is moved to a position 15 cm in front of the obstacle and then approaches the target point along a straight-line path.

(3) The gripper grasps the object and sets it down outside the scan area.

(4) The electromagnet sweeps over the scan area collecting spent ammunition and other light metal objects.

(5) The metal objects are set down outside the scan area.

(6) Stereovision is used to determine the position of a visual marker that has been placed over a buried mine (we are temporarily using visual markers to mark the location of mines in experiments because the detection sensor has not been completed yet).

(7) The air jet nozzle is positioned near the marker and releases a blast of air that blows away gravel and soil, exposing the mine.

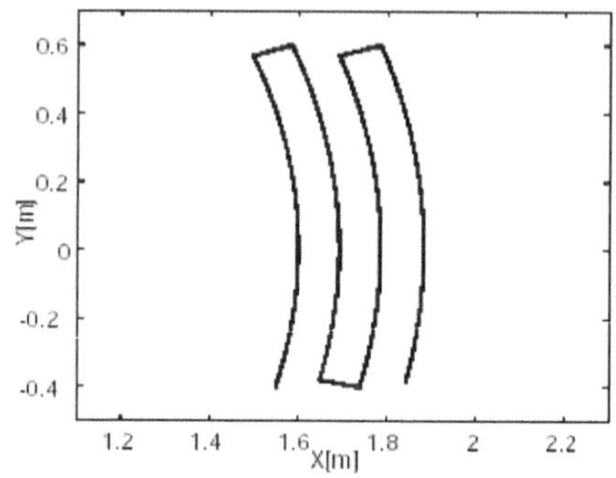

Figure 18. Orbit of electromagnet (*X–Y* plane).

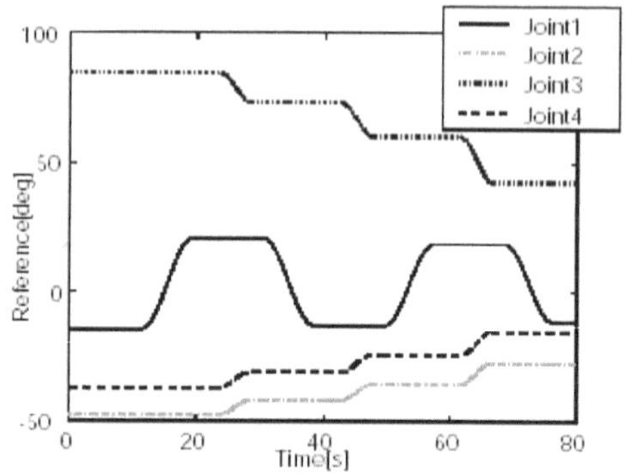

Figure 19. Orbit of electromagnet (reference for each joint).

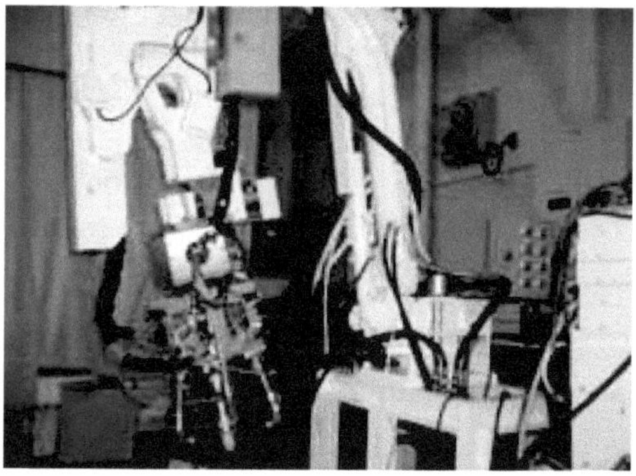

Figure 20. Overview of slave hand system for master-slave bilateral hand system for mine clearance.

7. MASTER-SLAVE BILATERAL CONTROL HAND SYSTEM

This section presents a master-slave system for the purpose of landmine and obstacle clearance. The master part of the system is a data-globe with an exo-skeleton while the slave part is a powerful hydraulic four-finger robot hand. This system was designed to be used in landmine fields but it can also be used in other human hostile environments. The total slave hand system with manipulator is shown in Figure 20.

This is the first master-slave system of this kind for the purpose of mine clearance [9]. This master slave systems uses a hydraulic robot hand as slave in contrast to other researchers who use electric motors. While the master hand is a human hand equipped with data glove the slave hand is of not anthropomorphic shape. This is not a common combination. The reason for that are

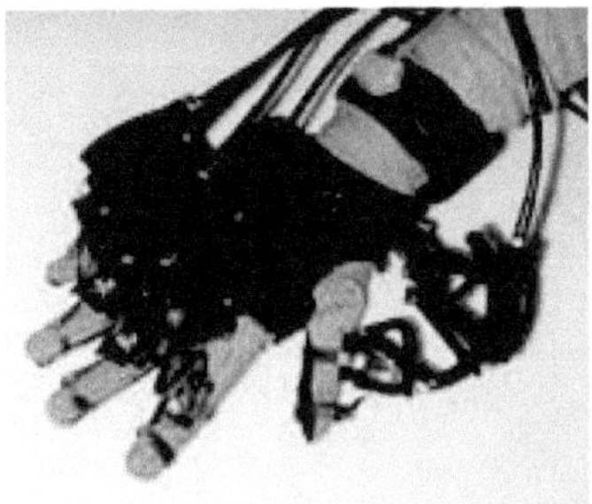

Figure 21. Master hand as human hand with a data-globe and with an exoskeleton.

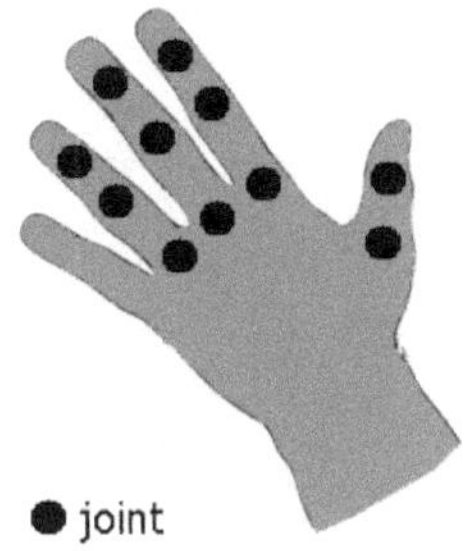

Figure 22. Data-glove with sensor.

derived from the applications and are described in this paper. On the other hand this kind of in homogeneity brings along master-to-slave mapping problems for that a solution is given.

This paper gives a description of the system. The design of the slave system is shown. The system modelling and identification attempts are described. In particular, the system nonlinearities are identified by means of experimental results. Control strategies for a joint which include pulse modulation techniques are presented. To deal with the strong temperature dependence adaptive control was designed. Compliance control is applied and experiments showing grasping of different objects are presented.

7.1 Master Hand System

The master hand system consists of a data-glove, an exoskeleton shown in Figures 21 and 22, and the control computer hardware. The data-glove has 22 film type position sensors. These sensors are measuring all angles of the fingers and some angles of the palm. The exoskeleton uses one pulley per finger tip for force feedback. The force signal coming from the slave hand is translated into a traction force giving the operator a sensation similar to touching an object with his finger tips. The master hand is also equipped with a 6-degree-of freedom sensor for detecting the hand position and orientation in space. This sensor will be used to control the manipulator movements. The master system is controlled by its own real time LINUX computer. The master system can be accessed through C++ routines. A virtual slave hand was programmed for visualization and VR (Virtual Reality).

7.2 Slave Hand System

The slave hand system is a powerful robot hand designed in association with the Fuji Heavy Industries Ltd. Figure 23 shows the hand holding an anti-personnel landmine of the type PMN. The hand has four fingers and two joints

Figure 23. Slave hand as a powerful hydraulic four-finger robot hand.

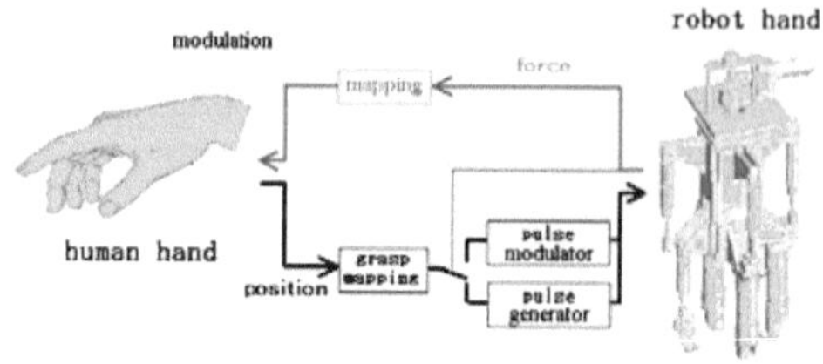

Figure 24. Overview of master-slave bilateral system.

per finger. The ranges of the proximal and distal joints are 40 degrees and 64 degrees respectively. The length of a finger is 240 mm and the maximal distance between two finger tips is 330 mm. The size of the robot hand is about two and a half times biger that the human hand. The whole hand structure can also rotate around an axis normal to the palm. The robot hand has all together nine degrees of freedom. Every joint of the hand is equipped with linear potentiometer as position sensor. Every finger tip has a force sensor. The hand is driven by hydraulic cylinders.

7.3 Master-Slave Bilateral Control Hand System

The overview of the master-slave bilateral system is shown in Figure 24. The position signal is generated by the human hand (master hand). It passes the grasp mapping unit until 7.4. The grasp mapping is used to allow the operator an intuitive usage of four-finger grasp and two-finger grasp. The modified position signal becomes the reference for the robot hand (slave hand). The force signal coming from the finger tips of the slave hand passes through the mapping unit and is fed back to the master hand exoskeleton. The operator using the master hand is able to natural interaction is given by the force feedback and an image of the environment displayed on the monitor as well as an ideal virtual environment constructed from the camera data and sensor data displayed on the computer screen.

Figure 25 shows the master-slave bilateral control. As shown in Figure 26, we first take the maximum value of two opposite finger tip forces then apply a dead-zone on the force signal so that only force signals larger that the trigger value have an effect on the control. This was done to cancel the effect of noise during non-contact manipulation. Next we use a low pass filter to cancel the noise during contact action. This modified force signal is weighted and added to the position loop influencing the reference position. The force signal changes only the x-value of the system which is the grasping direction. The y-coordinate which is the lifting direction remains unaffected by the force signal.

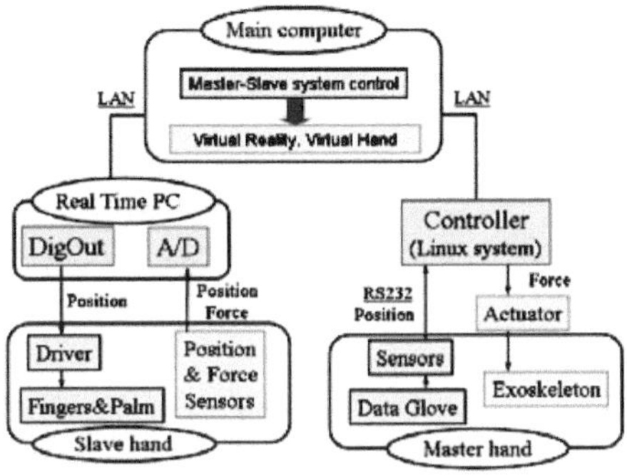

Figure 25. Master-slave bilateral control system.

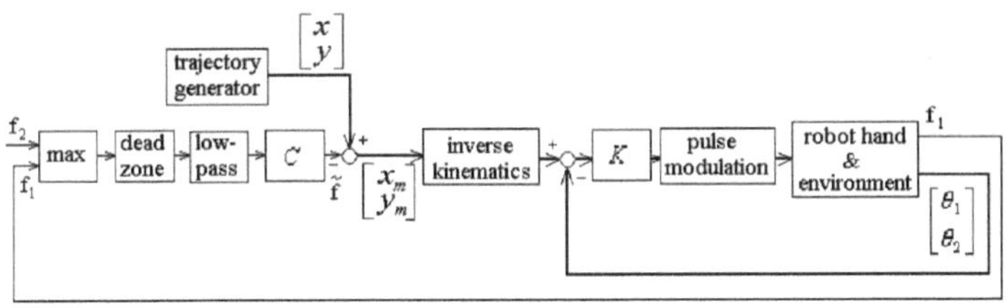

Figure 26. Force control block diagram.

If we know the size of the object the control performance can be improved. We use the knowledge of the object size. In our case only the x-dimension is relevant. Figure 27 shows the X–Y data of a hard object with known size. The reference in the X-dimension was chosen "tighter" than the real size of the object to enlarge the grasping force. Figure 28 shows grasping force during the manipulation. As can be seen the force in modified case is much smaller that in the case which the size of the object was unknown.

7.4 Grasp Mapping

The master hand and the slave hand have different structures. The master hand is a human hand equipped with a data glove with 22 position sensors, The slave hand is axial symametric and has nine d.o.f. The structural difference does not allow a direct mapping from human hand sensors to robot hand actuators. For this reason we designed a Neural Network (NN) grasp recoginition and grasp mapping unit as shown in Figure 29.

8. SUMMARY

We have discussed the current state of development of a anti-personnel mine exposure and clearance system, with emphasis on multi-joint control of a robot

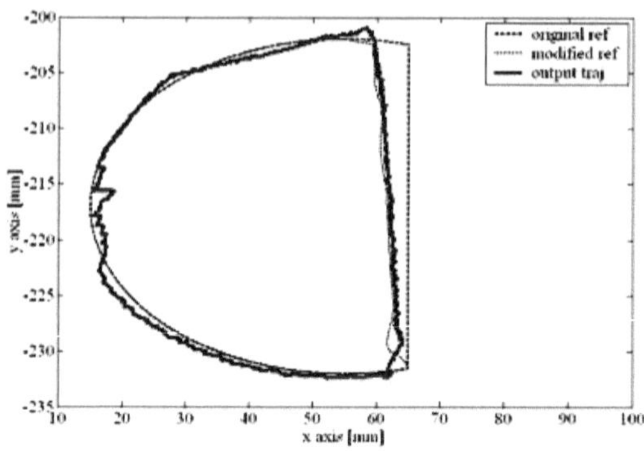

Figure 27. Bilateral force control – grasping a light and hard object; X–Y finger tip position.

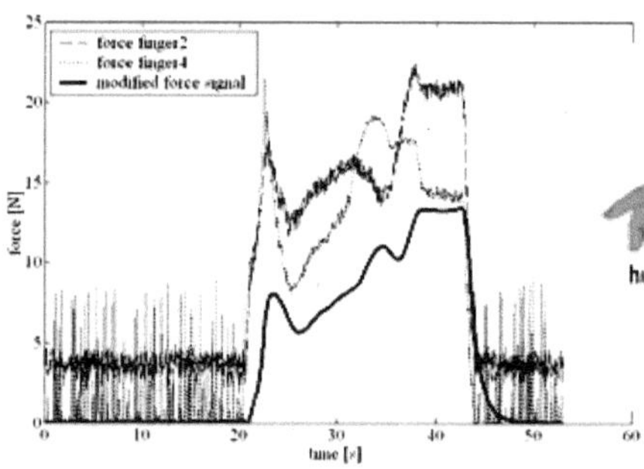

Figure 28. Bilateral force control – grasping a light and hard object; finger tip force.

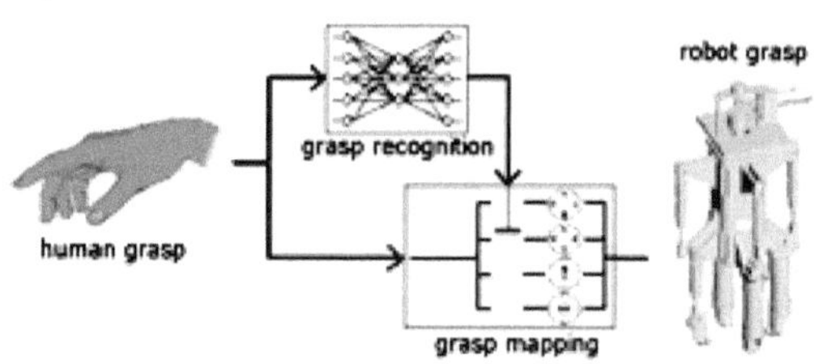

Figure 29. Grasp mapping.

arm. We have achieved the reasonable performance of 6-degree-of-freedom robot with multi-function tool by means of nonlinear control based on "LOOK AT TABLE" scheme. The focus of the next stage of development will be devising a means of joystick operation for the arm and tools. Research in the field of hydraulic robot arm control is extensive, but there is still plenty of room for improving performance. Our group is addressing a number of elements within the system at the same time with the aim of improving the system as a whole.

Currently the arm can only be operated by command instructions from the operator. For the future, various other command methods are being considered, such as joystick control, master-slave control, and other new human interface control ideas. We are also considering problems that may arise when the arm is mounted on the small vehicle. Presently the multifunction arm weighs 200 kg and the sensor armweighs 150 kg. The multifunction arm may need to be lightened to maintain a good balance. We will continue to develop the multifunction robot arm and work to improve performance in time for the planned field-test in Afghanistan.

REFERENCES

[1] Nonami, K. et al., Development and Control of Mine Detection Robot COMET-II and COMET-III, *JSME International Journal*, Ser. C, Vol. 46(3), 2003, 881–890.

[2] Research and Development of Sensing Technology and Access-and-Control Technology for Supporting Antipersonnel Mine Detection and Removal Activities from the Humanitarian Point of View: http://www.jst.go.jp/kisoken/jirai/index.html

[3] http://www.jst.go.jp/kisoken/jirai/kadai-frameset.html

[4] http://www.jst.go.jp/pr/jst-news/2004/2004-07/page16.html

[5] Sato, M., Fang, G. and Zeng, Z., Landmine Detection by a Broadband GPR System, in *Proc. IEEE Int. Symp. Geoscience and Remote Sensing*, IGARSS2003, 2003.

[6] Yamada,H., Takeichi, K. and Muto, T., Sliding Mode Control of Hydraulic Power Shovel, *Transactions of the Japan Society of Mechanical Engineers*, Vol. 68(670), 2002, 1768–1774.

[7] Humanitarian Demining-Developmental Technologies, 2000–2001 DoD Humanitarian Demining R&D Program.

[8] Nonami, K. et al., Preliminary Design and Feasiblity Study of 6DOF Robot for Excavation of Landmine, *Autonomous Robot*, Vol. 18, 2005, 293–301.

[9] Wojtara, T., Nonami, K. et al., Master-Slave Hand System of Different Strucure, Grasp Recognition by Neural Network and Grasp Mapping, *Robotica*, Vol. 22, 2004, 4.

[10] Nonami, K. et al., Research and Development of Mine Hunter Vehicle for Humanitarian Demining, in *Proc. IEEE Int. Advanced Robotics Program on Humanitarian Demining*, Tokyo, 2005.

EXCITATION AND CONTROL OF MICRO-VIBRO-IMPACT MODE FOR ULTRASONIC MACHINING OF INTRACTABLE MATERIALS

Vladimir Babitsky[1] and Vladimir Astashev[2]

[1]*Loughborough University, Wolfson School of Mechanical & Manufacturing Engineering, UK;*
[2]*Institute of Machine Studies Russian Academy of Sciences, Russia*

Abstract: The use of the nonlinear resonant vibration mode to develop mechanical work for machining of intractable materials is discussed. This can be achieved by condensing the vibration of a cutting tool into a sequence of microcollisions with a workpiece for impulsive action of the tool to the material being treated. To stabilise the nonlinear mode under unpredictable change of processing loads, an approach has been developed to design a cutting machine as an intelligent self-exciting oscillating system known as *autoresonant*. The paper is devoted to analytical consideration of the autoresonant ultrasonic machining system. The main qualitative features of the system dynamics are revealed and the basic design principles and parameter relations are obtained. The system was developed, tested and used for cutting of intractable materials. Improvements of machinig are obtained and described.

Key words: nonlinear dynamics and control, vibro-impact, ultrasonic machining, autoresonance.

1. INTRODUCTION

It has been shown that resonant excitation of the high-frequency nonlinear mode of the tool-workpiece interaction is the most effective method of ultrasonic influence on the dynamic characteristics of machining [1–4]. This discrete and repetitive influence imparts some unique properties to metal cutting, where the interaction between the workpiece and the cutting tool is transformed into *a micro-vibro-impact process* [4, 5].

Figure 1 demonstrates an experimental set-up of ultrasonically assisted cutting tools used in this work. The ultrasonic transducer consists of piezo ceramic rings within a package together with a wave guide (concentrator) and backing section. A cutting tip is fixed in the tool holder installed at the thin end of the concentrator. The transducer is fixed through its developed nodal cross section

H. Ulbrich and W. Günthner (eds), IUTAM Symposium on Vibration Control of Nonlinear Mechanisms and Structures, 55–67.

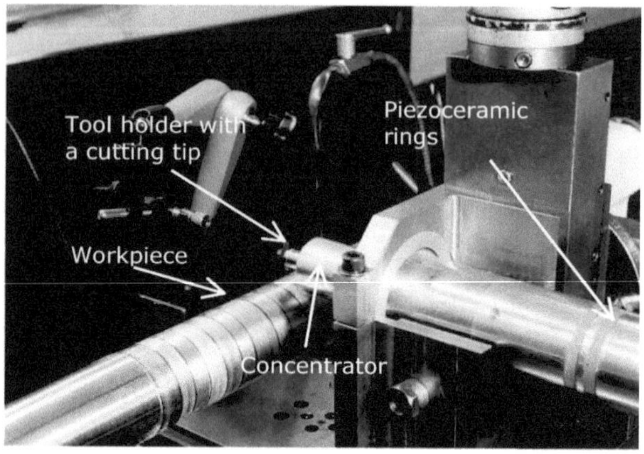

Figure 1. Experimental set-up of ultrasonically assisted machining system.

at the machine tool vertical slide. The workpiece is clamped by a three-jaw spindle chuck and rotates universally by a lathe drive.

When the high frequency electric impulses from an electronic amplifier are fed to the input of the piezo transducer it begins vibrating due to piezoelectric effect. The vibration excites the longitudinal waves in the concentrator (which intensifies the amplitude of vibration in the direction of the thin end) and through it vibration of the cutting tip.

The investigation of micro-vibro-impact mode tool-workpiece interaction needs an analysis of the nonlinear boundary problem for a rod-type transducer structure with the variable cross section. The nonlinear boundary condition reflects the complex process of tool–workpiece interaction during the cutting. Cutting force acting on the cutting edge of the tool is characterised by nonlinear dynamic characteristics of contact interaction, which is ill-defined and time varying. This describes both elasto-plastic deformation of the material due to cutting and following friction process.

The specific of such vibro-impact machining process, transforms the dynamical system of a machine in a processing regime into a nonlinear system with poorly predictable response to the excitation. Excitation, stabilisation and control of a nonlinear mode at the top intensity in such a system is an engineering challenge and needs a new method of adaptive control for its realisation.

To make the system adaptable to unpredictable variations of cutting conditions the new method for excitation and stabilisation of ultrasonic vibration was proposed, known as *autoresonance* [4–7]. In an autoresonant system (Figure 2) the amplified signal obtained from the performance sensor is fed to the piezo transducer by means of *a positive feedback*. This leads to dynamic instability of the acousto-mechanical system, which is controlled by intelligent

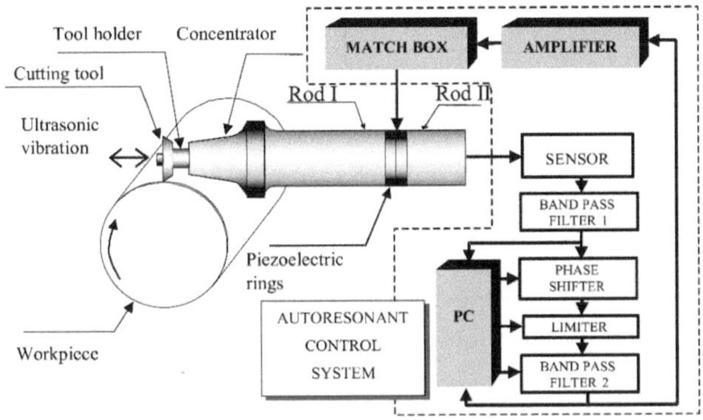

Figure 2. Schematic diagram of autoresonant ultrasonic cutting system.

tracing of the optimal relationship between phase shifting and limitation in the feedback circuitry.

Dynamics of the autoresonant ultrasonic machine was investigated and the results of analysis, design and experimentation are presented. The opportunity of application of the robust and high quality nonlinear resonant system under wide deviation of processing loads results in essential increase of machine productivity, accuracy, efficiency and improvement of design.

2. DYNAMIC MODEL OF ULTRASONICALLY ASSISTED MACHINING

The process of ultrasonic cutting is the accumulation of small residual deformations in every vibration cycle. Considering the cycling motion we can ignore during the vibration analysis the influence of slow average motion of the cutter relative to a workpiece and treat the cyclic motion as close to periodic. The cutting force f_l acting on the cutting edge of the tool can be described by a nonlinear dynamic characteristics of the contact interaction $f_l = F(u_l^l, \dot{u}_l^l)$, where $u_l^l(t)$ is the periodic motion of the cutting edge. This characteristic describes both the elasto-plastic deformation of the material due to cutting and the friction process [4].

Introducing the proper transfer functions for the electro-acousto-mechanical and the control system we can write the equation of cutter motion in the operator form

$$u_l^l(t) = W^{-1}(j\omega)\{N(j\omega)\Psi[T(j\omega)u_l^l(t)] - F[u_l^l(t), \dot{u}_l^l(t)] + M/r\}, \quad (1)$$

where for any operator $H(j\omega)$ and temporal function $x(t)$: $H(j\omega)x(t) \equiv \Im^{-1}\{H(j\omega)\Im[x(t)]\}$. Here $\Im$ is the operator of Fourier transform.

The transformation of the variables in (1) is represented by the block diagram in Figure 3. In Equation (1) is the force of the ultrasonic excitation of the

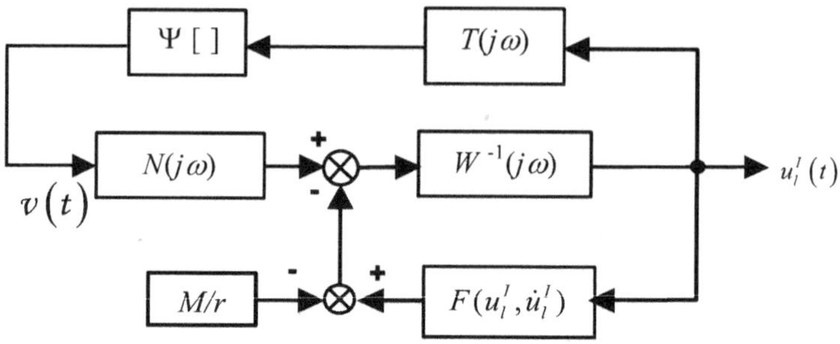

Figure 3. Block diagram of transformation of variables.

cutter, which is produced by an ultrasonic transducer with positive feedback. The feedback includes (see Figure 2) the sensor, the band pass filter 1 and the phase shifter with a common linear operator and the limiter with a nonlinear characteristic $\Psi[T(j\omega)u_l^l(t)]$; $N(j\omega)$ is a linear operator, which reflects the transfer of the ultrasonic excitation from the limiter to the cutter and includes the transfer functions of the band pass filter 2, the amplifier, the match box, the piezo transducer and the concentrator; $W(j\omega)$ is the reduced dynamic stiffness of the tool, which includes the proper operators for the concentrator and the tool; ω is the frequency of ultrasonic vibration, M is a torque of a lathe main drive; r is the radius of the workpiece; $j^2 = -1$.

The autoresonant excitation of the stable ultrasonic vibration corresponds to the existence and stability of the periodic solution of (1). Due to the strong filtering effect of the mechanical structure, the harmonic approximation of the solution can be found analytically by using the technique of harmonic linearisation [5]. We suppose that the periodic displacement of the cross-sectional area with the coordinate x of the rod i (i = I, II) (see Figure 2) can be presented as $u_x^i(t) = \tilde{a}_x^i e^{j\omega t}$, where $\tilde{a}_x^i$ is the complex amplitude of the vibration. The left end of the rod I having the coordinate $x = l$ is loaded with a non-linear cutting force $f_l(t) = F[u_l^l(t), \dot{u}_l^l(t)]$.

3. TRANSMISSIBILITY OF ULTRASONIC CUTTING UNIT UNDER LOAD

Using harmonic linearisation, we have

$$\begin{aligned} f_l = F(u_l^l, \dot{u}_l^l) &\approx [k(v, a_l^l) + j\omega b(v, a_l^l)]u_l^l + P_0(v, a_l^l) \\ &= \tilde{k}(v, a_l^l)u_l^l + P_0(v, a_l^l), \end{aligned} \tag{2}$$

where P_0 is the slow varying component of the load generated due to the ultrasonic cutting process (*vibro-induced force*). Here we took also into consideration an averaged movement of the cutter relative to the workpiece with a cutting speed v: $u_l^I(t) = vt + \tilde{a}_l^I e^{j\omega t}$. Following to the specific of the ultrasonically assisted cutting [1–4] we propose during the integration that $v < a\omega$.

Let us now write the equations for the complex vibration amplitudes of the transducer rods I and II clumping the piezo ceramic rings. For the cross-sectional areas of rod I with coordinates $x = 0$ and $x = l$ we have

$$\tilde{a}_0^I = L_{00}^I \tilde{F}_0 - L_{l0}^I \tilde{k} \tilde{a}_{1l},$$
$$\tilde{a}_l^I = L_{0l}^I \tilde{F}_0 - L_{ll}^I \tilde{k} \tilde{a}_{1l}. \quad (3)$$

From now on, $\tilde{F}_0$ is used to denote the complex amplitude of the force $f_0^I = f_0$, and $L_{sr}^i = L_{sr}^i(j\omega)$ is the dynamic compliance (receptance) connecting the force acting in the cross section with a coordinate s with the displacement of the cross section with a coordinate r. At the cross-sectional plane $x = 0$ of rod II we have

$$\tilde{a}_0^{II} = L_{00}^{II} \tilde{F}_0. \quad (4)$$

The state of the piezo ceramic elements is described by the equations that couple the forces applied with the deformation of the elements, voltage $\tilde{v}$ and charge $\tilde{q}$ on their plates,

$$-\tilde{F}_0 = K_1 \tilde{a} + \Phi \tilde{q}, \quad (5)$$

$$\tilde{v} = \Phi \tilde{a} + \frac{\tilde{q}}{C}, \quad (6)$$

where $K_1 = E_0 S / l_0$ is the stiffness of a piezo ceramic element without charge (short-circuited contacts), E_0 is the elastic modulus of the piezo ceramic, S is the area of a piezo ceramic plate,

$$\tilde{a} = (\tilde{a}_0^I + \tilde{a}_0^{II})/2 \quad (7)$$

is the complex amplitude of deformation for a single piezo ceramic plate, C is the capacitance of the piezo ceramic element without deformation ($\tilde{a} = 0$), Φ is the coefficient of electro-mechanic coupling, $\tilde{v}$ and $\tilde{q}$ are the complex amplitudes of voltage and charge.

As a result of the solution of the set of Equations (3)–(7), we obtained an equation (transmissibility) linking the amplitude of vibration $\tilde{a}_l^I$ for a cross section $x = l$ of rod I with the voltage $\tilde{v}$ applied to the piezo ceramic rings [6]:

$$\tilde{a}_l^I \triangleq \tilde{a}_l = \frac{-2\Phi C L_{0l}^I \tilde{v}}{2 + K_0 L_0 + (2L_{ll}^I + K_0 L_0 L_{ll}^I - K_0 (L_{0l}^I)^2)\tilde{k}}. \quad (8)$$

Here $L_0 = L_{00}^I + L_{00}^{II}$, $K_0 = K_1 - \Phi^2 C$.

The dynamic stiffness can be found from a solution of the boundary problem for steady-state waves in one-dimensional wave guides with variable cross section [5]. The equation of longitudinal vibration of the wave guide is

$$\rho \frac{\partial^2 u(x,t)}{\partial t^2} = \tilde{E}\left[\frac{\partial^2 u(x,t)}{\partial x^2} + \frac{S'_x}{S_x}\frac{\partial u(x,t)}{\partial x}\right], \tag{9}$$

where $u(x,t)$ is a motion of the cross section x, $\tilde{E} = E\left(1 + j\frac{\psi}{2\pi}\right)$, E is the elastic modulus, ψ is a coefficient of inner damping in material, ρ is the density of the material, $S(x)$ is an area of the cross section x, $S'_x(x) = dS(x)/dx$, $j^2 = -1$.

Let $x = s$ be a coordinate of the application of the exciting force $f_s(t) = \tilde{F}_s e^{j\omega t}$. For a wave guide of length l with free ends the boundary conditions are

$$\left.\frac{\partial u(x,t)}{\partial x}\right|_{x=l-s} = 0, \quad \tilde{E}S_x \left.\frac{\partial u(x,t)}{\partial x}\right|_{x=s} = (-1)^{(1-s/l)}\tilde{F}_s e^{j\omega t}. \tag{10}$$

Considering a harmonic vibration of the wave guide $u(x,t) = \tilde{a}_x e^{j\omega t} = a_x e^{j(\omega t - \varphi_x)}$, where a_x, φ_x are the amplitude and phase of the vibration of a cross section x, we have instead of (9) and (10):

$$\begin{aligned} &\tilde{E}[\tilde{a}''_x + (S'_x/S)\tilde{a}'_x] + \rho\omega^2\tilde{a}_x = 0, \\ &\tilde{a}'_x\big|_{x=l-s} = 0, \quad \tilde{E}S_x\tilde{a}'_x\big|_{x=s} = (-1)^{(1-s/l)}\tilde{F}_s. \end{aligned} \tag{11}$$

Further the solution depends on a form of the wave guide defined by $S(x)$. As an example, for a uniform wave guide $S(x) = S = \text{const.}$, we obtain from (11)

$$L_{sx}(j\omega) = W^{-1}(j\omega) = \frac{\tilde{a}_x}{\tilde{F}_s} = (-1)^{s/l}\frac{\lambda}{\omega^2\rho S}\frac{e^{\lambda x} + e^{-\lambda x}e^{2\lambda(l-s)}}{e^{\lambda s} - e^{-\lambda s}e^{2\lambda(l-s)}}, \tag{12}$$

where

$$\lambda = j\omega\sqrt{\frac{\rho}{\tilde{E}}} = \frac{j\omega}{c\sqrt{1 + j\psi/2\pi}}, \quad c = \sqrt{\frac{E}{\rho}}.$$

Usually wave guides are composed of parts with different properties. The parts may be made of different materials. The cross-sectional areas of these different parts may vary according to different forms. Some parts can be represented as elements with lumped parameters. Different compound concentrators with a tool attached used for ultrasonically assisted cutting belong to such systems. To describe the compound vibration system, it is practical to express the dynamic compliance of a concentrator (rod I, Figure 2) via the dynamic compliances of its parts.

Consider, as an example, its wave guide I (see Figure 2), which consists of two steps of length l_1 and l_2, where $l_1 + l_2 = l$, as is widely used in ultrasonic equipment. Dividing virtually the wave guide at the step, the interaction of its parts may be represented by a reaction force with a complex amplitude $\tilde{F}_{12}$. New coordinates $x_1 = x$ ($x_1 \in [0, l_1]$), $x_2 = x - l_1$ ($x_2 \in [0, l_2]$) may now be introduced. A complex displacement amplitude for the rod cross sections x_q ($q = 1, 2$) may be written for the system, which is under the action of force $\tilde{F}_0$ applied at the end $x = x_1 = 0$,

$$\begin{aligned} \tilde{a}_{x_1}^{(1)} &= L_{0x_1}^{(1)}(j\omega)\tilde{F}_0 - L_{l_1x_1}^{(1)}(j\omega)\tilde{F}_{12}, \\ \tilde{a}_{x_2}^{(2)} &= L_{0x_2}^{(2)}(j\omega)\tilde{F}_{12}, \end{aligned} \tag{13}$$

where the upper index (in brackets) denotes the number of the step to which the expression corresponds. Equality $\tilde{a}_{l_1}^{(1)} = \tilde{a}_0^{(2)}$ reflects the continuity of the system at cross section $x = l_1$ ($x_1 = l_1$; $x_2 = 0$), and allows the reaction amplitude to be found from (13)

$$\tilde{F}_{12} = L_{0l_1}^{(1)}(j\omega)\tilde{F}_0/[L_{l_1l_1}^{(1)}(j\omega) + L_{00}^{(2)}(j\omega)]. \tag{14}$$

Substituting (14) into (13), the dynamic compliance $L_{sx}(j\omega) = \tilde{a}_x/\tilde{F}_s$ of the compound system may be found. Its expressions for the end cross sections $L_{00}^{I}(j\omega)$ and $L_{0l}^{I}(j\omega)$ of the compound system are:

$$\begin{aligned} L_{00}^{I}(j\omega) &= L_{00}^{(1)}(j\omega) - \frac{L_{0l_1}^{(1)}(j\omega)L_{l_10}^{(1)}(j\omega)}{L_{l_1l_1}^{(1)}(j\omega) + L_{00}^{(2)}(j\omega)}, \\ L_{0l}^{I}(j\omega) &= \frac{L_{0l_1}^{(1)}(j\omega)L_{0l_2}^{(2)}(j\omega)}{L_{l_1l_1}^{(1)}(j\omega) + L_{00}^{(2)}(j\omega)}. \end{aligned} \tag{15}$$

Let, as usual, the steps be homogeneous rods with cross sections S_1 and S_2. We are assuming also that the steps are made from the same material. The dynamic compliances for the homogeneous rod under excitations acting at the end cross sections $s_i = l_i$ and $s_i = 0$ can be found from (12):

$$L_{l_ix_i}^{(i)}(j\omega) = -\frac{\lambda}{\rho S_i\omega^2}\frac{\operatorname{ch}\lambda x_i}{\operatorname{sh}\lambda l_i}, \tag{16}$$

$$L_{0x_i}^{(i)}(j\omega) = -\frac{\lambda}{\rho S_i\omega^2}\frac{\operatorname{ch}\lambda(l_i - x_i)}{\operatorname{sh}\lambda l_i}. \tag{17}$$

Substituting into (15) the proper expressions from (16), (17), we have

$$L_{00}^{I}(j\omega) = -\frac{\lambda}{S_1\rho\omega^2}\frac{S_1\operatorname{ch}\lambda l_1\operatorname{ch}\lambda l_2 + S_2\operatorname{sh}\lambda l_1\operatorname{sh}\lambda l_2}{S_1\operatorname{sh}\lambda l_1\operatorname{ch}\lambda l_2 + S_2\operatorname{sh}\lambda l_2\operatorname{ch}\lambda l_1}, \tag{18}$$

$$L_{0l}^{I}(j\omega) = -\frac{\lambda}{\rho\omega^2}\frac{1}{S_1\operatorname{sh}\lambda l_1\operatorname{ch}\lambda l_2 + S_2\operatorname{sh}\lambda l_2\operatorname{ch}\lambda l_1}, \tag{19}$$

Neglecting the dissipation in the material ($\psi = 0$), we can simplify expressions (18), (19) as follows

$$L_{00}^{l}(j\omega) = -\frac{l}{S_1 E \xi} \frac{S_1 \cos\xi_1 \cos\xi_2 - S_2 \sin\xi_1 \sin\xi_2}{S_1 \sin\xi_1 \cos\xi_2 + S_2 \sin\xi_2 \cos\xi_1}, \tag{20}$$

$$L_{0l}^{1}(j\omega) = -\frac{l}{E\xi} \frac{1}{S_1 \sin\xi_1 \cos\xi_2 + S_2 \sin\xi_2 \cos\xi_1}, \tag{21}$$

where $\xi = \omega l/c$ and $\xi_i = \omega l_i/c$.

With the use of the piezo ceramic transducer it is necessary to obtain a maximum dynamic loading of the ceramic rings. This corresponds to the fixed ends of the piezo transducer. When combined with a concentrator, this can be obtained by organisation of a node in a cross section $x = 0$. Condition of such tuning is

$$L_{00}^{l}(j\omega) = 0 \tag{22}$$

or according to (20)

$$S_1 \cos\xi_1 \cos\xi_2 - S_2 \sin\xi_1 \sin\xi_2 = 0. \tag{23}$$

It is possible to show that this expression defines the natural frequencies of a rod with a fixed end $x = 0$. In this case the concentrator works like a dynamic absorber of the ceramic ring vibration and the amplitude of the cross section $x = l$ can be found with the help of (21)

$$\tilde{a}_l^l \triangleq \tilde{a}_l = L_{0l}^{l}(j\omega)\tilde{F}_0, \tag{24}$$

whereas ξ_1, ξ_2 follow from (23).

Let us consider two special cases for condition (23):

$$1. \quad \cos\xi_1 = 0, \quad \sin_2 = 0, \tag{25}$$

or $\xi_1 = (2m-1)\pi/2$, $\xi_2 = n\pi$ $(m, n = 1, 2, \ldots)$. In this case the first step has a length equal to an odd number of quarter wave lengths: $l_1 = (2m+1)\pi c/2\omega$, but the second step's length is equal to an even number of quarter wave lengths: $l_2 = n\pi c/\omega$. From (24) with account of (21) and (25), we have an amplitude of a free end of the concentrator:

$$a_l = \frac{F_0 l}{E S_1 \xi}. \tag{26}$$

$$2. \quad \cos\xi_2 = 0, \quad \sin\xi = 0, \tag{27}$$

or $\xi_2 = m\pi$, $\xi_2 = (2n-1)\pi/2$. In this case the first step has a length equal to an even number of quarter wave lengths: $l_1 = m\pi c/\omega$, but a length of the second step is equal to an even number of a quarter wave lengths: $l_2 = (2n-1)\pi c/2\omega$.

From (24), with account of (21) and (27), we have an amplitude of a free end of the concentrator for this case:

$$a_l = \frac{F_0 l}{E S_2 \xi}. \tag{28}$$

For equal lengths of the concentrators, the second case gives increase of amplitude, which is used in practice.

4. FEEDBACK CONTROL

An effective influence of ultrasonic vibrations on cutting demands, as shown above, an excitation of ultrasonic vibrations with maximum amplitude. The complexity of the amplitude response4 makes the excitation and stabilisation of the resonant oscillation in conditions of variable technological loads difficult. The improvement of the excitation efficiency was achieved with the transition of the excitation into an autoresonant one (Figure 2) [4–7].

Filter action permits consideration of the main harmonic component only of the process in the feedback circuit. As a result, the voltage $\upsilon(t)$ applied to the piezo transducer can be described as

$$\upsilon(t) = \Psi[T(j\omega)u_l^l(t)]. \tag{29}$$

After a harmonic linearisation of expression (29) we have a complex amplitude of vibration

$$\tilde{\upsilon} = Z(a_l)\tilde{a}_l e^{j\varphi}, \tag{30}$$

where $Z(a_l)$ is a coefficient of the harmonic linearisation for the limiter, φ is a phase shift due to the action of the phase shifter. We suppose that a maximum response of the limiter is G, i.e.

$$\left|\Psi[T(j\omega)u_l^l(t)]\right| \le G. \tag{31}$$

The periodic function $\upsilon(t)$ at this case cannot have an amplitude of fundamental exceeding $4G/\pi$. As a result, it follows from (30)

$$Z(a_l)a_l \le \frac{4G}{\pi}. \tag{32}$$

After harmonic linearisation of (1) with account of (2) and (30) we have the following equations for the constant and periodic component of cutter vibration:

$$P_0(v, a_l) = M/r \tag{33}$$

$$1 + W^{-1}(j\omega)[\tilde{k}(v, a_l) - N(j\omega)Z(a_l)e^{j\omega}] = 0. \tag{34}$$

Equation (33) couples the vibration generated average force P_0 with an amplitude of the cutter vibration a_l and the force characteristics M of a machine-tool drive. Equating the real and imaginary part of (34) to zero, we find the conditions of a periodic self-sustained vibration

$$Z(a_l)\cos\varphi = U(\omega, a_l), \tag{35}$$

$$Z(a_l)\sin\varphi = V(\omega, a_l), \tag{36}$$

where $U(\omega, a_l) = \operatorname{Re} Q(j\omega, a_l)$, $V(\omega, a_l) = \operatorname{Im} Q(j\omega, a_l)$, $Q(j\omega, a_l) = [1 + W(j\omega)]\tilde{k}(v, a_l)/N(j\omega)$.

Excluding the trigonometric functions from (35) and (36), we have finally

$$a_l = a_l Z(a_l)[U^2(j\omega, a_l) + V^2(j\omega, a_l)]^{-1/2}. \tag{37}$$

Comparing (37) and (32) we can conclude that the maximum amplitude of the vibration is achieved when $a_l Z(a_l) = 4G/\pi$. This means that

$$Z(a_l) = 4G/\pi a_l. \tag{38}$$

Relationship (38) can be fulfilled by the use of a relay characteristic $\Psi(u_l)$ in feedback: $\Psi(u_l) \approx G \operatorname{sgn} u_l$. Due to the variation of the phase φ with the help of a phase shifter the amplitude of the cutter vibration a_l and its frequency ω, as follows from (35)–(37), belong to the amplitude-frequency characteristic $a_l = v_0[U^2(j\omega, a_l) + V^2(j\omega, a_l)]^{-1/2}$ of the *forced* vibration for the *loaded* transducer when $v_0 = 4G/\pi$.

In order to study the stability of the autoresonant regimes the method described in [5] can be used. It is assumed that in the vicinity of steady-state periodic solution the cutter vibration is of the form

$$\breve{u}_l(t) = vt + \breve{a}_l(t)\exp j\breve{\omega}(t)t, \tag{39}$$

where $\breve{a}_l$ and $\breve{\omega}(t)$ are slowly varying functions of time. Considering equality (36) as the reduced energy balance condition for the dissipative and exciting forces of the steady-state motion, the motion stability criterion with respect to small perturbation in amplitude, can be formulated as

$$\left(\frac{d \operatorname{Im} Q}{d\breve{a}_l}\right)_{\breve{a}_l = a_l} > 0. \tag{40}$$

Inequality (40) means that if the amplitude $\breve{a}_l$ derivates from a stationary value $a_l = \text{const.}$ due to perturbation, the energy is unbalanced in such a way that it compensates for the deviation. Using the notation introduced above, the condition (40) can be rewritten in the form

$$\frac{d}{d\breve{a}_l}[\omega b(v, \breve{a}_l) - K(\omega)Z(\breve{a}_l)\sin\varphi]_{\breve{a}_l = a_l} > 0,$$

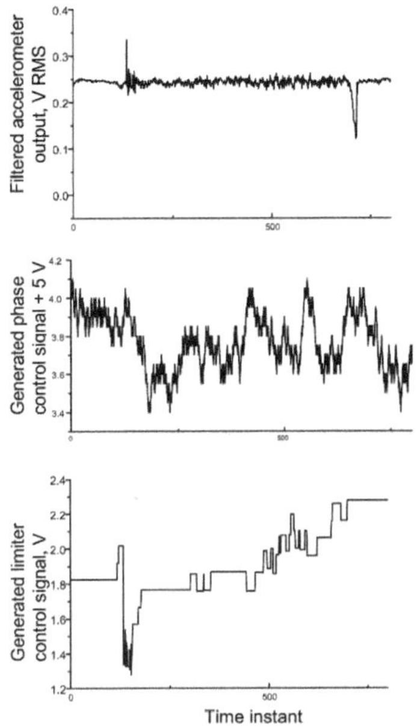

Figure 4. Autoresonant control of ultrasonic transducer.

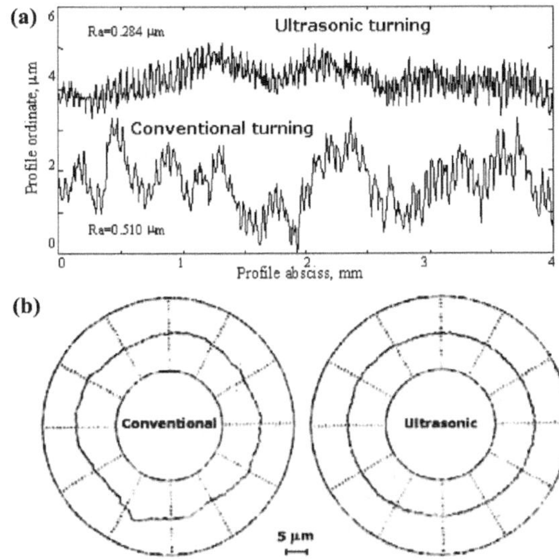

Figure 5. Comparative surface quality of Inconel -718 specimens machined conventionally and with ultrasonics vibration: a) axial surface profiles, b) roundness profiles.

where $K(\omega) = U_{sl}(\omega)/U_{ll}(\omega)$ is a coefficient that is dependent on the point at which the excitation force is applied and $\operatorname{sgn} K(\omega) \sin\varphi \geq 0$. According to (38), $[dZ/d\breve{a}_l]_{\breve{a}_l=a_l} < 0$, for real processing regimes $\omega b(v, \breve{a}_l) > 0$. From this it follows that under autoresonant excitation all points of the amplitude-frequency characteristic of the loaded transducer correspond to stable vibration as opposed to the forced excitation, when some branches of the characteristic are unstable due to the jump phenomena [4].

5. REAL TIME CONTROL

The hardware developed was fabricated and matched to provide the possibility for real time control. The phase shifter and limiter can be controlled from a computer on the basis of the reference signal properties (Figure 2). The principal aims of supervisory computer control are to provide the most efficient state of the electro-acousto-mechanical system and identify the control signal interval, if there have been any changes within the mechanical system (e.g. fitting a new cutting tip or a different tool holder). The control system keeps a stable level of vibration under variable cutting load [7]. Changing the phase shift within the autoresonant loop allows efficient control of the vibrating state. Changing the limiter threshold could control the vibration level.

The underlying principle of supervisory computer control is to acquire samples of the reference signal after band pass filtering, evaluate the efficiency of oscillations (the RMS value of the reference signal was used as a measure), elaborate new control values according to the specific algorithm, and set them up. Figure 4 demonstrates the operation of the autoresonant system in real working conditions. Here the horizontal axis corresponds to the time (each unit is about 50 ms which is the control cycle time). The vertical axis reflects (from top to bottom) the RMS value of the reference signal in V (the accelerometer was used), the phase shifter control signal in the control units, and the limiter control signal in V respectively. The autoresonant system keeps the stable vibration of the cutter under variable load.

Turning of some modern aviation materials was conducted with a developed system (Figure 5) [9]. Cutting conditions were typical of those used for manufacture. The roughness and roundness of the ultrasonically and conventionally machined workpieces were measured after machining and compared. An improvement of up to 50% had been achieved with superposition of ultrasonic vibration. A nanoindentation analysis revealed an essential reduction in the thickness of the modified material layer after cutting with ultrasonics.

6. CONCLUSIONS

The theory of machining unit for cutting with superimposing of ultrasonic vibration under autoresonant control is developed by methods of nonlinear dynamics. The autoresonant control provides the possibility of self-tuning and self-adaptation mechanisms for the system to keep the nonlinear resonant mode of oscillation under unpredictable variation of load. This allows simple program regulation of intensity of the process whilst keeping maximum efficiency at all times even for high Q-factor and nonlinear systems. Transient processes keep the resonant efficiency as well. The resulting solutions define the design parameters of the effective ultrasonic machining systems.

ACKNOWLEDGEMENTS

The authors thank Dr A. Kalashnikov and Dr A. Meadows for their valuable contributions to the development of the ultrasonic machining systems.

REFERENCES

[1] Cumabe, J., *Vibratory Cutting*, Dzikke Sjuppan, Tokyo, 1979 [in Japanese].

[2] Markov, A.I., *Ultrasonic Processing of Materials*, Mashinostroenie, Moscow, 1980 [in Russian].

[3] Astashev, V.K., Effect of ultrasonic vibration of a single-point tool on process of cutting, *Journal of Machine Manufacturing and Reliability*, 3, 1992, 65–70.

[4] Astashev, V.K. and Babitsky, V.I., Ultrasonic cutting as a nonlinear (vibro-impact) process, *Ultrasonics*, 36, 1998, 89–96.

[5] Babitsky, V.I., *Theory of Vibro-Impact Systems and Applications*, Springer, Berlin, 1998.

[6] Babitsky, V.I., Astashev, V.K. and Kalashnikov, A.N., Autoresonant control of nonlinear mode in ultrasonic transducer for machining applications, *Ultrasonics*, 42, 2004, 29–35.

[7] Babitsky, V.I., Kalashnikov, A.N. and Molodtsov, F.V., Autoresonant control of ultrasonically assisted cutting, *Mechatronics*, 14, 2004, 91–114.

[8] Astashev, V.K., Analysis of rod-type concentrator subjected to a nonlinear load, *Soviet Physics – Acoustics*, 27, 1981, 821–827.

[9] Babitsky, V.I., Kalashnikov, A.N., Meadows, A. and Wijesundara, A.A.H.P., Ultrasonically assisted turning of aviation materials, *Journal of Materials Processing Technology*, 132, 2003, 157–167.

STRUCTURE-CONTROL-OPTICS INTERACTION IN HIGH PRECISION TELESCOPES

H. Baier[1], M. Müller[1,2] and C. Zauner[1]
[1]*Institute of Lightweight Structures, TU Muenchen, D-85747 Garching, Germany*
[2]*Now European Southern Observatory ESO, Germany*

Abstract: Large precision telescopes require the consideration of the interaction of structural mechanics, control, and optics. Aspects for this are discussed from a simulation and technology point of view. In simulation, emphasis is put on handling and condensation of large structural dynamics models, while in the technology part the concept of smart structures for active damping and shape control is discussed.

Key words: integrated simulation, balanced model reduction, smart structures, shape control, active damping.

1. INTRODUCTION AND SUBJECT

High precision opto-mechanical systems such as large (earth or space) telescopes are characterized by significant interactions between structural, actuator and control dynamics and optics. This interaction has to be considered to properly minimize spurious vibrations within the optical path. In addition, quasistatic distortions e.g. due to temperature changes have to be counterbalanced. Such distortions in the sub-μm- or even nano-meter range put special requirements to the actuators and also to the sensing system.

In this paper, two aspects will be discussed, namely the model building and simulation of the interaction mentioned, and the investigation of smart structures technology for shape control and active damping. Integrated

H. Ulbrich and W. Günthner (eds), IUTAM Symposium on Vibration Control of Nonlinear Mechanisms and Structures, 69–79.

simulation is investigated for large earth bound telescopes, while smart structures are discussed within the context of different space applications.

2. INTEGRATED MODELLING

Within integrated models (see Figure 1) structural dynamics FE models with many hundred thousand degrees of freedom are to be handled. While in principle mechanical nonlinearities should be avoided, they might act locally such as in bearings and actuators. Another type of nonlinearity might come into play especially in space telescopes due to stick-slip effects in joints or due to "spontane" micro-stress relaxation in materials, which are often hard to predict.

Nonlinear optics is represented by ray-tracing techniques with possible linearization resulting into sensitivity matrices. More detailed discussion of the optical part can be found in [1], while the structural part is outlined in the following.

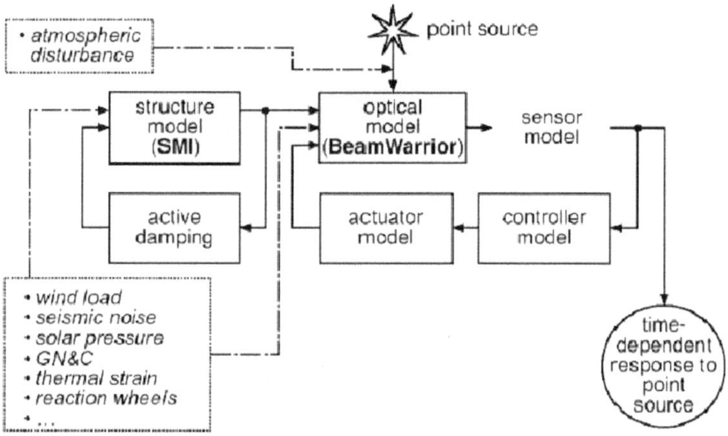

Figure 1. Elements of integrated modeling of telescopes.

2.1 The Structural Integrated Modeling Tool SMI

For the creation of (reduced) dynamic mechanical models the computational tool SMI (Structural Model Interface) has been developed. It uses modal data from an FEM-analysis and builds a linear time-invariant state-space model of a structure to be used within a time-dependent integrated simulation in a Matlab/Simulink environment. The basic software architecture is shown in Figure 2.

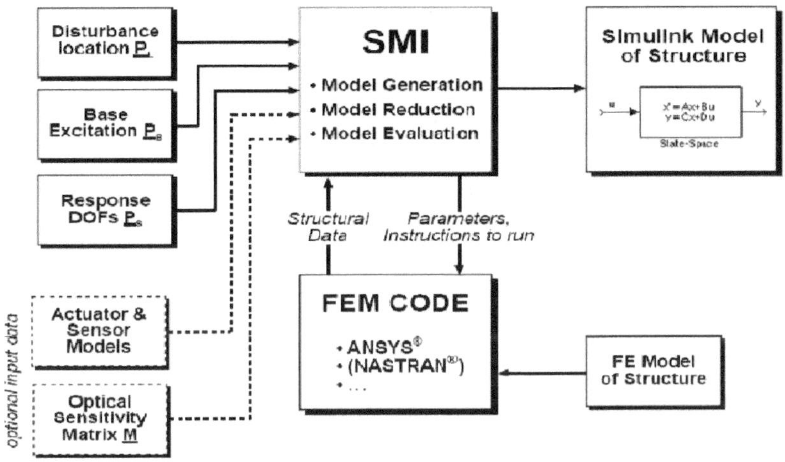

Figure 2. Architecture of SMI.

To define the system inputs and outputs, the degrees of freedom (dof) related to loads and displacement dofs relevant for optical performance must be defined. Typical dynamic loads from wind, rotating equipment or micro-seismic disturbances are in a frequency spectrum of 1-20 Hz. Quasi-static loads are caused by gravity (different view angles of the telescope) or temperature changes. A single load can have an arbitrary distribution on the nodes of the FEM structure with mutually independent time-histories (this includes also independent power spectral densities (PSDs)). The outputs can be combinations of responses of model nodes or the response of a virtual node defined in the FEM model by a constraint equation. In addition, atmospheric turbulence creates temporal and spatial fluctuations of the phase of the electric field of each interferometer beam. Their PSDs are converted into random time series, too. Higher-order modes are represented by time-varying Zernike coefficients.

A key feature of SMI are its quality-criteria based tools for manual and/or automatic model reduction required for integrated simulation and controller design in state space format. Automatic methods are based on effective modal mass tables, controllability/observability and balanced truncating. A system is called balanced if controllability and observability Gramians are equal and diagonal. For a balanced system the new state vectors are ranked according to their contribution to the Hankel singular values. This then allows to reduce states with little or no influence on observability and controllability. In order to take into account the expected frequency spectrum of the excitations, a frequency weighted balanced modal reduction has proven to be appropriate. For the balanced truncation the error bounds are displayed. To evaluate the model quality, effective mass tables can be also considered, and the transfer functions of full and reduced models can be compared. Moreover, dynamic models of actuators and sensors can be included.

2.2 Application Example

The usage of SMI was successfully applied to the Very Large Telescope VLT already under operation and recently within the concept study of the Overwhelmingly Large Telescope (OWL) of the European Southern Observatory (ESO), see Figure 3.

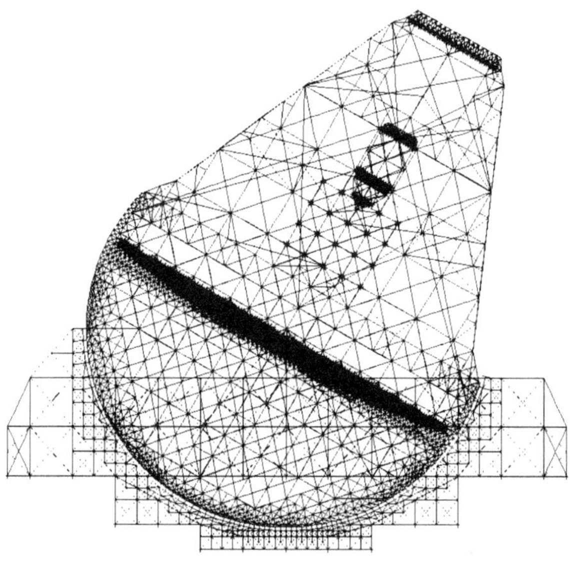

Figure 3. Overwhelmingly Large Telescope (OWL) FE-Model.

Figure 4 shows transfer functions for a full and a reduced model of OWL created via SMI. For a study of the friction effects and the control of the OWL bogies, reduced models were built for seven different configurations. From the detailed FE-model with about 135000 elements, 500 modes were used in a first modal model, which was reduced to 25 states in the end. Splitting the telescope into 15 sections, distributed wind loads are modelled as external disturbances. The model outputs were the telescope pointing, the mirror displacements and rotations, and for the friction simulation the load and displacement of each bogie.

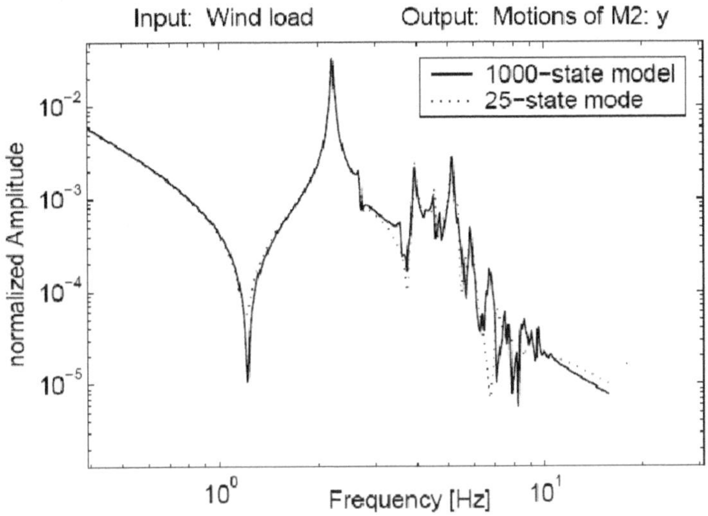

Figure 4. Transfer functions of wind load on OWL to the motions of mirror M2 in y-direction. Full state model (1000 states) in comparison with reduced state model (25 states).

Thus the final system consisted of 25 states, approximately 350 outputs and about 150 inputs.

3. SMART SPACE STRUCTURES

In-orbit requirements for mechanical systems of space telescopes and precision instruments are again related to very low distortions under thermal and dynamic excitations generated by the satellite platform and space environment. Basically these are met by proper design concepts and material selection such as CFRP with very low coefficients of expansion. Such "passive" approaches come to their limits for ever increasing requirements with distortion allowables in fractions of micro-meters or milli-arcseconds of pointing accuracy. Then active shape control and damping techniques come into play, with potential areas of applications being outlined in Figure 5.

Relevant types of actuators and control strategies will be briefly discussed together with applications for active damping of large solar arrays and shape control of precision reflectors.

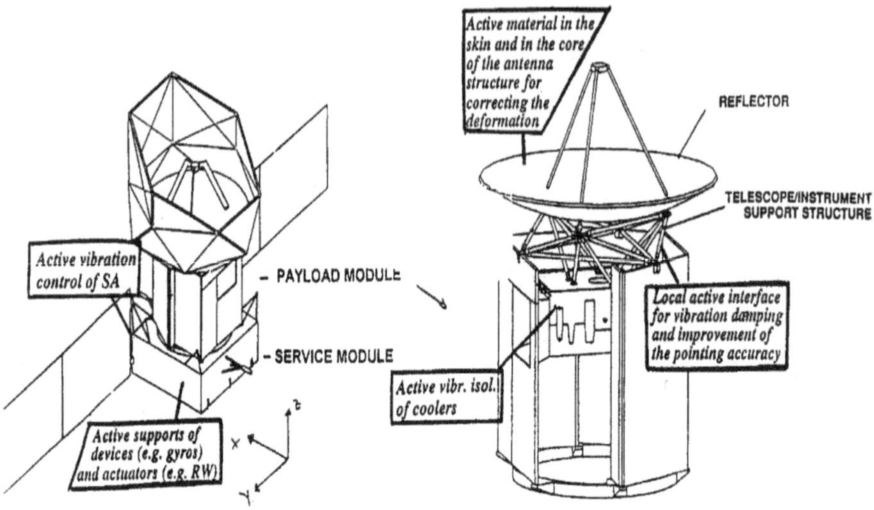

Figure 5. Application Scenario.

3.1 Relevant Smart Materials and Actuator Types

Actuators are an arrangement of power amplifiers together with electrical-mechanical power transfer mechanisms. For the latter, most often piezo-ceramic materials (especially PZT) are used. For precision active systems, power requirements are quite small due to the usually very small strokes involved. Depending on the type of application, piezo-electric materials are used in different actuator configurations such as

- linear actuators, where a stack of piezoceramics (PZT) is used to achieve linear actuation strokes typically in the sub-mm range. Industrial types of actuators are quite matured, but may be relatively heavy.
- PZT plates (planar actuation); also quite matured, but 2d actuation strains might be undesirable in some cases
- Piezo-fibres with thicknesses in the range of some μm up to sub-mm. Electrical contacting and integration of fibres in materials and components is one of the challenging tasks.
- Combinations and special forms such as proof mass actuators (PMA), where inertia forces generated by (small) masses accelerated e.g. by linear actuators are used as actuating lateral forces.

Typical properties of these smart materials are given in Table 1.

Table 1. Essential smart material data.

	piezo-ceramic (PZT)	piezo polymer (PVDF)	electro-strictive (PMN)	magneto-strictive (Terfenol D)	SMA (Nitinol)
Planar Max Strain	0.13%	0.07%	0.1%	0.2%	2%-8%
Modulus (GPa)	60.6	2	64.5	29.7	28-90
Density (kg/m3)	7500	1780	7800	9250	7100
Energy Density (J/kg)	6.8	0.28	4.1	6.4	250-4000
Hysteresis	10%	$> 10\%$	$< 1\%$	2%	High
Temp Range (°C)	-20 to 120	Low	0 to 40	High	-
Bandwidth	100 kHz	100 kHz	100 kHz	< 10 kHz	< 5 Hz

Generally speaking, piezoelectric (mostly ceramic) materials do have high dynamic bandwith, but only relatively low actuation strain and moderate force levels. Coefficients of thermal expansion (CTEs) are low but exceed those of very high precision structural materials. This can be at least partly compensated by proper combination with base materials having negative CTEs, such as special types of CFRP.

Transfer ratios from electrical to mechanical power are quite efficient and are easily exceeding 50 %. Play or slip stick is negligible, and most often lever of "gear" systems for stroke amplification can be avoided.

Temperature dependency of Piezo properties

Many space applications are in cold if not cryogenic environment. While reduction of CTEs can be observed at levels below –50°C and in many cases is beneficial, the possibly significant reduction of piezoelectric coefficient of expansion as outlined in Figure 6 has to be taken into account in cryogenic applications.

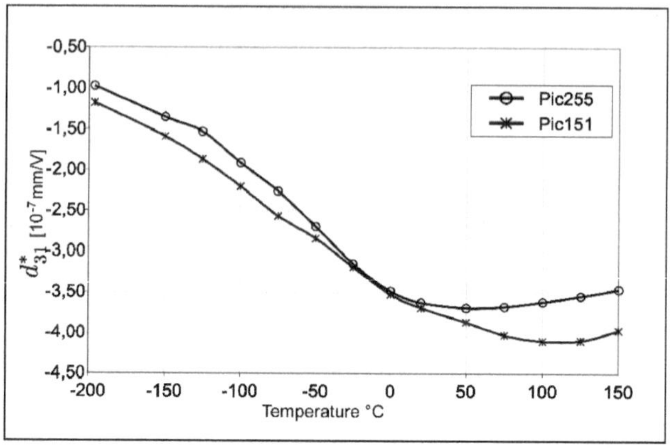

Figure 6. Achievable Piezo strain under varying temperature.

Material nonlinearity

When it comes to details, a non-linear behavior of relationship of strain vs. applied electric field has to be observed. Though practical consequences of this are small if sufficient stroke is available, this can lead to slightly smaller efficiency or slightly increased power consumption. Possibilities to compensate such nonlinearities are the use of different actuator control techniques which linearize the output vs. given input variations.

3.2 Control Laws

Control architecture can be differed between high authority control (HAC) and low authority control (LAC). Loosely speaking, in HAC relevant response data are measured at some points, while actuation can be triggered at other points. In addition, mathematical observer models which allow the estimation of the full system state from the usually small number of measured values are to be used. In LAC, data are measured at the actuator positions and more or less directly fed back to the actuators via different control laws. While HAC often may lead to better overall performance than LAC, it is much more complex and requires also certain provisions to keep the control loop stable. LAC is easier to implement, and control loop stability can be easier achieved. Most prominent examples of LACs are:

- direct velocity feedback (DVF), where the output of a velocity sensor is directly fed back to the actuator, multiplied by a gain factor

- acceleration, position and force feedback, where these data are measured and fed back via additional filters, i.e. with additional dynamics within the feedback loop.

An overview and short evaluation is given in Table 2.

Table 2. Different local authority controllers for active damping.

Control method	Direct Velocity Feedback (DVF)	Positive Position Feedback (PPF)	Acceleration Feedback (AF)	Integral Force Feedback (IF)
Feedback signal	velocity	displacement	acceleration	strut force
Controller type	P	PT_2	PT_2	I
Design parameters	gain	gain, frequency and damping of filter	gain, frequency and damping of filter	gain
Stability	for positive gain (no actuator / sensor dynamics!)	within parameter bounds	for positive gain	for positive gain
Influence on selected modes	not possible	possible by tuning of filter frequency	difficult	not possible

3.3 Applications and Lab Model Demonstrations

Solar arrays: active damping and satellite interface force minimization

Vibration damping of large solar arrays has to reject disturbances on the satellite platform with its precision and disturbance sensitive instruments, and may lead to simplifications of the satellite attitude control system AOCS. The AOCS is usually designed in such a way that the solar arrays cannot be excited in their eigenfrequency range. This superimposes constraints on the period length between two successive AOCS actions. The provision of a high damping of the solar array allows release of this constraint and consequently also leads to a saving of fuel consumption extending the satellite's lifetime.

A straightforward design solution is the implementation of an active interface between solar array and satellite platform.

Figure 7 shows the principle of such an interface for damping the low frequency bending modal shapes of large solar arrays in conventional configuration: the active interface is directly attached to the solar array pointing drive or another typical interface. The benefit of such an active interface is outlined in Figure 7 force (disturbance) amplitudes to the S/C become smaller and are damped out in significant shorter time period.

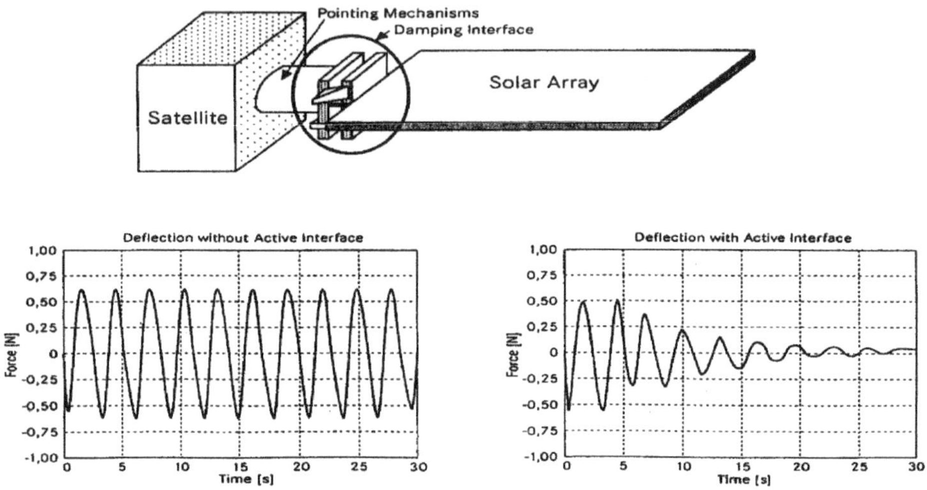

Figure 7. Active vibration attenuation (right) for a solar array interface force to the satellite.

Thermo-elastic shape control of precision reflectors

Design of shape controlled reflector means the determination of structural parameters (geometry, face sheet lay-up etc.) and control parameters (actuator positioning, size and geometry of piezo actuator layers, gain factors etc.).

Investigations carried out with PZT patches used as actuators on the reflector's rear side showed some benefit but also an inherent disadvantage of such plate or 2d actuators: piezoelectric or elastic coupling effects (e.g., Poisson's factor) also cause forced (control) strains in other directions orthogonal to those aimed for. So a decoupling into radial and circumferential actuation is desirable. This can be achieved for example by using radially or circumferentially integrated Piezo fibers, or orthogonally arranged linear actuators, or orthogonally arranged backside "stiffeners" with integrated actuating materials. A simulation result of such considerations is shown in Figure 8, where 40 actuators were "homogeneously" applied to a reflector's backside structure. Radially and circumferentially decoupled actuating forces show an improved rms error by a factor of 4 compared to that of the directionally coupled case when using 2d Piezo plates.

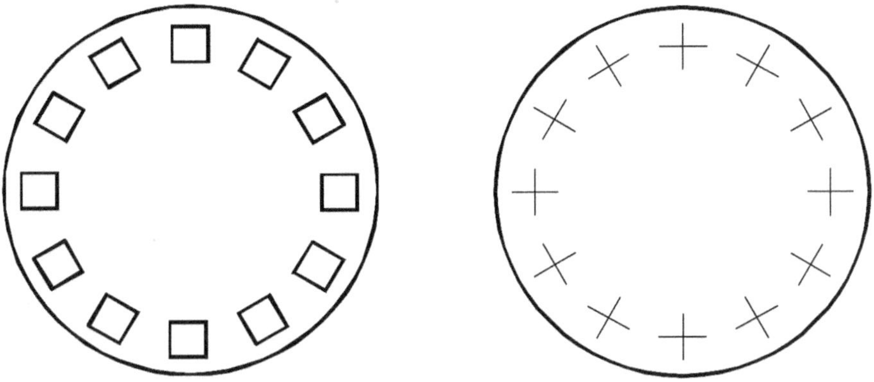

rms=1.0 rms= 0.25

Figure 8. Piezoplates vs. radial-circumferential actuators (piezofibres, orthogonal linear actuators) at reflectors and consequences for achievable rms shape deviation.

REFERENCES

1. Schoeller, M., Wilhelm, R. and Koehler, B., "Modeling the imaging process in optical stellar interferometers", Astron. & Astrophys. Suppl. Ser. 144, pp. 541–552, 2000.
2. Glindemann, A., "Very Large Telescope Interferometer: a status report", in Interferometry for Optical Astronomy II, W. A. Traub, ed., Proc. SPIE 4838, 2002.
3. Locatelli, G., Piezo-actuated Adaptive Structures for Vibration Damping and Shape Control, Fortschritts-Berichte VDI, Reihe 11, No. 303, VDI-Verlag, Düsseldorf, 2001.
4. Baier, H. and Scheulen, D., "Approaches and technologies for control-structure-interaction in smart structures", in Proc. ESTEC Conf. Spacecraft Structures and Mechanical Testing, 1998.
5. Müller, M., Baier, H., Langer, H., Helwig, G. and Wilhelm, R., "Structure-control-optics interaction and probabilistic analysis tools for high precision telescopes and mirrors", in Proc. European Conference on Spacecraft Structures, Materials and Mechanical Testing, 11–13.12.2002, Toulouse, France.
6. Baier, H. and Locatelli, G., "Microvibration control in telescope structures", in Proc. of SPIE Conference Astronomical Telescopes and Instrumentation 2000, Vol. 4004, Munich, March 2000.
7. Baier, H. and Müller, M., "Adaptive structures and mechatronic components for vibration and shape control of satellite precision payloads", in Proceedings of the 54nd International Astronautical Congress, Bremen, Germany, September 29–October 3, 2003.

ACCELERATION SENSOR BASED ON DIAMAGNETIC LEVITATION

François Barrot[12], Jan Sandtner[1] and Hannes Bleuler[1]
[1]EPFL (Ecole Polytechnique Fédérale de Lausanne), Laboratoire de Systèmes Robotiques, Station 9, CH-1015 Lausanne, Switzerland; [2]EMPA (Eidgenössische Materialprüfungs- und Forschungsanstalt), Structural Engineering Research Laboratoty, Überlandstrasse 129, CH-8600 Dübendorf, Switzerland

Abstract: A novel force feedback acceleration sensor concept, featuring levitation of the inertial mass, is presented in this paper.

Key words: sensor, diamagnetic levitation, accelerometer, seismometer.

1. INTRODUCTION

Acceleration sensors have a wide and growing spectrum of industrial applications, ranging from the monitoring of vibrations and accelerations (accelerometer, seismometers), to the control of speed and position in the ever expanding field of robotic applications.

Current acceleration sensors are based on the relative displacement between an inertial mass and the base of the instrument; however the existence of a material link between these two elements introduces uncontrollable distortions due to friction, stiffness changes and thermal differential dilation, which irremediably limit the precision of the measurements.

Ideally, all contact between the seismic mass and the base of the instrument should be avoided. We propose here the application of the diamagnetic effect for a totally contact-free levitation of the seismic mass.

H. Ulbrich and W. Günthner (eds), IUTAM Symposium on Vibration Control of Nonlinear Mechanisms and Structures, 81–90.

In conjunction with electrostatic actuation, this principle leads to an entirely new family of potentially low cost *high precision acceleration sensors.*

Furthermore, given the simplicity of their design, such sensors could be mass produced and at low cost.

2. DIAMAGNETIC LEVITATION

In 1842, Earnshaw [1] stated a theorem which stipulates that, with any combination of electrostatic, magnetostatic or gravitational fields, it is impossible to stabilize the levitation an object electrostatically charged or ferromagnetic.

However a stable levitation under external magnetostatic fields can be achieved by using diamagnetic materials [2, 3].

Diamagnetism is a magnetic property inherent to many materials such as gold, silver, bismuth, graphite and even water, wood or plastic.

Due to their characteristic negative magnetic susceptibility χ_m (Table 1) diamagnetic materials are repelled by magnetic fields and stabilized in the minimum flux density regions. The diamagnetic effect is however relatively weak in comparison to e.g. the flux-pinning or the Meissner-Ochsenfeld effect known for superconductors.

Table 1. Magnetic susceptibility of some diamagnetic materials.

Material	χ_m
Water	$-9.0 \cdot 10^{-6}$
Bismuth	$-1.5 \cdot 10^{-4}$
Graphite	$-1.6 \cdot 10^{-4}$
Pyrolitic graphite	$-4.5 \cdot 10^{-4}$
Superconductor	-1.0

Thanks to this negative magnetic susceptibility χ_m, the Laplacian of the potential interaction energy E_p between a diamagnetic substance and a magnetic field B can be made strictly positive (1) which proves that it is possible to trap a diamagnetic substance in a "potential bowl" region where a stable levitation takes place.

$$\Delta\left(E_p\right) \simeq \frac{-\chi_m}{\mu_0} \Delta\left(B^2\right) \geq 0 \tag{1}$$

Experimentally, although diamagnetic forces are very weak, it is indeed possible find diamagnetic materials light enough so that the diamagnetic effect may overcome gravity. We obtained such a stable equilibrium for a piece of diamagnetic material over a carefully optimized 2D arrangement of permanent magnets [4, 5].

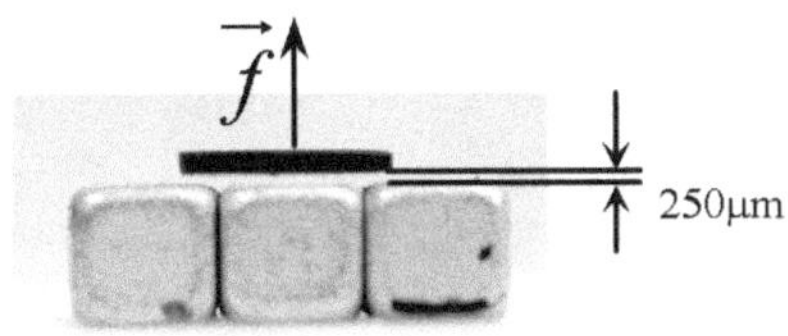

Figure 1. Diamagnetic material levitated over an array of permanent magnets.

Figure 1 shows a 7 mm diameter pyrolithic graphite disc floating motionless at room temperature, at about 250 μm above an array of NdFeB magnets. Small loads can be added to the levitated disc and, if more payload is needed, some weight compensating magnet can be added [5, 6].

The levitation force $\vec{F}$ (2) being proportional to the magnetic susceptibility χ_m, to the magnetic field $\vec{B}$ and to its gradient, a high diamagnetic susceptibility as well as high field gradients are sought.

$$\vec{F} = \frac{\chi_m}{2\mu_0}\vec{B}.\overrightarrow{grad}(B) \tag{2}$$

High diamagnetic susceptibility entails that the graphite must be oriented and this is why pyrolithic graphite (Table 1) is usually chosen. However high purity and high orientation graphite like the expensive "HOPG" (highly oriented pyrolithic graphite) used in scanning tunneling microscopy, is not needed in this case.

Ideally, in order to produce a *magnetic field with a steep gradient*, two neighboring magnets should have different polarities.

Accordingly several configurations for such magnet arrays are possible with a one-dimensional or a two dimensional pattern repetition[2]. Among these possibilities the "Halbach-2D" pattern (Figures 2 and 3) is the most efficient: it concentrates the magnetic field on only one side of the array while using a reduced number of magnets (the array is not entirely filled with magnets) and, in addition, it gives the highest diamagnetic levitation force [2, 5].

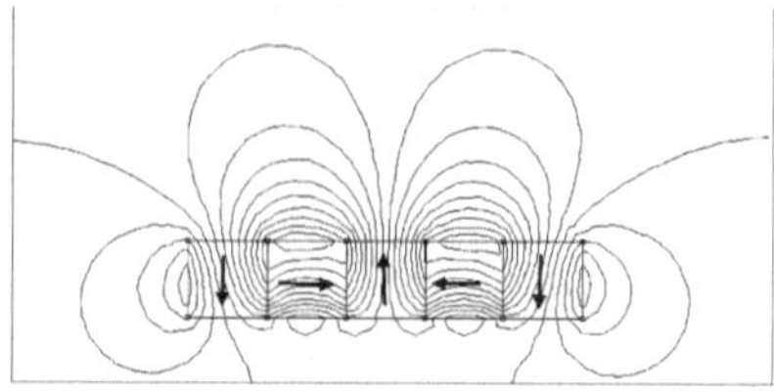

Figure 2. Halbach array: magnetic field contour lines.

However, the "opposite-2D" pattern (Figure 3), which is made of neighboring magnets with opposite polarities, is almost as effective and has the additional advantage of being easy to assemble since the magnets just stick together, without any bonding or gluing as opposed to the "Halbach-2D" configuration.

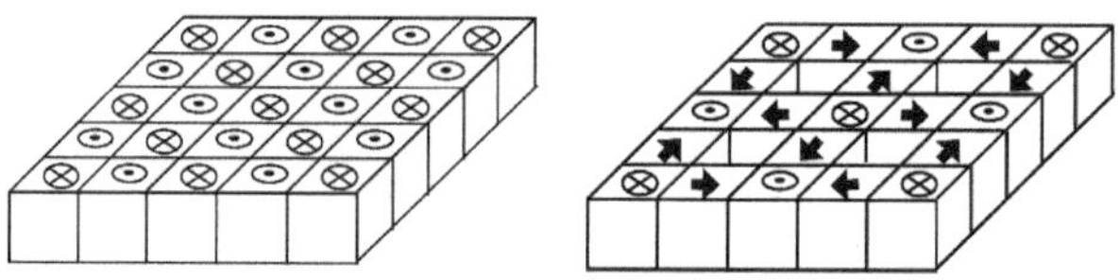

Figure 3. 2D patterns of magnets (left: Opposite 2D, right: Halbach 2D).

3. ACCELERATION SENSOR

3.1 2D-Acceleration Sensor

Combining the use of diamagnetic levitation and electrostatic actuators, highly sensitive contactless acceleration sensors, based on the magnetic levitation of a diamagnetic body over a two-dimensional permanent magnet array, can be designed.

The main advantage of such a design is the absence of contact between the levitated seismic mass and the base of the instrument which implies the absence of wear, friction or any material link subject to stiffness or nonlinear effects, thus resulting in a robust and potentially highly sensitive inertial sensor.

A prototype of such a sensor has been realized and is now described in some details.

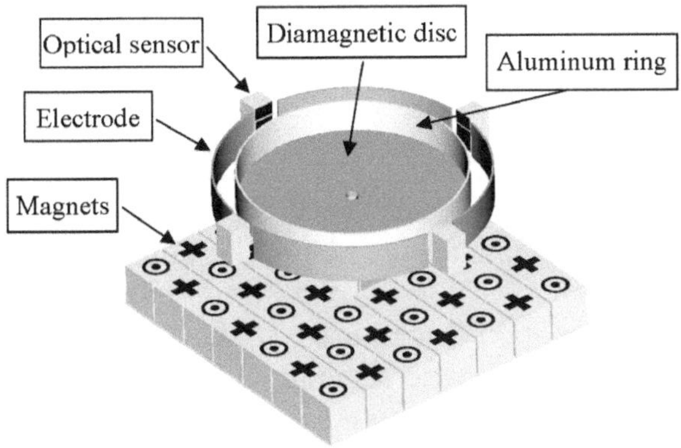

Figure 4. 2D acceleration sensor based on diamagnetic levitation.

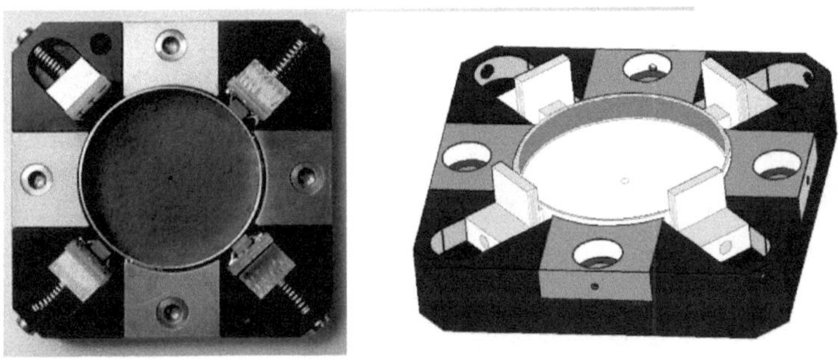

Figure 5. Version of the 2D accelerometer with four reflection position sensors.

A simplified 3D representation and a photograph of this prototype are shown in Figures 4 and 5. The seismic mass (1.68 g) is made of a pyrolithic graphite disc (28 mm diameter) surrounded by an aluminum ring. This seismic mass floats at 250 μm over a horizontal "Halbach-2D", or "Opposite-2D", arrangement of permanent magnets. The diamagnetic effect of this arrangement is strong enough to support the weight of the aluminium ring.

The relative horizontal position between the seismic mass and the base of the instrument is measured differentially. This position sensing technique, combined with the absence of mechanical contact, allows high accuracy measurements. Sensing can be achieved using:

either,

- one monolithic sensor consisting of a four segments photodiode as it is used for tracking and focusing in a CD player. An infrared LED shines through a small hole in the centre of the graphite disc and thus the radial (x-y) position of the disc is obtained with high accuracy (~100 nm) at very low cost

or

- two pairs of optical position sensors mounted on two orthogonal diameters axes and facing the aluminum ring (Figures 4 and 5).

The position information is then used in a feedback loop which includes the sensor(s) and an electrostatic actuator in order to hold the diamagnetic disc at a predefined position. The two pairs of electrodes of this actuator are also diametrically mounted on two orthogonal axes and exert radial forces on the aluminum ring (Figure 7).

For each actuator the resulting electrostatic force (3) is linearized (4) (Figure 6), with a bias voltage V_0, around the reference position.

$$f = \frac{\varepsilon_0 S}{2} \cdot \frac{V^2}{d^2} \tag{3}$$

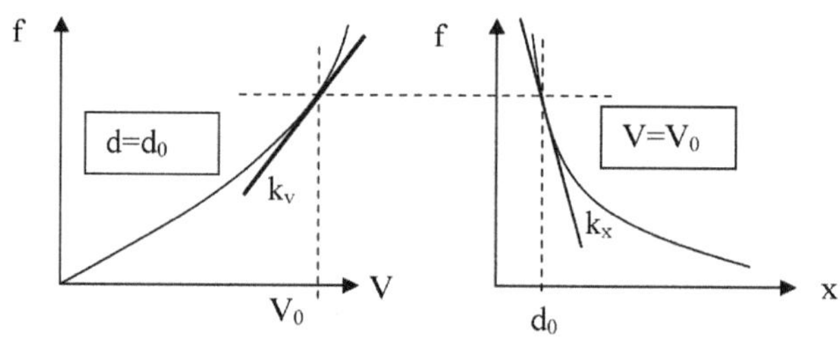

Figure 6. Linearization of the electrostatic force.

$$f = k_v V + k_x x \quad \text{with} \quad \begin{cases} k_v = 2\varepsilon_0 S \dfrac{V_0}{d_0^2} \\ k_x = 2\varepsilon_0 S \dfrac{V_0^2}{d_0^3} \end{cases} \tag{4}$$

In our case, the air gap d_0 being 300 μm, this results in a displacement smaller than 30 μm microns.

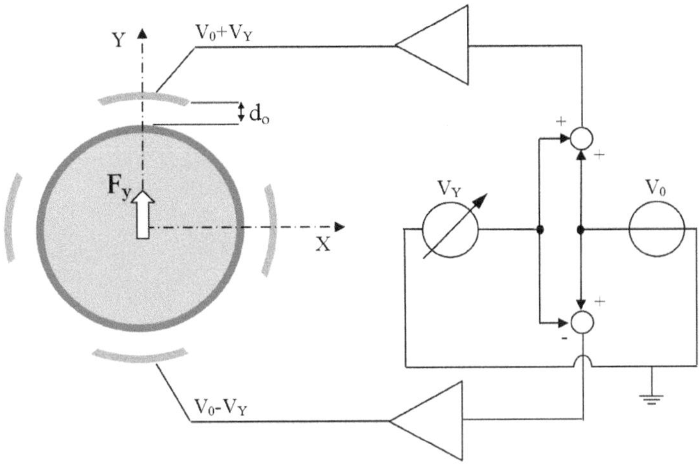

Figure 7. Differential control of the electrostatic bearing electrodes.

When subject to vibration of the base, the seismic mass is maintained at a reference position relative to this base and the acceleration is directly related to the voltage applied to the electrodes.

In addition, the stiffness and thus the frequency characteristics (bandwidth) of the actuator, *can be tuned electronically* by modifying the controller setting.

Regarding the power consumption of such a device, it is very low since the levitation of the seismic mass is passive (i.e., no energy consumption) and the electrostatic actuators do not require high power (few tens of μA at voltages in the range of typically up to a few hundreds Volts).

3.2 3D-Acceleration Sensor

This 2D-acceleration sensor measures horizontal accelerations. However, based on the same principles, a 3D acceleration sensor can also be designed by adding additional electrostatic actuators as well as additional differential position sensing systems for the vertical motion.

In this case, when the diamagnetic seismic mass moves vertically, the diamagnetic force (2) has a non linear variation [4, 5]. It is, nevertheless, still possible to precisely change stiffness and/or damping of the diamagnetic suspension in the vertical direction with an appropriate contact-free actuation similar to the radial system described earlier.

Indeed, using the diamagnetic disc as a common electrode located between a set of three pairs of horizontal electrostatic electrodes and, placing three comb shaped capacitive position sensors on each side of the diamagnetic disc [6], it is possible to control the vertical position of the seismic mass and thus measure its acceleration in the vertical direction.

Figure 8 shows how actuators and position sensors could be added to the previous design (Figure 4) in order to measure vertical acceleration. In this case, the comb shaped capacitive sensors could, for instance, be integrated on the same layer as the electrodes on a thin Flex-print.

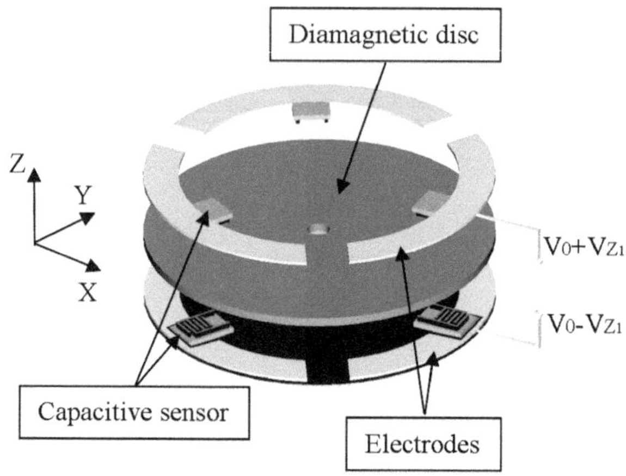

Figure 8. Placement of sensors/actuators for vertical acceleration sensing.

The current version of the 2D-acceleration sensor can experimentally undergo a peak acceleration of 0.5 g with a resolution in the μg range, making it particularly interesting for seismic sensing and instrumentation of civil engineering structures.

However, since, as shown hereunder, both diamagnetic levitation and the use of electrostatic forces are most relevant in small scale applications, it is possible to design a sensor with a broader sensing range by merely reducing its size.

Indeed, L being the dimensional factor:

- the ratio between the electrostatic force ($\sim L^{-2}$) and the weight ($\sim L^{3}$) of the inertial mass is proportional to L^{-5} ,
- the ratio between the diamagnetic levitation force ($\sim L^{2}$) and the weight ($\sim L^{3}$) of the inertial mass is proportional to L^{-1}.

Therefore scaling down the proposed acceleration sensors should result, in addition to a lower power consumption (less voltage needed for the electrostatic actuators) and the *possibility of measuring higher levels of g for use in robotic systems.*

In the proposed 2-D and 3-D implementations the seismic mass is a disc that can freely rotate around the vertical axis passing through its center. Hence, when the diamagnetic disc is maintained at its reference position, the actuators as well as the sensors may not always be facing the same part of the seismic mass and, the seismic mass being imperfectly cylindrical, it may induce errors in the case of high precision applications. A simple solution to block the free rotation of the seismic mass consists in using a non axisymmetrical seismic mass / actuators arrangement: i.e. a square or a triangular shaped diamagnetically levitated seismic mass facing a square or triangular shaped arrangement of electrodes.

3.3 Control Strategy

To keep the seismic mass at a reference position, the resulting force of each electrostatic actuator is linearized around this reference position and a PID controller is implemented. The equivalent damping and stiffness of each actuator can be tuned to modify the bandwidth and the sensitivity of the instrument.

One of the targeted applications of the proposed diamagnetically levitated acceleration sensor is seismic monitoring of civil engineering structures.

In this case the vibration signal of interest is stochastic and its dynamic range is huge, typically spanning from the nano *g* to a few *g*. However, seismic signals, though stochastic, are characterized by low amplitudes (low levels of *g*) at low frequencies and higher amplitudes at higher frequencies. Besides, the frequencies of seismic signals are low (0.1 Hz to 20 Hz) compared to the high sampling rate (5 kHz to 10 kHz) used for the position control of the levitated seismic mass.

Typically several kinds of acceleration sensors with different bandwidths and sensitivities are needed to cover the whole seismic spectrum. However, given the absence of friction and the ability to modify the equivalent damping and stiffness, the *diamagnetic acceleration sensor could be used as a single kind of acceleration sensor to cover a broad, maybe even the whole seismic spectrum*. Indeed, in this case, the modification of the bandwidth and sensitivity of the acceleration sensor does not require hardware changes but only parameter tuning in the digital controller. Several preset operating modes can thus be implemented on the sensor and then selected by the end-user, depending on the nature of the vibration to be recorded.

4. CONCLUSION

An entirely new type of accelerometer based on diamagnetic levitation and electrostatic actuation has been designed.

No mechanically highly precise parts are required at all, the *accuracy* of the sensor *depends essentially on the electronics*. Thus the actuator costs are potentially low in mass production.

The diamagnetic force has traditionally been thought of as such a weak force that its use at the macroscopic scale has not been considered for engineering applications.

However, recent advances in materials (mainly permanent magnets) lead to unexpected new possibilities in the design of contact-free systems based on diamagnetic levitation. This is opening up promising applications in the field of sensitive acceleration sensors and possibly of flywheel energy storage, to name only two.

REFERENCES

1. Earnshaw, S., On the Nature of the Molecular Forces which Regulate the Constitution of the Lumiferous Ether, *Trans. Camb. Phil. Soc.* 7, 1842, pp. 97–112.
2. Pelrine, R.E., Room Temperature Open-loop Levitation of Microdevices Using Diamagnetic Materials, in *Proc. IEEE Micro Electro Mechanical Systems*, 1990, pp. 34–37.
3. Simon, M.D., Heflinger, L.O. and Geim, A.K., Diamagnetically Stabilized Magnet Levitation, *Am. J. Phys.*, Vol. 69, No. 6, pp. 702–713.
4. Moser, R., F. Barrot, F. and Bleuler, H., Optimization of Two-dimensional Permanent Magnet Arrays for Diamagnetic Levitation, in *Proc. MAGLEV*, 2002.
5. Moser, R., Regamey, Y., Sandtner, J. and Bleuler, H., Passive Diamagnetic Levitation for Flywheels, in *Proc. 8th ISMB*, 2002.
6. Moser, R., Sandtner, J. and Bleuler, H., Diamagnetic Suspension Assisted by electrostatic Actuators, in *Proc. 6th ISMST*, 2001.

ACTIVE CONTROL STRATEGIES FOR VIBRATION ISOLATION

Brad M. Beadle[1], Stefan Hurlebaus[2], Lothar Gaul[1] and Uwe Stöbener[3]
[1]Institute A of Mechanics, University of Stuttgart, Allmandring 5 b,70550 Stuttgart, Germany;
[2]Dept. of Civil Engineering, Texas A&M University, College Station, TX, 77843-3136, USA;
[3]Halcyonics GmbH, Tuchmacherweg 12, 37079 Göttingen, Germany

Abstract: In the fields of high-resolution metrology and manufacturing, effective anti-vibration measures are required to obtain precise and repeatable results. This is particularly true when the amplitudes of ambient vibration and the dimensions of the investigated or manufactured structure are comparable, e.g. in sub-micron semiconductor chip production, holographic interferometry, confocal optical imaging, and scanning probe microscopy. In the active anti-vibration system examined, signals are acquired by extremely sensitive vibration detectors, and the vibration is reduced using a feedback controller to drive electrodynamic actuators. This paper deals with the modeling and control of this anti-vibration system. First, a six-degree-of-freedom rigid body model of the system is developed. The unknown parameters of the unloaded system, including actuator transduction constants, spring stiffness, damping, moments of inertia, and the vertical position of the center of mass, are determined by comparing measured transfer functions to those calculated using the updated model. Finally, two different strategies for actively controlling the vibration isolation system are considered.

Key words: active vibration isolation, MIMO control, parameter identification, SISO control, six-degree of-freedom rigid body.

1. INTRODUCTION

Isolating a piece of delicate equipment from the vibration of a base structure is of practical importance in a number of engineering fields. The quest for tighter production tolerances and higher resolution has led to more stringent requirements regarding ambient vibration levels. In the majority of cases, the base supporting a piece of precision equipment is flexible and vibrates with an unpredictable waveform containing significant energy over a broad range of frequencies. The active isolation of equipment from a vibrating base structure is considered in this paper. Passive anti-vibration mounts are

H. Ulbrich and W. Günthner (eds), IUTAM Symposium on Vibration Control of Nonlinear Mechanisms and Structures, 91–100.

widely used to isolate precision equipment from base vibration. Although conventional passive mounts offer good isolation at high frequencies, they suffer from vibration amplification at the mounted resonance frequency. Generally, the best isolation performance is achieved by using an active system in combination with a passive mount, whereby the fundamental mounted resonance can be actively controlled without compromising the high frequency performance.

Relevant active vibration isolation techniques can be found in several recent publications [1–5]. This paper investigates a commercially available six-degree-of-freedom (DOF) active vibration isolation system. Active vibration isolation is achieved in the as-built system using a decentralized control scheme consisting of independent, analog "SISO" (single input, single output) controllers. The ultimate goal of the study is to investigate alternative controllers, such as the "MIMO" (multiple input, multiple output) controller, for use in vibration isolation. However, successful implementation of such controllers first requires an accurate physical model of the vibration isolation system. In Section 2, a six-DOF rigid body model of the unloaded vibration isolation system is developed. In Section 3, model parameters associated with the unloaded vibration isolation system are determined using experimentally measured transfer functions. In Section 4, the SISO- and MIMO- control strategies are discussed.

2. MODELING

2.1 System Model

The vibration isolation system is essentially composed of two plates connected by springs and actuators. A physical model of the vibration isolation system is depicted in Figure 1. Assuming that the motion of the upper plate undergoes rigid body motion only, then its motion can be described using six coordinates. Referring to the upper plate, the coordinates x, y, and z are used to describe the linear motion of the plate, and the coordinates ϕ_x, ϕ_y, and ϕ_z are used to describe small rotations about the x, y, and z axes, respectively. The coordinate system is located at the center of mass of the plate and is aligned with the plate's principal axes. Due to plate symmetry, the principal axes align with the geometric axes of symmetry, and the x and y coordinates of the center of mass coincide with the geometric center of the plate. In a similar fashion, the motion of the lower plate is described using the coordinates x^{d}, y^{d}, z^{d}, ϕ_x^{d}, ϕ_y^{d}, and ϕ_z^{d}.

The upper plate is suspended above the lower plate with four springs, each of which is denoted by the stiffness k_{s}. The physical model contains additional stiffnesses, namely k_{t} and k_{r}, to model the transverse and rotational stiffness, respectively, associated with the suspension springs. Presumably, the motion of the lower plate is identical to the motion of the surface upon which the vibration isolation system rests. The coordinates describing the motion of the lower plate can therefore be viewed as disturbance inputs. The motion of the lower plate is coupled to the upper plate via the suspension, transverse, and rotational springs. Four vertical actuators, A_1-A_4, and four horizontal actuators, A_5-A_8, are rigidly attached to the upper plate, and interact with the lower plate via point contact. The vertical actuators can slide horizontally relative to the lower plate, and the horizontal actuators can slide vertically relative to the lower plate. Active vibration control is realized by driving the actuators with the appropriate control inputs. The coordinates L_1-L_3 are used to locate the suspension springs and the vertical actuators, and the coordinates L_4-L_6 are used to locate the horizontal actuators.

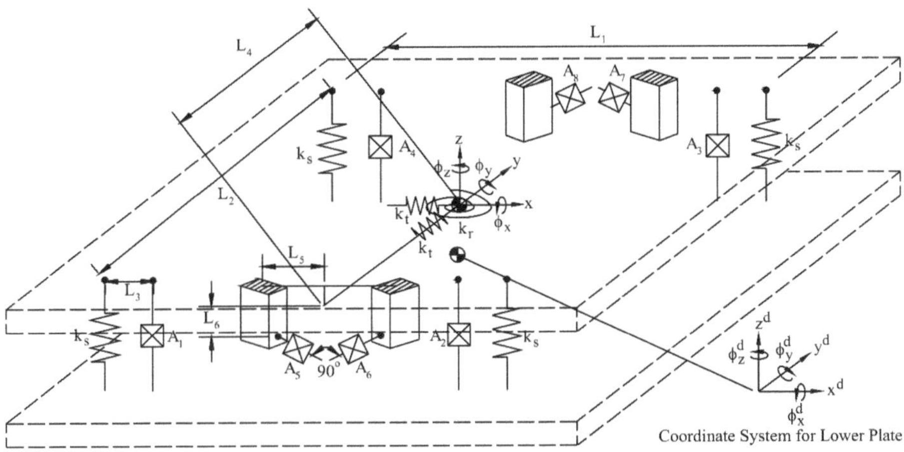

Figure 1. Physical model of the vibration isolation system.

2.2 Actuator Model

The physical model of an actuator is depicted in Figure 2. The main components of the actuator are a stationary permanent magnet and a coil-wrapped reaction mass m_a which moves relative to the permanent magnet. A damper having value d_a has been included to model energy losses, and a spring having stiffness k_a has been included to model the spring-like restoring force between the permanent magnet and the reaction mass. When a voltage V is applied to the coil, a current I will flow through it. An electromotive force F proportional to the current (and therefore to the applied voltage, assuming a purely resistive coil) is developed by the coil, thereby causing the reaction mass to accelerate.

A simplified actuator model is depicted to the right in Figure 2. This is the actuator model which is implemented in the overall model for the vibration isolation system. The reaction mass has been eliminated in the simplified actuator model since its motion is totally constrained in the vibration isolation system. Additionally, the damping parameter d_a has been excluded from the simplified actuator model since its effects are negligible in comparison to other system energy loss mechanisms, such as the friction occurring at the points of contact between the lower plate and the actuators.

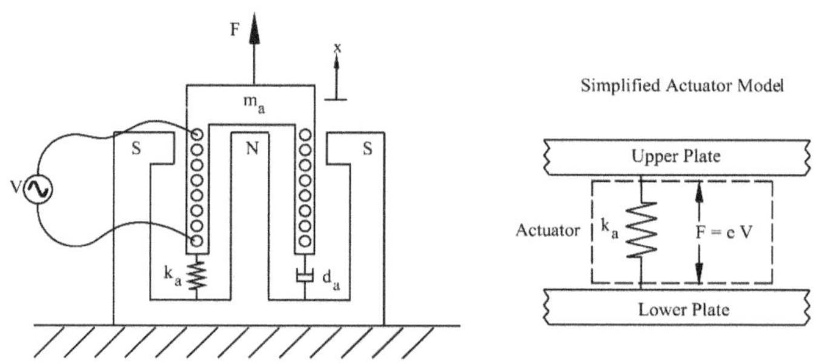

Figure 2. Physical model of an actuator.

2.3 Equations of Motion

The unloaded vibration isolation system is composed of two rigid bodies: the lower plate and the upper plate. The motion of the lower plate is presumably given, i.e., it has the same motion as the surface upon which it rests. The following equations of motion [6] can then be used to solve for the unknown motion of the upper plate:

$$\sum \overline{F} = m\overline{a}_{\mathrm{G}} \tag{1}$$

$$\sum \overline{M}_{\mathrm{G}} = \dot{\overline{H}}_{\mathrm{G}} \approx \begin{bmatrix} I_{xx} & 0 & 0 \\ 0 & I_{yy} & 0 \\ 0 & 0 & I_{zz} \end{bmatrix} \begin{Bmatrix} \ddot{\phi}_x \\ \ddot{\phi}_y \\ \ddot{\phi}_z \end{Bmatrix} \tag{2}$$

The force balance is stated in Eq. (1), where m is the mass of the upper plate, and $\overline{a}_{\mathrm{G}}$ is the acceleration of its mass center G. The moment balance is stated in Eq. (2), where $\dot{\overline{H}}_{\mathrm{G}}$ is the rate of change of angular momentum about the center of mass, and I_{xx}, I_{yy}, I_{zz} are the principle moments of inertia. The approximation in Eq. (2) is valid only for small angular displacements. The left hand side of Eqs. (1) and (2) are determined by summing force and moment contributions, respectively, from the various springs and actuators. The resultant equations of motion for the upper plate are:

$$[M]\ddot{\overline{w}} + [D]\dot{\overline{w}} + [K]\overline{w} = [D]\dot{\overline{w}}^{\mathrm{d}} + [K]\overline{w}^{\mathrm{d}} + [E]\overline{v}, \qquad (3)$$

$$\overline{w} = \{x \quad y \quad z \quad \phi_x \quad \phi_y \quad \phi_z\}^{\mathrm{T}},$$

$$\overline{w}^{\mathrm{d}} = \{x^{\mathrm{d}} \quad y^{\mathrm{d}} \quad z^{\mathrm{d}} \quad \phi_x^{\mathrm{d}} \quad \phi_y^{\mathrm{d}} \quad \phi_z^{\mathrm{d}}\}^{\mathrm{T}},$$

where $\bar{v}$ is the vector containing the input voltages applied to the actuators (the vector elements v_1-v_8 correspond to actuators A_1-A_8, respectively). Analysis details and expressions for the mass matrix $[M]$, damping matrix $[D]$, stiffness matrix $[K]$, and actuator transduction matrix $[E]$ are given by Beadle et al. [7]. The analysis assumes that the damping matrix is diagonal, so that the damping of a particular rigid body mode does not depend on the motion of another rigid body mode. The individual entries of the damping matrix are determined experimentally. The first two terms on the right hand side of Eq. (3) represent the disturbance excitation due to the motion of the lower plate. The third term on the right hand side of Eq. (3) is the excitation generated by the actuators.

The transfer function T relating displacement at an arbitrary location and in an arbitrary direction, to force input from the rth actuator has been derived by Beadle et al. [7]. The derivation assumes forced, harmonic, steady state motion and zero disturbance inputs, $\overline{w}^d = 0$.

3. PARAMETER IDENTIFICATION

3.1 Experimental Setup

The experimental setup for identifying the unknown parameters in the theoretical model is depicted in Figure 3. The unknown parameters include: spring stiffnesses, damping, actuator transduction constants, moments of inertia, and the vertical position of the upper plate's center of mass. The other constants required for the model are known: L_1=0.348 m, L_2=0.3 m, L_3=0.055 m, (L_4+L_5)=0.19 m, and mass m =19.3 kg. An HP Paragon measurement system is configured to drive one or more of the actuators using a random voltage excitation V. The excitation signal is amplified using the preamplifiers which are built into the commercial vibration isolation system. In the diagram, the vertical actuators are denoted by A_1-A_4, and the horizontal actuators are denoted by A_5-A_8. The response of the upper plate is measured using an accelerometer (Brüel & Kjær 4381) placed at one of the three measurement locations M_1-M_3. The vibration isolation system actually has built-in accelerometers at each of the actuator locations. In future efforts, these accelerometers will be used for the parameter identification procedure. The accelerometer used in the current study can be oriented to measure acceleration in any of the three coordinate directions. The acceleration signal is converted into displacement *X* using a signal conditioner before being received by the measurement system. The measurement system then computes the *X/V* transfer function from the random displacement and voltage waveforms. Division of the measured *X/V* transfer function by the appropriate actuator constant e_r results in the displacement-force transfer function *X/F*.

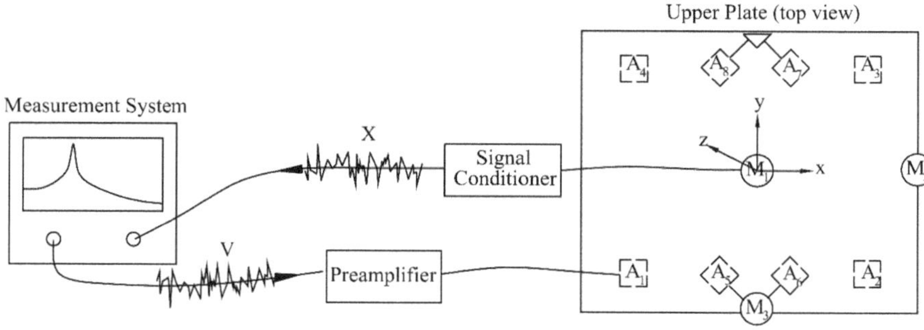

Figure 3. Experimental setup for determining model parameters.

3.2 Experimental Results

The results of the parameter identification procedure for the unloaded vibration isolation system are summarized in Table 1. In the table, the actuation constants are denoted by *e*, the damping coefficients are denoted by *d*, and the moments of inertia are denoted by *I*. A similar set of tests has been used to identify parameters for an arbitrarily loaded vibration isolation system [7]. For explicative purposes, refer to test 5 in Table 1. For this test, the four actuators A_1-A_4 were driven, and the accelerometer was used to measure the motion in the *z*-direction at measurement location M_3. Actuators A_3 and A_4 were driven 180° out of phase relative to actuators A_1 and A_2. This particular drive scheme excites pure rotational motion about the *x* axis. Hence, the unknown moment of inertia and rotational damping about this axis can be determined by comparing the measured transfer function with the theoretical transfer function. Specifically, the theoretical transfer function Z_3/F_1 is computed for several different values of I_{xx} and d_4. An error function given by

$$error = \left(\overline{T}_{\text{exp}} - \overline{T}_{\text{theo}}\right) \cdot \left(\overline{T}_{\text{exp}} - \overline{T}_{\text{theo}}\right)^* \tag{4}$$

is then calculated. Here, $\overline{T}_{\text{exp}}$ and $\overline{T}_{\text{theo}}$ are the experimental and theoretical transfer functions, respectively, and the * operator denotes complex conjugation of individual vector entries. The error function is calculated for values of the transfer functions in the frequency range 2.5-20 Hz. Finally, the parameter values which minimize the error function are determined.

Table 1. Parameter identification results for the unloaded vibration isolation system.

Test	Actuator	Displacement	Parameter	
1	A_1	z_1	$e_1 = 6.59$ N/V	
2	A_2	z_1	$e_2 = 7.85$ N/V	k_s=11300 N/m
3	A_3	z_1	$e_3 = 7.07$ N/V	d_3=275 Ns/m
4	A_4	z_1	$e_4 = 8.10$ N/V	
5	$(A_1+A_2)-(A_3+A_4)$	z_3	I_{xx}=0.340 kgm^2, d_4=6.74 Ns/rad	
6	$(A_1+A_4)-(A_2+A_3)$	z_2	I_{yy}=0.259 kgm^2, d_5=3.48 Ns/rad	
7	$(A_1+A_4)-(A_2+A_3)$	x_1	L_6=0.054 m	
8	A_5	x_1	$e_5 = 7.50$ N/V	
9	A_6	x_1	$e_6 = 7.89$ N/V	k_t=37600 N/m
10	A_7	x_1	$e_7 = 6.65$ N/V	d_1=279 Ns/m
11	A_8	x_1	$e_8 = 5.97$ N/V	
12	A_5+A_7	x_3	I_{zz}=0.500 kgm^2, k_r=1590 N/rad d_6=9.32 Ns/rad	

Comparison between the measured and theoretical transfer functions are depicted in Figures 4-7 for selected test cases. Overall agreement between experimental and theoretical curves is good, thereby giving confidence in the physical model and the parameter identification procedure. The deviation between the curves at low frequency is caused by ground-induced motion of the lower plate and/or the limited low frequency response of the accelerometer.

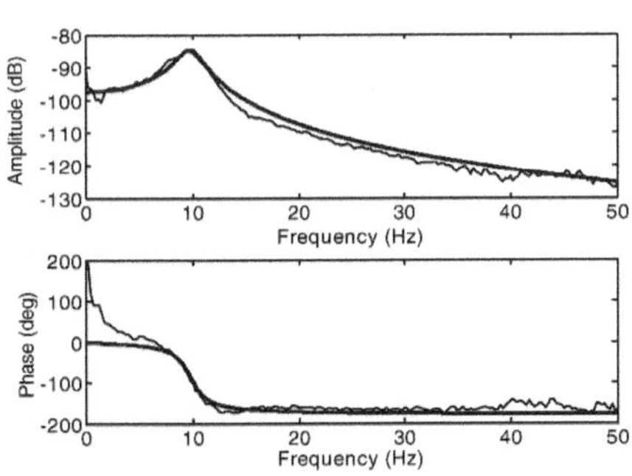

Figure 4. Experimental (—) and theoretical (**—**) transfer functions for test case 1.

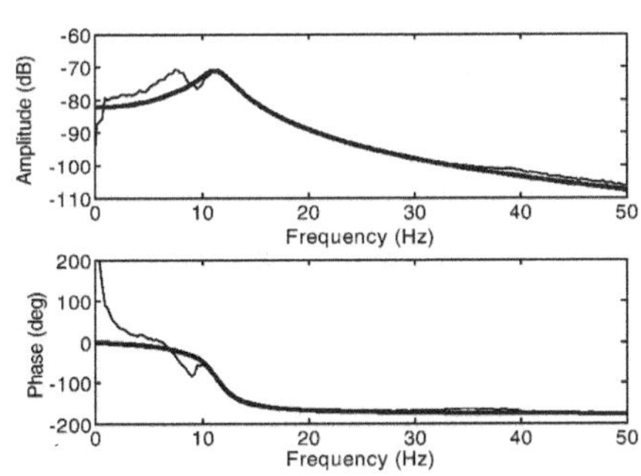

Figure 5. Experimental (—) and theoretical (**—**) transfer functions for test case 5.

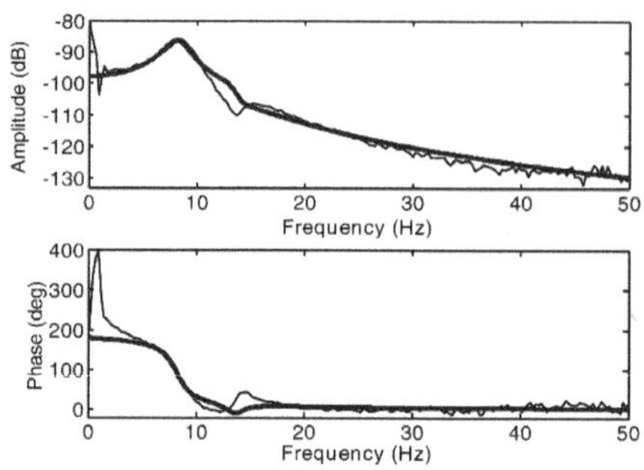

Figure 6. Experimental (—) and theoretical (**—**) transfer functions for test case 8.

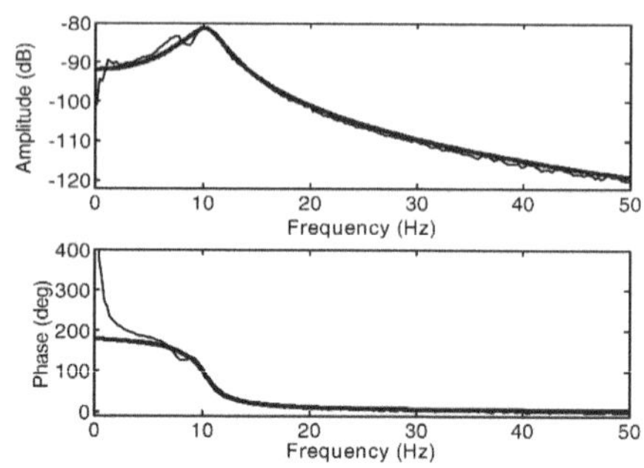

Figure 7. Experimental (—) and theoretical (**—**) transfer functions for test case 12.

4. CONTROL STRATEGIES

4.1 SISO Controller

The commercial vibration isolation system implements a decentralized SISO control strategy in which each of the actuators is controlled using feedback from a local acceleration sensor. The local controllers are proportional-derivative controllers, and they are realized using analog circuitry. The performance of the as-built vibration isolation system is assessed by measuring the transmissibilty. The experimental setup for measuring vertical transmissibilty is depicted in Figure 8. As shown, a large base plate is driven vertically using random excitation. The vertical motion Z_1 of the base structure and the vertical motion Z_2 of the upper plate in the vibration isolation system are measured using accelerometers. A measurement system is then used to compute the transmissibilty function Z_2/Z_1 from the measured displacements. The measured transmissibilty of the vibration isolation

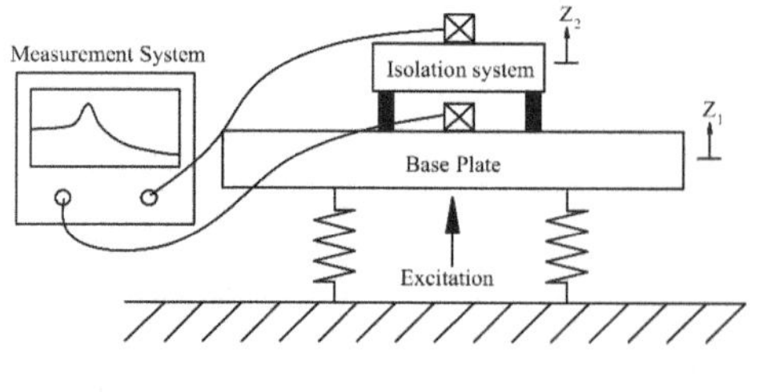

Figure 8. Experimental setup for measuring vertical transmissibility.

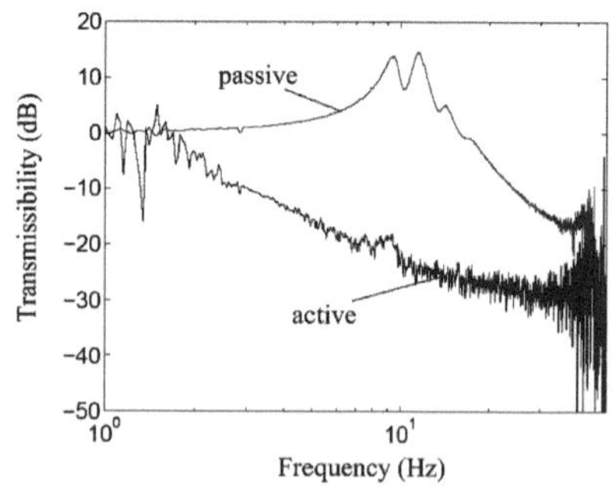

Figure 9. Measured transmissibility of the vibration isolation system with (active) and without (passive) control.

system with (active) and without (passive) control are depicted in Figure 9. Clearly, the vibration isolation system with analog SISO control effectively eliminates vibrations at the mounted vertical resonance frequency of ~10 Hz.

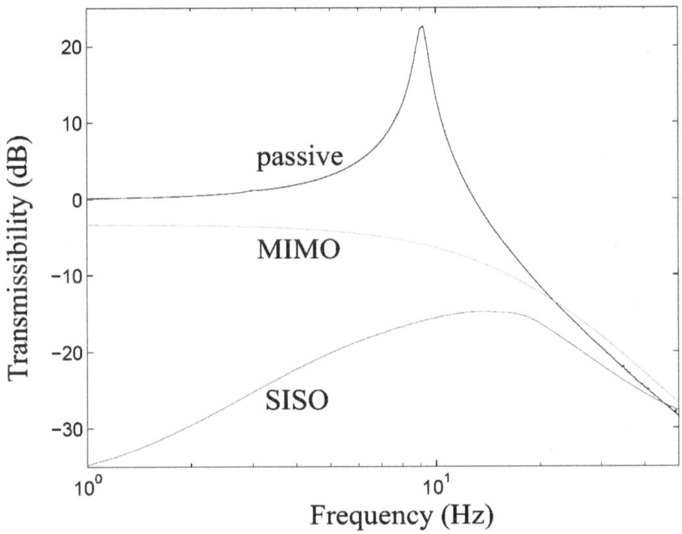

Figure 10. Simulated vertical transmissibilty curves for various control cases.

4.2 MIMO Controller

Ignoring the disturbance inputs, the equation of motion for the upper plate, Eq. (3), can be cast into state-space form:

$$\dot{\bar{x}} = [A]\bar{x} + [B]\bar{u}, \text{ with } \bar{x} = \begin{bmatrix} \bar{w} \\ \dot{\bar{w}} \end{bmatrix}, \quad \bar{u} = \bar{v} \,. \tag{5}$$

For the MIMO controller, the state feedback law $\bar{u} = -[K]\bar{x}$ is assumed. The elements in [*K*] are obtained using the pole placement method, and the elements of the state-space vector $\bar{x}$ are estimated from measured acceleration outputs using an observer [8]. Simulated transmissibility curves for passive, SISO-controlled, and MIMO-controlled systems are depicted in Figure 10. The SISO controller outperforms the MIMO controller for low-frequency vertical excitation. In future simulations, the performance of the controllers will be compared for arbitrary excitation composed of vertical, horizontal, and angular disturbance inputs. At higher frequencies, the structural response is dominated by inertial effects and is therefore only minimally influenced by disturbance and control inputs. Hence, the SISO-

and MIMO- controlled systems have a similar response to the passive system at higher frequencies.

5. CONCLUSIONS

A six-DOF rigid body model for a commercially available vibration isolation system has been developed. The parameters of the unloaded vibration isolation system, including actuator transduction constants, spring stiffness, damping, moments of inertia, and the vertical position of the center of mass, were determined by comparing theoretical and measured transfer functions. The responses predicted by the model agreed well with the experimental measurements, thereby giving confidence in the model. This work is important for model-based control techniques which require accurate identification of model parameters. Future work will involve modification of the current test bed so that vertical, horizontal, and angular disturbance inputs can be used to perturb the vibration isolation system. The SISO and MIMO controllers will be experimentally compared for the various disturbance input cases. Other control concepts, including adaptive and robust control algorithms, will also be implemented and compared.

REFERENCES

1. Fuller, C.R., Elliott, S.J. and Nelson, P.A., *Active Control of Vibration*, Academic Press, 1996.
2. Stöbener, U. and Gaul, L., "Piezoelectric Stack Actuator: FE Modeling and Application for Vibration Isolation," in *Proceedings of the NATO Advanced Study Institute on Responsive Systems for Active Vibration Control*, Ed. A. Preumont, Kluwer Academic Publishers, Dordecht, 2001.
3. Hurlebaus, S., *Smart Structures- Fundamentals and Applications*, Lecture Notes, Institute A of Mechanics, University of Stuttgart, 2005.
4. Huang, X., Elliott, S.J. and Brennan, M.J., "Active Isolation of a Flexible Structure from Base Vibration," *Journal of Sound and Vibration*, **263**, 357–376.
5. Riebe, S. and Ulbrich, H., "Modeling and Online Computation of the Dynamics of a Parallel Kinematic with Six Degrees-of-Freedom," *Archive of Applied Mechanics*, **72**, 2003, 817–829.
6. Ginsberg, J.H., *Advanced Engineering Dynamics*, 2nd edn., Cambridge University Press, New York, 1995.
7. Beadle, B.M., Hurlebaus, S., Stöbener, U. and Gaul, L., "Modeling and Parameter Identification of an Active Anti-Vibration System," SPIE International Symposia in Smart Structures & Materials/NDE, San Diego, March 2005.
8. Van de Vegte, J., *Feedback Control Systems*, 3rd edn., Prentice Hall, NJ, 1994.

DESIGN OF A LABORATORY CRANE FOR TESTING CONTROL APPROACHES

Dieter Bestle
Dept. of Eng. Mech. and Vehicle Dynamics, Brandenburg University of Technology, Germany

Abstract: The paper describes the development of a laboratory crane which is intended to serve as an experimental basis for testing different control approaches for under-actuated systems. The design steps consist of scaling a given simulation model, redesign according to the needs of electric drives, and parameter identification. A first control approach is based on ideas described in [1] where the control gains are found by optimization.

Key words: multibody dynamics, under-actuated systems, control, optimization.

1. INTRODUCTION

Control of under-actuated systems, where the number of control inputs is lower than the number of degrees of freedom, is a challenging task which has been investigated for a long time. In most cases, however, the studies remain academic based on simulations only. In order to account for real disturbances, there is a need for an experimental laboratory crane as a test platform for different control ideas like the one described in [1].

To understand the concept, a short review of this paper by Blajer and Kolodziejczyk shall introduce the control problem. A multibody system with n degrees-of-freedom summarized in a vector $\mathbf{y} \in I\!R^n$ of generalized coordinates may be described by second-order differential equations

$$\mathbf{M}(\mathbf{y})\ddot{\mathbf{y}} + \mathbf{k}(\mathbf{y},\dot{\mathbf{y}}) = \mathbf{q}(\mathbf{y},\dot{\mathbf{y}}) + \mathbf{B}^T(\mathbf{y})\,\mathbf{u}(t) \qquad (1)$$

where $\mathbf{M}$ is the $n \times n$-mass matrix, vector $\mathbf{k}$ summarizes the remaining inertia forces, vector $\mathbf{q}$ summarizes the applied forces, and the $n \times m$-matrix

H. Ulbrich and W. Günthner (eds), IUTAM Symposium on Vibration Control of Nonlinear Mechanisms and Structures, 101–110.

$\mathbf{B}^T$ distributes the m control inputs $\mathbf{u}$. If $m < n$, the system is called under-actuated.

The control inputs $\mathbf{u}(t)$ have to be specified in such a way, that some specific outputs $\hat{\mathbf{c}}(\mathbf{y})$ satisfy m prescribed motion constraints $\hat{\mathbf{c}}(\mathbf{y}(t)) = \breve{\mathbf{c}}(t)$ or implicitly

$$\mathbf{c}(\mathbf{y},t) := \hat{\mathbf{c}}(\mathbf{y}) - \breve{\mathbf{c}}(t) = \mathbf{0}. \tag{2}$$

Equations (1) and (2) look pretty much like the DAE-form of a closed-loop system, i.e.

$$\begin{aligned} &\mathbf{M}(\mathbf{y})\ddot{\mathbf{y}} + \mathbf{k}(\mathbf{y},\dot{\mathbf{y}}) = \mathbf{q}(\mathbf{y},\dot{\mathbf{y}}) + \mathbf{C}^T(\mathbf{y})\boldsymbol{\lambda} \\ &\mathbf{c}(\mathbf{y},t) = \mathbf{0} \end{aligned} \tag{3}$$

where $\mathbf{c}(\mathbf{y},t)$ represents m closing conditions of kinematical loops, $\mathbf{C} = \partial\mathbf{c}/\partial\mathbf{y}$ is the $m \times n$-closing Jacobian, and $\boldsymbol{\lambda} \in I\!R^m$ are Lagrange multipliers representing the constraint forces $\mathbf{C}^T\boldsymbol{\lambda}$, see e.g. [2]. As a result, there remain $f = n - m$ unconstrained degrees of freedom.

However, there is a major difference between the control problem (1), (2) and the closed-loop system (3). In case of the closed-loop system, a solution may be found by projecting the differential equation onto the tangential manifold of the closing conditions by a $f \times n$-matrix $\mathbf{D}$ orthogonal to $\mathbf{C}^T$, e.g. [3]. According to

$$\mathbf{D}\mathbf{C}^T = \mathbf{0} \tag{4}$$

the constraint forces can be eliminated from the differential equation (3) by pre-multiplication with $\mathbf{D}$ resulting in f ordinary differential equations

$$\mathbf{D}\mathbf{M}\ddot{\mathbf{y}} = \mathbf{D}(\mathbf{q} - \mathbf{k}). \tag{5}$$

For numerical integration they have to be accomplished by the second derivative of the closing conditions (2), i.e.

$$\mathbf{C}\ddot{\mathbf{y}} = -\dot{\mathbf{C}}\dot{\mathbf{y}} - \dot{\mathbf{c}}_t \tag{6}$$

where $\mathbf{c}_t = \partial\mathbf{c}/\partial t$. The reactions may be found from a projection of the differential equation onto the constrained subspace, which is the orthogonal complement to the tangential manifold, by multiplication with $\mathbf{C}\mathbf{M}^{-1}$, i.e.

$$\mathbf{C}\mathbf{M}^{-1}\mathbf{M}\ddot{\mathbf{y}}+\mathbf{C}\mathbf{M}^{-1}\mathbf{k}=\mathbf{C}\mathbf{M}^{-1}\mathbf{q}+\mathbf{C}\mathbf{M}^{-1}\mathbf{C}^{T}\boldsymbol{\lambda}\,. \tag{7}$$

In case of independent closing conditions the Jacobian, and thus $\mathbf{C}\mathbf{M}^{-1}\mathbf{C}^{T}$ have full rank. Therefore, equation. (7) can be solved as

$$\boldsymbol{\lambda}=\left(\mathbf{C}\mathbf{M}^{-1}\mathbf{C}^{T}\right)^{-1}\left[\mathbf{C}\mathbf{M}^{-1}(\mathbf{k}-\mathbf{q})-\dot{\mathbf{C}}\dot{\mathbf{y}}-\dot{\mathbf{c}}_{t}\right] \tag{8}$$

where the first term $\mathbf{C}\ddot{\mathbf{y}}$ has been substituted from (6).

Such a procedure doesn't work for the control system (1), (2). A projection with $\mathbf{C}\mathbf{M}^{-1}$ analogously to equation (7) yields

$$\mathbf{C}\mathbf{M}^{-1}\mathbf{M}\ddot{\mathbf{y}}+\mathbf{C}\mathbf{M}^{-1}\mathbf{k}=\mathbf{C}\mathbf{M}^{-1}\mathbf{q}+\mathbf{C}\mathbf{M}^{-1}\mathbf{B}^{T}\mathbf{u} \tag{9}$$

or after substituting the accelerations from (6):

$$-\dot{\mathbf{C}}\dot{\mathbf{y}}-\dot{\mathbf{c}}_{t}+\mathbf{C}\mathbf{M}^{-1}\mathbf{k}=\mathbf{C}\mathbf{M}^{-1}\mathbf{q}+\mathbf{C}\mathbf{M}^{-1}\mathbf{B}^{T}\mathbf{u}\,. \tag{10}$$

Since $\mathbf{B}^{T}$ is arbitrary and independent of the Jacobian $\mathbf{C}$ associated with motion constraints, the $m\times m$-matrix

$$\mathbf{P}=\mathbf{C}\mathbf{M}^{-1}\mathbf{B}^{T} \tag{11}$$

can have any rank $p\in[0,m]$. Thus, a solution of equation (10) for $\mathbf{u}$ is not possible in general. The different situations are discussed in [1] where also the following numerical solution to the problem is proposed.

Let us write problem (1), (2) in state space form by introducing the generalized velocities $\mathbf{z}=\dot{\mathbf{y}}$. These trivial equations combined with the equation of motion (1) projected onto the tangential manifold by $\mathbf{D}$ and equation (10), which is a projection onto the constrained space, as well as with (2) yield the following set of differential-algebraic equations:

$$\begin{aligned}\dot{\mathbf{y}}&=\mathbf{z},\\ \mathbf{D}\mathbf{M}\dot{\mathbf{z}}&=\mathbf{D}(\mathbf{q}-\mathbf{k})+\mathbf{D}\mathbf{B}^{T}\mathbf{u},\\ \mathbf{0}&=\mathbf{C}\mathbf{M}^{-1}(\mathbf{q}-\mathbf{k})+\mathbf{C}\mathbf{M}^{-1}\mathbf{B}^{T}\mathbf{u}+\dot{\mathbf{C}}\dot{\mathbf{y}}+\dot{\mathbf{c}}_{t},\\ \mathbf{0}&=\widehat{\mathbf{c}}(\mathbf{y})-\breve{\mathbf{c}}(t).\end{aligned} \tag{12}$$

These are $2n+m$ equations for $\mathbf{y}$, $\mathbf{z}$ and $\mathbf{u}$ which can be solved by an Euler-backward scheme for both the motion of the system and the required open-loop control. For closed-loop control the second derivative of the motion constraint used in equation (10) may be substituted by a stabilized form.

2. APPLICATION TO A CRANE MODEL

By simulation it was shown in [1] that the approach works quite well e.g. for a planar crane model consisting of a trolley and a load, Fig. 1. The system has 3 DOFs described by trolley position s, length l of the hoisting rope, and the swing angle φ. The trolley is controlled by a force F, the length of the rope by winch torque M_n, whereas the swing angle cannot be controlled directly. The control task is to force the load on a track $\breve{\mathbf{c}}(t)=[x(t)\ y(t)]^T$ in the plane. The equations (1), (2) read as

$$\begin{bmatrix} m_t+m & m\sin\varphi & ml\cos\varphi \\ m\sin\varphi & m+J/r^2 & 0 \\ ml\cos\varphi & 0 & ml^2 \end{bmatrix}\begin{bmatrix} \ddot{y} \\ \ddot{l} \\ \ddot{\varphi} \end{bmatrix}+\begin{bmatrix} 2m\dot{l}\dot{\varphi}\cos\varphi-ml\dot{\varphi}^2\sin\varphi \\ -ml\dot{\varphi}^2 \\ 2ml\dot{l}\dot{\varphi} \end{bmatrix}$$

$$=\begin{bmatrix} 0 \\ mg\cos\varphi \\ -mgl\sin\varphi \end{bmatrix}+\begin{bmatrix} 1 & 0 \\ 0 & 1/r \\ 0 & 0 \end{bmatrix}\begin{bmatrix} F \\ M_n \end{bmatrix},$$

$$\begin{bmatrix} s+l\sin\varphi \\ l_0-l\cos\varphi \end{bmatrix}-\begin{bmatrix} \Delta x\cdot p(t) \\ \Delta y\cdot p(t) \end{bmatrix}=\mathbf{0} \tag{13}$$

where the load has to move along the path with $p(t)=126\tau^5-420\tau^6+540\tau^7-315\tau^8+70\tau^9$, $\tau=t/T$, within a time interval $t\in[0,T]$.

2.1 Scaling

A first design decision is, how to scale down the model crane to laboratory size, Figure 2. The dynamics keeps invariant with respect to the following mapping between model data $\bullet$ to laboratory data $\bar{\bullet}$:

$$t=\alpha\bar{t}\,,\ M=\beta\overline{M}\,,\ L=\alpha^2\overline{L}\,,\ \frac{d}{dt}=\frac{1}{\alpha}\frac{d}{d\bar{t}}\,. \tag{14}$$

Solving the open-loop problem (12) for the scaled crane (13) yields identical shapes for motion and control as in [1] where force and torque scale down according to $\overline{F} = F/\beta$ and $\overline{M} = M/(\alpha^2\beta)$, respectively, see Figure 3 for $\alpha = 2$, $\beta = 20$.

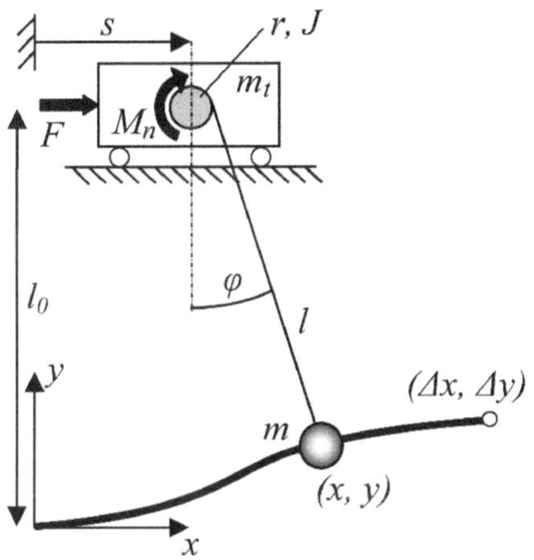

Figure 1. Crane model.

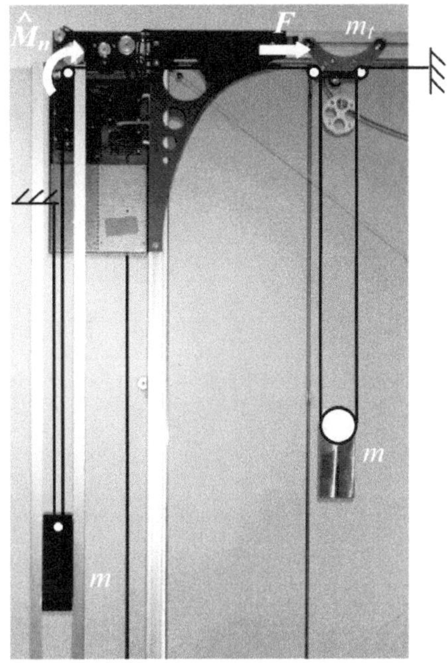

Figure 2. Laboratory crane.

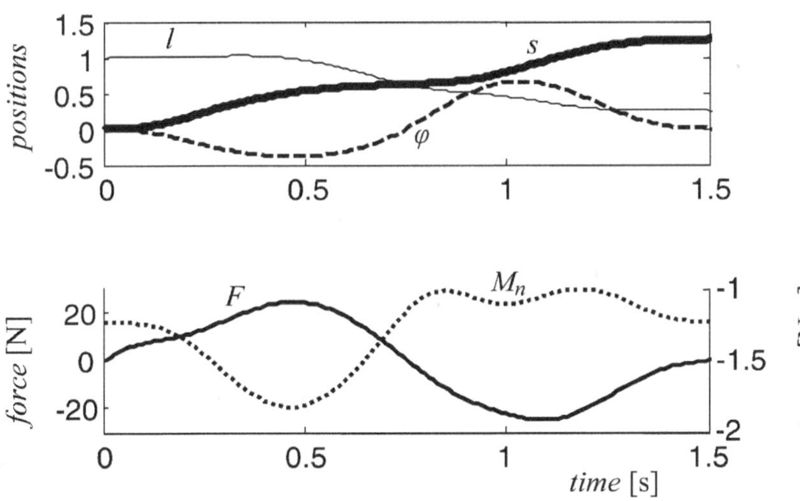

Figure 3. Open-loop control of the scaled crane model.

2.2 Drives

Based on these values, electric drives have been chosen for both trolley and winch. In order to keep the mass of the trolley as low as possible, it was decided to fix both drives to the inertial frame. Further the static load has to be taken from the electric winch drive by a counterweight, Figure 2. This, however, changes the equation of motion to

$$\begin{bmatrix} m_t + m & m\sin\varphi & ml\cos\varphi \\ m\sin\varphi & 2m + 4\hat{J}/r^2 & 0 \\ ml\cos\varphi & 0 & ml^2 \end{bmatrix} \begin{bmatrix} \ddot{y} \\ \ddot{l} \\ \ddot{\varphi} \end{bmatrix} + \begin{bmatrix} 2m\dot{l}\dot{\varphi}\cos\varphi - ml\dot{\varphi}^2\sin\varphi \\ -ml\dot{\varphi}^2 \\ 2ml\,\dot{l}\dot{\varphi} \end{bmatrix}$$

$$= \begin{bmatrix} 0 \\ mg(\cos\varphi - 1) \\ -mgl\sin\varphi \end{bmatrix} + \begin{bmatrix} 1 & 0 \\ 0 & 2/r \\ 0 & 0 \end{bmatrix} \begin{bmatrix} F \\ \hat{M}_n \end{bmatrix}. \tag{15}$$

By comparison with equation (13) it can be seen that dynamics will not change if we use $J = mr^2 + 4\hat{J}$ in the original model equation (13) for computing the winch control M_n and finally apply the torque $\hat{M}_n = M_n/2 + mgr/2$ to the laboratory crane. All other parameters are identical.

2.3 Parameters of the Laboratory Crane

Due to the counterweight the moment of inertia $\hat{J}$ couldn't be kept as low as required from scaling. Further the workspace limits have been changed to get a compact experimental setup. The final parameters of the laboratory crane according to equations (13) are

$$m_t = 0.5\,\mathrm{kg}\,,\; m = 3.235\,\mathrm{kg}\,,\; r = 0.0225\,\mathrm{m}\,,\; J = 1.66\cdot 10^{-3}\,\mathrm{kg\,m^2}\,,$$
$$l \in [0.4,\ 1.4]\,\mathrm{m}\,,\; s \in [0,\ 1.3]\,\mathrm{m}\,. \tag{16}$$

which have been identified by a procedure similar to the concept described in [4]. Friction of the belt drive is compensated by control which is realized by a real-time computing environment based on Matlab [5] and dSpace [6]. Figure 4 shows a comparison of the open-loop computation and the corresponding behavior of the laboratory crane controlled according to positions $s = s^0(t)$ and $l = l^0(t)$. According to control errors in these positions the swinging of the load pendulum is different which shows the need for a closed-loop concept including information on motion deviations.

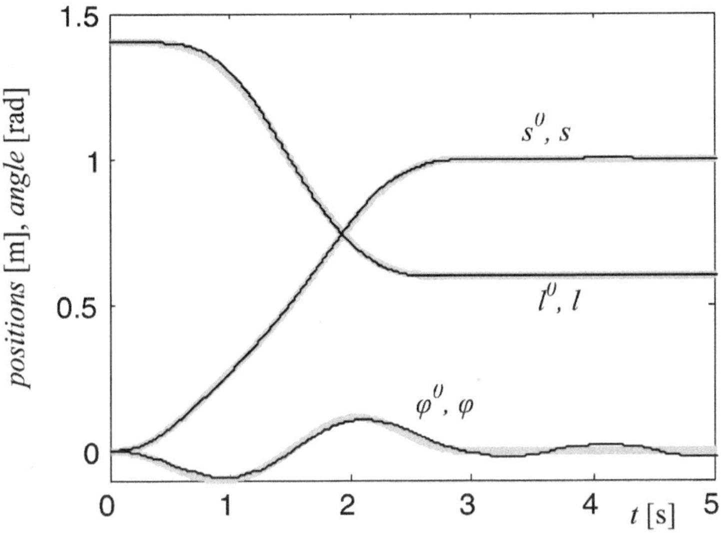

Figure 4. Position control of the laboratory crane: open-loop solution (grey: s^0, l^0, φ^0) and measurement (black: s, l, φ).

3. A FIRST CONTROL APPROACH

The control concept in [1] assumes that the system of nonlinear equations resulting from the Euler-backward scheme applied to equation (12) can be solved for both the control model and the system to be controlled. In reality, however, the behavior of the controlled system is not exactly predictable and the solution of nonlinear equations requires a Newton-like iteration procedure which cannot be guaranteed to converge in real-time. Therefore, a linearized modification of the approach shall be discussed.

Linearization of the closed-loop system (1), (2) with respect to the open-loop solution of equation (12), i.e. $\mathbf{u}(t) = \mathbf{u}^0(t) + \tilde{\mathbf{u}}$, $\mathbf{y}(t) = \mathbf{y}^0(t) + \tilde{\mathbf{y}}$ gives

$$\mathbf{M}^0(t)\,\ddot{\tilde{\mathbf{y}}} + \mathbf{P}^0(t)\,\dot{\tilde{\mathbf{y}}} + \mathbf{Q}^0(t)\,\tilde{\mathbf{y}} - \mathbf{B}^{0^T}(t)\,\tilde{\mathbf{u}} = \mathbf{0}, \tag{17a}$$

$$\tilde{\mathbf{c}}^0 := \mathbf{C}^0(t)\,\tilde{\mathbf{y}} = \mathbf{0} \tag{17b}$$

where $\mathbf{M}^0 = \mathbf{M}(\tilde{\mathbf{y}}^0)$, $\mathbf{P}^0 = [\mathbf{k} - \mathbf{q}]^0_{,\dot{y}}$, $\mathbf{Q}^0 = \left[\mathbf{M}\,\ddot{\mathbf{y}}^0 + \mathbf{k} - \mathbf{q} - \mathbf{B}^T\mathbf{u}^0\right]^0_{,y}$, and $\mathbf{C}^0 = \mathbf{C}(\tilde{\mathbf{y}}^0)$ can be obtained from NEWEUL [7]. Using the same concept as for equations (12), i.e. projections of (17a) with $\mathbf{D}^0$ and $\mathbf{C}^0\mathbf{M}^{0^{-1}}$, yields

$$\mathbf{D}^0\mathbf{M}^0\,\ddot{\tilde{\mathbf{y}}}+\mathbf{D}^0\left[\mathbf{P}^0\,\dot{\tilde{\mathbf{y}}}+\mathbf{Q}^0\,\tilde{\mathbf{y}}-\mathbf{B}^{0^T}\,\tilde{\mathbf{u}}\right]=\mathbf{0}\,, \tag{18a}$$

$$\mathbf{C}^0\,\ddot{\tilde{\mathbf{y}}}+\mathbf{C}^0\mathbf{M}^{0^{-1}}\left[\mathbf{P}^0\,\dot{\tilde{\mathbf{y}}}+\mathbf{Q}^0\,\tilde{\mathbf{y}}-\mathbf{B}^{0^T}\,\tilde{\mathbf{u}}\right]=\mathbf{0}\,. \tag{18b}$$

With (17b) also the derivatives and the integral are zero, why a linear combination will be zero as well:

$$\ddot{\tilde{\mathbf{c}}}^0+\boldsymbol{\alpha}\,\dot{\tilde{\mathbf{c}}}^0+\boldsymbol{\beta}\,\tilde{\mathbf{c}}^0+\boldsymbol{\gamma}\int\tilde{\mathbf{c}}^0\,dt=\mathbf{0}\quad\text{or}$$

$$\mathbf{C}^0\,\ddot{\tilde{\mathbf{y}}}+2\dot{\mathbf{C}}^0\,\dot{\tilde{\mathbf{y}}}+\ddot{\mathbf{C}}^0\,\tilde{\mathbf{y}}+\boldsymbol{\alpha}\left(\mathbf{C}^0\,\dot{\tilde{\mathbf{y}}}+\dot{\mathbf{C}}^0\,\tilde{\mathbf{y}}\right)+\boldsymbol{\beta}\,\mathbf{C}^0\,\tilde{\mathbf{y}}+\boldsymbol{\gamma}\int\mathbf{C}^0\,\tilde{\mathbf{y}}\,dt=\mathbf{0} \tag{19}$$

where $\boldsymbol{\alpha}$, $\boldsymbol{\beta}$ and $\boldsymbol{\gamma}$ are diagonal matrices of gain factors associated with the motion constraints. Equation (19) can be used for substituting the first term in (18b). After introducing $\tilde{\mathbf{z}}:=\dot{\tilde{\mathbf{y}}}$ and $\dot{\tilde{\mathbf{w}}}=\mathbf{C}^0\,\tilde{\mathbf{y}}$ we can summarize equations (18) and (19) as

$$\begin{aligned}
&\dot{\tilde{\mathbf{y}}}-\tilde{\mathbf{z}}=\mathbf{0}\,,\\
&\mathbf{D}^0\mathbf{M}^0\,\dot{\tilde{\mathbf{z}}}+\mathbf{D}^0\left[\mathbf{P}^0\,\tilde{\mathbf{z}}+\mathbf{Q}^0\,\tilde{\mathbf{y}}-\mathbf{B}^{0^T}\,\tilde{\mathbf{u}}\right]=\mathbf{0}\,,\\
&-2\dot{\mathbf{C}}^0\,\tilde{\mathbf{z}}-\ddot{\mathbf{C}}^0\,\tilde{\mathbf{y}}+\mathbf{C}^0\mathbf{M}^{0^{-1}}\left[\mathbf{P}^0\,\tilde{\mathbf{z}}+\mathbf{Q}^0\,\tilde{\mathbf{y}}-\mathbf{B}^{0^T}\,\tilde{\mathbf{u}}\right]\\
&\qquad=\boldsymbol{\alpha}\left(\mathbf{C}^0\tilde{\mathbf{z}}+\dot{\mathbf{C}}^0\,\tilde{\mathbf{y}}\right)+\boldsymbol{\beta}\,\mathbf{C}^0\,\tilde{\mathbf{y}}+\boldsymbol{\gamma}\,\tilde{\mathbf{w}}\,,\\
&\mathbf{C}^0\,\tilde{\mathbf{y}}=\mathbf{0}\,,\\
&\dot{\tilde{\mathbf{w}}}=\mathbf{C}^0\,\tilde{\mathbf{y}}\,.
\end{aligned} \tag{20}$$

These DAEs can be solved for a given predictive control time-step Δt by a semi-implicit Euler-scheme. The left hand side is assumed to be the result after such a time step which has to fulfill the dynamics and motion constraints, whereas the right hand side is interpreted as control error at a time point t_n before this step. After substituting the derivatives by finite differences we get a system of linear equations:

$$\begin{aligned}
&\tilde{\mathbf{y}}_{n+1}/\Delta t-\tilde{\mathbf{z}}_{n+1}=\tilde{\mathbf{y}}_n/\Delta t\,,\\
&\mathbf{D}^0\mathbf{M}^0\,\tilde{\mathbf{z}}_{n+1}/\Delta t+\mathbf{D}^0\left[\mathbf{P}^0\,\tilde{\mathbf{z}}_{n+1}+\mathbf{Q}^0\,\tilde{\mathbf{y}}_{n+1}-\mathbf{B}^{0^T}\,\tilde{\mathbf{u}}_{n+1}\right]=\mathbf{D}^0\mathbf{M}^0\,\tilde{\mathbf{z}}_n/\Delta t\,,\\
&-2\dot{\mathbf{C}}^0\,\tilde{\mathbf{z}}_{n+1}-\ddot{\mathbf{C}}^0\,\tilde{\mathbf{y}}_{n+1}+\mathbf{C}^0\mathbf{M}^{0^{-1}}\left[\mathbf{P}^0\,\tilde{\mathbf{z}}_{n+1}+\mathbf{Q}^0\,\tilde{\mathbf{y}}_{n+1}-\mathbf{B}^{0^T}\,\tilde{\mathbf{u}}_{n+1}\right]\\
&\qquad=\boldsymbol{\alpha}\left(\mathbf{C}^0\tilde{\mathbf{z}}_n+\dot{\mathbf{C}}^0\,\tilde{\mathbf{y}}_n\right)+\boldsymbol{\beta}\,\mathbf{C}^0\,\tilde{\mathbf{y}}_n+\boldsymbol{\gamma}\,\tilde{\mathbf{w}}_n\,,\\
&\mathbf{C}^0\,\tilde{\mathbf{y}}_{n+1}=\mathbf{0}\,,\\
&\tilde{\mathbf{w}}_{n+1}=\tilde{\mathbf{w}}_n+\Delta t\,\mathbf{C}^0\,\tilde{\mathbf{y}}\,.
\end{aligned} \tag{21}$$

Basically the equations serve for finding $\tilde{\mathbf{u}}_{n+1}$ in one solution step, and thus the control inputs $\mathbf{u} = \mathbf{u}^0(t_{n+1}) + \tilde{\mathbf{u}}_{n+1}$ to the system.

The gain matrices $\boldsymbol{\alpha}$, $\boldsymbol{\beta}$ and $\boldsymbol{\gamma}$ may be found by simulation and optimization for a given motion constraint. For this reason we disturb the initial position $\mathbf{y}(t_0)$ by a small vector $\delta\mathbf{y}_0$ and compute the motion from the nonlinear equations (1). The control inputs $\mathbf{u}(t)$ are computed from (21) and applied at equidistant time points $t_i = t_0 + i\,\delta t$, $i = 0,1,2,\ldots$. The time steps δt for control switching may be different from the control prediction time Δt and the simulation steps which typically are not equidistant. If we use the mean absolute motion errors

$$\boldsymbol{\mu} = \frac{1}{t_1 - t_0} \int_{t_0}^{t_1} |\mathbf{c}(\mathbf{y}, t)| \, dt \tag{22}$$

as criteria, a potential optimization problem reads as

$$\begin{aligned} &\min_{\boldsymbol{\alpha},\boldsymbol{\beta},\boldsymbol{\gamma}} \sqrt{\boldsymbol{\mu}^T \boldsymbol{\mu}} \\ &\text{subject to} \quad \text{(21) for computing } \mathbf{u}(t) \\ &\qquad\qquad\quad \text{(1) and (22) for computing } \mathbf{y}(t) \text{ and } \boldsymbol{\mu}. \end{aligned} \tag{23}$$

In order to avoid adapted control to a single situation, multiple disturbance combinations $\pm\delta y_{0_k}$ for the initial conditions y_{0_k} have been used to compute a single criterion vector $\boldsymbol{\mu}$. Figure 5 shows simulations for open-loop control and closed-loop control with optimized gain matrices for the crane (13) with parameters (16). Open-loop fails due to disturbances of initial conditions and interpolation of $\mathbf{u}^0(t)$, whereas closed-loop control can damp the load pendulum as well.

Implementation of the proposed control concept has to show how it works on the real laboratory crane. The goal, however, is to offer an experimental platform to the community, and everybody interested is invited to implement and test his approach.

ACKNOWLEDGEMENT

The author thanks W. Blajer for his help in understanding the control concept and performing some computations in the early design phase.

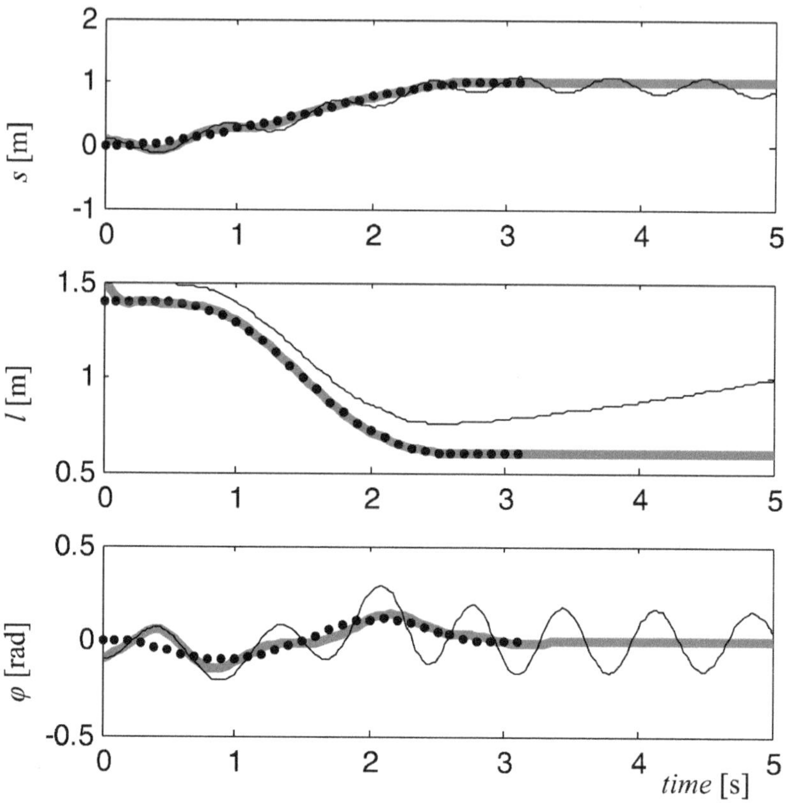

Figure 5. Open-loop behavior (nominal: dotted line, simulation: thin line) and closed-loop behavior (thick line) of the laboratory crane for optimized control gains.

REFERENCES

[1] Blajer, W. and Kolodziejczyk, K. (2004). A Geometric Approach to Solving Problems of Control Constraints – Theory and a DAE Framework. *Multibody System Dynamics* 11, 343–364.

[2] Bestle, D. (1994). *Analyse und Optimierung von Mehrkörpersystemen*, Springer, Berlin.

[3] Blajer, W., Bestle, D. and Schiehlen, W. (1994). An Orthogonal Complement Matrix Formulation for Constrained Multibody Systems. *J. Mech. Design* 116, 423–428.

[4] Bestle, D. and Glora, M. (2003): Optimization of Passive and Active Dynamic Systems. In: W. Schiehlen and M. Valasek (eds.), *Virtual Nonlinear Multibody Systems*, Kluwer, 297–316.

[5] *Matlab – The Language of Technical Computing* (2004). The MathWorks Inc., Natick.

[6] *ControlDesk – Experiment Guide* (2004). dSpace GmbH, Paderborn.

[7] Kreuzer, E. and Leister, G. (1992). *Programmiersystem NEWEUL*, Institute B of Mechanics, University of Stuttgart.

PREDICTION OF CONTROL OF OVERHEAD CRANES EXECUTING A PRESCRIBED LOAD TRAJECTORY

Wojciech Blajer and Krzysztof Kołodziejczyk
Technical University of Radom, Institute of Applied Mechanics, ul. Krasickiego 54, 26-600 Radom, Poland

Abstract: Manipulating payloads with overhead cranes can be challenging due to the underactuated nature of the system – the number of control inputs/outputs is smaller than the number of degrees-of-freedom. The control outputs (desired load trajectory coordinates), expressed in terms of the system states, lead to control constraints on the system, and the governing equations arise as index five differential-algebraic equations, transformed then to an index three form. An effective numerical code for solving the resultant equations is used. The feedforward control law obtained this way is then extended by a closed-loop control strategy with feedback of the actual errors to provide stable tracking of the required reference load trajectories in presence of perturbations.

Key words: cranes, dynamics, control, trajectory tracking, differential-algebraic equations.

1. INTRODUCTION

Overhead cranes belong to a broader class of underactuated systems – the controlled mechanical systems in which the number of control inputs/outputs is smaller than the number of degrees-of-freedom. The performance goal is a desired load trajectory, i.e. the control outputs are time-specified load coordinates $x_d(t)$, $y_d(t)$ and $z_d(t)$. The control inputs are the forces F_x and F_y actuating the trolley position and the winch torque M_n changing the rope length (see Fig. 1). The determination of control input strategy that force the system to complete the prescribed motion is a challenging problem, reflected in huge amount of research established hitherto.[1] The purpose of this study is to give a fresh view on the problem from the constrained motion perspective

H. Ulbrich and W. Günthner (eds), IUTAM Symposium on Vibration Control of Nonlinear Mechanisms and Structures, 111–120.

and to develop the mathematical tools for control design aimed at executing prescribed load trajectories with relative high speeds and without sway.

The control outputs, expressed in terms of the system states, are treated as *control constraints* on the system.[2] It is noticed, however, that control constraints differ from the classical *contact constraints* in several aspects. Mainly, they are enforced by means of available control forces (control inputs), which may have any directions with respect to the control constraint manifold, and in the extreme may be tangent. A specific methodology must then be developed to solve such 'singular' control problem. The initial governing equations arise as index five differential-algebraic equations (DAEs).[3] They are transformed then to an equivalent index three form, and an effective code for solving the resultant DAEs is proposed. The feedforward control law obtained this way is extended by a closed-loop control strategy with feedback of the actual errors to provide stable tracking of the required reference load trajectories in the presence of perturbations.

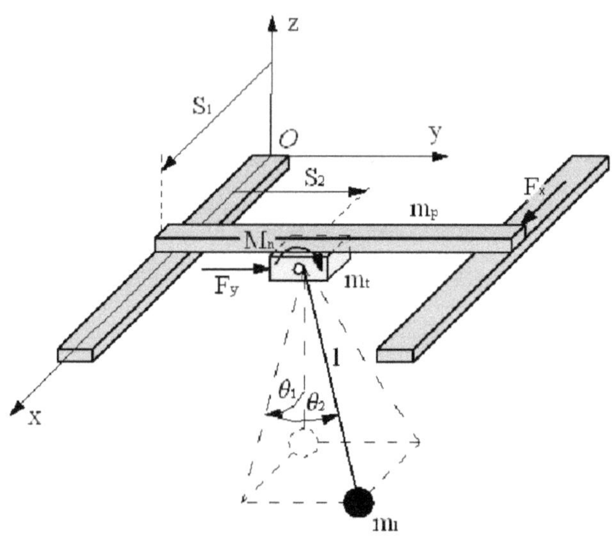

Figure 1. An overhead trolley crane.

2. MATHEMATICAL PRELIMINARIES

Consider a 5-degree-of-freedom ($n = 5$) overhead (gantry) crane seen in Figure 1, whose generalized coordinates are $\mathbf{q} = [s_1\ s_2\ l\ \theta_1\ \theta_2]^T$, and which is enforced by $m = 3$ actuators $\mathbf{u} = [F_x\ F_y\ M_n]^T$. The dynamic equations of the system can be written in the following generic form

$$\mathbf{M}(\mathbf{q})\ddot{\mathbf{q}} + \mathbf{d}(\mathbf{q}, \dot{\mathbf{q}}) = \mathbf{f}(\mathbf{q}, \dot{\mathbf{q}}, t) - \mathbf{B}^T(\mathbf{q})\,\mathbf{u}\ , \tag{1}$$

where $\mathbf{M}$ is the generalized mass matrix, $\mathbf{d}$ and $\mathbf{f}$ are the generalized dynamic and applied force vectors, and $\mathbf{B}^T$ is the matrix of influence of control inputs $\mathbf{u}$ on the generalized actuating force vector $\mathbf{f}_a = -\mathbf{B}^T\mathbf{u}$. Assumed the hoisting rope is massless, inextensible and flexible, and neglecting for simplicity all the forces associated with s_1, s_2 and l motions apart from the control inputs F_x, F_y and M_n, the components of dynamic equations are:

$$\mathbf{M} = \begin{bmatrix} m_b + m_t + m_l & \times & \times & \times & \times \\ 0 & m_t + m_l & \times & \times & \times \\ m_l \sin\theta_1 \cos\theta_2 & m_l \sin\theta_2 & m_l + J/r^2 & \times & \times \\ m_l l \cos\theta_1 \cos\theta_2 & 0 & 0 & m_l l^2 \cos^2\theta_2 & \times \\ -m_l l \sin\theta_1 \sin\theta_2 & m_l l \cos\theta_2 & 0 & 0 & m_l l^2 \end{bmatrix},$$

$$\mathbf{d} = \begin{bmatrix} 2m_l\dot{\theta}_1 \cos\theta_1(\dot{l}\cos\theta_2 - l\dot{\theta}_2 \sin\theta_2) - m_l \sin\theta_1[2\dot{l}\dot{\theta}_2 \sin\theta_2 + l(\dot{\theta}_1^2 + \dot{\theta}_2^2)\cos\theta_2] \\ m_l\dot{\theta}_2(2\dot{l}\cos\theta_2 - l\dot{\theta}_2 \sin\theta_2) \\ -m_l l(\dot{\theta}_1^2\cos^2\theta_2 + \dot{\theta}_2^2) \\ 2m_l l\dot{\theta}_1 \cos\theta_2(\dot{l}\cos\theta_2 - l\dot{\theta}_2 \sin\theta_2) \\ m_l l(l\dot{\theta}_2^2 \sin\theta_2 \cos\theta_2 + 2\dot{l}\dot{\theta}_2) \end{bmatrix},$$

$$\mathbf{f} = \begin{bmatrix} 0 \\ 0 \\ m_l g \cos\theta_1 \cos\theta_2 \\ -m_l g l \sin\theta_1 \cos\theta_2 \\ -m_l g l \cos\theta_1 \sin\theta_2 \end{bmatrix}, \qquad -\mathbf{B}^T = \begin{bmatrix} 1 & 0 & 0 \\ 0 & 1 & 0 \\ 0 & 0 & 1/r \\ 0 & 0 & 0 \\ 0 & 0 & 0 \end{bmatrix},$$

where m_b, m_t and m_l are the bridge, trolley and load masses, J and r are the moment of inertia and radius of the winch, and g is the gravitational acceleration, and $\times$ in the mass matrix $\mathbf{M}$ denotes a symmetric entry.

The performance goal is a desired load trajectory, i.e. the $m = 3$ outputs are time-specified load coordinates, $\boldsymbol{\gamma}_d(t) = [x_d(t)\ y_d(t)\ z_d(t)]^T$, equal in number to the number of control inputs $\mathbf{u}$. Expressed in terms of the system coordinates, the outputs lead to m *control constraints*[2] in the form

$$\mathbf{c}(\mathbf{q}, t) \equiv \mathbf{\Phi}(\mathbf{q}) - \boldsymbol{\gamma}_d(t) = \mathbf{0} \ . \tag{2}$$

The initial equations of control constraints are then twice differentiated with respect to time to obtain the constraint conditions at the acceleration level,

$$\ddot{\mathbf{c}} \equiv \mathbf{C}(\mathbf{q})\ddot{\mathbf{q}} - \boldsymbol{\xi}(\mathbf{q},\dot{\mathbf{q}},t) = \mathbf{0} \ , \tag{3}$$

where $\mathbf{C} = \partial\boldsymbol{\Phi}/\partial\mathbf{q}$ is the $m \times n$ constraint matrix, and $\boldsymbol{\xi} = \ddot{\boldsymbol{\gamma}}_d - \dot{\mathbf{C}}\dot{\mathbf{q}}$ is the constraint induced acceleration. For the crane shown in Figure 1 we have:

$$\mathbf{c} = \begin{bmatrix} s_1 + l\sin\theta_1\cos\theta_2 \\ s_2 + l\sin\theta_2 \\ -l\cos\theta_1\cos\theta_2 \end{bmatrix} - \begin{bmatrix} x_d(t) \\ y_d(t) \\ z_d(t) \end{bmatrix} = \mathbf{0} \ ,$$

$$\mathbf{C} = \begin{bmatrix} 1 & 0 & \sin\theta_1\cos\theta_2 & l\cos\theta_1\cos\theta_2 & -l\sin\theta_1\sin\theta_2 \\ 0 & 1 & \sin\theta_2 & 0 & l\cos\theta_2 \\ 0 & 0 & -\cos\theta_1\cos\theta_2 & l\sin\theta_1\cos\theta_2 & l\cos\theta_1\sin\theta_2 \end{bmatrix} ,$$

$$\boldsymbol{\xi} = \begin{bmatrix} \ddot{x}_d + l[(\dot{\theta}_1^2 + \dot{\theta}_2^2)\sin\theta_1\cos\theta_2 - 2\dot{\theta}_1\dot{\theta}_2\cos\theta_1\sin\theta_2] \\ -2\dot{l}(\dot{\theta}_1\cos\theta_1\cos\theta_2 - \dot{\theta}_2\sin\theta_1\sin\theta_2) \\ \\ \ddot{y}_d + l\dot{\theta}_2^2\sin\theta_2 - 2\dot{l}\dot{\theta}_2\cos\theta_2 \\ \\ \ddot{z}_d - l[(\dot{\theta}_1^2 + \dot{\theta}_2^2)\cos\theta_1\cos\theta_2 - 2\dot{\theta}_1\dot{\theta}_2\sin\theta_1\sin\theta_2] \\ -2\dot{l}(\dot{\theta}_1\sin\theta_1\cos\theta_2 + \dot{\theta}_2\cos\theta_1\sin\theta_2) \end{bmatrix} .$$

While Eq. (2) is mathematically equivalent to m rheonomic holonomic constraints $\mathbf{c}(\mathbf{q},t) = \mathbf{0}$, the resemblance of the trajectory control problem to the constrained motion case may be misleading. Assumed Eq. (2) represents contact constraints, $\mathbf{f}_a = -\mathbf{B}^T\mathbf{u}$ in Eq. (1) must be replaced by $\mathbf{f}_c = -\mathbf{C}^T\boldsymbol{\lambda}$, and by assumption the contact constraint reactions are orthogonal to the manifold of contact constrains. By contrast, the available control reactions may have arbitrary directions with respect to the control constraint manifold, and in the extreme some of the control reactions may be tangent. In the latter case, not all of the desired outputs $\boldsymbol{\gamma}_d$ can directly be actuated by the system inputs $\mathbf{u}$. A measure of the 'control singularity' is the rank of $m \times m$ matrix

$$\mathbf{P} = \mathbf{C}\mathbf{M}^{-1}\mathbf{B}^T \ , \tag{4}$$

which represents the inner product of the constrained and controlled subspaces.[4] For the case at hand, $\text{rank}(\mathbf{P}) = 1$, and this means that only one control input (M_n) actuates directly the control constraint conditions of Eq. (2), and the other two actuators (F_x and F_y) have no direct influence on realization of the control constraints.

3. GOVERNING EQUATIONS

The crane dynamic equations (1) can be projected into complementary constrained and unconstrained subspaces, defined by the 3×5 constraint matrix **C** and its orthogonal complement – a 5×2 matrix **D** such that[4]

$$\mathbf{C}\,\mathbf{D}=\mathbf{0} \quad \Leftrightarrow \quad \mathbf{D}^T\mathbf{C}^T=\mathbf{0}\ , \tag{5}$$

and for the crane considered **D** can be proposed as

$$\mathbf{D}^T=\begin{bmatrix} 1 & 0 & -\sin\theta_1\cos\theta_2 & -\dfrac{\cos\theta_1}{l\cos\theta_2} & \dfrac{\sin\theta_1\sin\theta_2}{l} \\ 0 & 1 & -\sin\theta_2 & 0 & -\dfrac{\cos\theta_2}{l} \end{bmatrix}.$$

The projection formula is

$$\begin{bmatrix} \mathbf{D}^T \\ \mathbf{C}\mathbf{M}^{-1} \end{bmatrix}\left(\mathbf{M}\,\ddot{\mathbf{q}}+\mathbf{d}=\mathbf{f}-\mathbf{B}^T\mathbf{u}\right), \tag{6}$$

and the governing equations can be manipulated to:

$$\dot{\mathbf{q}}=\mathbf{v}\ , \tag{7a}$$

$$\mathbf{D}^T\mathbf{M}\,\dot{\mathbf{v}}=\mathbf{D}^T(\mathbf{f}-\mathbf{d})-\mathbf{D}^T\mathbf{B}^T\mathbf{u}\ , \tag{7b}$$

$$\mathbf{0}=\mathbf{C}\mathbf{M}^{-1}(\mathbf{f}-\mathbf{d})-\mathbf{C}\mathbf{M}^{-1}\mathbf{B}^T\mathbf{u}-\boldsymbol{\xi}\ , \tag{7c}$$

$$\mathbf{0}=\boldsymbol{\Phi}(\mathbf{q})-\boldsymbol{\gamma}_d\ , \tag{7d}$$

where Eqs. (7b) and (7c) are the projections of Eq. (1) into the unconstrained and constrained subspaces, respectively.

While Eq. (7c) stands for $m=3$ algebraic equations, for the case at hand we have $\mathrm{rank}(\mathbf{C}\mathbf{M}^{-1}\mathbf{B}^T)=p=1$, and as such Eq. (7c) represents only one independent condition on **u** and $m-p=2$ restrictions on the crane motion, supplementary to original restrictions of Eq. (2). In this way, due to the mixed orthogonal-tangent realization[2] of control constraints, the total number of motion specifications is thus $m+m-p=5=n$, and as such the motion is fully specified. The situation corresponds to *flatness*[5] of the underactuated system in the partly specified motion.

4. SOLUTION CODE

For the case at hand, Eqs. (7) represent thirteen $(5+2+3+3)$ DAEs in ten states $\mathbf{q}$ and $\mathbf{v}$ and three control inputs $\mathbf{u}$. *Index* of the DAEs is three,[3] and they can be solved by using the simplest Euler backward difference approximation method. Representing Eqs. (7b), (7c) and (7c) symbolically as $\mathbf{H}(\mathbf{q})\,\dot{\mathbf{v}} = \mathbf{h}(\mathbf{q},\mathbf{v},\mathbf{u},t)$, $\mathbf{0} = \mathbf{b}(\mathbf{q},\mathbf{v},\mathbf{u},t)$ and $\mathbf{0} = \mathbf{c}(\mathbf{q},t)$, respectively, the solution code can be written as

$$\frac{\mathbf{q}_{n+1} - \mathbf{q}_n}{\Delta t} - \mathbf{v}_{n+1} = \mathbf{0} \ , \tag{8a}$$

$$\mathbf{H}(\mathbf{q}_{n+1})\frac{\mathbf{v}_{n+1} - \mathbf{v}_n}{\Delta t} - \mathbf{h}(\mathbf{v}_{n+1}, \mathbf{q}_{n+1}, \mathbf{u}_{n+1}, t_{n+1}) = \mathbf{0} \ , \tag{8b}$$

$$\mathbf{b}(\mathbf{v}_{n+1}, \mathbf{q}_{n+1}, \mathbf{u}_{n+1}, t_{n+1}) = \mathbf{0} \ , \tag{8c}$$

$$\mathbf{c}(\mathbf{q}_{n+1}, t_{n+1}) = \mathbf{0} \ . \tag{8d}$$

Given $\mathbf{q}_n$ and $\mathbf{v}_n$ at time t_n, Eqs. (8) represent thirteen nonlinear algebraic equations in $\mathbf{q}_{n+1}$, $\mathbf{v}_{n+1}$ and $\mathbf{u}_{n+1}$ at time $t_{n+1} = t_n + \Delta t$. By solving the equations, the simulation is advanced from t_n to t_{n+1}. In order to improve accuracy of the numerical solution, the rough Euler scheme can possibly be replaced by a higher order backward difference approximation method.[3] It may be worth noting that, due to the original control constraint equations $\mathbf{c}(\mathbf{q},t) = \mathbf{0}$ are involved in Eqs. (7), the solution is free from the constraint violation problem, and the truncation errors do not accumulate in time. The proposed simple code leads to reasonable and stable solutions.

5. SYNTHESIS OF CONTROL

As a solution to Eqs. (8), time-variations of state variables $\mathbf{q}(t)$ and $\mathbf{v}(t)$ in the prescribed motion and the control $\mathbf{u}(t)$ that assures the realization of the specified motion are obtained. The control obtained this way can be used as a feedforward control for the crane executing the load prescribed motion. It should then be enhanced by a feedback control to provide stable tracking of the load trajectory in the presence of perturbations. One possibility is to introduce, instead of Eq. (2), a stabilized form of the constraint equation at the acceleration level, $\ddot{\mathbf{c}} + \alpha\,\dot{\mathbf{c}} + \beta\,\mathbf{c} + \chi \int \mathbf{c}\,dt = \mathbf{0}$, where α, β and χ are gain values. The modification causes that Eq. (7c) is replaced with

$$\mathbf{0} = \mathbf{C}\mathbf{M}^{-1}(\mathbf{f} - \mathbf{d}) - \mathbf{C}\mathbf{M}^{-1}\mathbf{B}^T\mathbf{u} - \xi + \alpha\,\dot{\mathbf{c}} + \beta\,\mathbf{c} + \chi \int \mathbf{c}\,dt \ , \tag{9}$$

whose symbolic form is again $\mathbf{0} = \mathbf{b}(\mathbf{q}, \mathbf{v}, \mathbf{u}, t)$. In other words, the constraint induced accelerations $\boldsymbol{\xi} = \ddot{\boldsymbol{\gamma}}_d - \dot{\mathbf{C}}\dot{\mathbf{q}}$ are now modified to the stabilized form $\boldsymbol{\xi}_{stab} = \boldsymbol{\xi} - (\alpha\,\dot{\mathbf{c}} + \beta\,\mathbf{c} + \chi \int \mathbf{c}\,dt)$ by adding the correction terms due to the constraint violations. The hybrid control can then be synthesized from such modified Eqs. (7) using the code of Eq. (8). The idea for crane control with the use of the scheme is shown in Figure 2.

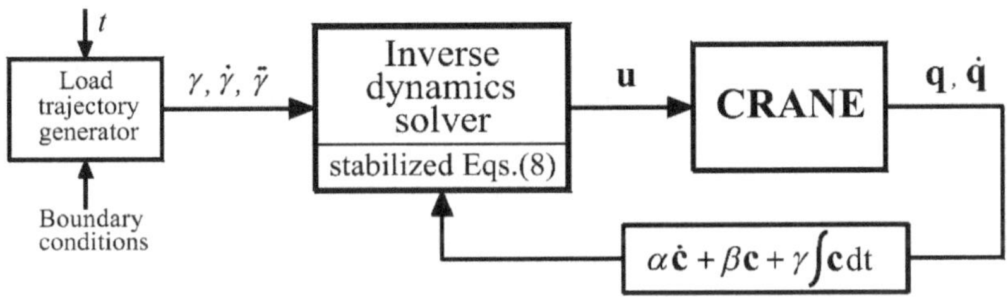

Figure 2. The hybrid control scheme.

6. NUMERICAL EXPERIMENTS

The crane data used in computations were: $m_b = 20\,\text{kg}$, $m_t = 10\,\text{kg}$, $m_l = 100\,\text{kg}$, $r = 0.1\,\text{m}$, and $J = 0.1\,\text{kg}\,\text{m}^2$. The control task was to move the load along a straight line following the rest-to-rest maneuver

$$\boldsymbol{\gamma}_d(t) = \boldsymbol{\gamma}_0 + (\boldsymbol{\gamma}_f - \boldsymbol{\gamma}_0)\,s(t), \quad \dot{\boldsymbol{\gamma}}_d = (\boldsymbol{\gamma}_f - \boldsymbol{\gamma}_0)\,\dot{s}, \quad \ddot{\boldsymbol{\gamma}}_d = (\boldsymbol{\gamma}_f - \boldsymbol{\gamma}_0)\,\ddot{s} \tag{10}$$

$$s(t) = 126\left(\frac{t}{\tau}\right)^5 - 420\left(\frac{t}{\tau}\right)^6 + 540\left(\frac{t}{\tau}\right)^7 - 315\left(\frac{t}{\tau}\right)^8 + 70\left(\frac{t}{\tau}\right)^9, \tag{11}$$

where $t \in < t_0, t_f >$, $\boldsymbol{\gamma}_0 = [x_0 \; y_0 \; z_0]^T$ and $\boldsymbol{\gamma}_f = [x_f \; y_f \; z_f]^T$ are the initial and final load positions at time t_0 and t_f, respectively, and $\tau = t_f - t_0$. For $x_0 = 0\,\text{m}$, $x_f = 5\,\text{m}$, $y_0 = 1\,\text{m}$, $y_f = 8\,\text{m}$, $z_0 = -5\,\text{m}$, $z_f = -1\,\text{m}$, $t_0 = 0\,\text{s}$, and $t_f = 6\,\text{s}$, the load motion specifications are illustrated in Figure 3.

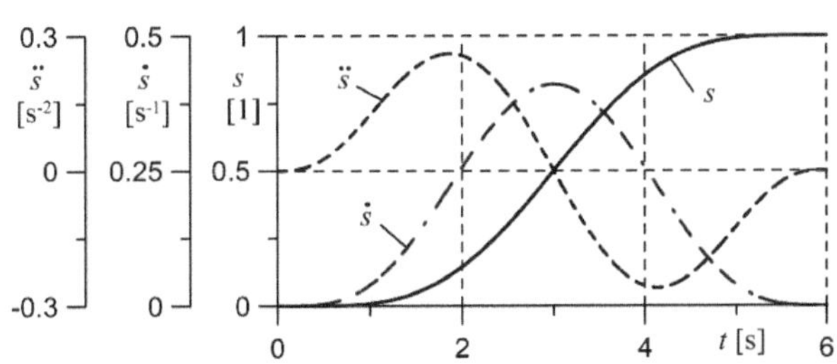

Figure 3. The load trajectory specifications according to Eqs. (10) and (11).

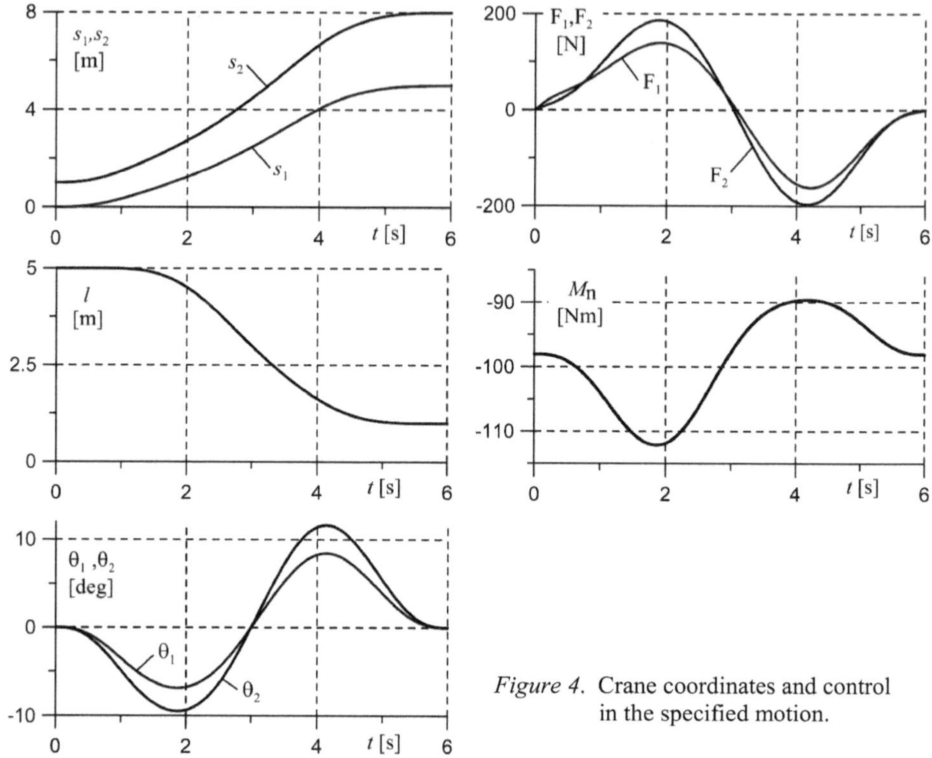

Figure 4. Crane coordinates and control in the specified motion.

The results of inverse simulation, i.e. the solution to the governing equations (7) by using the code (8), obtained for $\Delta t = 0.01\,\mathrm{s}$, are shown in Figure 4. The control rated this way can be used only as a feedforward control for the crane executing the prescribed load trajectory.

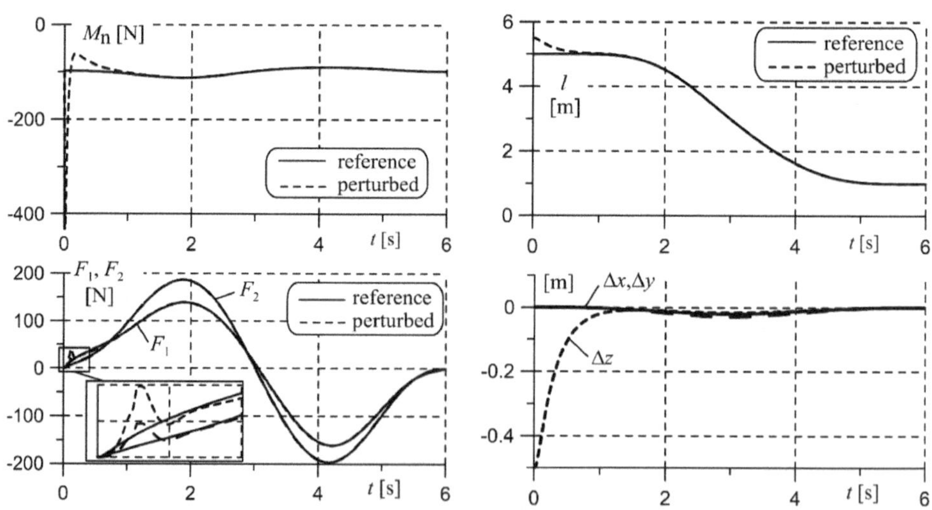

Figure 5. The hybrid control command in motion perturbed by inconsistent initial position.

The robustness of the hybrid control according to Eq. (9) (see Figure 2), was first tested by applying the inconsistent rest position at t_0 – the load was placed 0.5 m below its reference position, $l_0 = 5.5\,\text{m}$. The gain values were taken so that to assure the critical damping for the PID scheme,[6] i.e.

$$\alpha^2 = 8\beta\ , \qquad 32\chi = \alpha\,\beta \tag{12}$$

and a good choice for the integration time step $\Delta t = 0.01\,\text{s}$ was $\beta = 10$. The results of numerical simulations are shown in Figure 5. It can be seen that the system has a damped response about the reference trajectory.

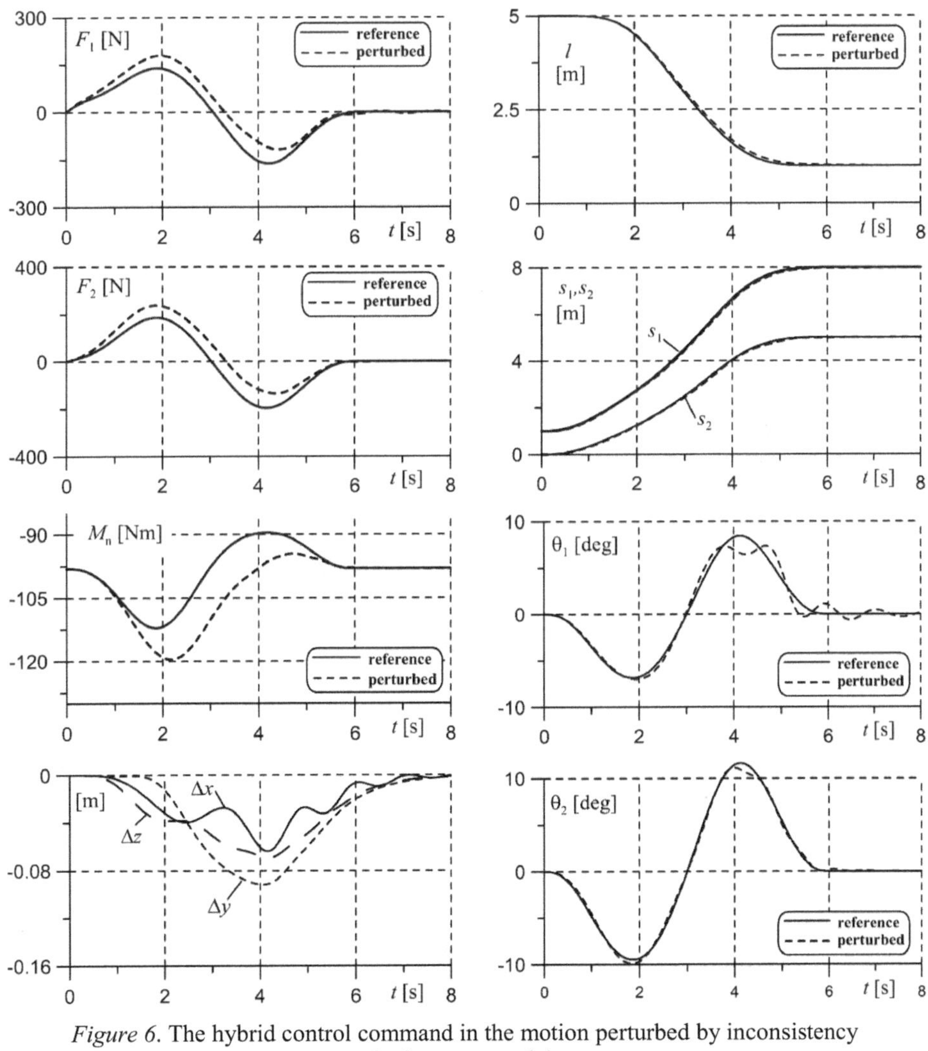

Figure 6. The hybrid control command in the motion perturbed by inconsistency in the crane model.

The other experiment consisted in checking the influence of modeling inconsistency. In the dynamic model used for the direct dynamic simulation, additional damping forces related to s_1, s_2 and l motions have been involved, not considered in the model used fort the determination of hybrid control. The additional forces were $-k_1\dot{s}_1$, $-k_2\dot{s}_2$ and $-k_3\dot{l}$, added respectively to the first, second and third entry of **f** described in Eq. (1), and the damping coefficient used were $k_1 = k_2 = 35\,[\mathrm{N\,s\,m^{-1}}]$ and $k_3 = 75\,[\mathrm{N\,s\,m^{-1}}]$. The motion disturbed this way was then stabilized along the reference motion by using the hybrid control. Some results of numerical simulations are shown in Figure 6. While the control characteristics are now decidedly different from the reference control (with no model inconsistencies), the motion of the load as well as the actual motion of the crane are very close to the reference motion characteristics. The simulation was extended over the end of the transfer maneuver (6s) up to 8 seconds, to show that the residual oscillations of the load are damped to the rest position as well.

7. CONCLUSION

A computational framework for control design of overhead cranes executing a prescribed load trajectory has been presented. The solution to the governing equations are the crane motion characteristics in the reference motion and the control required for its realization. The feedforward control scheme obtained this way is then enhanced by a feedback control, obtained by using the same governing equations in a slightly modified form.

REFERENCES

1. Abdel-Rahman, E.M., Nayfeh, A.H. and Masoud, Z.N., Dynamics and control of cranes: A review. *J. Vib. and Control* 9, 863–908 (2003).
2. Blajer, W. and Kołodziejczyk, K., A geometric approach to solving problems of control constraints: Theory and a DAE framework. *Multibody Syst. Dyn.* 11(4), 343-364 (2004).
3. Ascher, U.M. and Petzold, L.R., *Computer Methods for Ordinary Differential Equations and Differential-Algebraic Equations*, SIAM, Philadelphia (1998).
4. Blajer, W., A geometrical interpretation and uniform matrix formulation of multibody system dynamics. *ZAMM* 81(4), 247–259 (2001).
5. Fliess, M., Lévine, J., Martin, P. and Rouchon, P., Flatness and defect of nonlinear systems: Introductory theory and examples. *Int. J. Control* 21, 31–45 (1997).
6. Ostermayer, G.-P., On Baugarte stabilization for differential algebraic equations. In: *Real-Time Integration Methods for Mechanical System Simulation*, E.J. Haug and R.C. Deyo (eds.), NATO ASI Series, Vol. F69, Springer, Berlin, pp. 193–207 (1990).

HOISTING MANIPULATION BY MODAL COUPLING CONTROL FOR UNDERACTUATED CRANES

Andreas Bockstedte and Edwin Kreuzer
Mechanics and Ocean Engineering, TUHH – Hamburg University of Technology, 21071 Hamburg, Germany
bockstedte@tuhh.de, kreuzer@tuhh.de

Abstract: The reduction of payload pendulations is an important matter in the operation of cranes. Such oscillations can be excited by external influences like wind loads or by displacements of the support mechanism. Furthermore, for cranes with variant cable length oscillations are boosted by the hoisting of the payload.

The aim of the presented control strategy is to manipulate the desired hoisting velocity by superposition of a suitably modulated motion in order to reduce amplifications of the pendulations, in particular in absence of other effective control inputs.

Experimental results are presented for a simple pendulum. Numerical investigations extend the control strategy to an examplary 3-dimensional multibody system featuring the *flying-crane*-concept as designed for cargo airships.

Key words: crane dynamics, autoparametric resonances, modal coupling control.

1. INTRODUCTION

A variety of different cranes is used in technical applications, e.g. boom cranes, gantry cranes, crane ships and rescue helicopters. The payload of most cranes is subjected to excitations such as wind loads or motions of the support mechanism, forcing the pendulum formed by suspension cables and payload to oscillate.

In the past decades substantial effort has been invested in the analysis of crane dynamics and various control strategies have been developed to suppress oscillations of the payload [2]. Most of these techniques are based on a control input that displaces the support of the hoisting mechanism horizontally. However, controlled motion of the carrying structure might be not suitable or even impossible for some applications.

H. Ulbrich and W. Günthner (eds), IUTAM Symposium on Vibration Control of Nonlinear Mechanisms and Structures, 121–130.

Usually, cranes are equipped with winches to control the cable length. The actual cable length influences the dynamics of the pendulum significantly. Furthermore, payload oscillations are amplified by a decrease of the cable length during operation due to nonlinear effects. On the contrary, the possibility to influence and reduce pendulations by means of controlled variations of the cable length is hardly used in crane technology.

A control strategy where the cable length is adaptively adjusted with respect to the frequency of the excitation can be used to make the system less sensitive to disturbances [1].

In case the winches are capable of appropriate dynamical behavior the manipulation of the hoisting velocity can be used for an active reduction of the payload oscillations. Before this control strategy is discussed, two examplary crane models shall be introduced. Experimental results for a setup following this simple pendulum are presented in Section 5.

2. CRANE MODELS

The basic strategy of hoisting manipulation is illustrated for an *simple* elastic *pendulum*. The extension of the control strategy to more complex technical systems is demonstrated for a model of the *flying crane*.

The Simple Pendulum. The planar pendulum with variable rope length ℓ shown in Figure 1 is used as the most simple model of crane systems. Let the rope be visco-elastic, i.e. the force in the rope is

$$F_{\text{rope}} = F_0 + c \cdot (\ell - \ell_0) + d \cdot (\dot{\ell} - \dot{\ell}_0) \tag{1}$$

with stiffness parameter c and damping parameter d. The force F_0 and the length ℓ_0 denote the rope properties at equilibrium, where $\dot{\ell} = \dot{\ell}_0 = 0$.

With $F_0 = mg$ the equations of motion for the pendulum of mass m are

$$\ddot{\varphi} + 2\frac{\dot{\varphi}\dot{\ell}}{\ell} + \frac{g}{\ell}\sin\varphi = 0, \tag{2}$$

$$\ddot{\ell} - \dot{\varphi}^2\ell + g\,(1 - \cos\varphi) + \frac{d}{m}\left(\dot{\ell} - \dot{\ell}_0\right) + \frac{c}{m}\,(\ell - \ell_0) = 0. \tag{3}$$

Obviously, the angular and radial motion along $\mathbf{e}_\varphi$ and $\mathbf{e}_\ell$ are decoupled for the linearized equations of motion:

$$\ddot{\varphi} + \frac{g}{\ell}\varphi = 0, \tag{4}$$

$$\ddot{\ell} + \frac{d}{m}\left(\dot{\ell} - \dot{\ell}_0\right) + \frac{c}{m}\,(\ell - \ell_0) = 0. \tag{5}$$

Also, the controllability of the angular oscillation by means of a varying cable length is eliminated from the mathematical model by linearization. Thus, methods of nonlinear dynamics need to be used to enable successful control design.

The Flying Crane. A technical example for underactuated cranes is given by the cargo airship shown in Figure 2, designed for the transport of heavy payloads of up to 160 t. During the process of load exchange the *flying crane* constitutes a complicated pendulum. After unfastening the load frame from the ground the system is subjected to oscillations, amplified by hoisting. Due to the inertia of the buoyant hull, oscillations of the suspended payload cannot effectively be controlled by the thrusters mounted to the airship.

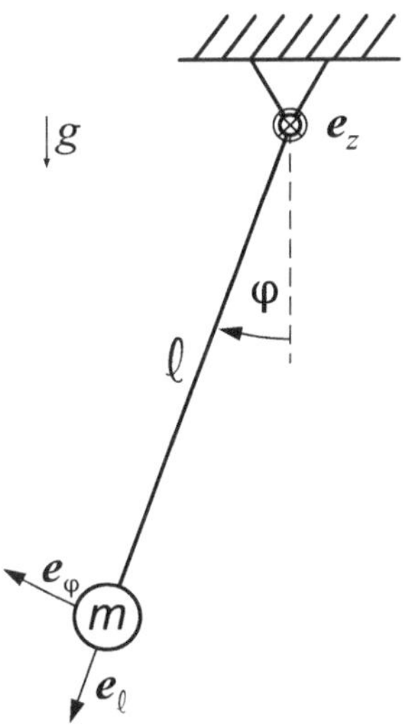

Figure 1. Simple pendulum.

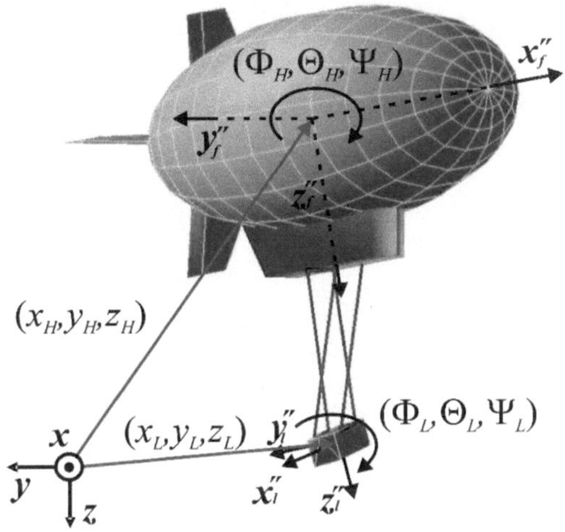

Figure 2. Flying crane.

The flying crane is modeled as multibody system with 12 degrees of freedom, including free motion of the hull without thruster interaction. The hull and the load frame have each three degrees of translatory freedom and three degrees of rotatory freedom. The position of hull and load frame are denoted by the position vectors of the envelope's center of volume $[\ x_{\mathrm{H}} \quad y_{\mathrm{H}} \quad z_{\mathrm{H}}\ ]^T$ and the center of mass of the load frame $[\ x_{\mathrm{L}} \quad y_{\mathrm{L}} \quad z_{\mathrm{L}}\ ]^T$ in Cartesian coordinates with respect to the inertial reference frame, see Figure 2. The rotation of the hull is described by *Cardan* angles Φ_{H}, Θ_{H}, Ψ_{H}. The angle Ψ_{H} denotes the rotation about the vertical z-axis, Θ_{H} determines the rotation about the new y'-axis and Φ_{H} denotes the rotation about the body fixed major axis (x''). The *Cardan* angles of the load frame are Φ_{L}, Θ_{L}, Ψ_{L}, respectively.

The vector of generalized coordinates for the system is therefore composed as follows:

$$\mathbf{y} = \left[\begin{array}{ccc|ccc|ccc|ccc} x_\mathrm{H} & y_\mathrm{H} & z_\mathrm{H} & \Phi_\mathrm{H} & \Theta_\mathrm{H} & \Psi_\mathrm{H} & x_\mathrm{L} & y_\mathrm{L} & z_\mathrm{L} & \Phi_\mathrm{L} & \Theta_\mathrm{L} & \Psi_\mathrm{L} \end{array}\right]^T . \tag{6}$$

Here, the suspension system consists of four cables driven by winches. The cables are assumed to be massless. The visco-elasticity of each cable is supposed to follow equation (1).

Computer algebra allows to derive the nonlinear equation of motion,

$$\mathbf{M}(\mathbf{y}) \cdot \ddot{\mathbf{y}} = \mathbf{f}(\mathbf{y}, \dot{\mathbf{y}}) \tag{7}$$

with the 12×12 mass matrix $\mathbf{M}(\mathbf{y})$ and the 12×1 vector of generalized forces $\mathbf{f}(\mathbf{y}, \dot{\mathbf{y}})$, and the system matrix $\mathbf{A}$ for the linearized dynamic system,

$$\dot{\mathbf{x}} = \mathbf{A}\mathbf{x}, \tag{8}$$

with state vector $\mathbf{x}^T = [\ \mathbf{y}^T \quad \dot{\mathbf{y}}^T\ ]$.

A more detailed description of the flying crane model is given in [3].

3. AUTOPARAMETRIC RESONANCES

The elasticity of the suspension cables enables vertical lift vibrations of the suspended payload. The natural frequency of vertical vibrations is affected by the stiffness of the cables, whereas the natural frequencies of pendulation oscillations are ruled by gravity.

Typically, the vertical vibrations are mathematically decoupled from pendulations in linearized models. However, weak nonlinear couplings between these oscillations occur in real systems. For certain values of the underlying parameters, i.e. basically the stiffness of the suspension cables in the case of cranes, these couplings become dominant for the system's behavior. This effect is called *autoparametric* or *internal resonance*. The necessary condition for autoparametric resonances requires integer multiples of the natural frequencies ω_i to commensurate approximately,

$$\sum k_i \omega_i \approx 0 \tag{9}$$

with integers k_i.

Typical autoparametric resonances for pendulums arise if the cable stiffness decreases until the natural frequency ω_lift of an elastic oscillation becomes just twice the value of the natural frequency ω_pend of some kind of pendulation,

$$\omega_\mathrm{lift} \approx 2 \cdot \omega_\mathrm{pend}. \tag{10}$$

Once the internal resonance is established, the amplitudes of involved oscillations can vary rapidly and energy is exchanged between these modes of oscillation. Examples for internal resonance have intensively been studied in the literature [5, 6].

In the case of the *flying crane*, distinctive couplings between the modes of oscillation can be observed depending on the geometry of the suspension system. Usually, these autoparametric resonances are regarded critical and undesired when high-frequency oscillations with moderate amplitude can suddenly turn into oscillations with low frequency but large amplitude. Here, physically relevant values of the cable stiffness turned out to be significantly higher than those critical values leading to autoparametric resonances. Hence, the autonomous suspension system is not subjected to critical internal resonances under typical conditions. This circumstance allows for extending considerations regarding control.

The phenomenon of energy exchange between linearly decoupled modes of vibration allows for the controlled energy transfer within the system and facilitates dissipation. Golnaraghi et al. suggest the method of *modal coupling control*, where the possibility of energy transfer between modes is utilized [4]. System properties are tuned by a control input such that autoparametric resonances are established and transfer energy into a motion where it can be dissipated more easily.

4. HOISTING MANIPULATION

Vertical motions of the suspended payload can easily be actuated, whereas pendulations are hardly accessible by means of the winch. However, some types of cranes lack efficient means to control the displacement of the suspension support. Therefore, a more sophisticated operation of the winches is desirable to improve the dynamic properties of the system.

Meeting the autoparametric resonance condition (10), hoisting interaction can suppress (and excite) oscillations of the pendulum. Vertical motions of the payload referring to the behavior of a tuned mass-spring-oscillator use the internal couplings to manipulate the energy transfer.

In a force-feedback loop a winch with appropriate dynamic behavior can emulate the visco-elasticity of a soft elastic pendulum. Choosing the control parameters accordingly to the internal resonance case establishes the excitation of a controlled periodic motion along the direction of the cables. Consequently, the amplitude of the pendulation is reduced. Emulation of viscosity adds active damping to the controlled heave motion, ensures energy dissipation, and is used to disable the converse excitation of the pendulation by the vertical oscillation.

The Simple Pendulum. Even tough the control strategy is based on nonlinear interactions, the design of the controller requires only the linearized model. Ironically, the designed controller would have no effect on the linear system.

The natural frequency for the angular oscillation follows from equation (4):

$$\omega_{\text{pend}} = \omega_\varphi = \sqrt{\frac{g}{\ell_0}}. \tag{11}$$

Equation (5) yields the resonant frequency $\Omega_{\text{res,lift}}$ and *Lehr*'s damping factor $D_{\text{res,lift}}$ for the radial motion:

$$\Omega_{\text{res,lift}} = \sqrt{\frac{c}{m} - \frac{d^2}{2m^2}} \quad , \quad D_{\text{res,lift}} = \frac{1}{2}\frac{d}{\sqrt{mc}}. \tag{12}$$

Even though the nonlinear couplings influence system dynamics even for a rough commensuration of frequencies following equation (9), close tuning of the frequencies is necessary for a controlled energy transfer. For the angular oscillation to initiate vertical oscillation, the lift frequency needs to be adjusted to the angular frequency,

$$\Omega_{\text{res,lift}} \approx 2 \cdot \omega_{\text{pend}}. \tag{13}$$

Depending on the desired cable length ℓ_0 and a desired damping factor $D_{\text{res,lift}} = D$, the emulated stiffness c and damping parameter d of the cable meet

$$c = \frac{4m}{1 - 2D^2} \cdot \frac{g}{\ell_0} \quad , \quad d = 2\sqrt{mc} \cdot D. \tag{14}$$

The damping factor D, $0 \leq D < 1/\sqrt{2}$, can be prescribed to control dissipation and excitation of the lift motion. Small values $D \approx 0$ allow for faster excitation of the lift oscillation and reduction of the angular oscillation, whereas the lift amplitude is kept smaller for larger values of D. If the desired cable length changes with time, $\dot{\ell}_0 \neq 0$, the tuned parameters are varied adaptively with respect to equations (14).

The Flying Crane. Usually, for more complicated technical systems such as the flying crane, algebraic manipulations neither on the linear nor on the nonlinear equations of motion are possible. Fortunately, the presented control strategy can be applied by numerical operations on the linear system (8).

Visco-elasticity following equation (1) is assumed for each cable j,

$$F_{\text{cable},j} = F_{0,j} + \frac{EA}{\ell_{0,j}} \cdot \left(\ell_j - \ell_{0,j}\right) + \frac{DA}{\ell_{0,j}} \cdot \left(\dot{\ell}_j - \dot{\ell}_{0,j}\right), \tag{15}$$

with cable stiffness EA and damping DA per unit length. At equilibrium, the particular cable lengths $\ell_{0,j}$ refer to the distance r_0 between the payload and the support of the suspension by geometry.

From the eigenvalues $\delta_k \pm i \cdot \omega_k$ of the linear system (8) follow the natural frequencies of the pendulum-like oscillations,

$$\omega_k = \omega_k(r_0, EA, DA), \tag{16}$$

the resonant frequencies of the lift oscillations, and *Lehr*'s damping factors, respectively,

$$\Omega_\ell = \sqrt{\omega_\ell^2 - \delta_\ell^2} = \Omega_\ell(r_0, EA, DA), \tag{17}$$

$$D_\ell = \frac{|\delta_\ell|}{\sqrt{\omega_\ell^2 + \delta_\ell^2}} = D_\ell(r_0, EA, DA). \tag{18}$$

Under the condition for resonance,

$$\Omega_\ell(r_0, EA, DA) = 2 \cdot \omega_k(r_0, EA, DA), \tag{19}$$

and damping,

$$D_\ell(r_0, EA, DA) = D, \tag{20}$$

the desired payload distance r_0 and damping factor D can be mapped to tuned stiffness and damping parameters, $EA_{\text{res}}(r_0, D)$ and $DA_{\text{res}}(r_0, D)$.

In order to emulate the visco-elasticity, each winch is operated with force-feedback according to

$$\dot{\ell}_j = \frac{\ell_{0,j}}{DA}\left(F_{\text{cable},j} - F_{0,j} - \frac{EA}{\ell_{0,j}} \cdot (\ell_j - \ell_{0,j})\right) + \dot{\ell}_{0,j}. \tag{21}$$

5. EXPERIMENTAL RESULTS FOR THE *SIMPLE PENDULUM*

The experimental setup for the simple pendulum is sketched in Figure 3. The rope is guided by pulleys via a force sensor to the winch. The maximum vertical length of the pendulum is 8.5 m. The winch is driven by a controlled motor. The signals of the force sensor and of a rotary sensor measuring the angle of the winch are fed back to a linear controller, setting the motor speed.

The controller is aimed to emulate the force-displacement-relation of a visco-elastic rope. Here, control design via pole placement was chosen, such that the eigenvalues $\lambda = \delta_\ell \pm i\omega_\ell$ referring to the lift oscillation meet the resonance condition,

$$\omega_\ell = 2 \cdot \sqrt{\frac{g(1 - D^2)}{\ell_0(1 - 2D^2)}} \quad , \quad \delta_\ell = -2 \cdot \sqrt{\frac{g \cdot D^2}{\ell_0(1 - 2D^2)}}. \tag{22}$$

For a third-order model of the plant with states ℓ, $\dot{\ell}$, and $\ddot{\ell}$, the third pole $\delta^\star$ is placed at the negative real axis, $\delta^\star \ll -\sqrt{\omega_\ell^2 + \delta_\ell^2}$.

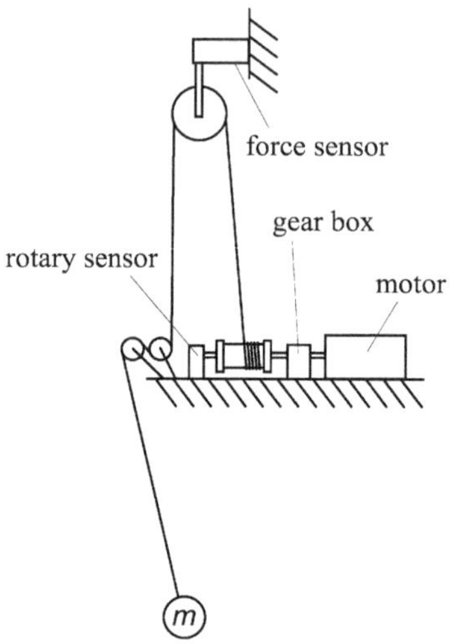

Figure 3. Experimental setup.

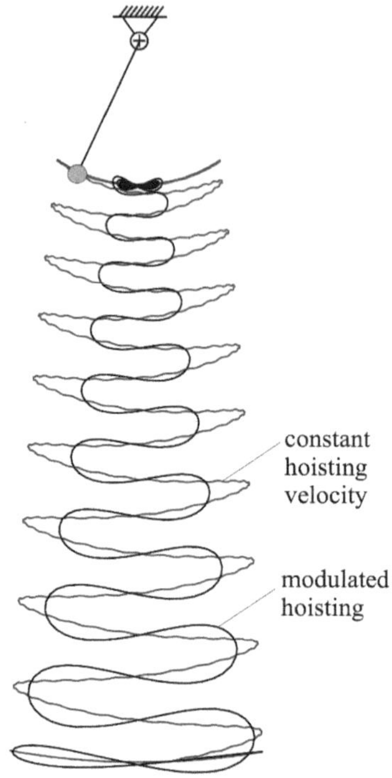

Figure 4. Simulation of modal coupling controlled hoisting.

A simulation of the elastic pendulum is shown in Figure 4. With constant hosting, $\dot{\ell} = \dot{\ell}_0 = \text{const}$, the angular oscillations of the pendulum are amplified. In contrast, the amplification can be reduced significantly by modulated hoisting.

The simulated scenarios were enacted in the experiment. Starting from an amplitude of 10°, the rope was hoisted from an initial length of 8 m to a final length of 1 m. Figure 5 compares hoisting of simulation and experiment and shows the specific force signals F_{rope}/m, respectively. The jerks induced by turning the controller on and off when hoisting starts and ends can be identified in the figure. Even though the force signal is interfered by noise from the rotating pulleys, the vertical oscillation is induced, leading to a reduction of the angular oscillation. After turning off the controller, the remaining force signal indicates the amplitude of the angular oscillation. (Note: The reduction of the rope weight due to hoisting was not reflected in the simulation.)

Along with the nonlinear interactions, the effect of vertical motion driven by the winch on the angular oscillation decreases for small angles. In general, total dissipation of the angular oscillation can not be expected. On the other

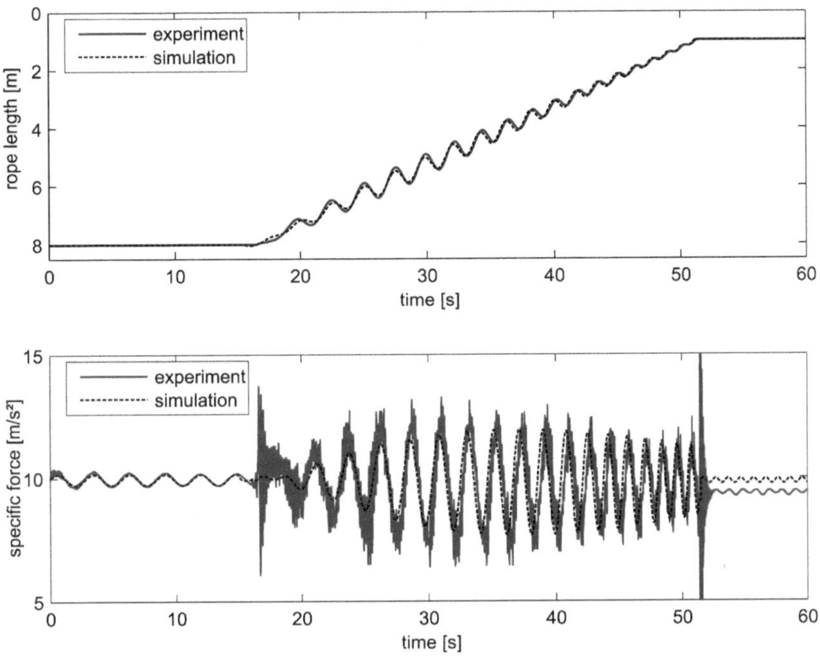

Figure 5. Comparison of experiment and simulation, $D = 0.03$.

hand, vertical motion that could excite angular oscillations is not induced for small angles.

The same holds for spherical oscillations of the pendulum. The gear angle of the pendulum can neither be measured nor be influenced by this controller. Starting from elliptical oscillations around the vertical axis, modulated hoisting can only dissipate energy until circular rotations with respect to the vertical axis arise, either.

6. NUMERICAL RESULTS FOR THE *FLYING CRANE*

In this case, the winches are to be operated with individual force feedback such that the internal resonance condition of the visco-elastic pendulum is emulated.

A simulation of payload oscillations along the local x-axis with constant and modulated hoisting is shown in Figure 6. Also in this case, payload oscillations are amplified at constant hoisting velocity until further hoisting is not expedient. In contrast, modulated hoisting keeps the oscillation smaller and allows for further, though slower, hoisting.

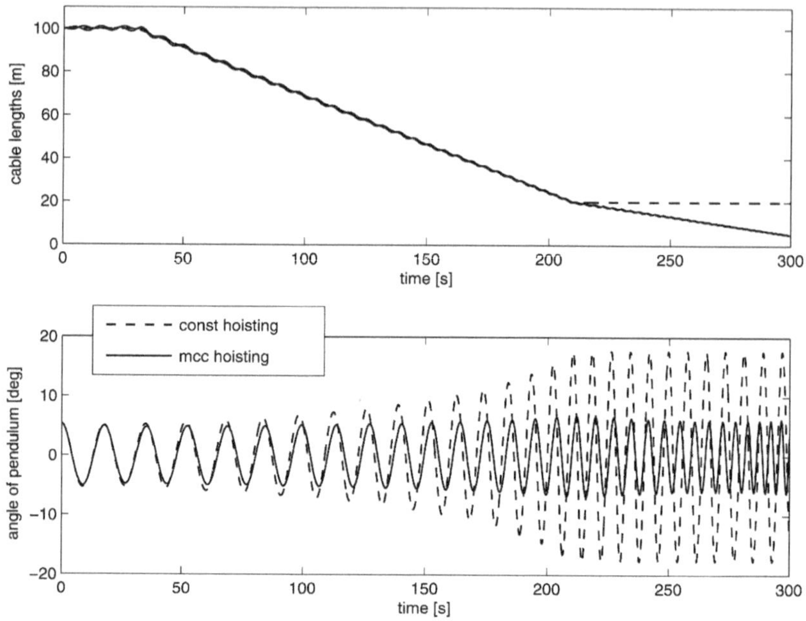

Figure 6. Simulation of constant and modulated (mcc) hoisting for the flying crane.

7. CONCLUSION

Instead of constant hoisting speeds the presented control strategy can modulate the hoisting velocity such that the amplification of oscillations due to hoisting can be reduced significantly. The application of modal coupling control appears particularly effective for cranes with varying cable length because of the direct intervention in the emergence of undesired amplification.

REFERENCES

[1] Abdel-Rahman, E.M. and Nayfeh, A.H. (2000). Cargo-pendulation reduction in boom cranes via cable-length manipulation. In *Proceedings of the 41th Structures, Structural Dynamics, and Materials Conference*, Atlanta, GA, USA.

[2] Abdel-Rahman, E.M., Nayfeh, A.H. and Masoud, Z.N. (2003). Dynamics and control of cranes: A review. *Journal of Sound and Vibration*, 9:863–908.

[3] Bockstedte, A. and Kreuzer, E. (2005). Multibody system model of a cargo airship to study the load exchange process. *Mathematical and Computer Modelling of Dynamical Systems*, 11.

[4] Golnaraghi, M.F. (1991). Regulation of flexible structures via nonlinear coupling. *Dynamics and Control*, 1:405–428.

[5] Nayfeh, A.H. (2000). *Nonlinear Interactions.* Wiley & Sons, New York.

[6] Tondl, A., Ruijgrok, T., Verhulst, F. and Nabergoj, R. (2000). *Autoparametric Resonance in Mechanical Systems.* Cambridge University Press, New York.

VIBRATION OF RESONANT GYROSCOPES

Application to the Gyros of Macro and Micro Sizes

Chia-Ou Chang and Chan-Shin Chou
Institute of Applied Mechanics, National Taiwan University, Taipei 106, Taiwan, ROC

Abstract: The vibration behavior of a hemispherical resonant gyro (HRG) and a micro-ring-gyro made of $\{111\}$ silicon wafer are studied. It was found that for HRG, the persisting vibration does not exist for the linear model but exists for the nonlinear one; whereas it exists in the linear model for micro-gyro, this is due to the size effect. The relationships between the angular velocity and the output vibrating magnitudes of both gyros are derived analytically by using perturbation technique.

Key words: nonlinear vibration, perturbation method, micro-gyro, shell theory.

1. INTRODUCTION

In working both the hemispherical resonant gyro (HRG) and micro-ring resonant gyro (MRRG) are excited into vibration at their lower natural frequencies. When these gyros are driven into rotation, their mode shapes will not rotate in the same rate as that of the gyros, then spin rates of the mode shapes are retarded by the Coriolis force. By measuring this backward angular velocity drift the spin rate of the gyro can be calculated from either the theoretical equation or the experimental curve. Although many papers studying the HRG of centimeter size were reported [1–3]. No paper derived analytically the explicit expressions for the equation which relates the angular velocity of spin to the modal precession rate. We used the thin shell theory[4] to establish the equations of vibration and the Galerkin method to obtain the amplitude equations. It was found from theoretical analysis that the vibration amplitude of the shell will be damped out due to the material damping from the analysis based on the **linear** equations of vibration, but analysis based on the non-linear equations of vibration can give the solution

H. Ulbrich and W. Günthner (eds), IUTAM Symposium on Vibration Control of Nonlinear Mechanisms and Structures, 131–140.

of non-vanishing vibration. The method of multiple scales and the center manifold theorem are used to get the precession rate of the mode shape in explicit form.

Due to the rapid development of MEMS technology many gyros of micrometer size were invented. Although there appeared many papers [5–9] and patents talking about ring micro-gyros, most of them gave only the conceptual designs and lack rigorous analysis. Gallacher et al. [10] seems to be the first to study the design of three-axis gyros. However, the mechanism of actuator, the effect of driving forces on the output amplitudes, and the coupling effects are not analyzed in their work.

In this paper the three-axis gyros excited by electrostatic field of capacitors are investigated. The nonlinear relationship between the three angular rates and the vibration amplitude of the ring are derived in closed form. The effect of the driving voltage and the gap of the capacitor surfaces on the output signal can be obtained through our analytical solution. The terms that cause linear instability for the HRG can be neglected for micro RRG.

2. ANALYSIS OF HRG

The geometry and coordinate system of a HRG is shown in Figure 1.

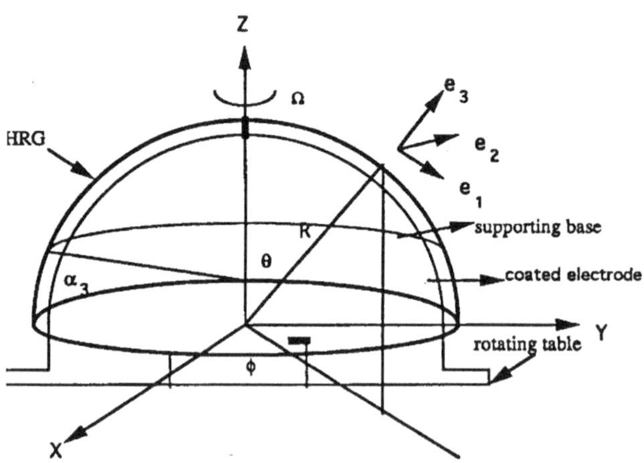

Figure 1. System configuration.

Let $\mathbf{u} = u^{\alpha}\mathbf{a}_{\alpha} + w\mathbf{a}_3$ (α=1,2) be the displacement of the middle surface. $\{\mathbf{a}_{\alpha}, \mathbf{a}_3\}$ are natural base vectors. We consider that the constitutive equation is linear and the strain-displacement is nonlinear, the nonlinearity comes from the moderate rotation [11].

$$\begin{aligned} E_{\alpha\beta} &= \tfrac{1}{2}(D_\alpha u_\beta + D_\beta u_\alpha) + \tfrac{1}{R} a_{\alpha\beta} w + \tfrac{1}{2}(w_{,\alpha} + b_\alpha^\lambda u_\lambda)(w_{,\beta} + b_\alpha^\delta u_\delta) \\ K_{\alpha\beta} &= D_\alpha D_\beta w - \tfrac{1}{R}(D_\alpha u_\beta + D_\beta u_\alpha) - \tfrac{1}{R^2} a_{\alpha\beta} w \end{aligned} \tag{1}$$

where $E_{\alpha\beta}$ is the in-plane strain tensor and $K_{\alpha\beta}$ the bending strain tensor, D_α the covariant differentiation. We use ring electrode to form a capacitor. If the applied voltage is $V = V_0 \cos\nu t$, the driving electric force is $P_e \equiv (1+\cos 2\nu t)(1+2w/\Delta_0)(\varepsilon_0 V_0^2/4\Delta^2{}_0)$, where Δ_0 is the gap between then electrode and the shell. The nonlinear equations of vibration can be obtained by using the principle of virtual work. At resonant frequency the displacement of shell can be put in the form

$$\begin{cases} u \equiv c_s u_s + c_a u_a = c_s \bar{u}(\theta)\cos n\phi + c_a \bar{u}(\theta)\sin n\phi \\ v \equiv c_s v_s - c_a v_a = c_s \bar{v}(\theta)\sin n\phi - c_a \bar{v}(\theta)\cos n\phi \\ w \equiv c_s w_s + c_a w_a = c_s \bar{w}(\theta)\cos n\phi + c_a \bar{w}(\theta)\sin n\phi \end{cases}, \tag{2}$$

where c_s and c_a are the amplitudes of symmetric and anti-symmetric modes, respectively. Using Galerkin method gives the amplitude equations

$$\begin{aligned} \ddot{c}_s + c_s + \varepsilon[2\varsigma^* \dot{c}_s + 2\Omega^* G \dot{c}_a + \alpha(c_s^3 + c_s c_a^2) - 2(h + h\cos 2\nu t)c_s] = 0 \\ \ddot{c}_a + c_a + \varepsilon[2\varsigma^* \dot{c}_a - 2\Omega^* G \dot{c}_s + \alpha(c_a^3 + c_a c_s^2) - 2(h + h\cos 2\nu t)c_a] = 0 \end{aligned} \tag{3}$$

We use the method of multiple scales to solve (3). Let $T_0 = t, T_1 = \varepsilon t$ and

$$x = x_0(T_0,T_1) + \varepsilon x_1(T_0,T_1) + O(\varepsilon^2),\ \ y = y_0(T_0,T_1) + \varepsilon y_1(T_0,T_1) + O(\varepsilon^2) \ldots \tag{4}$$

The linear part of Equation (3) is the type of the Mathieu's equations. It has the instability problem at primary resonance where $\nu = 2 + \grave{o}\sigma$ and the stability boundary is shown in Figure 2. Since the solution is damped out to zero in the stable region. We try to find the solution of the nonlinear equation (4) in the linearly unstable region to see whether a nonzero solution exists. Fortunately, the finite amplitude solution indeed exists. The first order solution is $x_0 = A_1(T_1)\exp(iT_0)$ and $y_0 = A_2(T_1)\exp(iT_0)$. Eliminating the secular terms yields the governing equations for unknown coefficients $A_1(T_1)$ and $A_2(T_2)$

$$\begin{aligned} -2iD_1A_1 - 2i\varsigma^* A_1 - 2i\Omega^* G A_2 + 2hA_1 + h\bar{A}_1 e^{i\sigma T_1} + [-3\alpha A_1^2 \bar{A}_1 - 2\alpha A_1 A_2 \bar{A}_2 - \alpha A_2^2 \bar{A}_1] = 0 \\ -2iD_1A_2 - 2i\varsigma^* A_2 + 2i\Omega^* G A_1 + 2hA_2 + h\bar{A}_2 e^{i\sigma T_1} + [-3\alpha A_2^2 \bar{A}_2 - 2\alpha A_2 A_1 \bar{A}_1 - \alpha A_1^2 \bar{A}_2] = 0 \end{aligned} \tag{5}$$

Let $A_1 = (a_1 + ib_1)\exp[i(\sigma T_1/2)]$ and $A_2 = (a_2 + ib_2)\exp[i(\sigma T_1/2)]$. Substituting these into (5) results in a set of equations having a pair of complex conjugate roots. Through the use of center manifold theorem [12] and coordinate transformation, we get

$$\dot{\mathbf{p}} = \begin{bmatrix} 0 & -\Omega^* G \\ \Omega^* G & 0 \end{bmatrix} \mathbf{p} + a_1(p_1^2 + p_2^2)\begin{pmatrix} p_1 \\ p_2 \end{pmatrix} + b_1(p_1^2 + p_2^2)\begin{pmatrix} -p_2 \\ p_1 \end{pmatrix} + H.O.T. \tag{6}$$

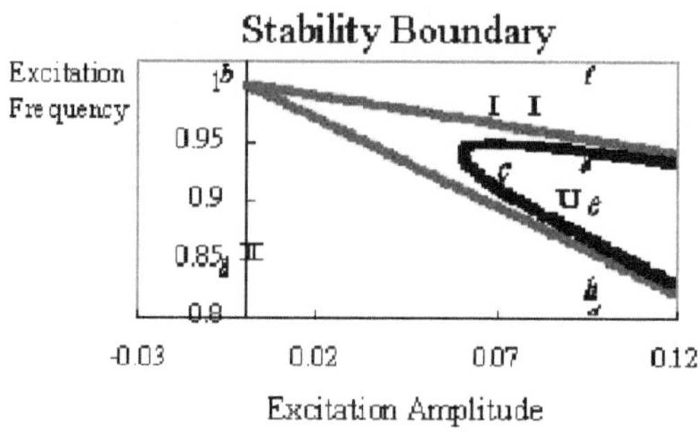

Figure 2. Stability boundary in amplitude-frequency plane.

Equation (6) is the Hopf singularity of type 1. The minimum-versal unfolding [12] needs only one parameter which is appropriate to represent the original three parameters h_r, σ, and h_0. The resultant equations are

$$\dot{\mathbf{p}} = \mathbf{Ep} + \mathbf{f}^*(\mathbf{p}, \mathbf{q}, \tilde{\mu}), \quad \dot{\mathbf{q}} = \mathbf{Fq} + \mathbf{g}^*(\mathbf{p}, \mathbf{q}, \tilde{\mu}), \quad \dot{\tilde{\mu}} = 0 . \tag{7}$$

where $h^2 - (\sigma + 2h_0)^2 \equiv \mu^{*2}$ and $\mu^* = 2\varsigma^* + \tilde{\mu}$. Since the matrix $\mathbf{E}$ has a zero eigenvalue. We can apply the second center manifold theorem again and express Eq.(7) in terms of polar coordinates by the transformation $p_1 = r\cos\phi$ and $p_2 = r\sin\phi$. According to Poincaré-Andronov-Hopf bifurcation, there exists an orbital solution of which the amplitude in closed form is

$$r = \sqrt{\bar{\mu} \Big/ \left[3\alpha \left(1 + 4\zeta^{*2} / \sigma_{2c}^2 \right) \right]}, \quad \tilde{\mu} = -\tilde{\mu}_1 (\sigma_c + 2h_0) / 2\xi^* . \tag{8}$$

The solution in the region far away the stability boundary cannot be solved through the use of the local theorem (i.e., the center manifold theorem). We assume that the coefficients are in the form

$$A_1 = R\cos\varphi \exp(i\frac{\sigma}{2}T_1 - i\theta_1), \quad A_2 = R\sin\varphi \exp(i\frac{\sigma}{2}T_1 - i\theta_2) . \tag{9}$$

The solution of amplitude is found to be in the closed form as

$$R_{0\pm} = \left[\frac{\sigma + 2h \pm \sqrt{h^2 - 4\varsigma^{*2}}}{3\alpha} \right]^{\frac{1}{2}}, \qquad \sigma_{c\pm} + 2h = \pm\sqrt{h^2 - 4\varsigma^{*2}} . \tag{10}$$

The normal displacement of the shell is

$$\begin{aligned} w &= c_s \cos n\phi + c_a \sin n\phi = 2R_0 \cos(n\phi - \Omega^* G T_1 - \varphi_0)\cos(T_0 + \sigma T_1/2 - \theta_0) \\ &= 2R_0 \cos(n\phi - \Omega G t - \varphi_0)\cos(t + \varepsilon\sigma t/2 - \theta_0) \end{aligned} . \tag{11}$$

From (11) we find that the mode shape drift with respect to the shell at the rate $\Omega^* G/2$ where $G = \int 2\bar{v}(\bar{u}\cos\theta + \bar{w}\sin\theta)\sin\theta d\theta$ is the sensing parameter. The angular drift of the vibrating mode relative to the spinning shell is shown in Figure 3.

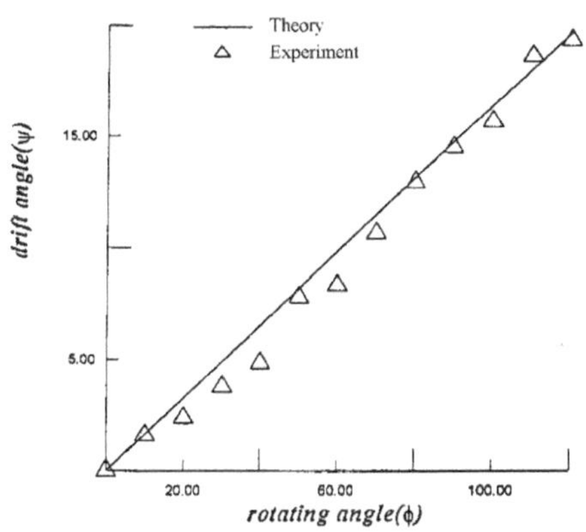

Figure 3. The precessional angle of vibrational mode (3,1) vs shell rotating angle.

3. ANALYSIS OF MRRG

The top view of the ring gyro is shown in Figure 4. The pairs of capacitors (PD1, PD3), (PD2,PD4) are used as actuators. The diameter, width, and thickness are denoted by a, h, and b respectively. Let u, v, w be the radial, tangential, and out-of-plane displacement of the neutral line, U_r, U_θ, U_z are the corresponding displacements of any point on the ring as shown in Figure 5. ϕ_i is the rotation angle about z-axis due to in-plane bending, ϕ_0 is the rotation angle about x-axis due to out-of-plane bending, and ϕ is the twist angle about the y-axis due to torsion.

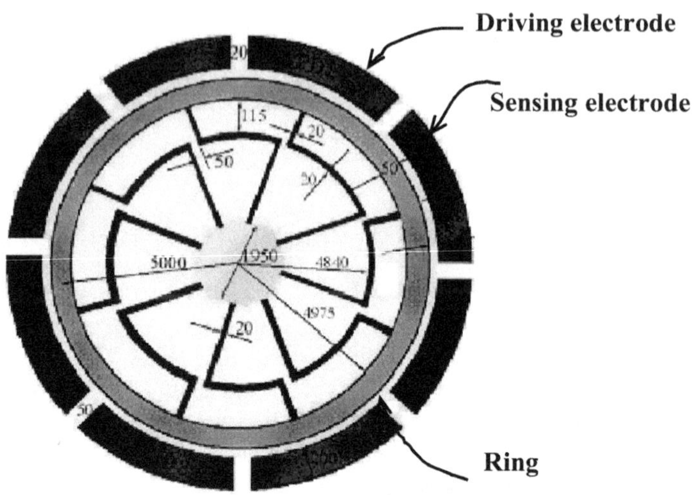

Figure 4. Top view of the ring.

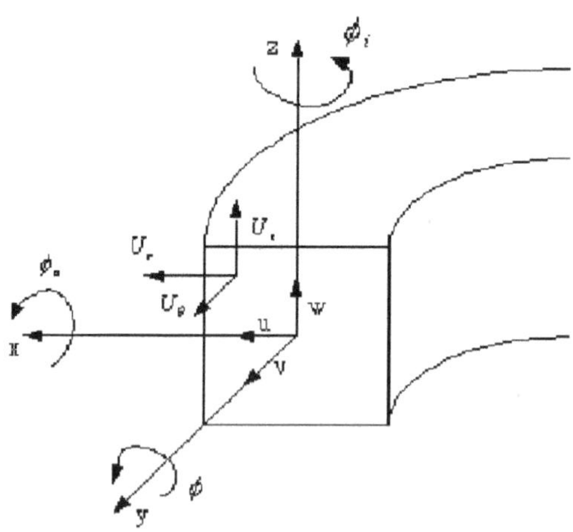

Figure 5. Displacements and coordinate system.

3.1 Natural Frequency Analysis

We adopt the Euler's beam theory, the displacements [14] are given by $U_r = u(\theta,t) + z\phi(\theta,t)$

$$U_r = u(\theta,t) + z\phi(\theta,t)$$
$$U_\theta = v(\theta,t) + r\phi_i - z\phi_o = v(\theta,t) + \frac{r}{a}\left(v(\theta,t) - \frac{\partial u(\theta,t)}{\partial \theta} \right) - \frac{z}{a}\frac{\partial w(\theta,t)}{\partial \theta} \tag{12}$$
$$U_z = w(\theta,t) - r\phi(\theta,t)$$

where $\phi_i = (v - \partial u/\partial\theta)/a$ and $\phi_0 = \partial w / a\partial\theta$. Using Hamilton's principle we get the equations of motion as

$$\begin{aligned}
&\ddot{u} + \frac{E}{a^2\rho}(u+v') + \frac{h^2}{12a^2}(\ddot{v}' - \ddot{u}'') + \frac{h^2 E}{12a^4\rho}(u'''' - v''') = 0\\
&\ddot{v} - \frac{12E}{12a^2\rho + h^2\rho}u' - \frac{h^2}{12a^2+h^2}\ddot{u}' - \frac{E}{a^2\rho}v'' + \frac{h^2E}{12a^4\rho + a^2h^2\rho}u''' = 0\\
&\ddot{w} - \frac{G}{12a^4\rho}w'' - \frac{(b^2G + h^2G + b^2E)}{12a^3\rho}\phi'' - \frac{b^2}{12a^2}\ddot{w}'' + \frac{b^2E}{12a^4\rho}w'''' = 0\\
&\ddot{\phi} + \frac{b^2E}{a^2(b^2+h^2)\rho}\phi - \frac{(b^2G + h^2G + b^2E)}{a^3(b^2+h^2)\rho}w'' - \frac{G}{a^2\rho}\phi'' = 0
\end{aligned} \tag{13}$$

We assume the solution in the form

$$u = Ae^{j(\omega_1 t + n\theta)},\ v = Be^{j(\omega_1 t + n\theta)},\ w = Ce^{j(\omega_2 t + m\theta)},\ \phi = De^{j(\omega_2 t + m\theta)}. \tag{14}$$

Substituting (14) into (13) yields in-plane natural frequencies

$$\omega_{11n}^2 = \frac{C_\omega - \sqrt{D_\omega}}{E_\omega} \quad \text{and} \quad \omega_{12n}^2 = \frac{C_\omega + \sqrt{D_\omega}}{E_\omega}, \tag{15}$$

where $C_\omega = E(12a^2(n^2+1) + h^2(2n^2+1)(n^2-1)$, $D_\omega = (12a^2 + h^2)[h^2(3n^2-1)^2 + 12a^2(n^2+1)^2]$, $E_\omega = 2a^2[12a^2 + h^2(n^2+1)]\rho$, ω_{11n} is the natural frequency of the nth inextensional mode and ω_{12n} is that of the nth extensional mode. Similarly, the frequencies of the m*th* out-of-plane modes are $\omega_{21m} = \omega_{21m}(E,G,a,h,b,m)$, $\omega_{22m} = \omega_{21m}(E,G,a,h,b,m)$.The geometric parameters of the system are designed so that the in-plane natural frequencies of the ring is tuned to be equal to the out-of-plane ones; therefore the output signal can be can be made of the same order as that of the input one, and this gyro can measure three-axis components of angular velocity. Consider the case where the data of silicon wafer are $E = 165$ GPa, $G = 67.6$ GPa, $\rho = 2330$ kg/m^3, $a = 4000\ \mu$m, $h = 100\ \mu$m, $b = 100\ \mu$m. In order to reconcile the in-plane and out-of-plane frequencies, that is, $\omega_{12n} = \omega_{22m}$ for $(n,m) = (2,3)$, we find that the geometric parameters are restricted to satisfy the equation $b = 0.34h$. And the radius has no effect on this reconciliation.

3.2 Analysis of Rotating Ring

The effect of Corioslis force due to the angular velocity input is considered. The eight supporting beam are included. Their strained energy is evaluated exactly, but their kinetic energy is approximated by considering their velocity as one half of the ring at the contact point. The sensing coefficients are derived, which are the key parameters that affect the performance of the gyros. The displacements can be expressed in terms of the eigenfunctions of the non-rotating ring as

$$\begin{aligned} u &= X_1(t)Cos(n\theta) + X_2(t)Sin(n\theta), \quad v = -\frac{1}{n}[X_1(t)Sin(n\theta) - X_2(t)Cos(n\theta)] \\ w &= X_3(t)Cos(m\theta) + X_4(t)Sin(m\theta) \\ \phi &= \frac{-m^2[(b^2+h^2)G + b^2m^2E] + a^2(12a^2 + b^2m^2)\rho\omega_2^2}{am^2[(b^2+h^2)G + b^2E]} \\ &\quad \times [X_3(t)Cos(m\theta) + X_4(t)Sin(m\theta)] \end{aligned} \tag{16}$$

where $X_i(t)$ $(i = 1,2,3,4)$ are generalized coordinates. The in-plane strain energy of the suspensions are evaluated in terms of generalized coordinates by using Castigliano theorem as

$$V_{si} = k_{ni}(X_1^2(t) + X_2^2(t)), \tag{17}$$

Similarly, the out-of-plane strain energy is $V_{si} = k_{ni}(X_3^2(t) + X_4^2(t))$. When the electrical force is applied, which is in the form

$$V_{e1}(t) = V_{e3}(t) = V_P Sin(vt), \quad V_{e2}(t) = V_{e4}(t) = V_P Cos(vt). \tag{18}$$

The equations of vibration in terms of the small parameter ε is

$$\ddot{\mathbf{X}} + \varepsilon \begin{bmatrix} \bar{\xi}_1 & -\lambda_1\bar{\Omega}_z & \lambda_2\bar{\Omega}_y & -\lambda_2\bar{\Omega}_x \\ \lambda_1\bar{\Omega}_z & \bar{\xi}_1 & \lambda_2\bar{\Omega}_x & \lambda_2\bar{\Omega}_y \\ -\lambda_3\bar{\Omega}_y & -\lambda_3\bar{\Omega}_x & \bar{\xi}_2 & -\lambda_4\bar{\Omega}_z \\ \lambda_3\bar{\Omega}_x & -\lambda_3\bar{\Omega}_y & \lambda_4\bar{\Omega}_z & \bar{\xi}_2 \end{bmatrix} \dot{\mathbf{X}} + \begin{bmatrix} 1+\varepsilon\bar{a}_1 & 0 \\ 0 & 1+\varepsilon\bar{a}_2 \\ 0 & 0 \\ 0 & 0 \end{bmatrix} \tag{19}$$

$$\begin{bmatrix} \cdots & 0 & 0 \\ \cdots & 0 & 0 \\ \cdots & \omega_r^2+\varepsilon^2\overline{a}_3[6\alpha+Sin(6\alpha)Cos(2\nu t)] & 0 \\ \cdots & 0 & \omega_r^2+\varepsilon^2\overline{a}_4[-6\alpha+Sin(6\alpha)Cos(2\nu t)] \end{bmatrix}\mathbf{X}=\varepsilon^3\begin{bmatrix}\overline{q}_1\\0\\0\\0\end{bmatrix}$$

where λ_i are the sensing coefficients of gyros and a_i is related to the applied voltages and the gap Δ_i of electrodes. They are

$$\lambda_1=\frac{48(2\pi\gamma\eta+2\kappa)}{60\pi\gamma\eta+9\pi\gamma^3\eta+60\kappa},\quad a_2=\frac{12a^2n^2\varepsilon_0V_p^2[-4\alpha+Sin(4\alpha)]}{\pi\rho h\Delta_i^3[h^2(n^2-1)^2+12a^2(n^2+1)]}\cdot \tag{20}$$

Assume the solution is in the form $\mathbf{x}(T)=\mathbf{Q}_1(T)+\varepsilon\mathbf{Q}_2(T)$, where $T=\omega_1 t$. The solution based on the perturbation technique of $\mathbf{Q}_1(T)$ is

$$Q_{11}(T)=A_{11}e^{jT},\ Q_{21}(T)=A_{21}e^{jT},\ Q_{31}(T)=A_{31}e^{jT},\ Q_{41}(T)=A_{41}e^{jT} \tag{21}$$

By eliminating the secular terms we can find solution for A_{11}, A_{21}, A_{31}, and A_{41}. Then the first order solutions $X_i(t)$ are obtained. We define the ratios of the output signal to the input one $X_1(t)$ as

$$\gamma_2=X_2(t)/X_1(t),\ \gamma_3=X_3(t)/X_1(t),\ \gamma_4=X_4(t)/X_1(t)\cdot \tag{22}$$

These equations, which relate the angular rates Ω_i to the displacements of the ring, are obtained explicitly in the form under the assumption that $\Omega_x, \Omega_y, \Omega_z \ll \omega$

$$\gamma_2=-\frac{4n\omega\Omega}{(n^2+1)(\xi_1\omega^2-ja_2)},\quad \gamma_3=(n-1)\left[-\frac{\Omega_y}{n\xi_2\omega}+\frac{4\Omega_x\Omega_z}{(n^2+1)\xi_1\xi_2\omega^2}\right]$$
$$\gamma_4=(n-1)\left[\frac{\Omega_x}{n\xi_2\omega}+\frac{4\Omega_x\Omega_z}{(n^2+1)\xi_1\xi_2\omega^2}\right] \tag{23}$$

Comparing equation (19) to equation (3), we find that the periodic coefficient terms for the MRRG is one order higher than those for the HRG. So the instability problem does not occur for MRRG. The nonlinear terms in (23) reveal the coupling effects. If these terms are dropped, they become

$$\gamma_2=\frac{4n\omega\Omega_z}{\sqrt{a_2^2+\xi_1^2\omega^4}},\ \gamma_3=-\frac{(n-1)\Omega_y}{n\xi_2\omega},\ \gamma_4=\frac{(n-1)\Omega_x}{n\xi_2\omega}. \tag{24}$$

It can be seen from (24) that the applied voltage and the electrode gap affect the measurement of Ω_2, and Ω_i are parameter-dependent, since ξ_1 and ξ_2 contain the geometrical size. If the displacement of X_1 is about $10^{-6}\mathrm{m}$, then the value of X_2 is around $10^{-8}\mathrm{m}$, the corresponding capacitance change is about $10^{-14} \sim 10^{-15}\mathrm{F}$. In order to measure this so small capacitance by a displacement-capacitor sensor, the measuring circuit and the vibrating ring must be made in the same wafer by CMOS processing.

4. CONCLUSION

Explicit solutions for the modal precession of HRG is obtained by solving the nonlinear equations of motion, while the displacement at the nodal point of MRRG are obtained by solving the linear ones. Although MRRG does not have the instability problem occurring in HRG, it has the problem of how to handle small signals.

REFERENCES

1. Loper, E. and Lych, D.D., The HRG: A new low noise inertial rate sensor, in: *Proceedings of the 16th JT Services Data Exchange for Inertial Systems*, Los Angles, 1982, pp. 16–18.
2. Matthews, A. and Varty, G.T., *Method for reading out a vibratory rotation sensor*, U.S. Patent 5,902,930, May 1999.
3. Burdess, J.S., Electrostatic Excitation of High Q Axisymmetric Shell Resonator, *J. Sound & Vibr.* **109**(3), 1986.
4. Niordson, F.I., *Shell Theory*, North-Holland, 1985.
5. Putty, W., *Microstructure for vibratory gyroscope*, U.S. Patent No. 5450751, 1995.
6. Hopkin, I.D., The performance and design of a silicon micromachined gyro, in *Symposium Gyro Technology*, Stuggart, Germany, 1997.
7. McNie, M.E., *Micro-machining of ring angular rate gyro*, Patent No. US6276205, 2001.
8. Fell, C.P., *Angular rate sensor*, Patent No. US 6282958, 2001.
9. Juneau et al., Dual axis operation of a micromachined rate gyroscope, in *Technical Digest of the 9th International Conference on Solid State Sensor & Actuator*, 1997, pp. 883–886.
10. Gallacher, B.J. et al., Principals of a three-axis vibrating gyroscope, *IEEE Transactions on Aerospace and Electronic Systems* **37**(4), 2001, 1333–1343.
11. Pietraszkiewicz, W., Lagrangian description and incremental formulation, *J. Non-Linear Mech.* **19**(2), 1983, 115–140.
12. Wiggins, S., *Introduction to Applied Nonlinear Dynamical Systems and Chaos*, Springer, 1990.
13. Kuznetsov, Y.A., *Elements of Applied Bifurcation Theory*, 2nd edition, Springer, 1998.
14. Kim, W. and Chung, J., Free non-linear vibration of a rotating thin ring with the in-plane and out-of-plane motions, *J. Sound & Vibr.* **258**(1), 2002, 167–178.

EXPERIMENTAL STUDY OF SNAKE-LIKE LOCOMOTION OF A THREE-LINK MECHANISM

Felix L. Chernousko[1], Friedrich Pfeiffer[2] and Nikolai A. Sobolev[1]
[1]*Institute for Problems in Mechanics of the Russian Academy of Sciences, Moscow, Russia;*
[2]*Technical University of Munich, Garching, Germany*

Abstract: Snake-like locomotion of a three-link mechanism equipped with actuators is considered. The results of experimental investigations are presented and compared with those prescribed by theory.

Key words: multibody system, mobile robot, dry friction, locomotion.

1. INTRODUCTION

The crawling motions of snakes and other limbless animals have always been of great interest for specialists in mechanics and biomechanics. Unlike walking and running creatures who alternate their supporting legs, snakes mostly keep permanent contact between their bodies and the ground. Although the friction force acting upon each moving segment of the body is directed against the velocity of the segment, the resultant of the friction forces, i.e., the thrust, should be directed along the velocity of the center of mass of the body. To explain this phenomenon, various models of snake-like locomotion have been proposed.

The motion of a snake inside a curved tube has been considered in [1]. It was shown that the required thrust can be created by the normal reactions of the tube. Locomotion of snakes have been described and classified into three classes in [2]. One of them, a rectilinear motion, requires the displacement of the snake's mass along its body, whereas motions of the other two classes (in which the snake twists its body) are possible only in the presence of vertical walls or other vertical of inclined objects. By pressing its body against these objects such as stones, grass, sand slopes, etc., the snake creates horizontal components of reactions along the direction of motion. Snakes always try to use vertical or inclined obstacles and avoid flat surfaces. Biomechanical aspects of the snake-like locomotion have been discussed in [3]. Mechanisms

H. Ulbrich and W. Günthner (eds), IUTAM Symposium on Vibration Control of Nonlinear Mechanisms and Structures, 141–150.

of the snake-like locomotion in the presence of obstacles have been considered in [4, 5].

Nonholonomic multilink snake-like robots have been designed and investigated in [6, 7]. These mechanisms consist of many elements equipped with passive wheels and connected by joints. By twisting, such robots can perform snake-like locomotion. In these motions, the wheels exert forces directed along their axes and thus produce the desired forces in the direction of motion. In fact, the wheels here play the role similar to that of vertical walls for snakes. Kinematics of snake-like locomotion of such nonholonomic mechanisms has been considered in [8–10].

In this paper, we consider a three-link mechanism, which can move along a horizontal plane in the presence of dry friction between the linkage and the plane. Particular attention is given to the experimental hardware implementation of a three-link snake-like mobile mechanism (robot).

2. THREE-LINK MECHANISM

2.1 Mechanical Model

Consider first a plane three-member linkage $O_1C_1C_2O_2$ moving along a fixed horizontal plane Oxy (Figure 1). For the sake of simplicity, we assume that the links O_1C_1, C_1C_2, and C_2O_2 are rigid massless bars, and the entire mass of the linkage is concentrated at the end points O_1, O_2 and joints C_1, C_2 which can slide along the plane. The mass of each of the end points is denoted by m_0, and the mass of each of the joints is denoted by m_1. Thus, the total mass of the linkage is $m = 2(m_0 + m_1)$. The length of the central link C_1C_2 is denoted by $2a$, and the lengths of the end links are denoted to l.

Denote by x, y the Cartesian coordinates of the middle of the central link, by θ the angle between this link and the x-axes, and by α_i the angles between the central link C_1C_2 and the end links O_iC_i, $i = 1, 2$ (Figure 1).

We assume that the dry friction forces acting between the points O_i, C_i, $i = 1, 2$, and the horizontal plane obey Coulomb's law. If a point mass m_0 (or m_1) moves, the friction force is directed opposite to the point velocity and equal to its weight m_0g (or m_1g) multiplied by the friction coefficient k_0 (or k_1). If the point mass m_0 (or m_1) is at rest, the friction force does not exceed m_0gk_0 (or m_1gk_1), and its direction can be arbitrary.

The control torques M_1 and M_2 about the vertical axes are created by actuators installed at the joints C_1 and C_2. We assume that these torques can provide a prescribed time-history for the angles $\alpha_i(t)$, $i = 1, 2$.

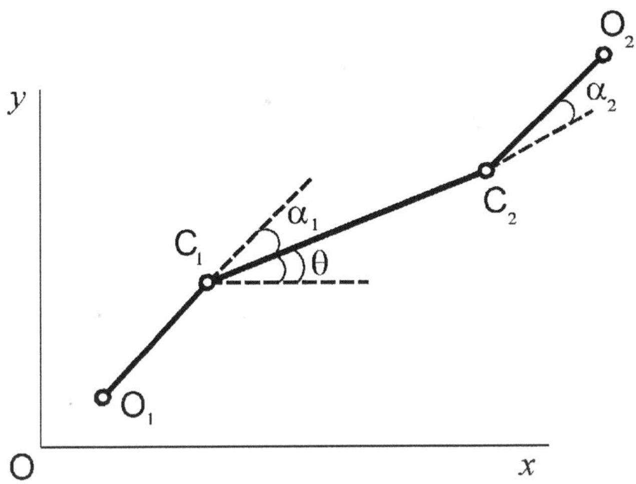

Figure 1. Three-member linkage.

2.2 Elementary Motions

We will construct various motions of the linkage as a combination of more simple motions, which we call elementary motions [11]. All elementary motions begin and end at states of rest of the linkage. In each elementary motion, the angles $\alpha_i(t)$, $i = 1, 2$, change within the interval $(-\pi, \pi)$ between the prescribed initial value α_i^0 and terminal value α_i^1; the time-histories $\alpha_i(t)$ can be more or less arbitrary. Either one or both angles α_1, α_2 can change in the elementary motion. In the latter case, they must change synchronously so that $\dot{\alpha}_2(t) = \pm\dot{\alpha}_1(t)$. Thus, the end links can rotate either in the same direction, or in the opposite directions.

Elementary motions are divided into slow and fast ones.

In *slow* motions, the values of the angular velocity $\omega(t) = |\dot{\alpha}_i(t)|$ and angular acceleration $\epsilon(t) = \dot{\omega}(t)$ are small enough, so that the central link stays at rest during these motions.

If both end links rotate in the same direction, then the central link stays at rest if [11–13]

$$m_0 l \left\{ \left[\omega_0^4 + \left(\epsilon_0 + g k_0 l^{-1}\right)^2 \right]^{1/2} + \left(\epsilon_0 + g k_0 l^{-1}\right) l a^{-1} \right\} \leq m_1 g k_1, \qquad (1)$$
$$\omega_0 = \max|\dot{\alpha}_i(t)|, \quad \epsilon_0 = \max|\ddot{\alpha}_i(t)|.$$

This condition remains valid for the case where only one end link rotates during the slow motion.

Condition (1) holds, if ω_0 and ϵ_0 are small enough and $m_0 k_0 (a + l) < m_1 k_1 a$. The last inequality can be easily satisfied by an appropriate choice of the lengths a and l and masses m_0 and m_1.

If the end links rotate in the opposite directions, condition (1) can be replaced by a weaker one [12, 13]

$$m_0 l \left[\omega_0^4 + \left(\epsilon_0 + g k_0 l^{-1}\right)^2\right]^{1/2} \le m_1 g k_1. \tag{2}$$

This condition holds, if ω_0 and ϵ_0 are small enough and $m_0 k_0 < m_1 k_1$.

In *fast* motions, the angular velocities and accelerations of the end links are sufficiently high, and the duration τ of this motion is much less than the duration T of the slow motion: $\tau \ll T$. The magnitudes of the control torques M_1 and M_2 during the fast motion are high compared with the torques due to friction forces

$$\begin{aligned} |M_i| \gg m^* g k^* l^*, \quad i = 1, 2, \quad m^* = \max(m_0, m_1), \\ k^* = \max(k_0, k_1), \quad l^* = \max(a, l). \end{aligned} \tag{3}$$

Hence, the friction can be neglected during the fast motion, and the conservation laws for the momentum and angular momentum hold in this motion. Therefore, the center of mass C of the linkage stays fixed, and its angular momentum is zero during the fast motion. Using these conservation laws, one can evaluate the terminal state of the linkage after the fast motion.

2.3 Periodic Locomotion

We will show how longitudinal, lateral, and rotational motions of the three-member linkage can be composed from elementary ones. Suppose that at the initial instant of time the linkage is at rest, and all its links are parallel to the x-axis. We have $\theta = \alpha_1 = \alpha_2 = 0$ in this state.

We denote slow and fast motions by the capital letters S and F, respectively. Let us indicate the initial and terminal values of the angles α_1, $i = 1, 2$, in each elementary motion by the respective superscripts ^{0} and ^{1}. Thus, we will describe the limits between which the angle α_i changes in an elementary motion as follows α_i: $\alpha_i^0 \to \alpha_i^1$, $i = 1, 2$.

Consider a sequence of elementary motions that results in the displacement of the robot by some distance along itself. First, the auxiliary slow motion is carried out: $S, \alpha_1 : 0 \to \gamma, \alpha_2(t) \equiv 0$ where $\gamma \in (-\pi, \pi)$ is a fixed angle. Then, the following four motions are performed:

$$\begin{aligned} &1) F, \alpha_1 : \gamma \to 0, \alpha_2 : 0 \to \gamma; \qquad &&2) S, \alpha_1 : 0 \to -\gamma, \alpha_2 : \gamma \to 0; \\ &3) F, \alpha_1 : -\gamma \to 0, \alpha_2 : 0 \to -\gamma; \qquad &&4) S, \alpha_1 : 0 \to \gamma, \alpha_2 : -\gamma \to 0. \end{aligned}$$

After stage 4, the configuration of the linkage coincides with its configuration before stage 1, so that stages 1–4 can be repeated any number of times. Thus, we obtain a periodic longitudinal motion, with each period (cycle) consisting of two fast and two slow phases. To return the linkage to its initial rectilinear configuration $\alpha_1 = \alpha_2 = 0$, one should perform the motion $S, \alpha_1 : \gamma \to 0, \alpha_2 \equiv 0$.

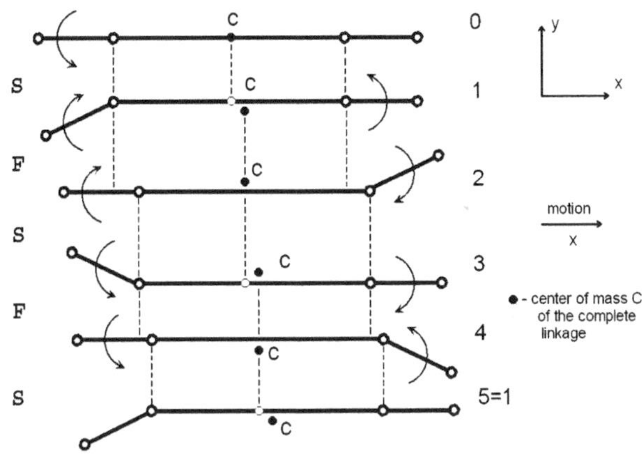

Figure 2. Longitudinal motion.

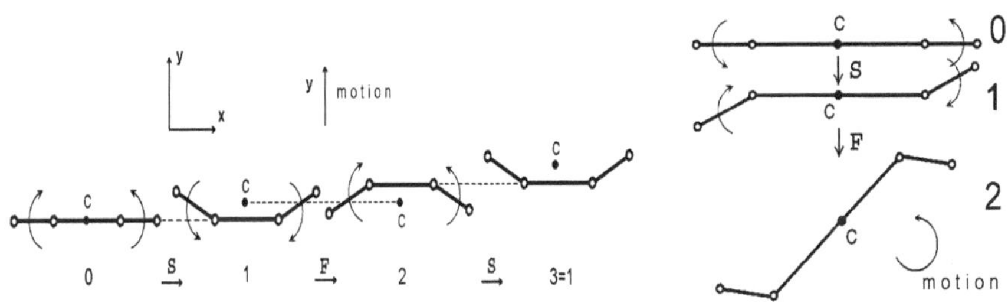

Figure 3. Lateral motion and rotation.

The displacement of the mechanism during one cycle and the average speed of the longitudinal motion are expressed by [11, 13]

$$\Delta x = 8m_0m^{-1}l\sin^2(\gamma/2), \quad v_1 = \Delta x(2T)^{-1} \tag{4}$$

By alternating fast and slow motions, one can also implement a lateral motion and a rotation of the robot. The sequences of elementary motions for the lateral motion and rotation are shown in Figure 3.

For the lateral motion mode, the distance travelled by the robot in one cycle and the average speed are given by [13]:

$$\Delta y = 4m_0m^{-1}l\sin\gamma, \quad v_2 = \Delta yT^{-1}. \tag{5}$$

For the rotation mode, the angle rotated by the robot in one cycle is given by [13]

$$\Delta\theta = \gamma + 2\frac{[m_0l^2 - (m_0+m_1)a^2]}{R}\arctan\left(\frac{R\tan(\gamma/2)}{m_0(a+l)^2 + m_1a^2}\right), \tag{6}$$

$$R = \left[m_0^2(l^2-a^2)^2 + m_1a^2(2m_0l^2 + 2m_0a^2 + m_1a^2)\right]^{1/2}.$$

Figure 4. The design of the three-link robot.

The average angular velocity of the rotation is defined as $\omega = \Delta\theta / T$.

Combining these three types of motions, the linkage can move from any initial position and configuration in the horizontal plane to any prescribed terminal position and configuration in this plane.

3. EXPERIMENTAL STUDY OF THE THREE-LINK MECHANISM

To validate the theory outlined above, an experimental model of the three-link mechanism (snake-like robot) was created. The design of this model was developed at the Munich Technical University and then modified at the Institute for Problems in Mechanics of the Russian Academy of Sciences. The sketch of the current design is shown in Figure 4.

The robot has three links — the central link and two end links. The central link consists of two active (actuated) joints connected by a pair of aluminum tubes. Each end link consists of a pair of aluminum tubes, two passive joints, and an end body (end mass). The active joints connect the central link to the end ones. Each joint of the central link is equipped with an electric actuator. To the active joints and the end bodies of the end links rigid plates are attached by means of which the robot contacts the supporting surface. To the bottoms of the contact plates, soft pads have been stuck.

Some structural features of the design are worth mentioning. The contact plates attached to the joints are connected to the rotating parts of the joints by means of bearings, which enables these plates to stay fixed during the slow motions in spite of the rotation of movable parts of the joints. This is necessary for the central link to stay fixed during the slow motion. The two passive joints of each end link provide two additional degrees of freedom for the end link to move in a vertical plane. This enables the robot to adapt to irregularities of the

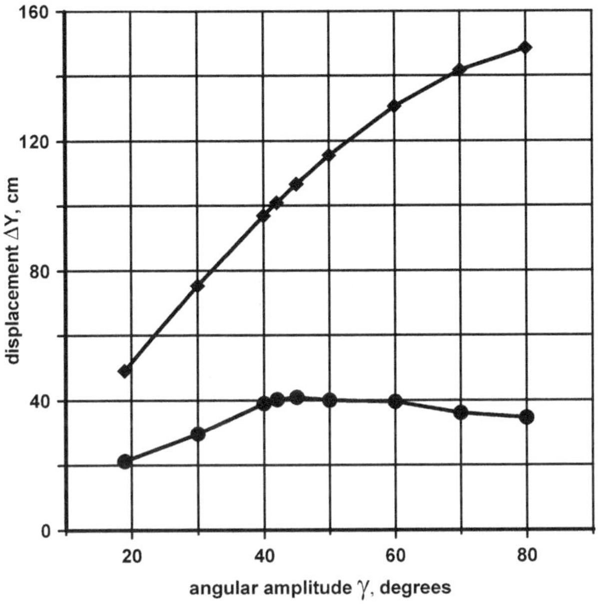

Figure 5. Theoretical and experimental curves for lateral motion.

surface along which it moves. The soft pads stuck to the bottoms of the contact plates are also needed for the adaptation to the surface irregularities.

A number of experiments with the snake-like robot performing longitudinal, lateral, and rotational motions were carried out. The major purpose of the experimental study was to verify the theoretical results, in particular, relations (4)–(6). For each motion mode, the parameters of the control law were chosen so as to make the central link stay fixed during the slow motions for all values of the angular amplitudes γ.

Lateral motion. The distance Δy travelled per cycle was investigated as a function of the angular amplitude γ of rotation of the end links (see (5)). For one measurement, 10 cycles were performed. For each value of γ, five measurements were carried out. The displacement $\Delta Y = 10\Delta y$ was calculated as the mean of the results of these five measurements. The results of these experiments are shown in Figure 5.

The curves marked with circles and squares correspond to the experimental and theoretical data, respectively. The mass and geometrical parameters of the robot utilized for the experiments related to the lateral motion were taken as follows: $m_0 = 820$ g, $m = 7070$ g, and $l = 32.5$ cm.

Longitudinal motion. Similar experiments were carried out for the longitudinal motion mode (see (4)). The results of these experiments are shown in Figure 6. As was the case for the lateral motion, we measured $\Delta X = 10\Delta x$, where

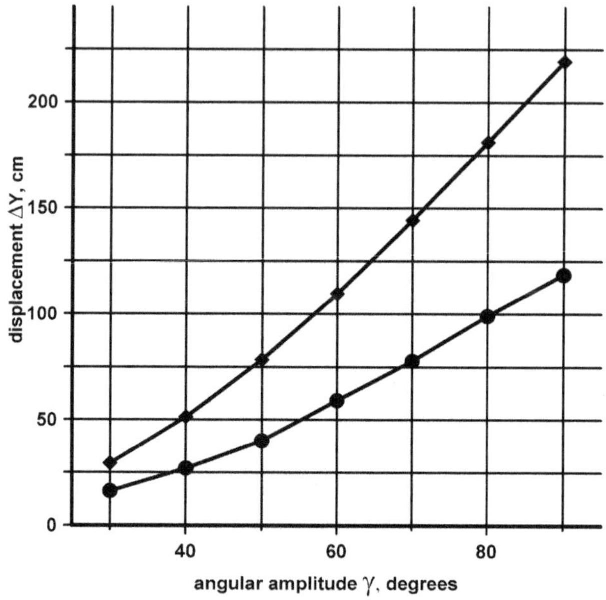

Figure 6. Theoretical and experimental curves for longitudinal motion.

Δx is the distance travelled by the robot during one cycle. The curve marked with circles is the experimental curve and the curve marked with squares is the theoretical one. The parameters of the robot utilized for the experiments with the longitudinal motion were taken as follows: $m_0 = 1220$ g, $m = 8010$ g, and $l = 36$ cm.

Rotation. The theoretical (squares, Eq. (6)) and experimental (circles) relations between the angular displacement $\Delta\Theta = 10\Delta\theta$, where $\Delta\theta$ is angle of rotation of the robot per one cycle, and angular amplitude γ are shown in Figure 7. For the rotational motion mode we utilized the robot with the following parameters: $m_0 = 820$ g, $m = 3715$ g, and $l = 32.4$ cm.

The experiments confirmed the applicability of the locomotion concept described in Section 2 to mobile mechanisms and validated major qualitative features of the motion predicted by the theory. However, Figures 5–7 demonstrate a significant quantitative disagreement between the experimental and theoretical data, especially for large γ. This can be accounted for by the fact that in the theory, the durations of the fast phases of the motion were assumed infinitely small and the action of friction at the fast phases was not taken into account. In the experiments, however, the ratio of the driving torques to the friction torques was not high enough to justify this hypothesis, and the effect of friction on the motion of the robot was noticeable. Hence, condition (3) was not satisfied as a strong inequality ($M \sim m^* g k^* l^*$ in some experiments).

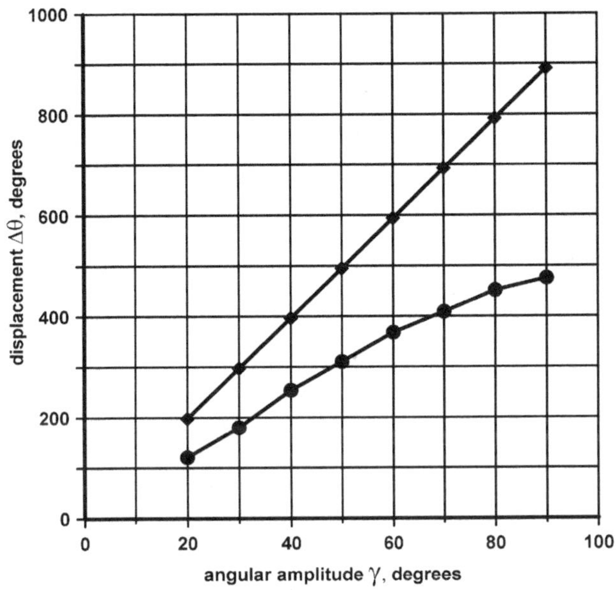

Figure 7. Theoretical and experimental curves for rotational motion.

The duration of the fast phases increases as the amplitude γ of rotations of the end links increases. This accounts for the increase in the departure of the experimental data from the theoretical ones as γ increases, as is apparent from Figures 5-7. To enable the theory to be suitable for more precise quantitative calculations, it is necessary to take into account the action of friction at the fast phases of the motion and the finite duration of these phases.

4. CONCLUSIONS

A physical model of a snake-like robot was created and experiments with this model were carried out. The experiments validated the concept of locomotion of snake-like mechanisms along rough surfaces by means of changing their configuration. This locomotion does not require wheels, legs, or protruding objects against which the mechanism could lean. The experiments confirmed qualitative features of the motion predicted by the theory but showed noticeable quantitative discrepancies. To reduce these discrepancies, the theory is being refined which takes into account the friction effect during the fast phases of the motion and the finite duration of these phases.

ACKNOWLEDGMENTS

The authors are grateful to M. Gienger and G. Mayr who took part in the design of the first experimental model, and also to K. Loeffler and K. Sorokin for the participation in experiments.

This research was supported by the Russian Foundation for Basic Research (Grants Nos. 04-01-04002 and 04-01-08046) and by the Grant for Russian Scientific Schools (NSh 1627.2003.1).

REFERENCES

Lavrentyev M.A. and Lavrentyev M.M. (1962) On one principle of creating the thrust force in motion, *Journal of Applied Mechanics and Technical Physics* (4), 6–9.

Gray J. (1968) *Animal Locomotion*, Weidenfeld & Nicolson, London.

Dobrolyubov A.I. (1987) *Travelling Waves of Deformation*, Nauka i Tekhnika, Minsk.

Bayraktaroglu Z.Y., Butel F., Pasqui V. and Blazevic P. (1999) A geometrical approach to the trajectory planning of a snake-like robot, in G.C. Virk, M. Randall, and D. Howard (eds.), *Proceedings of the Second International Conference on Climbing and Walking Robots CLAWAR 1999*, Portsmouth, pp. 851–856.

Bayraktaroglu Z.Y. and Blazevic P. (2000) Snake-like locomotion with a minimal mechanism, in M. Armada and P. Gonzales de Santos (eds.), *Proceedings of the Third International Conference on Climbing and Walking Robots CLAWAR 2000*, Madrid, pp. 201–207.

Hirose S. and Morishima A. (1990) Design and control of a mobile robot with an articulated body, *International Journal of Robotics Research* **9**(2), 99–114.

Hirose S. (1993) *Biologically Inspired Robots: Snake-like Locomotors and Manipulators*, Oxford University Press, Oxford.

Chirikjan G.S. and Burdick J.W. (1991) Kinematics of hyper-redundant robot locomotion with applications to grasping, in *Proceedings of the IEEE International Conference on Robotics and Automation*, Vol. 1, Sacramento, pp. 720–725.

Burdick J.W., Radford J. and Chirikjan G.S. (1993) A 'sidewinding' locomotion gait for hyper-redundant robots, in *Proceedings of the IEEE International Conference on Robotics and Automation* Vol. 3, Atlanta, pp. 101–106.

Ostrowski J. and Burdick J. (1996) Gait kinematics for a serpentine robot, in *Proceedings of the IEEE International Conference on Robotics and Automation*, Vol. 2, Minneapolis, pp. 1294–1300.

Chernousko F.L. (2000) The motion of a multilink system along a horizontal plane, *Journal of Applied Mathematics and Mechanics* **64**(1), 5–15.

Chernousko F.L. (2001) On the motion of a three-member linkage along a plane, *Journal of Applied Mathematics and Mechanics* **65**(1), 13–18.

Chernousko F.L. (2003) Snake-like locomotions of multilink mechanisms, *Journal of Vibration and Control* **9**(1–2), 235–256.

A PENALTY SHOOTING WALKING MACHINE

Dynamics and Control

Hubert Gattringer and Hartmut Bremer
Institute for Robotics, University of Linz, Austria
hubert.gattringer@jku.at; hartmut.bremer@jku.at

Abstract: In this paper, we consider the dynamical behavior and the control of a two legged Walking Machine. Special investigation is done in the evaluation of the equations of motion. Thus, for obtaining the accelerations of such a tree structured multi-body system, an O(n) algorithm is presented, enhancing the numerical stability. To fulfill the aim of modeling a gaitcycle, contact is opened and closed periodically. Therefore the O(n) algorithm is enhanced for this case. A controller is shown, that balances the robot when doing such highly dynamically processes like shooting a soccer penalty. Simulations and experimental results are presented.

Key words: two legged Walking Machine, Zero Moment Point, O(n) algorithm, unilateral contact.

INTRODUCTION

In the recent years Walking Machine design has become a more and more popular field of research. The main reason for this is, that the computational power to implement highly sophisticated control concepts has increased rapidly. Also the development of power drives, gears and sensors has progressed. Walking Machines have their advantage compared to wheel driven vehicles in their flexibility of acting in uncertain areas, see e.g. Hiller et al. (2004), where a four legged walker is proposed to achieve heavy forest work. There exist some highly developed two legged machines, for example Hirai et al. (1998) and Löffler et al. (2003). These robots have the same main degrees of freedom (DOF) like human beings. The here proposed machine has the same DOF for the legs, its mobility is thus the same. Much work is done in the efficient simulation of the equations of motion. An O(n) algorithm, using the tree structure of this multi-body system (MBS) has been implemented. The idea of solving the equations of motion recursively was first presented in Brandl et al. (1986).

H. Ulbrich and W. Günthner (eds), IUTAM Symposium on Vibration Control of Nonlinear Mechanisms and Structures, 151–159.

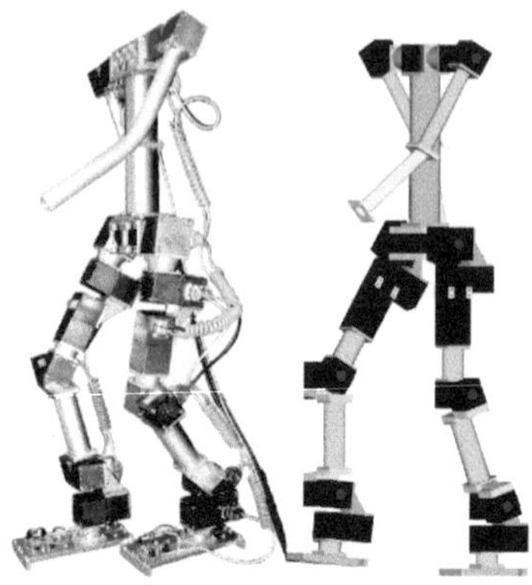

Figure 1. Walking Machine.

Unlike, here a very short and clear algorithm, obtained from the projection equation, see Bremer (2003), is presented. A main advantage of our algorithm is, that it makes no difference whether rigid or elastic bodies are under consideration, see Mitterhuber et al. (2004). In literature (Pfeiffer et al., 2000; Löffler et al., 2001), the contact problem is often solved as a Linear Complementary Problem. Here the Jacobi equation is used to obtain the contact forces which fit the O(n) algorithm requirements. The position control of the joint coordinates is realized with a feed-forward controller, that can be calculated very efficient, by again using the structure of the system, and a passive PD controller. There exist a lot of control strategies that use the Zero Moment Point(ZMP), which was first introduced by Vucobratovic et al. (1990), to balance the robot, see again Löffler et al. (2003) and Caballero et al. (2004). In this work, the ZMP is used to change the upper body orientation of the Walking Machine like humans do. Vucobratovic et al. (2004) gives a good summary about the ZMP. For the trajectory generation a simple Inverted Pendulum Mode (IPM) and a Two Masses Inverted Pendulum Mode (TMIPM) (Amos et al., 2003) are implemented. Shooting a soccer penalty is also done in this sense.

1. THE WALKING MACHINE

The two legged walking machine as shown in Figure 1 is under consideration. The robot consists of 14 joints, six for each leg and two for the arms. It is driven by 14 Amtec powercubes, where dc-motors, harmonic drive gears, microcontrollers and the power electronics are combined in one. These communicate with a dSpace1005 realtime control unit via CAN, and via RS232 for the evaluation of the special designed force moment sensors in the ankle of the robot, respectively. The sensors are used to evaluate the ZMP. The main control unit runs with a sampling time of 2 ms. For attaining as high flexibility as possible, a modular robotic system is chosen for the construction. Therefore the robot's configuration can be changed within minutes, concerning hardware as well as the corresponding software algorithms. It has a height of 1.8 m and a weight of 40 kg.

2. KINEMATIC MODEL

For the kinematic description of the robot, a flexible joint model is used. Therefore the relative angles between two links γ_A and the appropriate motor angle γ_M, see Figure 2 are chosen for minimal coordinates. The system has tree structure, so it makes sense, to formulate the kinematics in the form of a kinematical chain

$$\dot{\mathbf{y}}_n = \mathbf{T}_{np}\dot{\mathbf{y}}_p + \mathbf{F}_n\dot{\mathbf{s}}_n, \tag{1}$$

where

$$\mathbf{T}_{np} = \begin{bmatrix} \mathbf{A}_{np} & \mathbf{A}_{np}\ {}_p\tilde{\mathbf{r}}_{np}^T & \mathbf{0} & \mathbf{A}_{np}\ {}_p\tilde{\mathbf{r}}_{np}^T\ \mathbf{e}_R \\ \mathbf{0} & \mathbf{A}_{np} & \mathbf{0} & \mathbf{A}_{np}\mathbf{e}_R \\ \mathbf{0} & \mathbf{0} & 0 & 0 \\ \mathbf{0} & \mathbf{0} & 0 & 0 \end{bmatrix} \in \mathbb{R}^{8,8} \tag{2}$$

and

$$\mathbf{F}_n\dot{\mathbf{s}}_n = \begin{bmatrix} \mathbf{0} & \mathbf{0} \\ \mathbf{0} & \mathbf{0} \\ i_G & 0 \\ 0 & 1 \end{bmatrix} \begin{pmatrix} \dot{\gamma}_M \\ \dot{\gamma}_A \end{pmatrix}; \quad \mathbf{F}_n \in \mathbb{R}^{8,f_n}, \tag{3}$$

$f_n = 2$. Here, the describing velocities

$$\dot{\mathbf{y}}_n = [\mathbf{v_0}\ \boldsymbol{\omega}_0\ \Omega_M\ \Omega_A]^T \quad \in \mathbb{R}^8, \tag{4}$$

where $\mathbf{v_0}$, $\boldsymbol{\omega}_0$ are the "guiding velocities" and Ω_M, Ω_A are the relative motor and arm velocities, are suited for the proposed Walking Machine, see figure 2. To use this form for all bodies, the rotation axis $\mathbf{e}_R$ for each joint is defined. $\mathbf{A}_{np}$ is the corresponding rotation matrix, $\mathbf{r}_{np}$ is the connection vector and i_G specifies the transmission ratio of the harmonic drive gear. In this context, the describing velocity of body n is equal to the describing velocity of the predecessor p converted to the coordinate system of body n (via $\mathbf{T}_{np}$) with superimposed relative velocity $\mathbf{F}_n\ \dot{\mathbf{s}}_n$. For the evaluation of the predecessors, the topology vector of the system is used. The whole kinematical chain in matrix form can be expressed as

$$\begin{pmatrix} \dot{\mathbf{y}}_1 \\ \dot{\mathbf{y}}_2 \\ \\ \dot{\mathbf{y}}_N \end{pmatrix} = \underbrace{\begin{bmatrix} \mathbf{F}_1 & & & \\ \mathbf{T}_{21}\mathbf{F}_1 & \mathbf{F}_2 & & \\ \vdots & & \ddots & \\ \mathbf{T}_{N1}\mathbf{F}_1 & .. & .. & \mathbf{F}_N \end{bmatrix}}_{\mathbf{F}} \underbrace{\begin{pmatrix} \dot{\mathbf{s}}_1 \\ \dot{\mathbf{s}}_2 \\ \\ \dot{\mathbf{s}}_N \end{pmatrix}}_{\dot{\mathbf{s}}}. \tag{5}$$

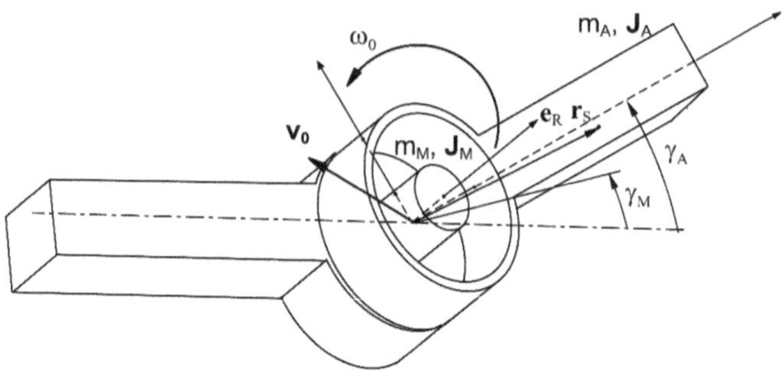

Figure 2. Subsystem.

This will be essential for formulating the O(n) algorithm.

Note that $\mathbf{T}_{ij} = \mathbf{T}_{i,p(i)}\text{x}...\mathbf{T}_{s(j),j}$ (p: predecessor, s: successor) is obtained recursively, without calculation as a whole.

3. DYNAMIC MODEL

3.1 Dynamic without Contact

For an efficient simulation of the whole system, an O(n) algorithm is derived. Starting point is the projection equation in terms of subsystems,

$$\sum_{n=1}^{N}\left(\frac{\partial \dot{\mathbf{y}}_n}{\partial \dot{\mathbf{s}}}\right)^T\left[\mathbf{M}_n\ddot{\mathbf{y}}_n + \mathbf{G}_n\dot{\mathbf{y}}_n - \mathbf{Q}_n\right] = \mathbf{0}, \tag{6}$$

or in matrix form

$$\begin{bmatrix} \mathbf{F}_1^T & \mathbf{F}_1^T\mathbf{T}_{21}^T & \dots & \mathbf{F}_1^T\mathbf{T}_{N,1}^T \\ & \mathbf{F}_2^T & & \mathbf{F}_2^T\mathbf{T}_{N,2}^T \\ & & \ddots & \vdots \\ & & & \mathbf{F}_N^T \end{bmatrix} \begin{pmatrix} \mathbf{M}_1\ddot{\mathbf{y}}_1 + \mathbf{G}_1\dot{\mathbf{y}}_1 - \mathbf{Q}_1 \\ \mathbf{M}_2\ddot{\mathbf{y}}_2 + \mathbf{G}_2\dot{\mathbf{y}}_2 - \mathbf{Q}_2 \\ \vdots \\ \mathbf{M}_N\ddot{\mathbf{y}}_N + \mathbf{G}_N\dot{\mathbf{y}}_N - \mathbf{Q}_N \end{pmatrix} = \mathbf{0}. \tag{7}$$

$\mathbf{M}_n$, $\mathbf{G}_n$, $\mathbf{Q}_n$ are the corresponding subsystem-matrices related to the describing velocities $\dot{\mathbf{y}}_n$. Exemplarily, the mass matrix for this case (with the describing velocities of Eq. (4)) reads

$$\mathbf{M}_n = \begin{pmatrix} m\mathbf{E} & m\tilde{\mathbf{r}}_s^T & \mathbf{0} & m\,\tilde{\mathbf{r}}_s^T\mathbf{e}_R \\ m\tilde{\mathbf{r}}_s & \mathbf{J}_A^0 + \mathbf{J}_M & \mathbf{J}_M\mathbf{e}_R & \mathbf{J}_A^0\mathbf{e}_R \\ \mathbf{0} & \mathbf{e}_R^T\mathbf{J}_M & \mathbf{e}_R^T\mathbf{J}_M\mathbf{e}_R & 0 \\ m\,\mathbf{e}_R^T\tilde{\mathbf{r}}_s & \mathbf{e}_R^T\mathbf{J}_A^0 & 0 & \mathbf{e}_R^T\mathbf{J}_A^0\mathbf{e}_R \end{pmatrix}_n \in \mathbb{R}^{8,8}, \tag{8}$$

where m is the mass, $\mathbf{J}_A$, $\mathbf{J}_M$ are the inertia tensors of arm and motor. The equation of motion can be calculated in minimal configuration. However, with the help of the kinematic chain, the global functional matrix $\mathbf{F}$ which is built with the Jacobians $\partial\dot{\mathbf{y}}_\mathbf{n}/\partial\dot{\mathbf{s}}$ has a lower triangular structure, see Eq. (5). Therefore the accelerations $\ddot{\mathbf{s}}$ can be calculated like a Gauss elemination process. From the last block Eqs. (7)

$$\mathbf{F}_N^T \ (\mathbf{M}_N\ddot{\mathbf{y}}_N + \mathbf{G}_N\dot{\mathbf{y}}_N - \mathbf{Q}_N) = \mathbf{0}, \tag{9}$$

the accelerations

$$\begin{aligned}\ddot{\mathbf{s}}_N &= -\left[\mathbf{F}_N^T\mathbf{M}_N\mathbf{F}_N\right]^{-1} \\ &\quad \left\{\mathbf{M}_N\left(\mathbf{T}_{N,N-1}\ddot{\mathbf{y}}_{N-1} + \dot{\mathbf{T}}_{N,N-1}\dot{\mathbf{y}}_{N-1}\right) + \mathbf{G}_N\dot{\mathbf{y}}_N - \mathbf{Q}_N\right\}\end{aligned} \tag{10}$$

can be computed, but they contain the unknown accelerations of the predecessors via $\ddot{\mathbf{y}}_{N-1}$. By substituting $\ddot{\mathbf{s}}_N$ into the prelast block Eqs. (7), and by rearranging terms, we obtain

$$\mathbf{F}_{N-1}^T \left(\mathbf{M}_{N-1}^*\ddot{\mathbf{y}}_{N-1} + \mathbf{G}_{N-1}^*\dot{\mathbf{y}}_{N-1} - \mathbf{Q}_{N-1}^*\right) = \mathbf{0}. \tag{11}$$

This procedure can be repeated till the first body. Body 1 has no predecessor, therefore the accelerations $\ddot{\mathbf{s}}_1$ can be evaluated. Thus, with known $\ddot{\mathbf{s}}_1$ all other accelerations can be computed in a forward step. An O(n) algorithm can be formulated in three steps.

1st Step: Calculate $\mathbf{T}_{ip}$, $\dot{\mathbf{T}}_{ip}$ (forward step).

2nd Step: Calulate the modified system matrices (backward step)

$$\mathbf{M}_p^* = \mathbf{M}_p + \sum_{i\in\{s_j(p)\}} \mathbf{T}_{ip}^{*T}\mathbf{M}_i^*\mathbf{T}_{ip} \tag{12}$$

$$\mathbf{G}_p^* = \mathbf{G}_p + \sum_{i\in\{s_j(p)\}} \mathbf{T}_{ip}^{*T}\left(\mathbf{G}_i^*\mathbf{T}_{ip} + \mathbf{M}_i^*\dot{\mathbf{T}}_{ip}\right) \tag{13}$$

$$\mathbf{Q}_p^* = \mathbf{Q}_p + \sum_{i\in\{s_j(p)\}} \mathbf{T}_{ip}^{*T}\left(\mathbf{Q}_i^* - \mathbf{G}_i^*\mathbf{F}_i\dot{\mathbf{s}}_i\right), \tag{14}$$

where $s_j(p)$ are the successors of p, known from the topology matrix, and the abbreviations read

$$\begin{aligned}\mathbf{M}_{Ri} &:= \left[\mathbf{F}_i^T\mathbf{M}_i^*\mathbf{F}_i\right], & \mathbf{N}_i &:= \left[\mathbf{E} - \mathbf{M}_i^*\mathbf{J}_i\right], \\ \mathbf{T}_{ip}^* &:= \mathbf{N}_i^T\mathbf{T}_{ip}, & \mathbf{J}_i &:= \mathbf{F}_i\mathbf{M}_{Ri}^{-1}\mathbf{F}_i^T.\end{aligned}$$

3rd Step: Calculate $\ddot{\mathbf{s}}_i$ (forward step)

$$\ddot{\mathbf{s}}_i = -\mathbf{M}_{Ri}^{-1}\mathbf{F}_i^T\left\{\mathbf{M}_i^*\left(\mathbf{T}_{ip}\ddot{\mathbf{y}}_p + \dot{\mathbf{T}}_{ip}\dot{\mathbf{y}}_p\right) + \mathbf{G}_i^*\dot{\mathbf{y}}_i - \mathbf{Q}_i^*\right\}. \tag{15}$$

Note that only a matrix of rank f_n has to be inverted which corresponds to the n-th subsystem dimension ($f_n = 2$ in the present case of the Walking Machine).

3.2 Contact

During walking, ground contact is closed and opened periodically. This can be achieved by adding additional forces $\boldsymbol{\lambda}$ to the minimal representation $\mathbf{M}\ddot{\mathbf{s}} + \mathbf{G}\dot{\mathbf{s}} - \mathbf{Q} - (\partial\dot{\boldsymbol{\Phi}}/\partial\dot{\mathbf{s}})^T\boldsymbol{\lambda} = \mathbf{0}$. The force $\boldsymbol{\lambda}$ is also known as Lagrange Parameter. It can be calculated with the help of the second time derivative of the implicit constraint equation

$$\ddot{\boldsymbol{\Phi}} = \left(\frac{\partial\dot{\boldsymbol{\Phi}}}{\partial\dot{\mathbf{s}}}\right)\ddot{\mathbf{s}} + \left[\frac{d}{dt}\left(\frac{\partial\dot{\boldsymbol{\Phi}}}{\partial\dot{\mathbf{s}}}\right)\right]\dot{\mathbf{s}} = 0 \tag{16}$$

yielding

$$\mathbf{A}\boldsymbol{\lambda} + \mathbf{b} = \mathbf{0}, \tag{17}$$

where

$$\mathbf{A} = \left[\left(\frac{\partial\dot{\boldsymbol{\Phi}}}{\partial\dot{\mathbf{s}}}\right)\mathbf{M}^{-1}\left(\frac{\partial\dot{\boldsymbol{\Phi}}}{\partial\dot{\mathbf{s}}}\right)^T\right], \tag{18}$$

$$\mathbf{b} = \left(\frac{\partial\dot{\boldsymbol{\Phi}}}{\partial\dot{\mathbf{s}}}\right)\mathbf{M}^{-1}(\mathbf{G}\dot{\mathbf{s}} - \mathbf{Q}) - \left[\frac{d}{dt}\left(\frac{\partial\dot{\boldsymbol{\Phi}}}{\partial\dot{\mathbf{s}}}\right)\right]\dot{\mathbf{s}}. \tag{19}$$

In this form, the total mass matrix has to be inverted and the recursive formulation gets lost. However, usually the constraints arise in only one of the subsystems (e.g. in the swinging foot of the Walking Machine). The constraint Eq. (16) can be rewritten in terms of the describing velocities and accelerations

$$\ddot{\boldsymbol{\Phi}} = \left(\frac{\partial\dot{\boldsymbol{\Phi}}}{\partial\dot{\mathbf{y}}_C}\right)\ddot{\mathbf{y}}_C + \left[\frac{d}{dt}\left(\frac{\partial\dot{\boldsymbol{\Phi}}}{\partial\dot{\mathbf{y}}_C}\right)\right]\dot{\mathbf{y}}_C = \mathbf{0}, \tag{20}$$

where C is the body which undergoes constraints. Inserting successively $\ddot{\mathbf{s}}_i$ from the recursion procedure, one again obtains

$$\mathbf{A}\boldsymbol{\lambda} + \mathbf{b} = \mathbf{0}, \tag{21}$$

but now

$$\mathbf{A} = \left(\frac{\partial\dot{\boldsymbol{\Phi}}}{\partial\dot{\mathbf{y}}_C}\right)\left[\sum_{i=1}^{C}\mathbf{T}^*_{C,i}\mathbf{J}_i\mathbf{T}^{*T}_{C,i}\right]\left(\frac{\partial\dot{\boldsymbol{\Phi}}}{\partial\dot{\mathbf{y}}_C}\right)^T, \tag{22}$$

$$\begin{aligned}\mathbf{b} \; = \; & \left[\frac{d}{dt}\left(\frac{\partial\dot{\boldsymbol{\Phi}}}{\partial\dot{\mathbf{y}}_C}\right)\right]\dot{\mathbf{y}}_C + \\ & + \left(\frac{\partial\dot{\boldsymbol{\Phi}}}{\partial\dot{\mathbf{y}}_C}\right)\sum_{i=1}^{C}\mathbf{T}^*_{C,i}\left[\mathbf{J}_i\left(\mathbf{Q}^*_i - \mathbf{G}^*_i\dot{\mathbf{y}}_i\right) + \mathbf{N}^T_{i+1}\dot{\mathbf{T}}_{i+1,i}\dot{\mathbf{y}}_i\right].\end{aligned} \tag{23}$$

Notice that the only inversion needed is $\mathbf{A}^{-1}$ which has the dimension of $\boldsymbol{\lambda}$ (e.g. $\boldsymbol{\lambda} \in \mathbb{R}^1$ for the Walking Machine). Once $\boldsymbol{\lambda}$ is known, the recursive calculation is started, where the impressed force $\mathbf{Q}_C$ is completed with $\left(\partial\dot{\boldsymbol{\Phi}}/\partial\dot{\mathbf{y}}_C\right)^T \boldsymbol{\lambda}$.

4. CONTROL

Any prescribed motion $\mathbf{y}_{id} = \mathbf{y}_{id}(t),\ i = 1..N$ may be looked at as a response of imposed rheonomic constraints. The corresponding constraint forces Q_{Ci} enter the motion Eqs. (7) via $\mathbf{Q}_{Cp} = \mathbf{T}_{ip}^T \mathbf{Q}_{Ci}$ yielding

$$\begin{pmatrix} \mathbf{Q}_{C1} \\ \mathbf{Q}_{C2} \\ \vdots \\ \mathbf{Q}_{CN} \end{pmatrix} = \begin{bmatrix} \mathbf{I} & \mathbf{T}_{21}^T & \cdots & \mathbf{T}_{N,1}^T \\ & \mathbf{I} & \ddots & \mathbf{T}_{N,2}^T \\ & & \ddots & \vdots \\ & & & \mathbf{I} \end{bmatrix} \begin{pmatrix} \mathbf{M}_1\ddot{\mathbf{y}}_{1d} + \mathbf{G}_1\dot{\mathbf{y}}_{1d} - \mathbf{Q}_1 \\ \mathbf{M}_2\ddot{\mathbf{y}}_{2d} + \mathbf{G}_2\dot{\mathbf{y}}_{2d} - \mathbf{Q}_2 \\ \vdots \\ \mathbf{M}_N\ddot{\mathbf{y}}_{Nd} + \mathbf{G}_N\dot{\mathbf{y}}_{Nd} - \mathbf{Q}_N \end{pmatrix}, \quad (24)$$

for calculation of $(\mathbf{Q}_{C1}(t)...\mathbf{Q}_{CN}(t))$. Provided that the system is controllable, the motor torques can always be construed as $\mathbf{Q}_{mot,i} = \mathbf{F}_i^T \mathbf{Q}_{Ci}$, which fades the noninteracting constraints out. The desired trajectories $\mathbf{y}_{id}, \dot{\mathbf{y}}_{id}, \ddot{\mathbf{y}}_{id}$ depend on $\mathbf{q}_d, \dot{\mathbf{s}}_d, \ddot{\mathbf{s}}_d$ and are easily calculated with the kinematical chain, see Eq. (1). Since the motor coordinates γ_{Mi} do not contribute to position kinematics, one may set $\gamma_{Mi} = \gamma_{Ai} + \Delta\gamma_i$ and evaluate the open loop control neglecting $\Delta\gamma_i$ which has then to be considered within the closed loop part. To compensate parameter inaccuracies and $\Delta\gamma$, a passive PD-controller is used to stabilize the system. In Figure 3 it is combined in the block "Joint Control". The determination of the velocities $\dot{\mathbf{s}}$ is done with a high gain observer. To stabilize the gait of the Walking Machine, a ZMP Controller is used to change the orientation (Kardan angles $\alpha_H, \beta_H, \gamma_H$) of the upper part of the body, if the ZMP tends to leave the foot sole.

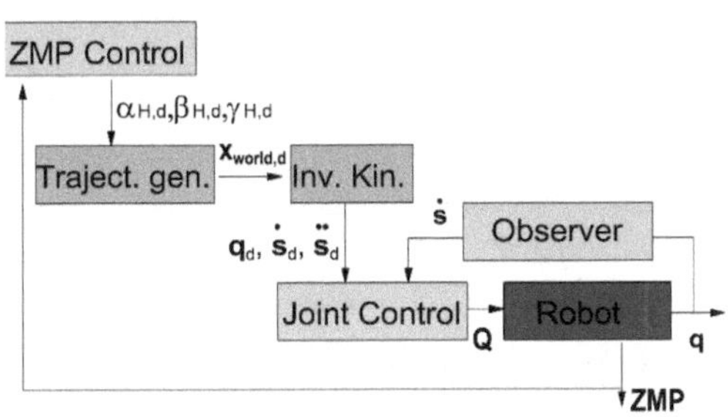

Figure 3. Control concept.

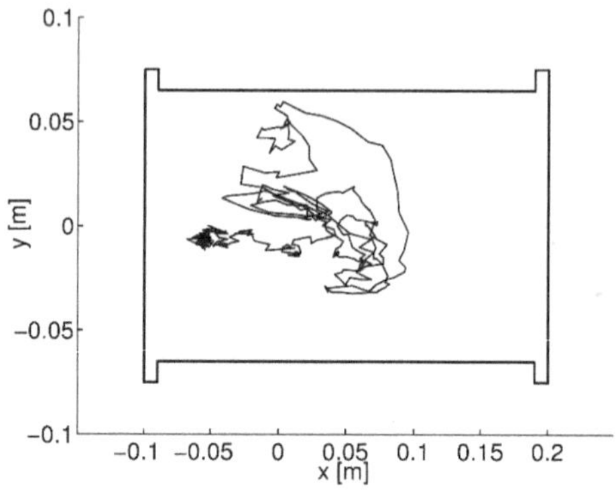

Figure 4. Zero Moment Point during penalty shooting (measurement).

5. TRAJECTORY GENERATION – PENALTY SHOOTING

To achieve the required tasks like walking in three dimensional areas or shooting a soccer penalty, the trajectories have to be computed in realtime. For the feet, these are normally polynomials of degree 5. For the Center of Gravity, a lumped mass model is used to calculate the trajectories in the sense of an inverted pendulum (IPM). Tests on the real platform have shown, that the TMIPM brings no advantages in walking stability. A robot is said to walk dynamically stable as long as the ZMP stays within the sole of the supported foot. The ZMP is measured with the help of special designed torque sensor in the ankle of the Walking Machine. Figure 4 shows the ZMP on the supporting foot during shooting the penalty. It stays within the convex hull of the foot and therefore the robot will not fall. The orientation of the feet and the hip is specified by means of Kardan angles (α, β, γ). For each trajectory, the second time derivative has to be steady to use them as input for the feed-forward controller. Therefore the joint coordinates have to be calculated. Because of the special buildup of the Walking Machine, there is no analytical solution for the inverskinematic problem. A Newton algorithm is used to solve this problem numerically. Hence, the joint angles can be calculated for each leg separately, the computational efford is not very high.

In the context of the ZMP also the trajectories for shooting the penalty are generated. To get a steady motion, polynomials of degree 6 are evaluated. Figure 5 shows a sideview of the robot during shooting. The ball with a mass of 0.2 kg reaches a start velocity of about 1.5 m/s. It does not influence the dynamic of the Walking Machine, and was therefore neglected.

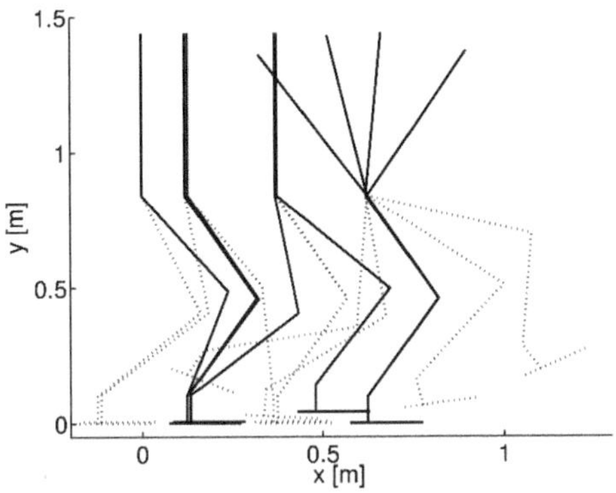

Figure 5. Trajectories for penalty shooting.

REFERENCES

Amos, A., Gerth, W. (2003) *Analytic Path Planning Algorithms for Bipedal Robots without a Trunk*, Journal of Intelligent and Robotic Systems, Kluwer Academic Publishers, Netherlands, pp. 109-127.

Brandl, H., Johanni, R., Otter, M. (1986) *A very efficient algorithm for the simulation of robots and similar multibody systems without inversion of the mass matrix*, Proceedings of the IFAC/IFIP/IMACS International Symposium on Theory of Robots, Vienna.

Bremer, H. (1988) *Dynamik und Regelung mechanischer Systeme*, B.G.Teubner Verlag, Stuttgart.

Bremer, H. (2003) *On the use of nonholonomic variables in robotics,*World Scientific, Singapore, pp. 1-48.

Caballero, R., Armada, M. A., Akinfiev T. (2004) *Robust Cascade Controller for Nonlinearly Actuated Biped Robots: Experimental Evaluation*, The International Journal of Robotics Research, Sage Publications, pp. 1075-1095.

Hiller, M., Germann, D., Morgado de Gois, J.A. (2004) *Design and control of a quadruped robot walking in unstructered terrain*, IEEE Conference on Control Applications, CCAA 2004, Taiwan.

Hirai, K.,Hirose, M., Takenaka, T. (1998) *The development of Honda humanoid robot*, In Proc. IEEE Int. Conf. on Robotics and Automation, Leuven, Belgium, pp. 160-165.

Löffler, K., Gienger, M., Pfeiffer, F. (2001) *Simulation and control of a biped jogging robot*, In 4th International Conference on Climbing and Walking Robots, Professional Engineering Publishing Limited, London, pp. 867-874.

Löffler, K., Gienger, M., Pfeiffer, F. (2003) *Sensors and Control Concept of Walking "Johnnie"*, The International Journal of Robotics Research, Sage Publications, pp. 229-239.

Mitterhuber, R., Gattringer, H., Bremer H. (2004) *Dynamische Modellierung hybrider Mehrkörpersysteme unter Berücksichtigung numerischer Aspekte*, PAMM-Proc. Appl. Math. Mech., Vol. 4, Issue 1, pp. 161-162.

Pfeiffer, F., Glocker, C. (2000) *Multibody Dynamics with unilateral Contacts*, Springer-Verlag, Wien New York.

Vucobratovic, M., Borovac, B., Surla, D., Stokic, D. (1990) *Biped locomotion: dynamics, stability, control and applications*, Springer-Verlag, Berlin.

Vucobratovic, M., Potkonjak, V., Tzafestas, S. (2004) *Human and Humanoid Dynamics*, Journal of Intelligent and Robotic Systems, Kluwer Academic Publishers, Netherlands, pp. 65-84.

DIFFERENT CONTROL STRATEGIES FOR THE ACTIVE SUPPRESSION OF BRAKE SQUEAL

Peter Hagedorn and Daniel Hochlenert
Technische Universität Darmstadt, Department of Applied Mechanics, Germany
hagedorn@mechanik.tu-darmstadt.de, hochlenert@mechanik.tu-darmstadt.de

Abstract: The present paper is devoted to the modeling and the active suppression of brake squeal. Two different control strategies are described and tested on the basis of a four degree of freedom disk brake model. The first control law is based on optimal control and includes a KALMAN–BUCY filter to estimate the system's state. The second law maximizes the energy dissipated by the friction forces between the disk and the brake pads. Both control strategies are compared with respect to their practical application and implemented in a floating caliper disk brake at a test rig.

Key words: brake squeal, active control.

INTRODUCTION

Brake squeal is a high frequency noise (1–12 kHz) of brake systems. It does not affect the performance of the brake but is detrimental the passenger's comfort and the subjective quality of the vehicle. Due to the fact that the source and mechanism of generating brake squeal is not completely understood yet, the design of disk brakes of modern passenger cars demands a significant amount of costly experiments. Active control of brake squeal could be a helpful tool for the development and experimental evaluation of new brake systems.

1. MODELING OF BRAKE SQUEAL

It is commonly accepted by engineers and scientists working in the field of brake noise, that brake squeal in a disk brake is initiated by an instability due to the friction forces, leading to self-excited vibrations. The self-excited brake system then oscillates, reaching a limit cycle. The reason for the onset of instability has been put forward on three possible grounds, namely, the change of the friction characteristic with the speed of the vehicle [1], change of the relative orientation of the disk and the friction pads leading to a modification

H. Ulbrich and W. Günthner (eds), IUTAM Symposium on Vibration Control of Nonlinear Mechanisms and Structures, 161–171.

of the friction force [2] and/or a flutter instability which is found even with a constant friction coefficient [3]. Most of these models and mechanisms have however not been validated by physical experiments. It is of course well known that a negative slope in the friction characteristic leads to instability and self-excitation. On the other hand, the friction properties in a brake depend strongly on the temperature and wear (actually on the whole history of temperature and wear) and also on the pressure, humidity, etc. It is known from experience with brake squeal, that there may be instability and self-excitation even in the absence of the negative slope. Obviously other mechanisms for instability are also important and relevant to the problem, even if the friction characteristic with negative slope seems to be an obvious culprit. The authors believe that the flutter instability, which may even occur with constant coefficient of friction, is a more realistic candidate in most cases. This instability mechanism can be studied on systems having two ore more degrees of freedom. Models for brake squeal often are based on multi-body systems having up to 20 degrees of freedom. A typical model with 14 degrees of freedom is described in [4].

The model-based real time active control of squeal requires low degree of freedom models, so that the control laws can be implemented in real time on present day computers. Figure 1 shows a new four degree of freedom model, including the in-plane motion of the idealized brake pads and a wobbling brake disk.

The brake disk is fixed at its center of mass and modelled as a rotating, elastically supported (rotational stiffness k_t), rigid disk with thickness h. A Newtonian (inertial) coordinate system is given by $\boldsymbol{n}_i$ and the disk-fixed coordinate system by $\boldsymbol{d}_i$, $i=1,2,3$. The orientation of the disk is described by the CARDAN-angles q_1, q_2, q_3 and intermediate coordinate systems are specified by the unit vectors $\boldsymbol{a}_i$ and $\boldsymbol{b}_i$, $i=1,2,3$. The angular velocity of the disk with respect to the Newtonian frame can be written as

$$^N\boldsymbol{\omega}^D = \dot{q}_1\boldsymbol{n}_1 + \dot{q}_2\boldsymbol{a}_2 + \dot{q}_3\boldsymbol{b}_3.$$

The central inertia dyadic of the disk is

$$\boldsymbol{\Theta} = \Theta\ \boldsymbol{d}_1 \otimes \boldsymbol{d}_1 + \Theta\ \boldsymbol{d}_2 \otimes \boldsymbol{d}_2 + \Phi\ \boldsymbol{d}_3 \otimes \boldsymbol{d}_3.$$

The disk is driven by a torque $\boldsymbol{M}_A = M_A\boldsymbol{n}_3$ such that the nonholonomic constraint

$$^N\boldsymbol{\omega}^D \cdot \boldsymbol{n}_3 = \Omega = \text{const.} \tag{1}$$

is fulfilled. Two idealized pads (point mass m_p each) with preload N_0 are in contact (point contact) with the disk. The preload is considered to be large enough so that there is always contact between the pads and the disk. The position vectors from the center of the disk to the upper and lower contact point

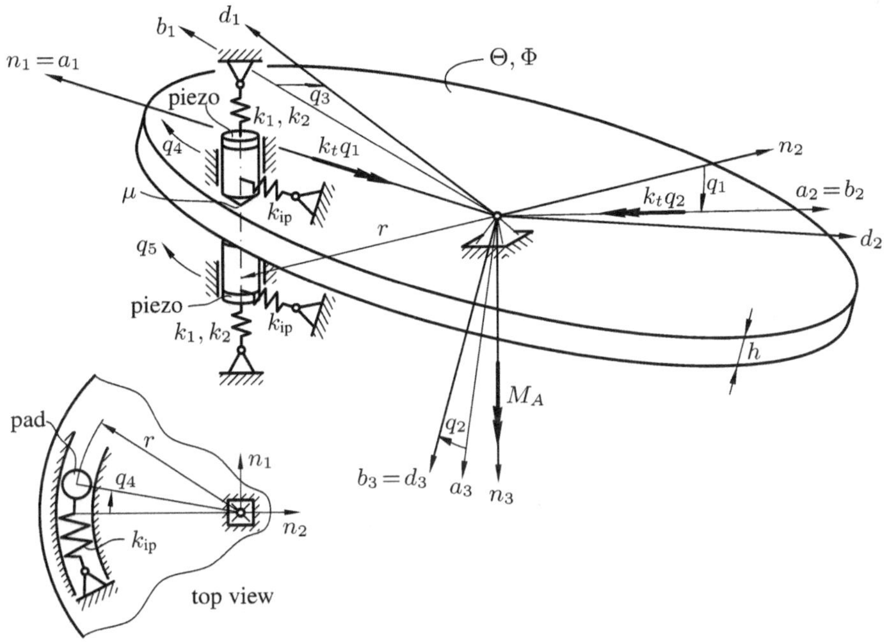

Figure 1. Disk brake model for control.

are $\boldsymbol{p}_1$ and $\boldsymbol{p}_2$ respectively. Piezoceramic actuators are located at the back of each pad, generating two independent control forces. The pads can move in the $\boldsymbol{n}_3$-direction and on a circle with radius r from the center of the disk in the $\boldsymbol{n}_1, \boldsymbol{n}_2$-plane. Both pads are supported by a linear spring with stiffness k_{ip} and a nonlinear spring representing the typical behavior of the friction material. The force of the nonlinear spring is given by $k_1 x + k_2 x^3$, where x is the spring's elongation. The in-plane motions of the pads on the circle with radius r are described by the two angles q_4 and q_5.

COULOMB friction is assumed between the pads and the disk, the magnitude of the friction force R_i being related to the magnitude of the normal force N_i by $R_i = \mu N_i$ for $i = 1$ (upper pad) and $i = 2$ (lower pad). The direction of the normal forces is given by $\boldsymbol{d}_3$. The friction forces are orthogonal to the normal forces and therefore have to lie in the $\boldsymbol{d}_1, \boldsymbol{d}_2$-plane. Let $\boldsymbol{r}_i$ $(i = 1, 2)$ be the unit vectors in the direction of the relative velocities between the contact points of the pads and the disk, the the contact forces acting on the disk are

$$\boldsymbol{F}_{d1} = -R_1 \boldsymbol{r}_1 + N_1 \boldsymbol{d}_3,$$
$$\boldsymbol{F}_{d2} = -R_2 \boldsymbol{r}_2 - N_2 \boldsymbol{d}_3.$$

The resulting torque acting on the disk with respect to its center is given by the torque due to the elastic suspension, the constraint torque M_A and the torque of the contact forces with respect to the center of the disk. It can be written as

$$\boldsymbol{M} = -k_t q_1 \boldsymbol{n}_1 - k_t q_2 \boldsymbol{a}_2 + M_A \boldsymbol{n}_3 + \boldsymbol{p}_1 \times \boldsymbol{F}_{d1} + \boldsymbol{p}_2 \times \boldsymbol{F}_{d2}.$$

The system's equations of motion follows from the law of angular momentum for the disk

$$\frac{{}^N\mathrm{d}}{\mathrm{d}t}\left(\boldsymbol{\Theta}\,{}^N\boldsymbol{\omega}^D\right) = \boldsymbol{M},$$

and NEWTONS's law for each of the pads

$$m_p \ddot{\boldsymbol{p}}_i = \boldsymbol{F}_{pi} \qquad i = 1, 2,$$

where $\boldsymbol{F}_{pi}$ is the sum of the forces acting on the corresponding pad due to the support, the contact with the disk, the linear and nonlinear spring as well as the piezo-ceramic actuators.

The nonlinear equations of motion were derived in symbolic form using the commercial software `Autolev` based on KANE's algorithm [5]. The nonholonomic constraint (1) is used to eliminate the degree of freedom corresponding to the variable q_3 and to determine the driving torque M_A (a constraint torque). The equations of motion can be written as

$$\boldsymbol{M}\ddot{\boldsymbol{q}} + \boldsymbol{D}\dot{\boldsymbol{q}} + \boldsymbol{C}\boldsymbol{q} + \boldsymbol{n}(\boldsymbol{q}, \dot{\boldsymbol{q}}, \boldsymbol{u}) = \boldsymbol{f}(t), \tag{2}$$

where the vector of generalized coordinates is $\boldsymbol{q} = \begin{bmatrix} q_1 & q_2 & q_4 & q_5 \end{bmatrix}^{\mathrm{T}}$, the matrices are

$$\boldsymbol{M} = \begin{bmatrix} \Theta + 2m_p r^2 & 0 & 0 & 0 \\ -\mu m_p h r & \Theta & 0 & 0 \\ -\mu m_p r^2 & 0 & m_p r^2 & 0 \\ \mu m_p r^2 & 0 & 0 & m_p r^2 \end{bmatrix}, \quad \boldsymbol{D} = \begin{bmatrix} \frac{\mu N_0 h^2}{2\Omega r} & \Omega\Psi & 0 & 0 \\ -\Omega\Psi & 0 & 0 & 0 \\ 0 & 0 & 0 & 0 \\ 0 & 0 & 0 & 0 \end{bmatrix},$$

$$\boldsymbol{C} = \begin{bmatrix} k_t + N_0 h + 2kr^2 & \frac{1}{2}\mu N_0 \frac{h^2}{r} & \frac{1}{2}\mu N_0 h & -\frac{1}{2}\mu N_0 h \\ -\mu(2N_0 r + k_1 h r) & k_t + (1+\mu^2) N_0 h & N_0 r & -N_0 r \\ -\mu k_1 r^2 & (1+\mu^2) N_0 r & k_{\mathrm{ip}} r^2 & 0 \\ \mu k_1 r^2 & -(1+\mu^2) N_0 r & 0 & k_{\mathrm{ip}} r^2 \end{bmatrix}$$

and $\boldsymbol{n}(\boldsymbol{q}, \dot{\boldsymbol{q}}, \boldsymbol{u})$ includes all nonlinear terms due to the kinematics and the nonlinear springs. The right hand side of the equation of motion is

$$\boldsymbol{f}(t) = \begin{bmatrix} -r & r \\ \frac{1}{2}\mu h & -\frac{1}{2}\mu h \\ \mu r & 0 \\ 0 & \mu r \end{bmatrix} \kappa \begin{bmatrix} u_1(t) \\ u_2(t) \end{bmatrix},$$

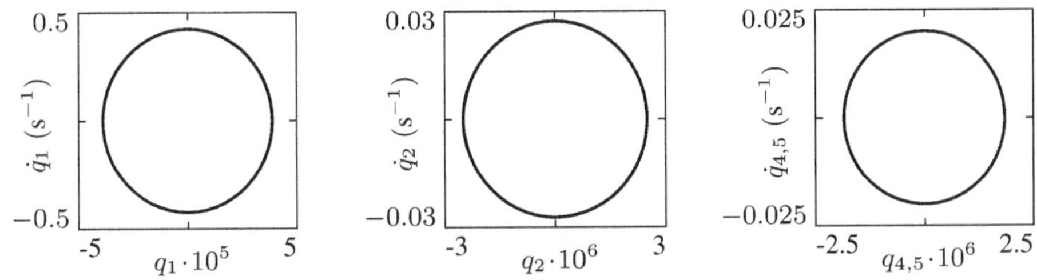

Figure 2. Limit cycles of q_i ($i = 1, 2, 4, 5$).

where $\boldsymbol{u} = \begin{bmatrix} u_1 & u_2 \end{bmatrix}^{\mathrm{T}}$ is the control vector (input voltages applied to the piezoceramic actuators at the back of the brake pads). The inertia and damping of the piezoelectric actuators are neglected, so that the control forces are proportional to the applied voltages (constant of proportionality κ).

The parameters were chosen to fit the floating caliper disk brake presently used at the test rig in Darmstadt. The following parameters are used for the simulations described in the next sections:

$$\begin{aligned}
\Theta &= 0.156 \text{ kgm}^2, & \Psi &= 2\Theta, & m_p &= 0.4 \text{ kg},\\
h &= 0.02 \text{ m}, & r &= 0.13 \text{ m}, & \mu &= 0.6,\\
k_t &= 1.34 \cdot 10^7 \text{ Nm}, & k_1 &= 2.0 \cdot 10^8 \text{ N/m}, & \Omega &= 5\pi \text{ s}^{-1},\\
k_{\text{ip}} &= 1.00 \cdot 10^6 \text{ N/m}, & k_2 &= 3.1 \cdot 10^{17} \text{ N/m}, & N_0 &= 3000 \text{ N},\\
\kappa &= 1.6 \text{ N/V}.
\end{aligned}$$

The asymmetries in the matrices M and C for the chosen parameters lead to an instability of the trivial solution (eigenvalues of the linearized system with positive real part) and therefore to self excited vibrations, which can be interpreted as squeal. The amplitudes of the nonlinear system are bounded and tend towards the limit cycles shown in Figure 2.

The relative velocity of the material point of the disk and the material point of the upper pad being in actual contact can be roughly estimated by $\boldsymbol{v}_{\text{rel}} \approx (\Omega r - \dot{q}_2 h/2 - \dot{q}_4 r)\boldsymbol{n}_1$. It is obvious that stick-slip does not occur in the presented model because $\boldsymbol{v}_{\text{rel}}$ will never vanish for the given parameters and the amplitudes of the limit cycles shown in Figure 2.

2. CONTROL STRATEGIES

2.1 Optimal Control

In developing a first control law, optimal control with an LQR formulation is used, so that the linear part of the equations of motion (2) are written in the standard state space form

$$\begin{aligned} \dot{\boldsymbol{x}}(t) &= \boldsymbol{A}\boldsymbol{x}(t) + \boldsymbol{B}\boldsymbol{u}(t), \\ \boldsymbol{y}(t) &= \boldsymbol{C}\boldsymbol{x}(t) + \boldsymbol{D}\boldsymbol{u}(t), \end{aligned} \tag{3}$$

where $\boldsymbol{x}^{\mathrm{T}}(t) = \left[\boldsymbol{q}^{\mathrm{T}} \; \dot{\boldsymbol{q}}^{\mathrm{T}}\right]$ is the state vector, $\boldsymbol{u}^{\mathrm{T}}(t) = \left[u_1 \; u_2\right]$ the control vector, $\boldsymbol{y}(t)$ the measurement vector, $\boldsymbol{A}$ the system matrix, $\boldsymbol{B}$ the input matrix, $\boldsymbol{C}$ the measurement matrix and $\boldsymbol{D}$ the direct transmission matrix (no confusion should occur with the matrices $\boldsymbol{C}$ and $\boldsymbol{D}$ from (2)). The system should be controlled by state feedback of the form

$$\boldsymbol{u}(t) = -\boldsymbol{K}\boldsymbol{x}(t)$$

with the constant feedback matrix $\boldsymbol{K}$. In the context of optimal control, the matrix $\boldsymbol{K}$ is found by minimizing the cost function

$$J = \int_0^\infty \boldsymbol{x}^{\mathrm{T}}(t)\boldsymbol{Q}\boldsymbol{x}(t) + \boldsymbol{u}^{\mathrm{T}}(t)\boldsymbol{R}\boldsymbol{u}(t)\,\mathrm{d}t, \tag{4}$$

where $\boldsymbol{Q}$ and $\boldsymbol{R}$ are positive semidefinite and positive definite, constant weighting matrices for the states and the inputs respectively. They are chosen as diagonal matrices such that the closed loop meets the desired dynamic behavior within the scope of the used piezoelectric actuators and power amplifiers. It is possible and useful to weight only the generalized velocities because the control may be used for models having more degrees of freedom or especially for a real brake where $\dot{\boldsymbol{q}} \equiv \boldsymbol{0}$ (no squeal) does not necessary imply $\boldsymbol{q} \equiv \boldsymbol{0}$. The cost function (4) is minimized by solving the appropriate matrix RICCATI equation.

Given that the complete state vector cannot be measured in the experiment, a KALMAN–BUCY filter is used to estimate the state vector. It is assumed that the states of the system $\boldsymbol{x}(t)$ and the measurement signals $\boldsymbol{y}(t)$ are superposed by white noise. The state observer should provide an optimal estimate $\hat{\boldsymbol{x}}$ for the state vector $\boldsymbol{x}$ and consequently minimize the error

$$\boldsymbol{e}(t) = \boldsymbol{x}(t) - \hat{\boldsymbol{x}}(t). \tag{5}$$

The differential equation for the estimation of the state vector with a KALMAN–BUCY filter is

$$\dot{\hat{\boldsymbol{x}}}(t) = \boldsymbol{A}\hat{\boldsymbol{x}}(t) + \boldsymbol{B}\boldsymbol{u}(t) + \boldsymbol{L}\big(\boldsymbol{y}(t) - \hat{\boldsymbol{y}}(t)\big),$$

where

$$\hat{\boldsymbol{y}}(t) = \boldsymbol{C}\hat{\boldsymbol{x}}(t) + \boldsymbol{D}\boldsymbol{u}(t)$$

is the estimated measurement signal. The matrix $\boldsymbol{L}$ is found by minimizing the expected value of the error (5) for known covariance matrices of $\boldsymbol{\epsilon}$ and

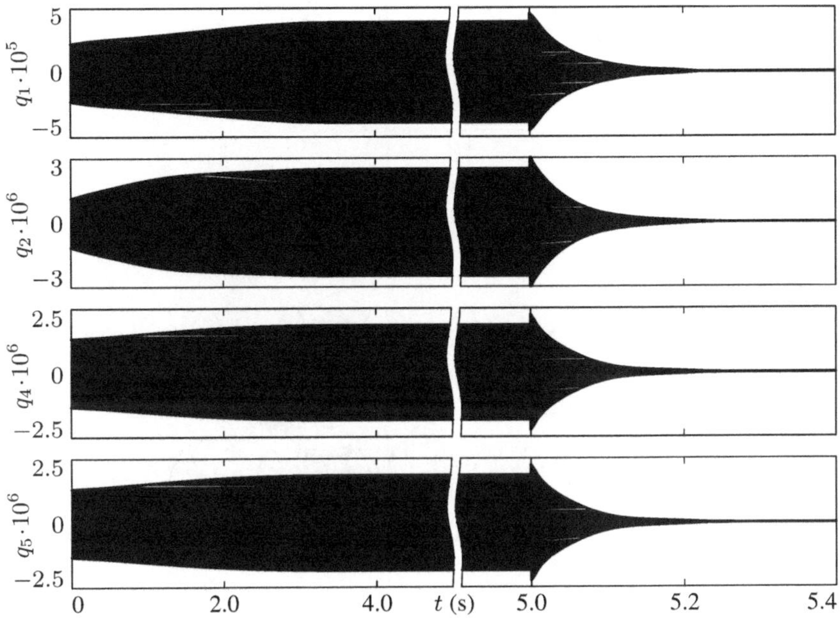

Figure 3. Optimal control with KALMAN–BUCY filter.

$\boldsymbol{\nu}$. The minimization leads to an appropriate matrix RICCATI equation for the observer matrix $\boldsymbol{L}$.

Figure 3 shows the response of the controlled nonlinear system. Only the acceleration $\ddot{q}_4$ (measured by means of an corresponding accelerometer at the back plate of the pad) is used here as an input to the KALMAN–BUCY filter. The controller was switched on after 5 s. It is possible to quench the vibrations and thus the squealing within approximately 300 ms. The increasing vibrations shortly after 5 s are due to transient oscillations of the KALMAN–BUCY filter because it starts with zero initial conditions. The voltage applied to the piezo-ceramic actuator at the upper pad is shown in Figure 4. The input signal of the second actuator is similar. Due to the limits of the power amplifiers, the controller is designed such that the maximum input voltage does not exceed 100 V. Since the input signals are proportional to the states of the system, the voltage decreases with decreasing vibration amplitudes.

Even though the squealing was quenched within only 300 ms, the vibration amplitudes vary only quasi-statically at the time scale of the period of the vibrations. This can be seen from Figure 5, where the phase shift of the coordinates q_4, q_5 (in-plane motion of the pads) and the corresponding input signals u_1, u_2 is shown. The minimization of the cost function (4) and therefore the optimal placement of the eigenvalues of the closed loop obviously leads to a phase shift

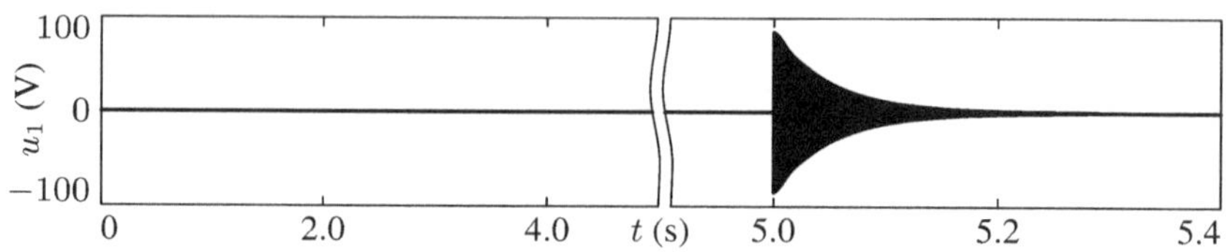

Figure 4. Voltage applied to the actuator at the upper pad.

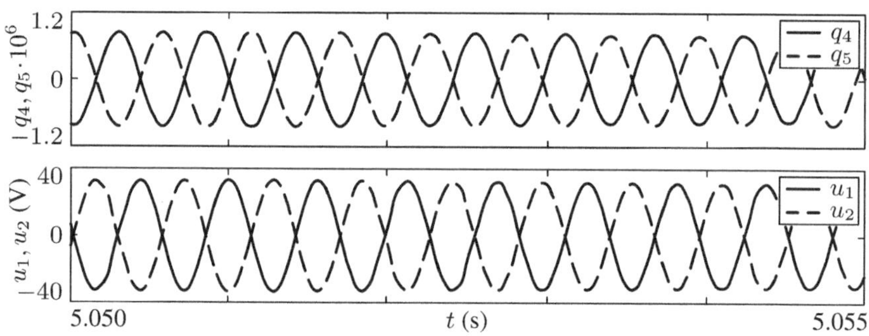

Figure 5. Phase shift of q_4, q_5 and the input signals u_1 and u_2.

of approximately $\pi/2$ between the in-plane motion of the pads and externally applied voltage. This circumstance will recur in the next section.

2.2 Control by Maximum Dissipation

The instability of the trivial solution is a result of an energy transfer from the driving torque M_A to the vibrations of the disk, the energy being transferred via the friction forces. This leads to the idea of controlling the system by minimizing the energy transferred via the friction forces. Such a control algorithm for the suppression of friction induced vibrations was proposed by POPP in [6] for a simpler model. The following calculations lead to similar results for the described four degree of freedom model.

For the linearized version of the model, the power of the friction forces at the brake pads is a second order polynomial in q_i $(i = 1, 2, 4, 5)$ and its derivatives. The expression for the upper pad can be written as

$$\begin{aligned} P_{R_1} &= R_1 \boldsymbol{r}_1 \cdot \dot{\boldsymbol{p}}_1 \\ &= \mu r \kappa u_1 \dot{q}_4 + \mu r \big(N_0 - \mu N_0 q_2 + k r q_1 + m_p r \ddot{q}_1\big) \dot{q}_4 + \mu r N_0 q_2 \dot{q}_1. \end{aligned}$$

The squealing brake vibrates in a limit cycle as shown in Figure 2. All coordinates oscillate quasi-harmonically with circular frequency ω and some constant relative phase shift. The origin of time is chosen such that

$$q_4(t) = Q_4 \cos \omega t$$

holds for the in-plane motion of the upper pad. Furthermore it is assumed that the voltage applied to the upper pad is

$$u_1(t) = U_1 \cos(\omega t - \varphi_1),$$

where U_1 is a constant amplitude and φ is a phase shift between the voltage and the in-plane motion of the pad. As mentioned above, the amplitudes of the vibrations are reduced only quasi-statically at the time scale of the period of the limit cycle because of the limited power of the piezo actuators. Therefore the energy transferred by the friction force during one period of vibration is approximately

$$E_{R_1} = \int_0^{\frac{2\pi}{\omega}} P_{R_1} \, \mathrm{d}t = -\pi \mu r Q_4 \kappa U_1 \sin \varphi_1 + \text{terms independent of } u_1.$$

Consequently minimum energy is transferred or, in other words, maximum energy is dissipated by the friction force for $U_1 = 100$ V (limit of the amplifier) and $\varphi_1 = \pi/2$. Thus the control law simply is

$$u_1(t) = 100 \cos(\omega t - \pi/2) \text{ V}. \tag{6}$$

Similar calculations for the lower pad lead to an optimal phase shift of $\varphi_2 = \pi/2$ between the in-plane motion of the lower pad $q_5(t)$ and the voltage $u_2(t)$.

It should be mentioned that the work of the friction forces acting at the disk is irrelevant in this minimization process. The material point of the disk in actual contact with the brake pad has a dominant velocity in the direction of the rotation of the disk (no stick-slip). Thus a fluctuating contact force due to the fluctuating input voltage does not affect the energy transferred to the disk during one period in the linear case. On this account the described control approach does not reduce the braking power of the system.

Figure 6 shows the response of the nonlinear system with this control law. The controller was switched on after 5 s. It reduces the amplitudes of the oscillations significantly, stopping the squeal. The input signal (6) is not proportional to the state of the system; it is applied even when the oscillation has been quenched. Therefore small in-plane oscillations of the brake pads remain for $t > 5.2$ s. A diagram showing the input signal and the phase shift between q_4 and the input signal is omitted, because all these relations are clearly defined by (6).

2.3 Comparison of the Control Strategies

Both control strategies are able to suppress brake squeal within the scope of the described model. Furthermore they were successfully used in experiments

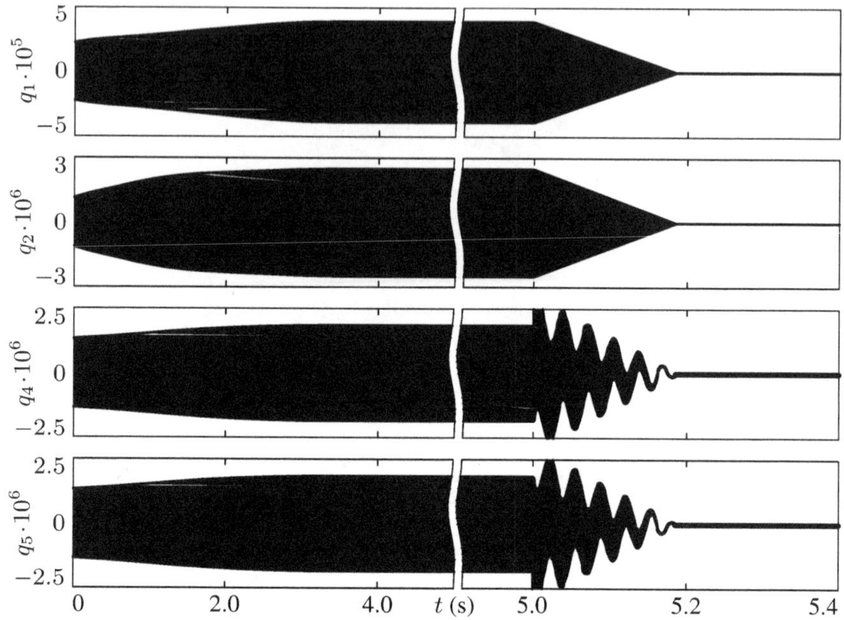

Figure 6. Control by maximum dissipation.

with a real disk brake at a brake test rig in Darmstadt and also in an industrial laboratory. The time histories of the control variables obtained by the two concepts are quite similar, but the practical implementation differs a lot.

The optimal control with a KALMAN–BUCY filter requires a large amount of real time calculations. A system of eight first order ODEs has to be integrated with frequencies up to 60 kHz at the test rig. The sampling frequency can be reduced to 15 kHz by a time-discrete design of the controller. However the code for the real time computations has to be well organized and optimized. Last but not least the efficiency of the control strongly depends on the parameters of the observer. The main advantage of the optimal control is that only one sensor e.g. the in-plane acceleration of one brake pad is needed for the control and that the input signal is proportional to the state of the system. This means that there is almost no voltage needed to maintain the system quiet once the squeal has been suppressed.

The control by maximum dissipation is applicable without any detailed knowledge of the parameters of the system. Only the approximate frequency of squeal has to be known in advance to design appropriate filters to realize the control law (6). This is important, since the parameters of a real brake system change with temperature, wear and time. In contrast to the optimal control two sensors (one for each pad) are needed. A disadvantage of this control is the constant amplitude of the input signals. Even when the squealing has been stopped, small in-plane oscillates of the brake pads remain, due to the fact that the actuators are still operating, which could be changed by a state dependent amplitude of the control function (6).

For the time being in our laboratory in Darmstadt the control law (6) is realized with a zero-crossing detection based on the band-pass filtered signal of the in-plane motion. The generation of a signal with constant amplitude and defined phase shift to a reference signal however is a typical application of a phase locked loop (PLL). This and the implementation of different control laws (e.g. a passivity based control) are the a subject of the present and future work.

REFERENCES

[1] Ouyang, H. and Mottershead, J.E., "Friction-induced parametric resonances in disc: Effect of a negative friction-velocity relationship", *Journal of Sound and Vibration*, 209(2), 1998, 251–264.

[2] Millner, N., "An analysis of disc brake squeal", SAE Technical Paper Series, 780332, 1978.

[3] Chakraborty, G., Jearsiripongkul, T., von Wagner, U. and Hagedorn, P., "A new model for a floating caliper disk brake", VDI-Tagung Reibung und Schwingungen in Fahrzeugen, Maschinen und Anlagen, Hannover (Germany) November 26–27, 2002.

[4] Jearsiripongkul, T., Hochlenert, D, von Wagner, U. and Hagedorn, P., "A nonlinear floating caliper disk brake model to incorporate squeal", in *Proceedings of SAE 23nd Annual Brake Colloquium & Exhibition*, 2005, submitted.

[5] Kane, T.R. and Levinson, D.A., *Dynamics: Theory and Applications*, McGraw-Hill, 1985.

[6] Popp, K., Rudolph, M., Kröger M. and Lindner, M., "Mechanisms to generate and avoid friction induced vibrations", VDI-Bericht 1736, 2002, 1–15.

MODELLING AND IDENTIFICATION OF ROBOTS WITH JOINT AND DRIVE FLEXIBILITIES

Toon Hardeman, Ronald Aarts and Ben Jonker
Faculty of Engineering Technology, University of Twente, The Netherlands
t.hardeman@ctw.utwente.nl

Abstract: This paper deals with modelling and identification of flexible-joint robot models that can be used for dynamic simulation and model based control of industrial robots. A nonlinear finite element based method is used to derive the dynamic equations of motion in a form suitable for both simulation and identification. The latter requires that the equations of motion are linear in the dynamic parameters. For accurate simulations of the robot tip motion the model should describe the relevant dynamic properties such as joint friction and flexibilities. Both the drive and the joint flexibilities are included in the model. Joint friction is described by means of a static friction model, including Coulomb and viscous friction components. The dynamic parameters describing mass, inertia, stiffness, damping and friction properties are obtained from a least squares solution of an over determined linear system assembled from closed loop identification experiments. In the identification experiment the robot moves along a prescribed trajectory while all joint angles, flexible deformations and driving torques are recorded. In order to excite joint vibrations during the identification feed forward torques at frequencies above the bandwidth of the control system are superposed on the joint torques. The applicability of the method is demonstrated in a numerical study of a four link industrial robot.

Keywords: parameter identification, flexible/elastic-joint robots, nonlinear finite element method.

INTRODUCTION

Dynamic simulation and model based control of industrial robots require a correct description of the equations of motion and accurate knowledge of the dynamic parameters. Determination of the dynamic parameters by using CAD data may not yield a complete representation as it is difficult to include dynamic parameters like joint friction, masses introduced by additional equipment and stiffness parameters. Experimental parameter identification using

H. Ulbrich and W. Günthner (eds), IUTAM Symposium on Vibration Control of Nonlinear Mechanisms and Structures, 173–182.

the assembled robot may be the only practical method to determine accurate values of the dynamic parameters.

The dynamic parameters can be identified by using a linear least squares method provided that all degrees of freedom are measured during the identification experiment. Industrial robots are usually only equipped with motor encoders that measure joint angles. This implies that additional sensors are required to measure flexible joint degrees of freedom, e.g. by means of acceleration sensors (Pham et al., 2002), link position and/or velocity sensors (Tsaprounis and Aspragathos, 2000; Huang, 2003) or torque sensors (Albu-Schäffer and Hirzinger, 2001). Östring et al. (2003) and Pham et al. (2001) have tried to avoid the use of additional sensors. Their approach is restricted to linear models, so nonlinear terms like Coulomb friction and gravitation forces can not be included in the dynamic model. Omitting these forces is not realistic since joint friction accounts for a significant part of the motor torques of industrial robots. Therefore application of additional sensors seems to be the most promising method to identify the dynamic parameters of flexible-joint robots. Within the scope of this paper it is assumed that all flexible deformations can be measured for example by means of a camera system as used by Yoshikawa et al. (2001).

In most of the literature concerning flexible-joint robots only dive flexibilities are considered. Flexible joints are modelled as torsional springs (Spong, 1987). From static experiments on an industrial robot in our laboratory it appeared that the joint stiffness introduced by the bearings is of the same order as the drive stiffness. The problem of combined joint and drive flexibilities has not received much attention so far. However for realistic dynamic simulations of robot tasks that demand high tracking accuracy combined with high operational speeds the effects of both joint and drive flexibilities should be included in the model.

In this paper a nonlinear finite element method is used to generate dynamic equations of motion in terms of a mixed set of generalized coordinates describing joint angles and deformation parameters that characterize both joint and drive flexibilities. Joint friction is described by means of a static friction model, including Coulomb and viscous friction components. The equations are also written in a form linear in the dynamic parameters. This is to facilitate the use of linear least squares techniques for the identification of the parameters. Measurement data is collected from closed loop identification experiments where the robot is programmed to follow a prescribed trajectory. In order to excite the flexibilities feed forward torques with frequencies above the bandwidth of the feedback controller will be applied. The identification method will be demonstrated on simulated responses of a four link flexible-joint manipulator. The results are validated by comparing computed tip motions of the original and the identified model.

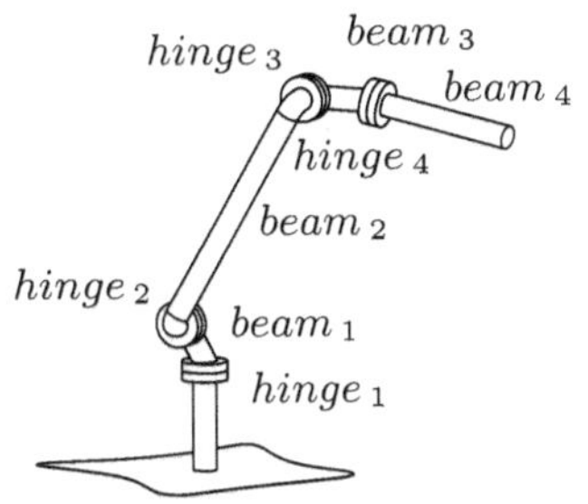

Figure 1. Model 4 DOF manipulator.

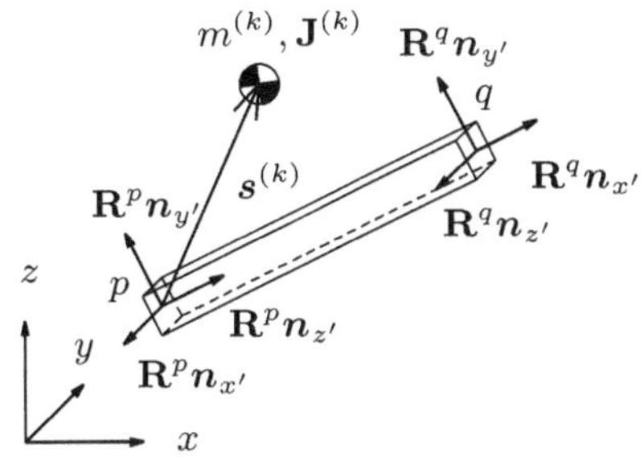

Figure 2. Link k with lumped mass $m^{(k)}$ and rotational inertia $\mathbf{J}^{(k)}$.

1. ROBOT MODEL

In the presented finite element method a manipulator is modelled as an assembly of finite elements (Jonker, 1990). This is illustrated in Figure 1 where a 4 DOF manipulator is modelled with two different types of elements. The manipulator links are modelled as beam elements and the joints are modelled by hinge elements. The location of each element is described relative to a fixed inertial coordinate system by a set of nodal coordinates $x_i^{(k)}$. Some coordinates may be Cartesian coordinates of the end nodes, while others describe the orientation of orthogonal triads, rigidly attached to the element nodes. The superscript k is added to show that a specific element k is considered. With respect to some reference configuration of the element, the instantaneous values of the nodal coordinates represent a fixed number of deformation modes for the element. The deformation modes are specified by a set of deformation parameters $\boldsymbol{e}^{(k)}$, some of which are associated with large relative displacements and rotations between the element nodes, while others describe small elastic deformations of the element and will be denoted by $\boldsymbol{\epsilon}^{(k)}$. The number of deformation parameters is equal to the number of nodal coordinates minus the number of degrees of freedom of the element as a rigid body. Note that rigid body motions of the elements are characterized by displacements and rotations of the nodal points for which all deformations are zero. The components of the vector of deformation parameters $(\boldsymbol{e}^{(k)}, \boldsymbol{\epsilon}^{(k)})$ can be expressed as analytic functions of the vector of nodal coordinates $\boldsymbol{x}^{(k)}$. In this way for each element k a vector function $(\boldsymbol{e}^{(k)}, \boldsymbol{\epsilon}^{(k)}) = \boldsymbol{\mathcal{D}}^{(k)}(\boldsymbol{x}^{(k)})$ can be defined.

1.1 Link Model

The manipulator links are modelled by beam elements. The configuration of the beam element is determined by the position vectors $\boldsymbol{x}^p$ and $\boldsymbol{x}^q$ of the end nodes and the angular orientation of orthogonal triads $(\boldsymbol{n}_{x'}, \boldsymbol{n}_{y'}, \boldsymbol{n}_{z'})$ rigidly attached to the element nodes as shown in Figure 2. The triads coincide with the axis pq and the principle axes of its cross section. The rotation part of the motion of the beam element is described by rotation of the triads $(\boldsymbol{n}_{x'}, \boldsymbol{n}_{y'}, \boldsymbol{n}_{z'})$ which are determined by rotation matrices $\mathbf{R}^p$ and $\mathbf{R}^q$, respectively. Six deformation parameters are defined for the beam element. As the links are rigid all deformations are prescribed zero. For a detailed description of the deformation functions $\mathcal{D}^{(k)}(\boldsymbol{x}^{(k)})$ the reader is referred to Jonker (1990).

The assumption of rigid links enables the use of a lumped mass description, in which the inertia and mass properties of a link are modelled by means of a lumped body attached at node p of the beam element, see Figure 2. The dynamic properties of each body are described by a symmetric rotational inertia matrix $\mathbf{J}^{(k)}$, a mass $m^{(k)}$ and a vector $\boldsymbol{s}^{(k)}$ defining the center of gravity with respect to the corresponding element node at which the body is lumped. For each link element a lumped parameter vector $\boldsymbol{p}^{(l,k)}$ is defined as,

$$\boldsymbol{p}^{(l,k)} = (m, m\, s_{x'}, m\, s_{y'}, m\, s_{z'}, J_{x'x'}, J_{x'y'}, J_{x'z'}, J_{y'y'}, J_{y'z'}, J_{z'z'})^{(k)} \quad (1)$$

containing the components of $m^{(k)}$,$\boldsymbol{s}^{(k)}$ and $\mathbf{J}^{(k)}$ respectively. A detailed description can be found in Hardeman et al. (2005).

1.2 Joint Model

Each robot joint is modelled by two cylindrical hinge elements, a motor hinge element and a flexible hinge element, describing the joint and drive flexibilities, see Figure 3. Each cylindrical hinge element hereafter called hinge element has two nodes p and q at the element axis, see Figure 4. The configuration of the element is determined with two orthogonal triads $(\boldsymbol{n}_{x'}, \boldsymbol{n}_{y'}, \boldsymbol{n}_{z'})$ rigidly attached to each end node. For the hinge element three deformation functions are defined, namely a large relative rotation about the element axis denoted $e_1^{(k)}$ and two small orthogonal bending deformations denoted $\epsilon_2^{(k)}$ and $\epsilon_3^{(k)}$ respectively. For a detailed description of the hinge element the reader is referred to Geradin et al. (1986) and Schwab and Meijaard (1999).

The motor hinge represents the driving system which generates the net driving torque $\tau^{(k)}$. The relative rotation $e_1^{(k)}$ is referred to as the joint angle. The bending deformations $\epsilon_2^{(k)}$ and $\epsilon_3^{(k)}$ of the motor hinge are prescribed zero. The net driving torque is the driving torque of the motor $\tau^{(m,k)}$ minus the torque to accelerate a motor inertia $J^{(m,k)}$ and a joint friction torque $\tau^{(f,k)}$

$$\tau^{(k)} = \tau^{(m,k)} - J^{(m,k)}\ddot{e}_1^{(k)} - \tau^{(f,k)}. \quad (2)$$

Joint friction is described using a static friction model containing Coulomb $c^{(c,k)}$ and viscous $c^{(v,k)}$ friction, i.e.

$$\tau^{(f,k)} = c^{(c,k)}\mathrm{sign}(\dot{e}_1^{(k)}) + c^{(v,k)}\dot{e}_1^{(k)}. \quad (3)$$

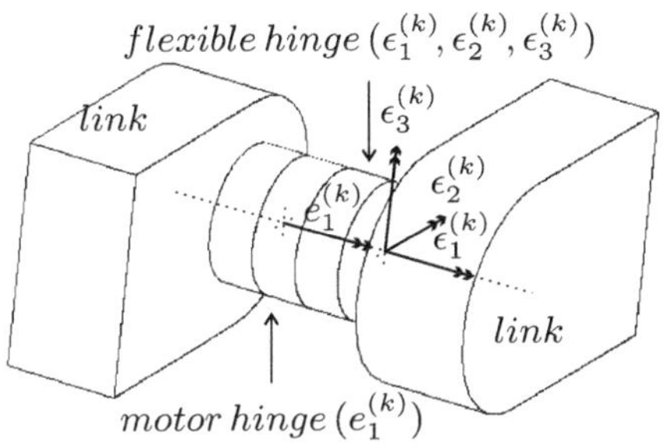

Figure 3. Joint model.

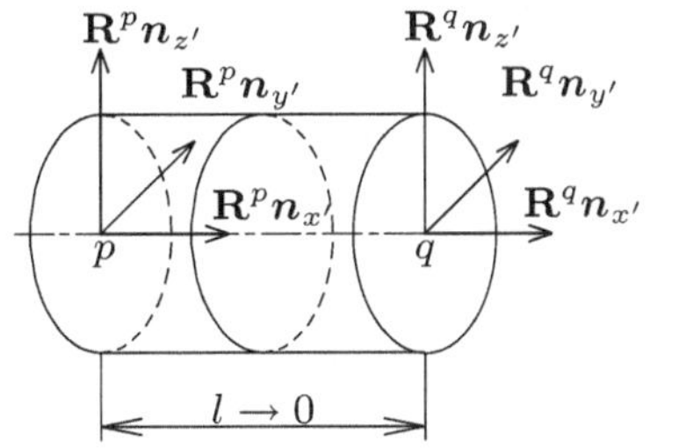

Figure 4. Hinge element.

The joint and drive flexibilities are described by the flexible hinge element. The elastic deformations of the hinge are assumed to be small and are denoted by $\boldsymbol{\epsilon}^{(k)} = (\epsilon_1^{(k)}, \epsilon_2^{(k)}, \epsilon_3^{(k)})$ for respectively the drive flexibility and the two orthogonal bending modes of the joint. The vector of associated stress resultants $\boldsymbol{\sigma}^{(k)}$ is linearly related to the deformations $\boldsymbol{\epsilon}^{(k)}$ and their time derivatives $\dot{\boldsymbol{\epsilon}}^{(k)}$ by the constitutive equation

$$\boldsymbol{\sigma}^{(k)} = \mathbf{K}^{(k)}\boldsymbol{\epsilon}^{(k)} + \mathbf{D}^{(k)}\dot{\boldsymbol{\epsilon}}^{(k)}, \tag{4}$$

where $\mathbf{K}^{(k)}$ and $\mathbf{D}^{(k)}$ are diagonal matrices containing respectively the stiffness parameters $\boldsymbol{k}^{(k)}$ and structural damping parameters $\boldsymbol{d}^{(k)}$.

1.3 Equations of Motion

A manipulator model is built up with finite elements by letting them have nodal points in common. The assembly of finite elements is realized by defining a vector $\boldsymbol{x}$ of nodal coordinates for the entire manipulator mechanism. The motion of the manipulator mechanism is described by relative degrees of freedom, which are the joint rotations, denoted $\boldsymbol{e}^{(m)}$, as well as the elastic deformation parameters of the joints denoted by $\boldsymbol{\epsilon}^{(m)}$. By means of the geometric transfer function $\mathcal{F}^{(x)}$, the nodal coordinates $\boldsymbol{x}$ are expressed as functions of the generalized coordinates $\boldsymbol{q} = (\boldsymbol{e}^{(m)}, \boldsymbol{\epsilon}^{(m)})$, so

$$\boldsymbol{x} = \mathcal{F}^{(x)}(\boldsymbol{e}^{(m)}, \boldsymbol{\epsilon}^{(m)}) = \mathcal{F}^{(x)}(\boldsymbol{q}). \tag{5}$$

Differentiating the transfer function with respect to time gives

$$\dot{\boldsymbol{x}} = \mathrm{D}\mathcal{F}^{(x)}\dot{\boldsymbol{q}}, \tag{6}$$

where the differentiation operator D represents partial differentiation with respect to the degrees of freedom. Again differentiating with respect to time yields the accelerations

$$\ddot{\boldsymbol{x}} = \mathrm{D}\mathcal{F}^{(x)}\ddot{\boldsymbol{q}} + (\mathrm{D}^2\mathcal{F}^{(x)}\dot{\boldsymbol{q}})\dot{\boldsymbol{q}}. \tag{7}$$

The geometric transfer functions $\mathcal{F}^{(x)}$ and their derivatives $\mathrm{D}\mathcal{F}^{(x)}$ and $\mathrm{D}^2\mathcal{F}^{(x)}$ are computed from the deformation functions of the finite elements and are determined numerically in a iterative way, see Jonker (1990).

Let $\mathbf{M}$ be the global mass matrix and $\boldsymbol{f}$ be the vector of nodal forces including the velocity dependent inertia forces and the gravitational forces obtained by assembling the contributions of the lumped bodies of the links. Let $\boldsymbol{\tau}^{(m)}$ be the vector of driving torques, $\boldsymbol{\tau}^{(f)}$ be the global vector of friction torques and $\boldsymbol{\sigma}$ be the global vector of stress resultants obtained by assembling the contributions of equation (2) and equation (4) respectively. Then the equations of motion described in terms of the degrees of freedom $\boldsymbol{q}$ are given by

$$\begin{bmatrix} \bar{\mathbf{M}}^{ee} & \bar{\mathbf{M}}^{e\epsilon} \\ \bar{\mathbf{M}}^{\epsilon e} & \bar{\mathbf{M}}^{\epsilon\epsilon} \end{bmatrix} \begin{bmatrix} \ddot{\boldsymbol{e}}^{(m)} \\ \ddot{\boldsymbol{\epsilon}}^{(m)} \end{bmatrix} + \begin{bmatrix} \mathrm{D}_e\mathcal{F}^{(x)T} \\ \mathrm{D}_\epsilon\mathcal{F}^{(x)T} \end{bmatrix} \left[\mathbf{M}\left(\mathrm{D}^2\mathcal{F}^{(x)}\dot{\boldsymbol{q}}\right)\dot{\boldsymbol{q}} - \boldsymbol{f}\right] + \begin{bmatrix} \boldsymbol{\tau}^{(f)} \\ \boldsymbol{\sigma} \end{bmatrix} = \begin{bmatrix} \boldsymbol{\tau}^{(m)} \\ \mathbf{0} \end{bmatrix}, \tag{8}$$

where

$$\begin{aligned} \bar{\mathbf{M}}^{ee} &= \mathrm{D}_{ee}\mathcal{F}^{(x)T}\mathbf{M}\mathrm{D}_{ee}\mathcal{F}^{(x)} + \mathrm{diag}(\boldsymbol{J}^{(m)}), & \bar{\mathbf{M}}^{e\epsilon} &= \mathrm{D}_{e\epsilon}\mathcal{F}^{(x)T}\mathbf{M}\mathrm{D}_{e\epsilon}\mathcal{F}^{(x)}, \\ \bar{\mathbf{M}}^{\epsilon e} &= \mathrm{D}_{\epsilon e}\mathcal{F}^{(x)T}\mathbf{M}\mathrm{D}_{\epsilon e}\mathcal{F}^{(x)}, & \bar{\mathbf{M}}^{\epsilon\epsilon} &= \mathrm{D}_{\epsilon\epsilon}\mathcal{F}^{(x)T}\mathbf{M}\mathrm{D}_{\epsilon\epsilon}\mathcal{F}^{(x)}, \end{aligned}$$

are the reduced mass matrices. Note that matrix $\bar{\mathbf{M}}^{ee}$ also includes the vector of motor inertias $\boldsymbol{J}^{(m)}$. The above formulation of the equations of motion in acceleration linear form will be used for the forward dynamic simulations.

The equations of motion can also be written in a parameter linear form which will be used to estimate the dynamic parameters

$$\Phi(\boldsymbol{q}, \dot{\boldsymbol{q}}, \ddot{\boldsymbol{q}})\boldsymbol{p} = \begin{bmatrix} \boldsymbol{\tau}^{(m)} \\ \mathbf{0} \end{bmatrix}, \tag{9}$$

where

$$\Phi = \begin{bmatrix} \Phi^{(l)} & \mathrm{diag}(\ddot{\boldsymbol{e}}^{(m)}) & \mathrm{diag}(\dot{\boldsymbol{e}}^{(m)}) & \mathrm{diag}(\mathrm{sign}(\dot{\boldsymbol{e}}^{(m)})) & \mathbf{0} & \mathbf{0} \\ & \mathbf{0} & \mathbf{0} & \mathbf{0} & \mathrm{diag}(\boldsymbol{\epsilon}^{(m)}) & \mathrm{diag}(\dot{\boldsymbol{\epsilon}}^{(m)}) \end{bmatrix} \tag{10}$$

and parameter vector $\boldsymbol{p}$ is defined as

$$\boldsymbol{p} = (\boldsymbol{p}^{(l)}, \boldsymbol{J}^{(m)}, \boldsymbol{c}^{(c)}, \boldsymbol{c}^{(v)}, \boldsymbol{k}, \boldsymbol{d}). \tag{11}$$

Vector $\boldsymbol{p}^{(l)}$ includes the lumped mass and inertia parameters of the robot links. Furthermore $\boldsymbol{J}^{(m)}, \boldsymbol{c}^{(c)}, \boldsymbol{c}^{(v)}, \boldsymbol{k}, \boldsymbol{d}$ represent the parameter vectors containing motor inertias, Coulomb friction, viscous friction, stiffness and structural damping coefficients, respectively. For a detailed description of the matrix $\Phi^{(l)}$ the reader is referred to Hardeman et al. (2005).

2. IDENTIFICATION METHOD

The linear least squares technique will be used to identify the unknown dynamic parameters of a robot. During an identification experiment the robot is moved along a prescribed trajectory while the joint driving torques $\boldsymbol{\tau}^{(m)}$ and the degrees of freedom $\boldsymbol{q}$ are measured. The velocities $\dot{\boldsymbol{q}}$ and accelerations $\ddot{\boldsymbol{q}}$ will be obtained afterwards by numerical differentiation. The driving torques $\boldsymbol{\tau}^{(m)}(t_i)$ and the system matrices $\Phi(t_i)$ at all time steps $t_i = 1 \cdots N$ are stacked in vector $\boldsymbol{y}$ and regression matrix $\mathbf{A}$ respectively,

$$\boldsymbol{y} = \begin{bmatrix} \boldsymbol{\tau}(t_1) \\ \mathbf{0} \\ \vdots \\ \boldsymbol{\tau}(t_N) \\ \mathbf{0} \end{bmatrix}, \quad \mathbf{A} = \begin{bmatrix} \Phi(t_1) \\ \vdots \\ \Phi(t_N) \end{bmatrix}, \tag{12}$$

yielding the equations

$$\boldsymbol{y} = \mathbf{A}\boldsymbol{p} + \boldsymbol{\rho}, \tag{13}$$

where the vector $\boldsymbol{\rho}$ represents the residual torques arising from measurement noise and model errors. Given the measurements $\boldsymbol{y}$ and the regression matrix $\mathbf{A}$ the dynamic parameters are identified by solving the linear least squares problem

$$\hat{\boldsymbol{p}} = \arg\min_{\boldsymbol{p}} ||\boldsymbol{y} - \mathbf{A}\boldsymbol{p}||^2. \tag{14}$$

The solution can be found by using the concept of the pseudo-inverse

$$\hat{\boldsymbol{p}} = \mathbf{A}^+\boldsymbol{y}, \tag{15}$$

where $\mathbf{A}^+$ represents the pseudo-inverse of $\mathbf{A}$ that can be calculated from a Singular Value Decomposition of the regression matrix $\mathbf{A}$, see Lawson and Hanson (1974).

Not all parameters of the vector $\boldsymbol{p}$ can be identified since some of the parameters do not affect the dynamic response or affect the dynamic response in linear combinations with other parameters. Application of the Singular Value Decomposition gives insight in the maximum number of identifiable parameters and parameters that show up in linear combinations. Both depend on the actual model, the size of the parameters, the excitation signal and the amount of measurement noise.

The design of an identification experiment includes the choice of input and output signals. With the experimental limitations in mind, the following approach is suggested. Parameters related to position dependent behavior like gravitation can only be identified when measurement data is obtained from sufficiently distinguishable robot configurations. Therefore the robot is programmed to follow a trajectory $\boldsymbol{r}$ along various configurations in its work envelope. For stability and safety reasons the experiments will be carried out in closed loop. Problems arise for the identifiability of the dynamic parameters of the flexible joints because the controller is designed to damp out the flexibilities

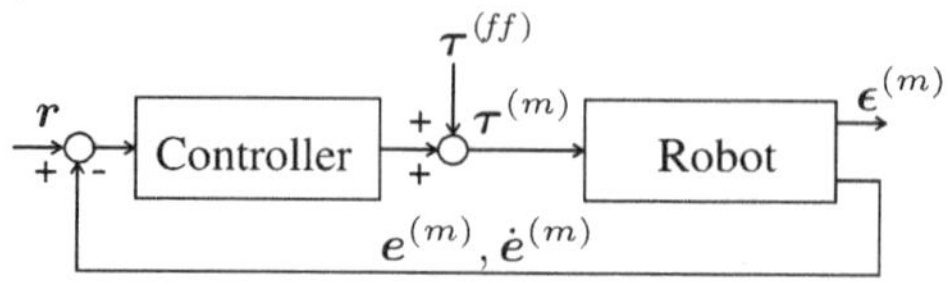

Figure 5. Closed loop system.

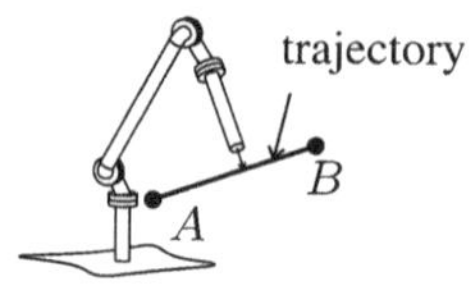

Figure 6. Validation trajectory.

of the robot joints. Therefore during the identification the driving torques are perturbed with feed forward torques $\boldsymbol{\tau}^{(ff)}$ having a frequency spectrum beyond the bandwidth of the closed-loop system, see Figure 5.

3. RESULTS AND DISCUSSION

The identification method will be demonstrated on the basis of simulated responses of the four link manipulator presented in Figure 1. The dynamic parameters of the robot model are adopted from a real industrial robot. The mass and inertia parameters are obtained from CAD data. Friction parameters are found by measuring the joint torque during constant velocity motion. The stiffness parameters are estimated by measuring the static deformations of the joints as functions of static loads applied on the robot arms. A relative damping equal to 1% is assumed for each vibrational mode. The dynamic responses are computed in a closed-loop simulation with a model of a real digital PID controller, see Figure 5. The bandwidth of the closed loop system is in the order of 15 Hz.

During the identification the robot performs a trajectory where each joint angle varies between 0 and $\frac{1}{2}\pi$ according to a sine function with a period of 4.1 s. The joint flexibilities are excited by means of a feed forward torque described by a multi sine signal containing 204 frequencies ranging from 0.12 Hz to 50 Hz. A multi sine signal has been used in order to obtain a sharp cutoff for frequencies above 50 Hz. This is required because in the actual controller the torque and position data are sampled at 250 Hz without any aliasing filter. The velocities and accelerations are obtained by means of numerical differentiation of the position signals using an eight order central difference scheme given by Khan and Ohba (2003).

From the simulated responses the regression matrix **A** and the measurement vector $\boldsymbol{y}$ are constructed. The parameter vector $\boldsymbol{p}$ contains 76 parameters, namely for each link 10 lumped inertia parameters $\boldsymbol{p}^{(l,k)}$ and for each joint 1 motor inertia, 2 friction parameters, 3 stiffness and 3 damping parameters. Using a Singular Value Decomposition of the regression matrix, a parameter set with 65 linearly independent parameters has been identified from the data.

Simulated tip responses of both the original and the identified model are used for validation of the identified parameters. The robot is programmed to

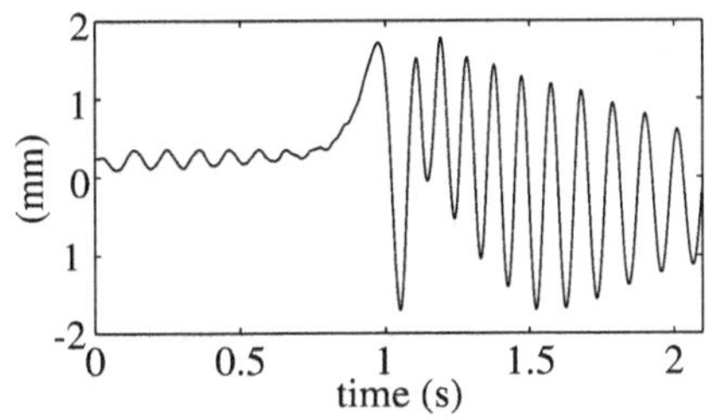

Figure 7a. Vertical deviation robot tip original model.

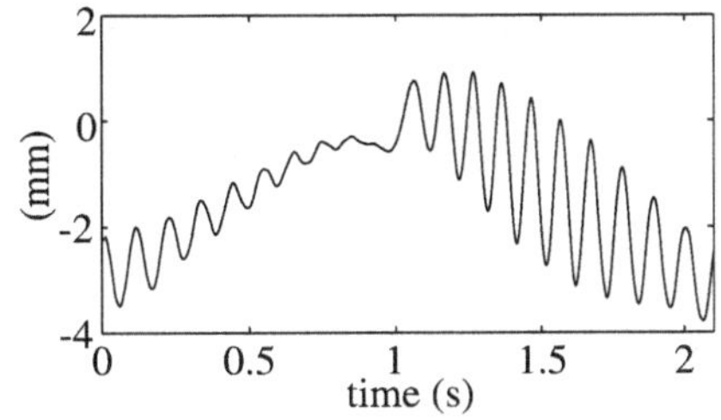

Figure 7b. Horizontal deviation robot tip original model.

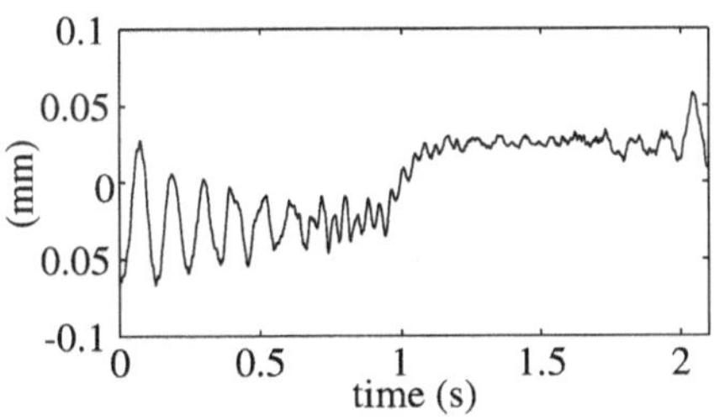

Figure 7c. Difference between tip motion of original and identified model in vertical direction.

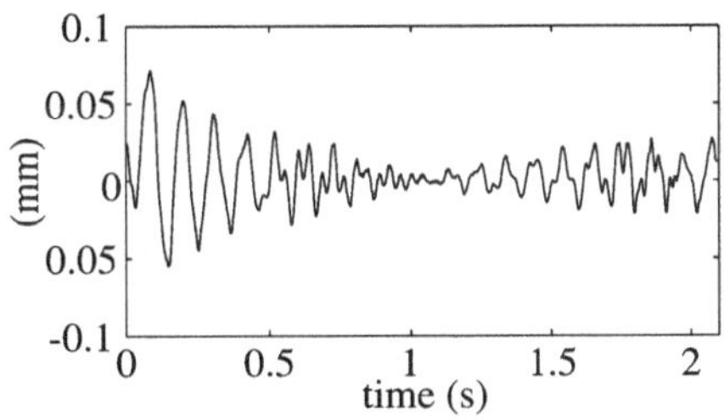

Figure 7d. Difference between tip motion of original and identified model in horizontal direction.

perform a tip motion along a straight line of 1.7 m in the horizontal plane as illustrated in Figure 6 with starting point A and endpoint B. Due to the flexibilities and tracking errors of the control system the robot tip will deviate from the straight line. The tip deviations of the original model are shown in Figures 7a and 7b in vertical direction and horizontal direction, respectively. Figures 7c and 7d show the differences between the tip motion of the original and the identified model again in vertical and horizontal direction. These differences appear to be at least an order of magnitude smaller than the amplitudes of the tip deviations of the original model. Similar results are found when other quantities are compared, e.g. the driving torques. It can be concluded that there is a good agreement between both models.

In this paper we have provided a dynamic robot model describing both drive and joint flexibilities. A nonlinear finite element based method is used to derive the dynamic equations of motion in a form suitable for both simulation and identification. The presented numerical example of the four link manipulator showed that the proposed identification method gives useful results, at least in simulations. In a future research the identification will be carried out with experimental data from the real industrial robot.

ACKNOWLEDGMENTS

This research was carried out under project number MC8.03161 in the framework of the Strategic Research programme of the Netherlands Institute for Metals Research in the Netherlands (www.nimr.nl).

REFERENCES

Albu-Schäffer, A. and Hirzinger, G. (2001). Parameter identification and passivity based joint control for a 7dof torque controlled light weight robot. In *IEEE International Conference on Robotics and Automation*, pp. 2852–2858.

Geradin, M., Robert, G. and Buchet, P. (1986). *Finite Element Methods for Nonlinear Problems*, chapter "Kinematic and Dynamic Analysis of Mechanisms a Fininite Element Approach Based on Euler Parameters". Springer-Verlag, Berlin.

Hardeman, T., Aarts, R.G.K.M. and Jonker, J.B. (2005). A finite element formulation for dynamic parameter identification of robot manipulators. Submitted to *Multibody System Dynamics*.

Huang, J.T. (2003). A new approach to parametric identification of a single-link flexible joint manipulator. *Jounal of Intelligent and Robotic Systems*, 37:273–284.

Jonker, J.B. (1990). A finite element dynamic analysis of flexible manipulators. *International Journal of Robotic Research*, 9:59–74.

Khan, I.R. and Ohba, R. (2003). Taylor series based finite differentce approximations of higher-degree derivatives. *Journal of Computational and Applied Mathematics*, 154:115–124.

Lawson, C.L. and Hanson, R.J. (1974). *Solving Least Squares Problems*. Prentice-Hall, Englewood Cliffs, NJ.

Östring, M, Gunnarsson, S. and Norrlöf, M. (2003). Closed-loop identification of an industrial robot containing flexibilities. *Control Engineering Practice*, 11:291–300.

Pham, M.T., Gautier, M. and Poignet, Ph. (2001). Identification of joint stiffness with bandpass filtering. In *IEEE International Conference on Robotics and Automation*, pp. 2867–2872.

Pham, M.T., Gautier, M. and Poignet, Ph. (2002). Accelerometer based identification of mechanical systems. In *IEEE International Conference on Robotics and Automation*, pp. 4293–4298.

Schwab, A.L. and Meijaard, J.P. (1999). The belt, gear, bearing and hinge as special finite elements for kinematic and dynamic analysis of mechanisms and machines. In *Dynamics, Nonlinear Oscillations, Rotor Dynamics and Software Development*.

Spong, M.W. (1987). Modeling and control of elastic joint robots. *Journal of Dynamic Systems, Measurement, and Control*, 109:310–319.

Tsaprounis, C.J. and Aspragathos, N.A. (2000). Adaptive tracking controller for rigid-link elastic-joint robots with link acceleration astimation. *Journal of Intelligent and Robotic Systems*, 27:68–83.

Yoshikawa, T., Ohta, A. and Kanaoko, K. (2001). State estimation and parameter identification of flexible manipulators based on visual sensor and virtual joint method. In *IEEE International Conference on Robotics and Automation*, pp. 2840–2845.

OPTIMAL ROBUST CONTROLLERS FOR MULTIBODY SYSTEMS

Petko Kiriazov
Institute of Mechanics, Bulgaria

Abstract: The proposed paper addresses control design problems for multibody systems (MBS) like robot manipulators or mechanical structures with active vibration damping. When performing motion tasks, such systems may be subject to even severe disturbances. A control design approach for motion stabilisation has to meet the increasing demands for faster response, higher position accuracy, and reduced energy consumption. A central role in solving such a complicated control optimisation problem plays the matrix that transfers the control inputs into mechanical accelerations. For MBS having as many control forces as controlled outputs, simple conditions on that matrix are found to be necessary and sufficient for such systems to be controllable in the presence of bounded random disturbances. There are proposed optimal trade-off relations for designing decentralised controllers with maximum degree of robustness. An interesting extension of these concepts to the important class of over-controlled MBS is proposed. Examples with a car body suspension and an elastic-joint manipulator are presented to show how the proposed control design approach can be applied and developed.

Key words: arbitrary disturbances, robustness, decentralised control, optimisation, vibration damping, active suspension, elastic-joint manipulators.

1. INTRODUCTION

The proposed paper addresses control design problems for multibody systems (MBS) like robot manipulators or mechanical structures with active vibration damping. MBS of micro- and nano-scale that may be used in LCD or semiconductor manufacturing, optical-electronic industry or microbiology are also of great interest as vibration isolation for such systems is a

H. Ulbrich and W. Günthner (eds), IUTAM Symposium on Vibration Control of Nonlinear Mechanisms and Structures, 183–192.

significant problem, too. When performing motion tasks, MBS may be subject to severe disturbances that may appear arbitrary in time and location.

Usually, active vibration isolation systems are based upon a hybrid concept, combining a passive isolator for higher frequencies and a servo control system for lower frequencies. Actuators for vibration control may be of various-type, e.g., electrical motors, electro-hydraulic cylinders, piezo-electric, electro-magnetic actuators, etc. In practice, servo-controllers of decentralised structure are preferred, as they are simpler, faster and more reliable than centralised ones.

A control design approach for MBS motion stabilisation has to meet the increasing demands for faster response, higher position/orientation accuracy, and reduced energy cost. Classical control design techniques, as verified in Keel and Bhattacharyya (1997), cannot ensure optimality and robustness even for linear systems if substantial errors in their modelling exist.

In optimal feedback control problems, one has in principle to optimise between error and energy for finding the most appropriate control gains. In the well-known control methods using linear-quadratic or H-infinity optimisation functionals (Amirifar and Sadati, 2004), the weight matrices are critical towards the required system performance. Choosing their coefficients only by trial-and-error, however, may not guarantee achievement of the required motion accuracy within the existing control power limits. Moreover, optimality and robustness cannot be guaranteed if a control design method does not take into account a realistic estimate for modelling errors and system disturbances.

Our intention is to propose and develop a control design approach, which has the potential to overcome any of the above problems in an efficient way. MBS of various type can be considered with the premise that they can be equipped with a sufficiently great number of actuators. Our concept is that optimal and robust, to arbitrary-in-place external disturbances, system performance is hard to provide if the actuator number is smaller than the number of the controlled performance variables. The proposed approach is based on a diagonal dominance property of the MBS control transfer matrix, which is proved to be the necessary and sufficient condition for robust decentralised controllability (Kiriazov, 1994; Kiriazov et al., 1997). Using that property, optimal trade-off relations between bounds of disturbances and control power limits can be derived. In other words, they make it possible to design decentralised controllers with maximum degree of robustness. Two important case studies: a car body suspension and an elastic-joint manipulator will be presented to show how our control design approach can be applied and developed. A very important feature of this approach is that its concepts and design criteria can be directly used for efficient overall optimisation of controlled MBS.

2. MATHEMATICAL MODELLING

First, it is very important to properly define the controls, the performance variables to be controlled, and the structure of the input-output relations (Skogestad and Postlethwaithe, 1996). A general mathematical model of a flexible, in general, mechanical system driven by control forces $u \in R^m$ is

$$\ddot{q} = f(p,t,q,\dot{q},u,d) \tag{1}$$

where $q \in R^n$ are the displacements (relative or absolute), p - model parameters, and d - arbitrary, but of course, bounded disturbances.

We have to find however reduced-order, nominal dynamic models that are appropriate for both, accurate parameter identification and robust control design (Skelton, 1999). Any mechanical system can be approximated, in general, by a composition of rigid bodies connected by joints, springs, dampers, and driving forces (Schiehlen, 1990). Applying the Lagrange formalism, the dynamic performance of such a MBS can be described by

$$M(q)\ddot{q} = \tau - v(q,\dot{q}) - s(q) - g(q) \tag{2}$$

The main dynamic terms are the inertia $M(q)\ddot{q}$, generalised τ, velocity $v(q,\dot{q})$, stiffness $s(q)$, and gravity $g(q)$ forces. The generalised forces can be related to the actuator forces/torques u by $\tau = Bu$, where B is a matrix representing the actuator locations/gains.

Thus for MBS with elastic elements, dynamic model structures can be described by a relatively small number of parameters. All coefficients in (2) can be identified in a systematic way as proposed in Kiriazov (2001).

In principle, using both feedforward and feedback control functions is needed for optimal performance to be achieved (Riebe and Ulbrich, 2004). With a feedforward control, we can compensate, to some extent, for the gravitation, stiffness and other forces. Then, for the purpose of feedback control design, the following simple, but a sufficiently general, model for the error dynamics can be used:

$$\ddot{e} = A(q)u + d \tag{3}$$

where $e = q - q^{ref}$; vector d stands for the model imprecision and the uncompensated terms as well as for the measurement and environment noises; $A = M^{-1}(q)\ B$ is the control transfer matrix which will play a central role in the next control design considerations.

3. CONTROL DESIGN APPROACH

With passive vibration damping we may provide sufficiently good performance for higher frequencies. However, passive isolation system has bad performance around natural frequencies. Active vibration damping has to be used to improve the isolation performance for lower frequencies.

A control design approach for MBS motion stabilisation has to meet the increasing demands for faster response, higher position/orientation accuracy, and reduced energy cost. In the face of bounded-in-magnitude but arbitrary-in-time/place disturbances, our objective is to design *n* properly bounded decentralised feedback controllers that have maximum capability to achieve the required motion accuracy.

As a measure of tracking precision, we take the absolute value of $s = \dot{e} + \lambda e$, $\lambda > 0$. We will use decentralised controllers, i.e., the stabilising control force u_i depends solely on the corresponding controlled output s_i. As usual, robotic systems have decentralised controllers at the joint level, which are simpler, faster and more reliable than centralised ones.

DEFINITION: A decentralised controller is robust against random disturbances d with known upper bounds $\hat{d}$ if it gets the local subsystem state $(\dot{q}, q)$ at each joint to track the desired state $(\dot{q}^{ref}, q^{ref})$ with maximum allowable absolute values $\hat{s}$ of the errors s.

As we stated, robust to arbitrary-in-place external disturbances system performance can be provided if the number of the actuators *m* is not smaller than the number of the performance variables *n*. When *m* equals *n* and we have a square control transfer matrix, analytical guarantees for optimal and robust feedback control of such MBS are next proposed. The case of over-controlled MBS, i.e. *m>n,* will be considered in the next subsection.

3.1 Control Design with Square Control Transfer Matrix

For MBS with square control transfer matrix A, we will use the following key result (Kiriazov, 1994; Kiriazov et al., 1997):

THEOREM: A necessary and sufficient condition for a MBS to be robustly controlled by a decentralised controller is that A be a generalised diagonal dominant (GDD) matrix.

When matrix A is GDD, the non-negative matrix theory (Lunze, 1992) ensures that there always exists a positive vector $\hat{u}$ of control force magnitudes solving the following system of equations

$$A_{ii}\hat{u}_i - \sum_{j \neq i} | A_{ij} | \hat{u}_j = \hat{d}_i , \; i = 1,...,n \tag{4}$$

Equations (4) present optimal trade-off relations between the bounds of model uncertainties and the control force limits. Then, with known coefficients of *A*, we can design optimal robust controllers. To do that, we can use sliding-mode controllers or position-derivative (PD) controllers with bounded control functions, (Slotine and Shastry, 1983; Kiriazov, 1994):

$$\begin{aligned} &u_i(s_i) = -\hat{u}_i \, \mathtt{sat}(s_i / \delta_i), i = 1,\dots,n \\ &\mathtt{sat}(y) = y \text{ for } |y| < 1 \text{ and } \mathtt{sat}(y) = \mathtt{sgn}(y) \text{ for } |y| \geq 1 \end{aligned} \tag{5}$$

Parameters $\delta_i > 0$ in (5) are needed to provide the control functions with some smoothness as required in practice. The design relations (4) give the minimum control force magnitudes $\hat{u}$ that can overcome the disturbances. In what follows, we relate them with the physical actuator limits when the number of actuators is greater than the number of the controlled outputs.

3.2 Over-Controlled Mechanical Systems

The proposed control design concepts can also be developed and applied to over-controlled mechanical systems, i.e., when their actuators are more than the controlled performance variables. Such systems are, e.g., full-car body models, redundant manipulators, or, more generally, flexible mechanical structures represented by MBS that are over-controlled.

We consider again the dynamic model (2) and assign τ to be the control forces/torques in this case. The corresponding control transfer matrix is $M^{-1}(q)$ and we can be in position to design robust decentralised controllers for τ, using the optimal trade-off relations (4), if this matrix is a GDD one. In most practical cases, such condition can be fulfilled if we consider the problem of motion control of the MBS in the real space. For example, this is the case with the three-degree-of-freedom platform, Figure 1.

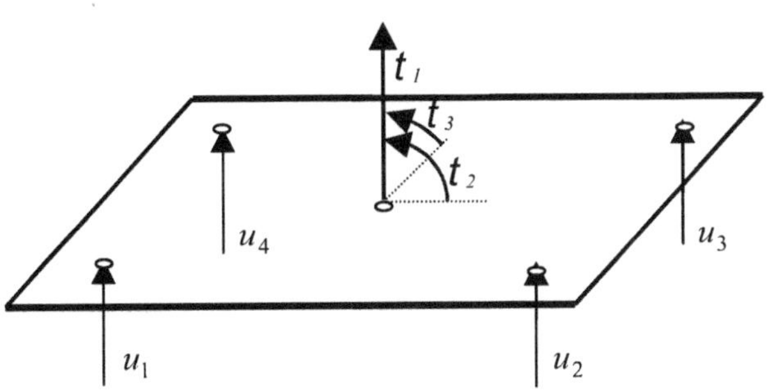

Figure 1: Bounce/pitch/roll control of a platform.

The reason to assign four single-axis force actuators to stabilise its bounce/pitch/roll motion comes naturally from its rectangular shape. Various controlled mechanical systems used on the earth (e.g., ground vehicles) and in the space (stations, platforms) may have such a shape and such a number of actuators.

In order to estimate the capability of the control functions τ for robust stabilisation, we have to take into consideration the power limits of the real actuators u, based on the relationship $\tau = Bu$. Optimal design relations regarding the magnitudes of τ -controllers are next proposed.

We use the Moore-Penrose pseudo inverse of matrix B, i.e., B^+. Then

$$u = B^+\tau \tag{6}$$

The fact that the decentralised τ -controllers, during the motion stabilisation task, may take either positive or negative values independently, means that the following inequalities have to be satisfied:

$$\sum_{j=1}^{n} \left|B_{ij}^+\right| \hat{\tau}_j \le \hat{u}_i, \forall i = 1,\ldots,m \text{ where } \max\left|u_i\right| = \hat{u}_i \text{ and } \max\left|\tau_j\right| = \hat{\tau}_j \tag{7}$$

These relations can be useful, in fact, when designing decentralised τ -controllers for any MBS having GDD inverse inertia matrices.

4. CASE STUDIES

As we see, the proposed approach is quite a generic one and can be applied in various cases: vehicle suspensions, manipulators with elastic joints, Stewart- (Riebe and Ulbrich, 2004) and other type platforms (Amirat et al., 1996). To illustrate its main features in specific cases, we consider two simple models having different dynamic structures: a half-car body with hybrid (passive and active) suspension and a two-link manipulator model with elastic joints.

4.1 Half-Car Body Model

Active vibration damping for minimising vertical car body acceleration and oscillation is especially important for off-road vehicles. We consider the hybrid suspension concept where the active mount is modelled as a single-axis force actuator in parallel with a passive spring and damper.

When displacements are relatively small, the two motions, pitch/bounce and roll/bounce, are weakly coupled (Takezono et al., 1999) and can be considered separately (Figure 2).

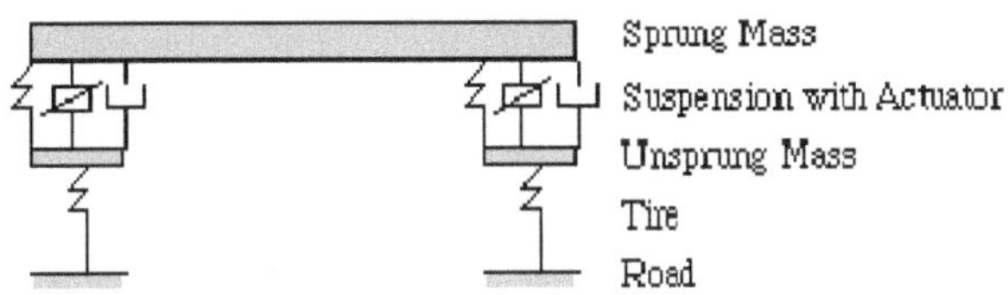

Figure 2: Sky-hook damping half-car model.

Such controlled dynamic systems satisfy the GDD-condition and our control design approach can be applied. Bounded PD controllers (5) are designed according to the optimal trade-off relations (4). In the numerical considerations, a mathematical model similar to that given in Krtolica and Hrovat (1992) is used and random disturbances with non-zero mean value are applied. It is found that the motion stabilisation control can be robust even when the disturbance magnitudes take values up to 80% of the nominal accelerations (produced by the nominal control forces). The robust transient behaviour, starting from a non-zero initial state, is depicted in Figure 3.

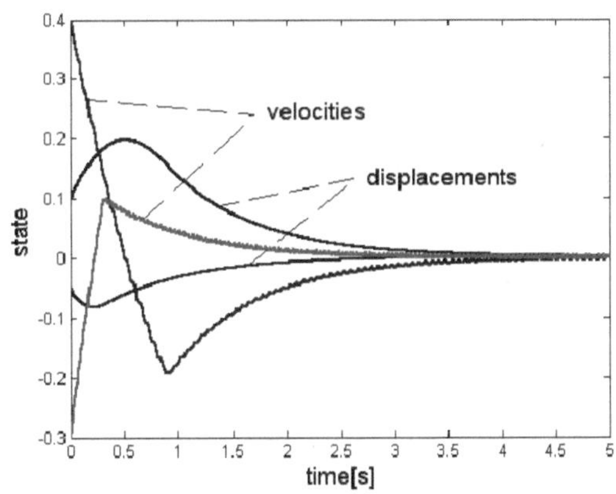

Figure 3: Stabilisation with hybrid suspension control.

Interesting application of our control design approach can be done also for driver seats with hybrid suspension. In general, implementing hybrid suspension control can be useful not only for vehicles moving on rough terrain. Any vehicle (on roads or rails) needs some active control for better performance when making turns.

4.2 Manipulator with Elastic Joints

The dynamic performance of a rigid-link elastic-joint manipulator can be described, by a system of differential equations, as in Lim et al. (1997)

$$\begin{aligned} M(q)\ddot{q} &= K(q_{mot} - q) - v(q,\dot{q}) - g(q) \\ J(q)\ddot{q}_{mot} &= u - K(q_{mot} - q) - C\dot{q}_{mot} \end{aligned} \tag{8}$$

Here, u are the motor torques, q_{mot} and q represent the motor and the corresponding link rotations, and the diagonal constant matrices *K, J,* and *C* describe the elasticity, inertia, and damping of the joint actuators, respectively. The motors do not, however, directly drive the manipulator. According to Eq. $(8)_1$, its driving torques are $K(q_{mot} - q)$, generated due to the joints elasticity. Thus, we have a complicated MBS to control but we can take advantage of the fact that the response of the actuators can be much faster than that of the manipulator.

Our control design approach is applied in the case of two rotational joints. Using design relations (4), we find first the optimal ratio between the magnitudes of the driving torques. Then, we design a stabilising controller composed of two PD controllers: one is to provide the necessary magnitudes for the driving torques and the other is for tracking the required accuracy:

$$u = -\hat{u}\,\mathrm{sat}(a(q_{mot} - q + l^{(1)}\dot{q}_{mot} + b\,\mathrm{sign}(q + l^{(2)}\dot{q}))) \tag{9}$$

A computer simulation is done in the worst case of stretched-out manipulator configuration. It is found that, even with substantial external disturbances, the link position errors converge to zero, as shown in Figure 4.

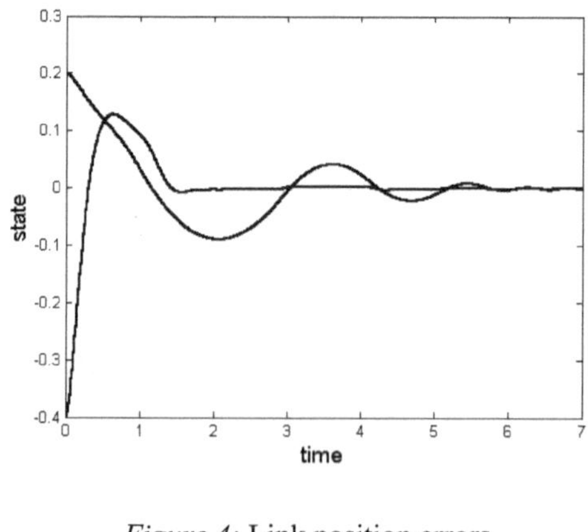

Figure 4: Link position errors.

5. INTEGRATED STRUCTURE-CONTROL DESIGN

Usually, control design methods assume that all hardware components of the mechanical system (especially, the mechanical structure incl. its motors) have been designed except the control component. But it is important to know how its structure should be designed in order for the system to have a higher capability for better dynamic performance. Structure design considerations will be next proposed for MBS with GDD control transfer matrices. Numerous verifications (Kiriazov, 1997; Kiriazov et al., 1997; Kiriazov and Schiehlen, 1997) show that the GDD-condition is not so restrictive but rather a reasonable one. Such a condition can be fulfilled for most properly designed controlled MBS. It has been verified that this requirement is even in accordance with the basic, strength requirements.

If we again consider Eq. (4), we can see that the greater the determinant D of this system of linear equations, the less control effort will be required to overcome the disturbances. In other words, the index D can quantify the capability of MBS to be robustly controlled in a decentralised manner. It can be used first to optimise the control system structure, e.g., finding optimal number and locations for their actuators and sensors. Then we can follow the following

Structure Design Optimisation Scheme

1. *Mechanical Structure*
   - *design parameters*: mass/inertia parameters;
   - *constraints*: strength requirements and GDD-condition;
   - *design objective*: maximise D;
2. *Actuators*
   - *design parameters*: actuator sizes;
   - *constraints*: strength requirements and Eq. (4);
   - *design objective*: minimise actuator power needed;

For over-controlled MBS, we have to add relations (7) to the constraints for the actuator sizes.

6. CONCLUSION

For MBS having as many control forces as controlled outputs, simple (GDD) conditions on the control transfer matrix are found to be necessary and sufficient for such systems to be controllable in the presence of arbitrary-in-time/place but bounded-in-magnitude disturbances. Optimal trade-off relations are proposed that can be used for designing decentralised controllers with maximum degree of robustness. An interesting extension of these concepts to the important class of over-controlled MBS is presented. Numerical examples with a car body suspension and an elastic-joint manipulator are given to show how efficiently the proposed control design approach can be applied. Moreover, its concepts and design criteria can be directly used for direct overall optimisation of controlled MBS.

REFERENCES

Amirat, Y., Francois, C., Fried, G., Pontnau, J. and Dafaoui, M., 1996, Design and control of a new six dof parallel robot: application to equestrian gait simulation, *Mechatronics*, Vol. 6, pp. 227–239.

Amirifar, R. and Sadati, N., 2004, A low-order H-infinity controller design for an active suspension system via linear matrix inequalities, *Journal of Vibration and Control*, Vol. 10(8), pp. 1181–1197.

Keel, L.H. and Bhattacharyya, S.P., 1997, Robust, fragile, or optimal?, *IEEE Transactions on Automatic Control*, Vol. 42(8), pp. 1098–1105.

Kiriazov, P., 1994, Necessary and sufficient condition for robust decentralized controllability of robotic manipulators, *American Control Conference*, Baltimore MA, pp. 2285–2287.

Kiriazov, P., 1997, Robust decentralized control of mechanical systems, In: *Solid Mechanics and Its Applications*, Vol. 52, Ed. D. van Campen, Kluwer Acad. Publ., pp. 175–182.

Kiriazov, P., 2001, Efficient approach for dynamic parameter identification and control design of structronic systems, In: *Solid Mechanics and Its Applications*, Vol. 89, Eds. U. Gabbert and H. Tzou, Kluwer Acad. Publ, pp. 323–330.

Kiriazov, P., Kreuzer, E. and Pinto, F., 1997, Robust feedback stabilization of underwater robotic vehicles, *J. Robotics and Autonomous Systems*, Vol. 21, pp. 415–423.

Kiriazov P. and Schiehlen, W., 1997, On direct-search optimization of biped walking. In *CISM Courses*, Vol. 381, Eds. A. Morecki et al., Springer, Wien/NewYork, pp. 134–140.

Krtolica, R. and Hrovat, H., 1992, Optimal active suspension control based on a half-car model: an analytical solution, *IEEE Trans. on Automatic Control,* No. 4, pp. 528–532.

Lim, S.Y., Dawson, D.M., Hu, J. and de Queiroz, M.S., 1997, An adaptive link position tracking controller for rigid-link flexible-joint robots without velocity measurements, *IEEE Transactions on Systems, Man, and Cybernetics-Part B*, Vol. 27, No. 3, pp. 412–427.

Lunze, J., 1992, *Feedback Control of Large-Scale Systems*, Prentice Hall, UK.

Riebe, S. and Ulbrich, H., 2004, Model-based vibration isolation of a hexapod-system using a combined feedforward-feedback control concept, *Proceedings of the IEEE Conference on Mechatronics and Robotics*, Aachen, pp. 400–405.

Schiehlen, W. (Ed.), 1990, *Multibody Systems Handbook*. Springer, Berlin.

Skelton, R.E., 1999, System design: The grand challenge of system theory, *Plenary lecture of the 1999 American Control Conference*, San Diego, California, USA.

Slotine, J.-J. and Shastry, S.S., 1983, Tracking control of nonlinear systems using sliding surfaces with application to robotic manipulators, *Int. Journal of Control*, No. 2, 465–492.

Skogestad, S. and Postlethwaithe I., 1996, *Multivariable Feedback Control: Analysis and Design*, Wiley.

Takezono, S., Minamoto, H. and Tao, K., 1999, Two-dimensional motion of four-wheel vehicles, *Vehicle System Dynamics*, Vol. 32, No. 6, pp. 441–458.

CONTROL APPROACH FOR STRUCTURED PIEZO-ACTUATED MICRO/NANO MANIPULATORS

Kostadin Kostadinov[1], Roland Kasper[2] and Muhammed Al-Wahab[1]
[1]Institute of Mechanics, Bulgarian Academy of Sciences, Acad. G. Bonchev St., Block 4, Sofia 1113, Bulgaria; [2]Institute of Mechatronics and Drives, Otto-von-Guericke University, Hermann-Gruson-Gebäude, 2 Universitaetsplatz, D-39106 Magdeburg, Germany

Abstract: This paper considers a new method for integration of piezo-electric actuators and motion amplification elements (MAE) into one component, subjected here to be telecontrolled as micro/nano manipulator. In this case development of a control approach for such micro and nano manipulators is the object of this paper. The used piezo-ceramic disc is segmented according the desired task space. It is characterized with non-linear behavior while the segments are controlled either separately or in combination. The main problems concerning the integrated structure-control design optimization approach is defined and considered in order to realize a telemanipulation control of the desired micro/nano operations. An appropriate hybrid telemanipulation approach is developed to improve the efficiency of any micro- and nano manipulation and operation for which the operator has neither experience nor knowledge. It combines the advantages of the known two teleoperated control approaches – direct and task-oriented teleoperation.

Key words: piezo-structured ceramics, micro/nano manipulator, control approach, non-linear mechanic structure, telemanipulation control, impedance scaling.

1. INTRODUCTION

The effective piezo-electric materials together with powerful electronic driving systems pushed the attractiveness of piezo-electric actuators in the past years significantly for bride specter of applications. Compared to competitive actuators they offer substantial advantages e.g. large force, high

H. Ulbrich and W. Günthner (eds), IUTAM Symposium on Vibration Control of Nonlinear Mechanisms and Structures, 193–202.

frequency and a small size [1-4]. Nevertheless, many practical applications need larger motion, realizable using amplification elements only. But it results in bigger actuator size and more difficult its integration and fitting.

In the second part of this paper utilizing of a new method for integrate piezo-electric actuators and motion amplification elements (MAE) into one component [5], which in our case is a micro/nano manipulator. Based on a mechanical structuring process of the piezo-electric material, these actuators offer an improved adaptation of actuator force and motion and facilitate new applications in industrial and consumer products. To adapt actuator characteristics to the requirements of applications, MAS is utilized [5]. In our preliminary study this focus is set on mechanical solutions [6], as they offer a better way of integration into ceramic structures. They offer a simple mechanical construction, but have problems with friction, backlash and stiffness. To avoid these drawbacks and to ease handling, frame construction is used, which integrate several tasks like structure optimization, motion amplification, pre-stress generation and safe mounting. Unfortunately, all known MAS increase size and moved masses of the actuator. The actuator dynamics with MAS is worsened while the control requirements are increased. This is a contradiction to an optimized actuator system, fine tuning all its components, required for today's micro- and nano-applications.

A further goal in addition to the generation of the movement parameters is the integration of the piezo-ceramics into the mechatronic handling device used as a micro/nano manipulator. The structured piezo-ceramics makes possible the micro- and nano-positioning. The mechanism for the macro positioning must possess interfaces and survey, in order to take up structured ceramics. The piezo-ceramics must likewise make possible the admission in its structure. Here above all adjustment work is necessary. The used piezo-ceramic disc is segmented according the desired task space. It is characterized with non-linear behavior while the segments are controlled either separately or in combination, which has to be analyzed and overcame.

The main problems concerning the integrated structure-control design optimization approach are defined and considered in the third part in order to realize a telemanipulation control of desired micro/nano operations. Proper hybrid telemanipulation approach to improve efficiency of any micro/ nano manipulation and operation for which the operator has neither experience nor knowledge is developed in the forth part combining the advantages of the known two teleoperated control approaches – direct and task-oriented teleoperation [7]. Two prospective applications are considered in the fifth part. In the last part some conclusions and future prospects are presented.

2. MICRO/NANO MANIPULATOR BASED ON MECHANICALLY STRUCTURED PIEZO-CERAMICS

An alternative approach uses (3,1)-effect like a bender actuator to generate a large motion due to the large effective length of a planar actuator. Integration of a MAE into this planar actuator can be done in several ways. An important distinguishing feature of an MAE is the integration in the same plane as the actuator or generating motion perpendicular to this plane.

Planar systems offer the advantage that all components are located in the same physical plane. As a consequence they can be produced from piezo-ceramic plates or discs within one structuring process. Prototypes shown on Table 1 are manufactured and tested [8].

We consider a planar piezo-ceramic disk structured into 3 segments as it is shown in Figure 1. This allows realizing the micro/nano manipulator by MAS characterizing with parallel closed structure. The obtained micro/nano manipulator has 3 degree of freedom (DOF) – 2 rotations around axes X and Y and one translation along the Z axis. It also can be considered and used as a local structure of robot with more degree of freedom.

The base vibration noise passed through the manipulator mechanical structure has to be eliminated to the end-effector. The possible vibrations generated by the non-linear mechanical structure also have to be damped in addition of generation and control of any desired motions of the manipulator end-effector.

Table 1. Actuator prototypes utilizing in micro/nano manipulators.

Planar Systems	Spatial Systems			
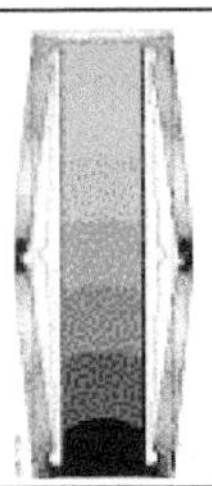	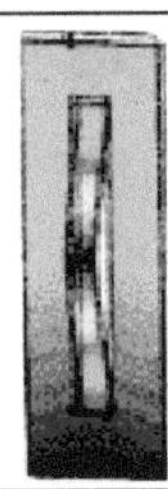		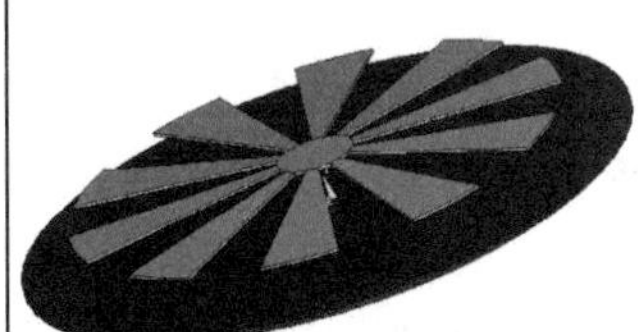	
a) Two-sided lever system	b) One-sided lever system	c) Hybrid one-sided lever system	d) Hybrid one-sided disc system	
2x88	64	281	511	Motion[μm]
4	1	20	40	Force[N]
70x25	70x25	70x25	$\pi 35^2$	Area

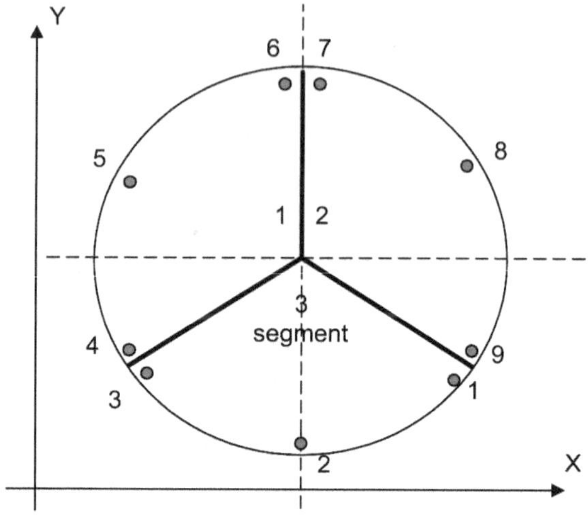

Figure 1. One-sided disc piezo system structured into 3 segments with measurement points.

3. INTEGRATED STRUCTURE-CONTROL DESIGN OPTIMIZATION APPROACH

The resulting control problem of the complete manipulator with 6 DOF consisting of regional structure with 3 DOF and ceramic piezo-structure with 3 DOF is very complex. Development of the integrated structure-control design approach is consisted of the following tasks and steps comprising all problems significantly influenced and which have to be taken into account to the control performance of the manipulator end-effector operation:

3.1 Definition of the Performance Optimization Problem and Its Transformation into Integrated Structure-Control Design Optimization Task

System requirements are used to formulate an adequate structure control problem for the micro/nano manipulator. Formulation of design criteria will include:

- strength/load capacity and force of operation/ manipulation;
- positioning accuracy, respectively stiffness of the active and passive components of the piezo-structure and MAS;
- positioning speed of manipulator regional and local piezo-structure;
- restrictions in control variables;
- level and specter of base vibrations.

3.2 Decomposition of the Design Optimization Task into Design Problems for the Micro/Nano Manipulator Systems and Overall Control

The complete control problem is divided in a problem of coarse positioning associated to the regional manipulator structure and a problem of fine positioning associated to the piezo-ceramic structure utilized here as the local manipulator structure. An overall controller will coordinate both sub-controllers to realize a suitable interface for a superior operation unit.

3.3 Design of Manipulator Structures Control

Control is realized on the PC based control system with controllers for the structured piezo-ceramics and for the regional structure actuated by other types such as step or DC motors. The whole system is integrated in the micro processor control and communication unit specially developed [7]. In the control synthesis appropriate sensor systems is subject to be integrated including positioning, vibration, and temperature sensors.

3.4 Design of Piezo-Structure Control

Elastic deformations of any one of the segments of piezo-structured ceramics are transformed to the movement of the end effector via the MAS in a nonlinear way (Figure 2). The model used is a SOLID5 which has a 3-D magnetic, thermal, electric, piezo-electric and structural field capability with limited coupling between the fields. The element has eight nodes with up to six degrees of freedom at each node. As a consequence there is a nonlinear transfer function from control inputs (segment voltages) to the controlled variables (position of the end effector). A nonlinear feed forward compensation method is used to realize a nearly linear transfer function. Segment voltages are controlled in a feedback control loop utilizing analogue drivers. This is possible because the piezo-actuator's capacity is very small and losses can be neglected.

4. TELEMANIPULATION CONTROL APPROACH

For the control of micro and nano manipulators two approaches are utilized: teleoperation approach or automatic manipulation. To improve efficiency of any micro- and nano manipulation and operation realized by the micro/nano manipulator it is necessary to develop appropriate telemanipulation approach

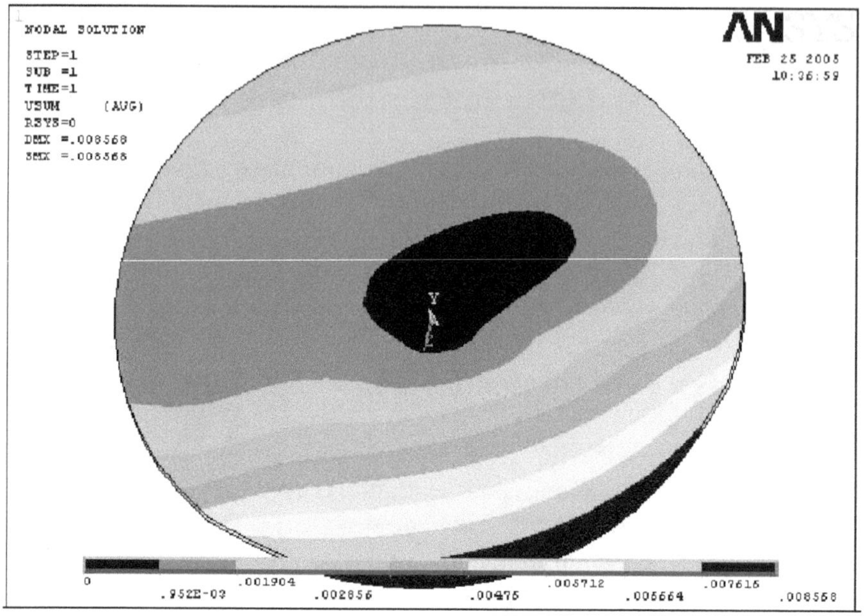

Figure 2. FEM results for the piezo-structure with controlled 3rd segment.

and appropriate automatic control in addition keeping the vibrations of the end-effector under the desired level. Telemanipulation is a process where the operator has to do some tasks at the far environment where he cannot be physically. Here the most important tasks to be decided are how stable and high performance control can be obtained in spite of highly variable human operator and environment dynamics (Figure 3.), time delays in communication channel, and other effects such as hysteresis, interaction between separate segments, etc.

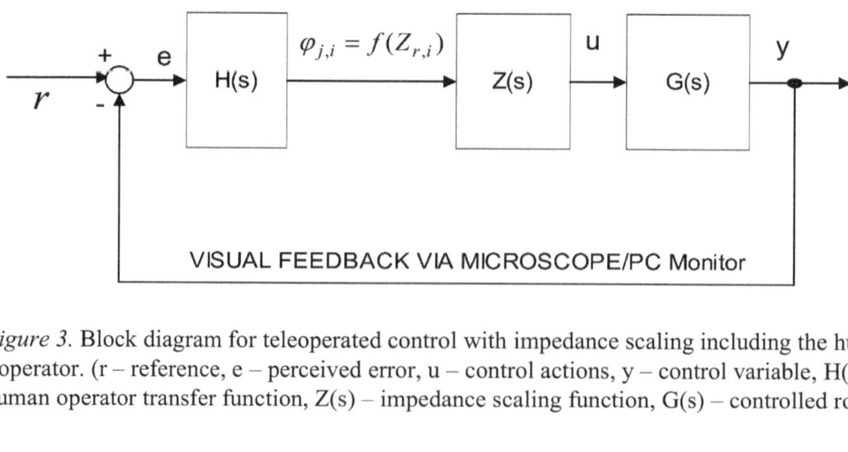

Figure 3. Block diagram for teleoperated control with impedance scaling including the human operator. (r – reference, e – perceived error, u – control actions, y – control variable, H(s) – human operator transfer function, Z(s) – impedance scaling function, G(s) – controlled robot).

Since the object of this paper is development of a manipulator for micro and nano operations it is necessary to develop appropriate telemanipulation approach to improve efficiency of any micro- and nano manipulation and operation for which the operator has neither experience nor knowledge.

Two teleoperated control approaches are known – direct teleoperation and task-oriented teleoperation [7]. In the first one the human wealth of experience in manipulation tasks is used. The operator manipulates directly in the robot control-loop using a man-machine interface. Dexterous telemanipulation enables a user to interact with the environment from a real world into the micro or sub micro world. In the case when humans use telemanipulation device only to move objects than the reaction force from the far environment has no significant effect on the performance, then measurement of the operator's position and visual feedback may be enough.

Direct teleoperation approach can realize tasks requiring high-level intelligence and flexibility. It is slow, not so precise, not exactly repeatable and engaged in many complex and challenging scaling problems. However, the task–oriented approach avoids those problems by executing only the given task in closed-loop autonomous control [9]. In the automatic control approach the robot has a closed loop control using sensor information without user intervention. The problems characterizing the automatic control in the micro and nano world are not reliable at present due to complexity of the micro and nano dynamics, difficulties in nano positioning and real time visual feedback, changing and uncertain physical parameters, and insufficient models and intelligent control strategies [10].

To meet these requirements the hybrid teleoperation approach is developed as a combination of a direct teleoperation and a task-oriented teleoperated manipulation using haptic interface via a master joystick, PC keyboard and visual feedback.

In any micro- or nano-teleoperation task the physical size of the environment and slave differs greatly from that of the human operator (master joystick), therefore it is necessary to scale up and down the information (forces and motions) exchanged between them. In the case of cell micro and nano manipulations the force of interaction is not important. This is why the force information is not foreseen to be exchanged between the parts of the telemanipulation system. In this case the human operator has no perception about the manipulator dynamics and environment where he manipulates. But as the most important requirement in the case of cell micro and nano manipulations is to do in a minute the whole operation the haptic interface has to provide the operator with the feeling that he knows the dynamics of the manipulator and object to be manipulated. Since the design of the teleoperation robot system must be based on human perceptual capabilities [11], it is necessary to quantify human perceptual capabilities, and to have means to incorporate them into the control design (design methodology, tools, and proper formulation). This could be realized by introducing a mechanical impedance characterization, which is nonlinear scaling with corresponding constant. It is also called impedance scaling. The manipulator dynamics can be assigned to virtual coupling mechanical impedance Z, appropriate for the operator to manipulate in the micro/nano world as in familiar for him real world, as follows:

$$M\ddot{x}_{R,i} + B\dot{x}_{R,i} + Kx_{R,i} = F_i .$$

In this case the macro joystick angular positions $\varphi_{j,i}$ define the force F_i by which the manipulator axes characterizing by their impedance parameters assigned preliminary will be driven, i.e.:

$$F_i = z_{s,i} \cdot \varphi_{j,i}$$

Reference mechanical impedance $z_{s,i}$ is determined by the desired dynamic interaction, i.e. the dynamic accuracy [12]:

$$[\Delta E] = \sum_{1=0}^{2} \{[p^i x_r] - [p^i x_a]\} = \Delta x_d + \Delta v_d + \Delta a_d = [F][Z^{-1}]$$

where $[p^i x_r]$ and $[p^i x_a]$- reference and actual motion parameters; Δx_d, Δv_d and Δa_d - position, velocity and acceleration errors, p - Laplace's operator.

In this way the manipulator will operate with the micro/nano object with its dynamics appropriate for the operator. This dynamics of interaction between the human operator and the manipulator can be modified and adjusted at any time. Hence, in this way the teleoperation control can realize a desired dynamic coupling between the macro and nano world.

5. APPLICATIONS

The work started with a requirement analysis for mechatronic handling devices for two applications as micro/nano manipulator:

- for cell manipulations: Typical manipulations as cell penetration, cell sorting and moving or treating microorganisms will be analyzed in order to specify requirements to mechatronic handling device for bio research applications. Critical issues here are speed of cell penetration, necessary stroke for penetration or operation and orientation of the end-effector, working space and space of the device itself, sensing and approach for teleoperation control able to offer a minute performance of the desired operation.

- for electrochemical impedance application techniques: The desired operation of the probe for electrochemical impedance technique is analyzed in order to achieve positioning and dynamic requirements for the right measurement task execution. They are consisted of 2 scanning rotation motion and translational one for fine adjustment of the probe. Here the regional coarse scanning robot structure with 3 DOF (translation) is necessary.

Requirements defined here build the reference tasks for synthesis, optimization, control and test operation.

The experimental results for segment motion generation and its control have been obtained. They confirmed the FEM results shown in Figure 2. On that basis the transfer functions used in the control synthesis have been determined.

6. CONCLUSION AND FUTURE PROSPECT

Based on an analysis of the pros and cons of today's piezo-actuators, this paper presents a new construction method and control approach for these actuators that allows a better scaling and adaptation of actuators characteristics to given applications. The fundamental idea of a better integration of actuator's function elements like force/motion generator, MAS and housing, is realized by mechanically structuring piezo-ceramics into different number of segments depending of the desired manipulation task and additional elements. In future project steps, functional elements of special application fields also will be integrated in the mechanical structure.

The prototypes discussed here cover an economically important segment of actuator force and motion that cannot be served by classical stack or bender actuators. Compared with competing magnetic or electromechanical systems, these actuators offer a very simple mechanical construction, which results in very low cost. As the development of this type of actuator is just at the beginning, the experimental and FEM results achieved with first prototypes are promising significant improvements in the future research. On one hand this can be achieved by optimizing the structure and geometry of existing actuators to match new applications. On the other hand, more kinematical structures and principles of solid mechanics can be used and will lead to new and better actuator structures. All structures put plenty of problems to the proposed control approach which are the object of further development.

ACKNOWLEDGEMENTS

The authors gratefully acknowledge the partial support of Bulgarian National Found "Scientific Research" under the Contract *Ro-MiNa* TH-1308/2003 and the project *MeCHaPiCS*/2005.

REFERENCES

[1] Patentschrift der Festo AG & Co., EP 1104863 A2, 06.06.2001.

[2] Patentschrift Hoerbiger Fluidtechnik GmbH, AT 396392 B, 25.8.1993, Piezo-Ventil.

[3] Lubitz, K., Schuh, C., Steinkopff, T. and Wolff, A. (2000) Properties and Reliability of large PZT multilayer actuators, in *ACTUATOR 2000, 7th International Conference on New Actuators*, June 2000, Bremen, Germany, pp. 58–61.

[4] Linden, D. (1999) Hydraulisches Piezoservoventil NG10, *O+P Ölhydraulik und Pneumatik* 43, Nr. 7, pp. 538–543.

[5] Kasper, R., Mechatronisches Design integrierter Piezoaktoren, 21–22. März 2002, ISOM 2002 *Innovative Antriebstechnik*, ISBN 3-00-007504-6. Chemnitz, pp. 379–392.

[6] Kasper, R., Heinemann, W. and Wagner, A. (1998) Modelling and control for piezoelectric actuators for high speed applications, in *The Fourth International Conference on Motion and Vibration Control, MOVIC*, 25–28.8.1998, Volume 1, pp. 231– 236, ETH Zürich, Switzerland.

[7] Ionescu, Fl., Kostadinov, K., Hradynarski, R. and Vuchkov, I. (2003) Teleoperated control for robot with 6 DOF for micro and nano manipulations, in *Intelligente technische Systeme und Prozesse*, R. Kasper (ed.), Logos Verlag, Berlin, ISBN 3-8325-0338-2, pp. 127–132.

[8] Kasper, R. and Al-Wahab, M. (2004) Mechanically structured piezoelectric actuators, in *ACTUATOR 2004, 9th Int. Conference on New Actuators*, 14–16 June 2004, ISBN 3-933339-06-5, Bremen.

[9] Sitti M. and Hashimoto, H. (2000) Two-dimensional fine particle positioning under optical microscope using a piezoresistive cantilever ass a manipulator, *Journal of Micromechanics*, Vol. 1, No. 1, pp. 25–48.

[10] Sitti M. and Hashimoto, H. (1999) Teleoperated nano scale object manipulation" in *Recent Advances on Mechatronics*, O. Kaynak et al. (eds), Springer Verlag, Singapore, pp. 322–335.

[11] Cavusoglu, M.C., Cenk, M., Sherman, A. and Tendick, F. (2001) Bilateral controller design for telemanipulation in soft environments, in *Proceedings of the IEEE International Conference on Robotics and Automation (ICRA 2001),* Seoul, Korea, May 21–26, 2001.

[12] Kostadinov K., Accommodation control in the drive dynamic accuracy for positioning robot, in *38 Internationales wissenschaftliches Kolloquium*, Tagungsband, Ilmenau, 20–23.09.1993, pp. 100–108.

ACTIVE AND SEMI-ACTIVE CONTROL OF ELECTRORHEOLOGICAL FLUID DEVICES

Andreas Kugi, Klaus Holzmann and Wolfgang Kemmetmüller
Chair of System Theory and Automatic Control, Saarland University, Building 13, 66123 Saarbrücken, Germany
andreas.kugi@lsr.uni-saarland.de

Abstract: This paper is devoted to two different applications of electrorheological (ER) fluid devices. The first application deals with the modeling and nonlinear control of an active actuator consisting of a double-rod cylinder and four ER valves arranged in a full bridge configuration. Secondly, a semi-active shock absorber system is designed by utilizing the special properties of ER valves. The latter application is also intended to demonstrate the benefits of a mechatronic design approach, where the control strategy and the system components are designed simultaneously. Measurement results prove the feasibility of the proposed ER devices.

Key words: electrorheological fluids, ER actuator, ER shock absorber, active and semi-active control, nonlinear control.

1. INTRODUCTION

This contribution deals with the mathematical modeling and the control of active and semi-active electrorheological (ER) fluid devices in flow-operation mode, see, e.g., Butz and von Stryk (2002) for the classification of the different operation principles. In general, an ER fluid is a suspension of solid particles in a fluid phase which experiences dramatic changes in its rheological properties (e.g. the apparent viscosity) when subjected to a sufficiently strong electric field. Although the exact mechanisms behind the ER effect are not fully understood, two facts are generally accepted (see, e.g., Eckart, 2000): (i) the particles form chains upon application of sufficiently large electric fields and (ii) these chains are responsible for the change in the rheological properties. The ER effect is known to be essentially reversible and it exhibits a very rapid response time upon application or removal of the electric field in the order of a few milliseconds as shown e.g. by Whittle et al. (1996). The ability of purposefully altering the rheological properties of ER fluids motivates the

H. Ulbrich and W. Günthner (eds), IUTAM Symposium on Vibration Control of Nonlinear Mechanisms and Structures, 203–212.

design of new types of active and semi-active actuators. Examples for these applications are clutches, shock absorbers, dampers, and servo drives (see, e.g., Fees, 2001; Gavin, 2001; Hoppe et al., 2000). The central part of the applications presented in this contribution is the ER valve. Roughly speaking, an ER valve comprises two concentric cylindrical electrodes forming an annular channel where the ER fluid is passing through. Applying a voltage to the inner cylindrical electrode, with the outer electrode being earthed, an electric field perpendicular to the direction of the flow is generated. If a constant volume flow of ER fluid is maintained through the ER valve, an increase in the voltage (in the electric field) causes an increase in the apparent viscosity and hence also in the pressure drop across the valve. Thus, the resistance of the ER fluid to flow through the valve can be controlled by the voltage. This way of using the ER effect is also known as the flow- or valve-operation mode.

Based on the observations about the mechanism of the ER effect mentioned above there are a number of approaches for the modeling of ER fluids in the literature. They can roughly be divided into microscopic and macroscopic models. The microscopic modeling approaches try to describe the movement and aggregation of particles under the influence of an external electric field, see, e.g., Parthasarathy and Klingenberg (1996) for an overview. The major drawback of the microscopic modeling approach is that in general it can only be used to model the behavior of a limited number of particles. Therefore, a straightforward application of these microscopic models to describe the flow through the gap of an ER valve is not possible. In the design of technical applications macroscopic models are more suitable. Apart from the pure phenomenological models describing the input-output behavior of ER devices (see, e.g., Butz and von Stryk, 2002), there also exist macroscopic models in the context of continuum mechanics, mainly based on the so-called generalized Cauchy stress tensor (see, e.g., Eckart, 2000; Rajagopal and Wineman, 1992; Růžička, 2000). In this approach the ER suspension is treated as a homogenous continuum. Furthermore, it is assumed that changes in the electric field take effect instantaneously and that there are no memory effects. The usage of a generalized Cauchy stress tensor has at least the advantage that under certain assumptions the constitutive equations of the ER fluid satisfy some general principles of material behavior, i.e. the material frame-indifference and the principle of dissipation (Clausius–Duhem inequality). For this reason we will use this approach in the following mathematical modeling.

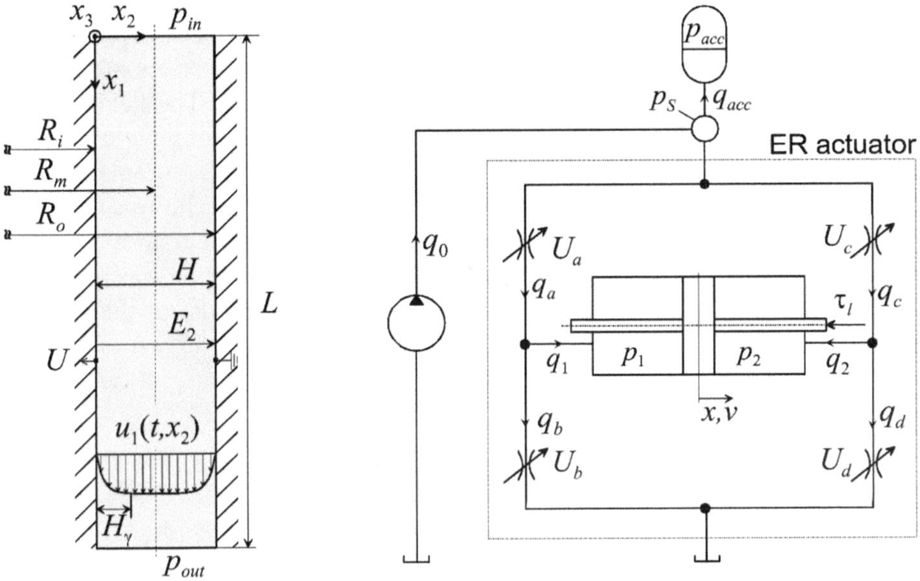

Figure 1. (a) Longitudinal section of an ER valve. (b) Schematic diagram of the ER actuator test stand.

2. MATHEMATICAL MODELING OF ER COMPONENTS

In this section the mathematical model of an ER valve and a double-rod cylinder is briefly derived. For a more detailed treatment the reader is referred to e.g., Kemmetmüller and Kugi (2004b).

Figure 1(a) shows the longitudinal section of an ER valve. The annular channel between the two cylindrical electrodes with inner radius R_i and outer radius R_o has a height of $H = R_o - R_i$ and a length L. Since the channel gap is small compared to the mean radius $R_m = (R_o + R_i)/2$, the ER valve can be modeled as a flat gap with the dimensions H, L and width $W = 2R_m\pi$. Under the assumptions that (i) the flow in the gap remains laminar, (ii) the fluid in the gap is incompressible and (iii) the temperature is constant, the constitutive equations based on the generalized Cauchy stress tensor due to e.g., Eckart (2000), Rajagopal and Wineman (1992) and Růžička (2000), can be simplified to a "Bingham-like" material model

$$\sigma_{12} = \tau_0 (E_2) \operatorname{sign} (\gamma) + \eta\gamma \qquad \text{for} \qquad \gamma \neq 0 \tag{1}$$

with the shear stress σ_{12}, the dynamic viscosity η, the electric field $E_2 = U/H$, the field dependent yield stress $\tau_0 (E_2)$ and the shear rate $\gamma = \partial u_1/\partial x_2$. Furthermore, $u_1(t, x_2)$ denotes the velocity profile in the gap of and U is the voltage applied to the ER valve, cf. Figure 1(a). For $|\sigma_{12}| < \tau_0(E_2)$ $(\gamma = 0)$ the ER fluid behaves like a linear elastic solid with the shear stress $\sigma_{12} = G\varepsilon_{12}$ depending on the shear modulus G and the shear strain ε_{12}. The field dependent yield stress grows quadratically for low electric field strength and shows almost a linear behavior for higher electric field strength. Measurements show that an approximation of the form $\tau_0(E_2) = a_2E_2^2 + a_3E_2^3$, with real constants a_2, a_3, can be used to describe the behavior in the domain of operation. Based on this model the velocity profile $u_1(t, x_2)$ for non-vanishing electric fields $E_2 \neq 0$ comprises a field dependent plug zone in the middle of the gap, i.e. $\gamma = 0$ for $H_\gamma \leq x_2 \leq H - H_\gamma$, and a parabolic profile in the rest of the gap. The volume flow q through the ER valve is calculated by integration over the velocity profile $u_1(t, x_2)$ and yields

$$q = \frac{W(\tau_0(E_2) + PH)(-2\tau_0(E_2) + PH)^2}{12P^2\eta} \quad \text{for} \quad |P| > \tau_0(E_2)\frac{2}{H} \tag{2}$$

with $P = (p_{in} - p_{out})/L$. If $|P| < 2\tau_0(E_2)/H$ the plug zone covers the whole gap and the ER valve is closed, i.e. $q = 0$.

The compressibility of the ER fluid cannot be neglected in the double-rod cylinder of the active ER actuator, cf. Figure 1(b). Therefore, we take account of it by means of the bulk modulus $\beta = \rho\partial p/\partial\rho$ with the mass density ρ and the pressure p. Neglecting the leakage flows, we get the mathematical model of the double-rod cylinder in the form (see, e.g., Merritt, 1967)

$$\begin{aligned} \dot{x} &= v, & \dot{p}_1 &= \frac{\beta}{V_1 + Ax}(-Av + q_1) \\ \dot{v} &= \frac{1}{m}(A(p_1 - p_2) - \tau_l), & \dot{p}_2 &= \frac{\beta}{V_2 - Ax}(Av + q_2) \end{aligned} \tag{3}$$

with the piston position x and the velocity v, the chamber pressures p_1 and p_2 and the volume flows into the two chambers, q_1 and q_2, respectively. Furthermore, V_1 and V_2 denote the chamber volumes for $x = 0$, A is the effective piston area, m denotes the mass of the piston and all masses rigidly connected to the piston and τ_l stands for the external load force including friction.

3. APPLICATION: ACTIVE ER ACTUATOR

The active ER actuator as designed and manufactured by FLUDICON GmbH (2001) consists of a double-rod cylinder and four ER valves arranged in a full bridge configuration. The schematic diagram of the ER actuator and the test stand is depicted in Figure 1(b). In principle, the pressure supply consists of a gear pump with a constant volume flow q_0, a connection block with pressure p_S and a hydraulic accumulator with pressure p_{acc} and volume flow q_{acc}. The flows q_i through the four ER valves can be controlled within certain limits by means of four independent high voltage amplifiers applying the voltages U_i, $i \in \{a, b, c, d\}$, to the ER valves. Moreover, for control purposes the test stand is equipped with three pressure transducers for the supply pressure p_S and the two chamber pressures p_1 and p_2 and a sensor for the piston position x.

The primary objective of the controller design is to track a desired trajectory x_d of the piston position x, with the voltages U_i, $i \in \{a, b, c, d\}$, as the control inputs. The control concept is based on a cascaded structure with an inner control loop for the chamber pressures and an outer control loop for the piston position. If we consider the volume flows into the two cylinder chambers, $q_1 = q_a - q_b$ and $q_2 = q_c - q_d$, as control inputs, then we have shown in Kemmetmüller and Kugi (2004b) that the nonlinear control law

$$\begin{aligned} q_1 &= Av + \frac{V_1 + Ax}{\beta}(-\delta_p(p_1 - p_{1,d}) + \dot{p}_{1,d}), & p_{1,d} &= \frac{p_{S,d}}{2} + \frac{\tau_d}{2A} \\ q_2 &= -Av + \frac{V_2 - Ax}{\beta}(-\delta_p(p_2 - p_{2,d}) + \dot{p}_{2,d}), & p_{2,d} &= \frac{p_{S,d}}{2} - \frac{\tau_d}{2A} \end{aligned} \tag{4}$$

and

$$\tau_d = m\left(\ddot{x}_d - \delta_{x,2}\dot{e}_x - \delta_{x,1}e_x - \delta_{x,0}\int e_x \mathrm{d}t\right) \tag{5}$$

yields an exponentially stable dynamics for the piston position error $e_x = x - x_d$ for suitable positive parameters δ_p, $\delta_{x,j}$, $j = 0, 1, 2$, and sufficiently smooth desired trajectories x_d. Here $p_{S,d}$ denotes the desired average value of the supply pressure p_S. Clearly, the volume flows q_1 and q_2 from Eq. (4) do not uniquely determine the volume flows q_i, $i \in \{a, b, c, d\}$, through the ER valves. However, the remaining degrees-of-freedom can be advantageously used to cope with the following problems: (i) the ER valves show undesirable hysteresis for small volume flows and therefore, the control strategy should circumvent a complete closing of the valves; (ii) the peak volume flows of q_1 and q_2 needed for fast trajectories are much higher than the constant volume flow q_0 provided by the gear pump. In such cases, the high volume flows must be made available by the accumulator. An optimal operation of the accumulator requires that the supply pressure p_S is kept within certain limits. Therefore, a controller for the average value of the supply pressure is included in the control concept (see Kemmetmüller and Kugi, 2004b). All the demands can be fulfilled by dividing the flows q_1 and q_2 from Eq. (4) into the volume flows q_i, $i \in \{a, b, c, d\}$, in the following form

$$q_a = \mathrm{sg}\,(q_1) + \bar{q}, \quad q_b = \mathrm{sg}\,(-q_1) + \bar{q}, \quad q_c = \mathrm{sg}\,(q_2) + \bar{q}, \quad q_d = \mathrm{sg}\,(-q_2) + \bar{q} \tag{6}$$

with $\bar{q} = \frac{q_0 + q_S^m}{2}$ and $\mathrm{sg}\,(q) = q$ for $q > 0$ and $\mathrm{sg}\,(q) = 0$ for $q \leq 0$. Furthermore, it can be easily seen that with this choice the volume flow q_S^m has no

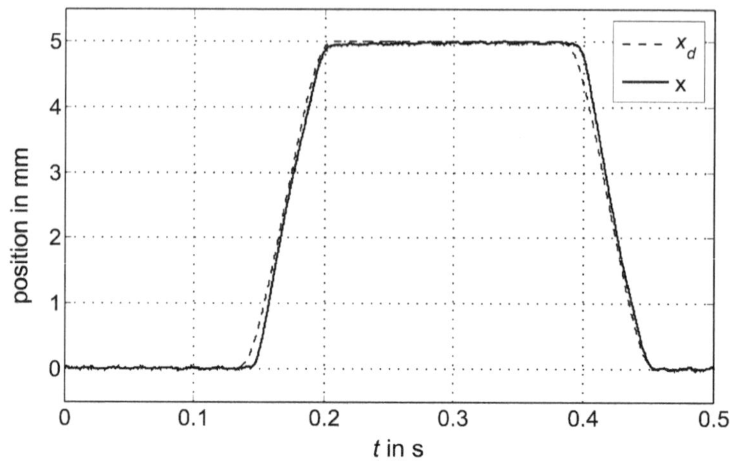

Figure 2. Measured piston position x for a rectangular-like reference trajectory x_d.

influence on q_1 and q_2 and therefore can serve as a control input for the supply pressure controller. More details on the supply pressure controller and a rigorous proof of the stability can be found in Kemmetmüller and Kugi (2004a). The voltages U_i, $i \in \{a, b, c, d\}$, can then be directly calculated by an analytic inversion of Eq. (2) with q_i from Eq. (6). The control concept was implemented with a sampling time of 200 μs by means of the real-time hardware DS1103 from DSPACE together with MATLAB/SIMULINK. Figure 2 shows the typical response of the closed-loop system for a rectangular-like trajectory with an amplitude of 5 mm.

4. APPLICATION: SEMI-ACTIVE SHOCK ABSORBER

The second application is a shock absorber system as depicted in Figure 3. The objective of this device is to significantly reduce the acceleration a_1 induced on the platform by a high external excitation (shock) and to assure a fast and accurate repositioning of this platform after the shock. Furthermore, for low excitations the platform should be well damped in order to prevent oscillations. A schematic diagram of the construction is shown in Figure 3(a).

Thereby, the acceleration a_2 of the shock desk is enforced by the environment. The variables x_1, v_1, a_1 and x_2, v_2, a_2 denote the position, the velocity and the acceleration of the platform and the shock desk, respectively. The platform and the shock desk are connected to each other via an air spring with the pressure p_c and the piston of a damping cylinder with the chamber pressures p_1 and p_2. The two cylinder chambers are connected to a bypass throttle and an ER valve. A voltage U applied to the ER valve acts as the control input

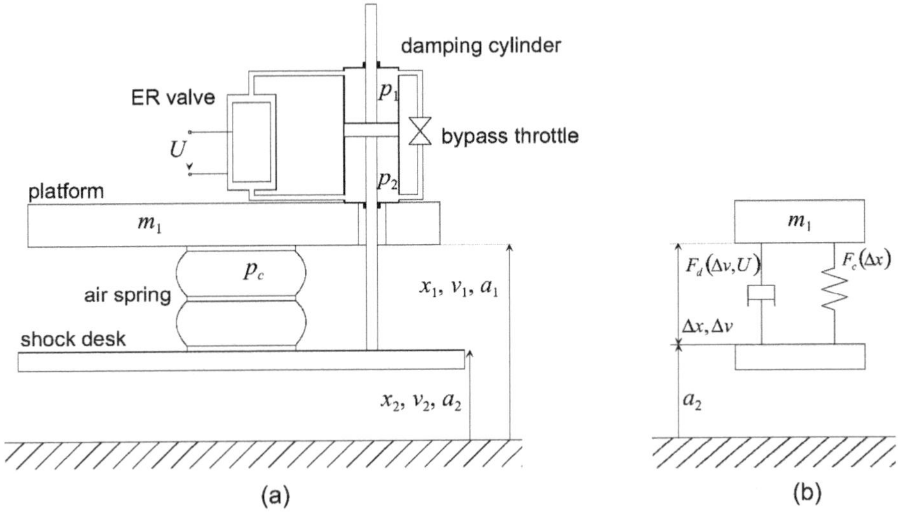

Figure 3. (a) Schematic diagram of the shock absorber system. (b) Simplified equivalent spring-mass-damper system.

to the system. The platform is assumed to be ideally stiff and hence can be modeled as a rigid body. Furthermore, it is assumed that the excitation a_2 acts only vertically. Since we are interested in the absolute value of the acceleration a_1 of the platform but not in the absolute values of its velocity or position, it is reasonable to use the relative position $\Delta x = x_1 - x_2$ and the relative velocity $\Delta v = v_1 - v_2$ as state variables for the mathematical model.

In addition to the standard ER components as described in Section 2 two further components are important in this application, namely the air spring and the piping of the system. For the air spring it is assumed that no displacement in transverse direction occurs and that the movement of the air spring is sufficiently fast. The last assumption incorporates that the heat exchange with the environment can be neglected. Therefore, the thermodynamic process can be regarded as isentropic (adiabatic and reversible). With the precharge pressure p_0 of the air spring and the initial air volume V_0, the pressure in the air spring can be calculated in the form $p_c(\Delta x) = p_0(V_0/V_c(\Delta x))^\kappa$, with the air volume $V_c(\Delta x)$ and the isentropic coefficient κ. The resulting force of the air spring reads as

$$F_c(\Delta x) = p_c(\Delta x)\, A_c(\Delta x) \tag{7}$$

with the effective cross sectional area $A_c(\Delta x)$ of the spring, which is in general a nonlinear function of Δx. Due to the high accelerations and thus the large changes in the volume flows in the system, the piping has an essential influence on the system dynamics. Due to space limitations we will not go into the details of the mathematical model here. Furthermore, due to the high accelerations cavitation can occur in the damping cylinder. Thus, the assumption of a constant bulk modulus in the mathematical model Eq. (3) has to be released, see, e.g., Franc et al. (1995) for more details on an extension to the case including cavitation. For designing a control strategy, we may regard the hydraulic part of the shock absorber system in a quasi-static way. Doing so, we get the following relationship for the damping cylinder together with the bypass throttle and the ER valve

$$F_d(\Delta v, U) = \left[p_{1,s}(\Delta v, U) - p_{2,s}(\Delta v, U)\right] A_d \tag{8}$$

where F_d is a nonlinear function of the relative velocity Δv and the voltage U applied to the ER valve. Here A_d describes the effective cross sectional area of the damping cylinder and $p_{1,s}$ and $p_{2,s}$ are the quasi-static pressures in the cylinder chambers. Then, the equations of motion for the shock absorber system take the simple form

$$\Delta\dot{x} = \Delta v, \qquad \Delta\dot{v} = \frac{1}{m_1}\left(F_d(\Delta v, U) + F_c(\Delta x) - m_1 g\right) - a_2 \tag{9}$$

with m_1 the mass of the platform and all parts mounted on it and the gravitational constant g. Clearly, the acceleration a_1 of the platform according to Figure 3(b) is given by $a_1 = \frac{1}{m_1}\left(F_d(\Delta v, U) + F_c(\Delta x) - m_1 g\right)$.

The control task as described above can be viewed as a mechanical impedance matching problem with the voltage U as the control input. It is immediately clear that with the ER actuator as used in the shock absorber system of Figure 3 it is only possible to control the (nonlinear) damping characteristics of the system. The (nonlinear) spring characteristics is defined by the choice of the air spring. Nonetheless, in the sense of a mechatronic design of the shock absorber system the geometric design of the spring can be regarded as a part of the controller design. After optimizing the air spring with respect to the preload pressure p_0, the suspension travel and the maximum spring force, we try to control the damping force F_d of Eq. (8) in accordance with the following requirements: (i) under normal operation conditions, i.e. when the excitation a_2 is small, the damping should be high in order to avoid undesired oscillations of the platform and (ii) in case of a shock, i.e. when the excitation a_2 exceeds a certain threshold, the damping has to be sufficiently small to assure a minimum acceleration of the platform. In exploiting the special properties of the ER valve we are able to satisfy all these demands even without any sensor information. We simply apply a sufficiently large voltage to the ER valve and use the fact that the ER valve opens if the pressure drop exceeds a certain limit, cf. Eq. (2).

Figure 4 shows that for high mass flows (in the shock case) the additional pressure difference Δp_{add} in the damping cylinder caused by the ER valve

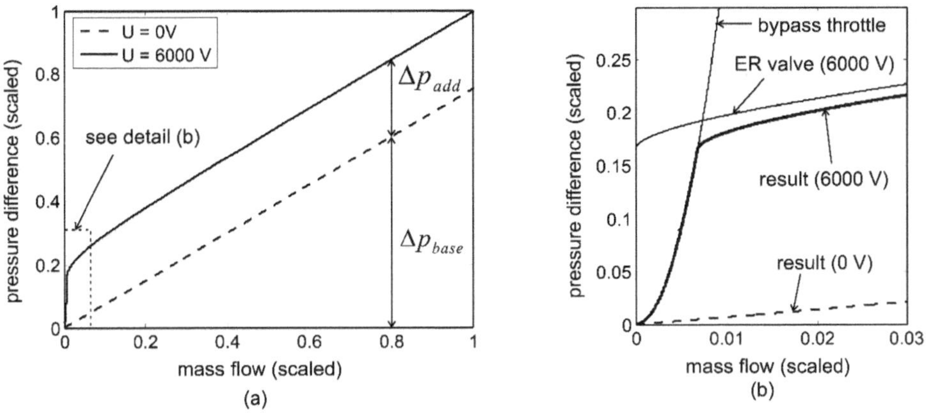

Figure 4. Pressure difference across the damping cylinder in case of a typical shock for different values of the applied voltage U.

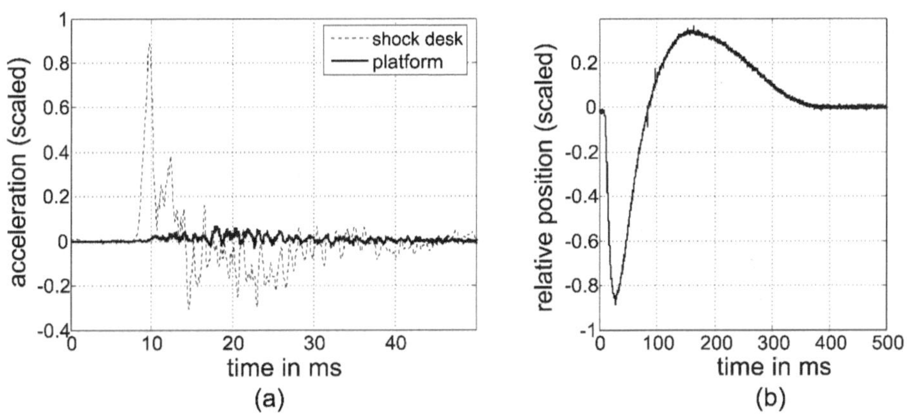

Figure 5. (a) Scaled accelerations a_1 and a_2. (b) Scaled relative position Δx.

with $U = 6000$ V is small compared to the pressure difference Δp_{base} with no voltage applied to the ER valve. In contrast to this the damping is dominated by the bypass throttle for small mass flows under normal operation conditions. Figure 5(a) shows scaled measurement results of the accelerations a_1 and a_2 of the shock desk and the platform, respectively. It can be seen that the acceleration of the platform a_1 is significantly reduced. The corresponding relative position Δx is depicted in Figure 5(b) to demonstrate the fast repositioning of the platform after the shock.

ACKNOWLEDGEMENTS

The authors would like to thank FLUDICON GmbH, in particular Dr. R. Adenstedt, Dr. H. Rosenfeldt and Dipl.-Ing. M. Stork for the fruitful co-operation.

REFERENCES

Butz, T. and von Stryk, O. (2002). Modelling and Simulation of Electro- and Magnetorheological Fluid Dampers, *Z. Angew. Math. Mech.*, Vol. 82, No. 1, 3–20.

Eckart, W. (2000). Phenomenological Modeling of Electrorheological Fluids with an Extended Casson-Model, *J. of Cont. Mechanics and Thermodynamics*, Vol. 12, 341–362.

Fees, G. (2001). Statische und dynamische Eigenschaften eines hochdynamischen ER-Servoantriebes, *Ölhydraulik und Pneumatik*, Vol. 45, 45–48.

FLUDICON (2001). Rheact, Technical Product Description 1, 1–13.

Franc, J.P. et al. (1995). *La Cavitation Mécanismes Physiques et Aspects Industriels*, Presses Universitaires de Grenoble.

Gavin, H.P. (2001). Poiseuille flow of ER and MR Materials, *J. of Rheology*, Vol. 45, No. 4, 983–994.

Hoppe, R.H.W., Mazurkevitch, G., von Stryk, O. and Rettig, U. (2000). Modeling, simulation, and control of electrorheological automobile devices, in *Proc. Conf. Int. Symp. SFB 438*, Munich, Germany, June 30–July 2, 1999, pp. 251–276.

Kemmetmüller, W. and Kugi, A. (2004a). Nonlinear Control in Electrorheological Fluid Devices, in *Proc. of the 3rd European Conference on Structural Control*, Vienna, Austria, July 12–15, 2004, Vol. 1, pp. M9-9–M9-12.

Kemmetmüller, W. and Kugi, A. (2004b). Modeling and Control of an Electrorheological Actuator, Preprints of the *3rd IFAC Symposium on Mechatronic Systems*, Sydney, Australia, September 6–8, 2004, pp. 271-276.

Merritt, E.M. (1967). *Hydraulic Control Systems*, John Wiley, New York.

Parthasarathy, M. and Klingenberg, D.J. (1996). Electrorheology: mechanisms and models, *Material Science and Engineering*, R17, 57–103.

Rajagopal, K.R. and Wineman, A.S. (1992). Flow of Electrorheological Materials, *Acta Mechanica*, Vol. 91, 57–75.

Růžička, M. (2000). *Electrorheological Fluids: Modeling and Mathematical Theory*, Springer, Berlin.

Whittle, M., Atkin, R.J. and Bullough, W.A. (1996). Dynamics of an electrorheological valve, *Int. J. of Modern Physics B*, Vol. 10, 2933–2950.

PNEUMATIC ZERO-COMPLIANCE MECHANISM USING NEGATIVE STIFFNESS

Takeshi Mizuno[1], Masaya Takasaki[1] and Masato Murashita[2]
[1]*Saitama University, Shimo-Okubo 255, Sakura-ku, Saitama 338-8570, Japan;* [2]*Suzuki Motor Corporation, Japan*

Abstract: This paper investigates the applicability of pneumatic actuator to zero-compliance mechanism for vibration isolation. A control method of realizing prescribed negative stiffness by a pneumatic actuator is shown. Its efficacy is verified in a fabricated single-axis apparatus. It was also demonstrated that the equalization of the amplitude of negative and positive stiffness enables the system to have zero-compliance to direct disturbance.

Key words: vibration isolation, pneumatic actuator, negative stiffness, zero compliance.

1. INTRODUCTION

Demands for high-performance vibration isolation systems have been increasing in various fields [1]. There are two kinds of vibration to be reduced by vibration isolation system. One is vibration transmitted from ground through suspension (spring). The other is vibration caused by direct disturbance acting on the isolation table. Lower stiffness of suspension is better for reducing the former while higher stiffness is better for suppressing the latter. A trade-off between them is inevitable in conventional passive-type vibration isolation systems so that their performances are limited.

The author has proposed a new approach to breaking through the trade-off [2, 3]. In the proposed system, a suspension with negative stiffness is connected with a suspension with positive stiffness in series through a middle mass. It enables the system to have good characteristics in reducing vibration transmitted from ground and in suppressing the effects of direct disturbance.

H. Ulbrich and W. Günthner (eds), IUTAM Symposium on Vibration Control of Nonlinear Mechanisms and Structures, 213–222.

The first developed system uses zero-power magnetic suspension as negative spring [2]. The second system used a voice coil motor instead of a hybrid magnet for realizing a negative spring [3]. However, pneumatic-type actuator is superior in supporting heavy tables and reducing costs. In this work, a zero-compliance suspension mechanism is developed which uses a pneumatic actuator for realizing a suspension with negative stiffness.

2. PRINCIPLE

First it will be shown that infinite stiffness can be realized by connecting a normal spring with a spring that has negative stiffness. When two springs with spring constants of k_1 and k_2 are connected in series, the total stiffness is given by

$$k_c = \frac{k_1 k_2}{k_1 + k_2} . \quad (1)$$

This equation shows that the total stiffness becomes lower than that of each spring when normal springs are connected. However, if one of the springs has negative stiffness that satisfies

$$k_1 = -k_2 , \quad (2)$$

the resultant stiffness becomes infinite, that is

$$|k_c| = \infty . \quad (3)$$

This research applies this principle to developing a zero-compliance mechanism for vibration isolation.

Figure 1a shows the general configuration of the proposed mechanism. A middle mass m_1 is connected to the base through a spring k_1 and a damper c_1 that work as a conventional vibration isolator. A linear actuator is attached to the middle mass. The actuator suspends and drives an isolation table m_2.

The linear actuator is controlled for the suspension system to have negative stiffness. In the initial steady states in Figure 1b, the distance between the table m and the base is kept to be L. When the mass of the table increases by Δm, the distance increases by ΔL. The displacement of the table is in the direction opposite to the added force Δmg so that the static stiffness of the suspension is given by

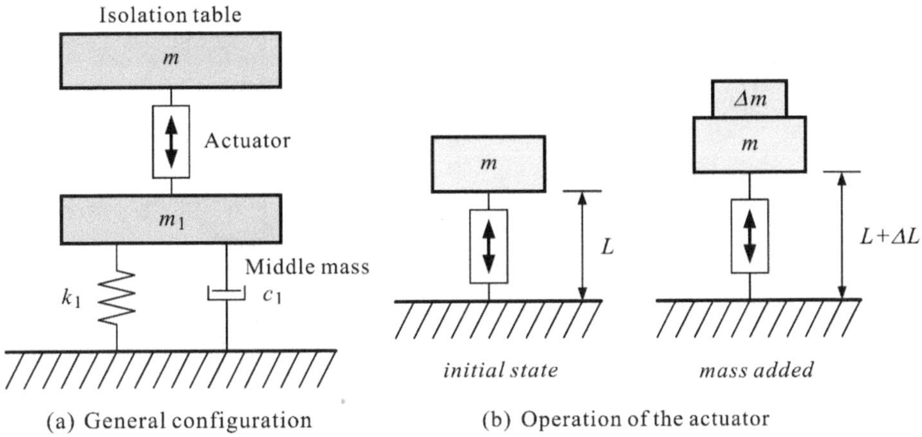

Figure 1. General structure of zero-compliance system using negative stiffness.

$$-\frac{\Delta mg}{\Delta L}(\equiv -k_n). \tag{4}$$

When the actuator is controlled in this way, the system shown by Figure 1a behaves as follows. When the isolation table m_2 is subject to downward force f_0, the table would move upwards by f_0 / k_n if the middle mass were fixed. Meanwhile, the middle mass moves downwards by f_0 / k_1 because of the increase of the suspension force. If $k_1 = k_n$, the increase of the length of the actuator is cancelled by the downward displacement of the middle mass. Thus the isolation table is maintained at the same position as before so that the system has zero compliance to direct static disturbance.

3. REALIZATION OF NEGATIVE STIFFNESS BY PNEUMATIC ACTUATOR

3.1 Basic Model

A pneumatic cylinder of a diaphragm type is developed for the realization of suspension with negative stiffness. Figure 2 shows its photograph and schematic drawing. This type of cylinder is characterized by short stroke and small friction. The pneumatic pressure applied to the diaphragm creates the driving force for the mover.

Figure 3 shows an analytical model for discussing control system design. The cylinder moves a table m, which is connected, to the base through a spring k and a damper c. The equation of motion is given by

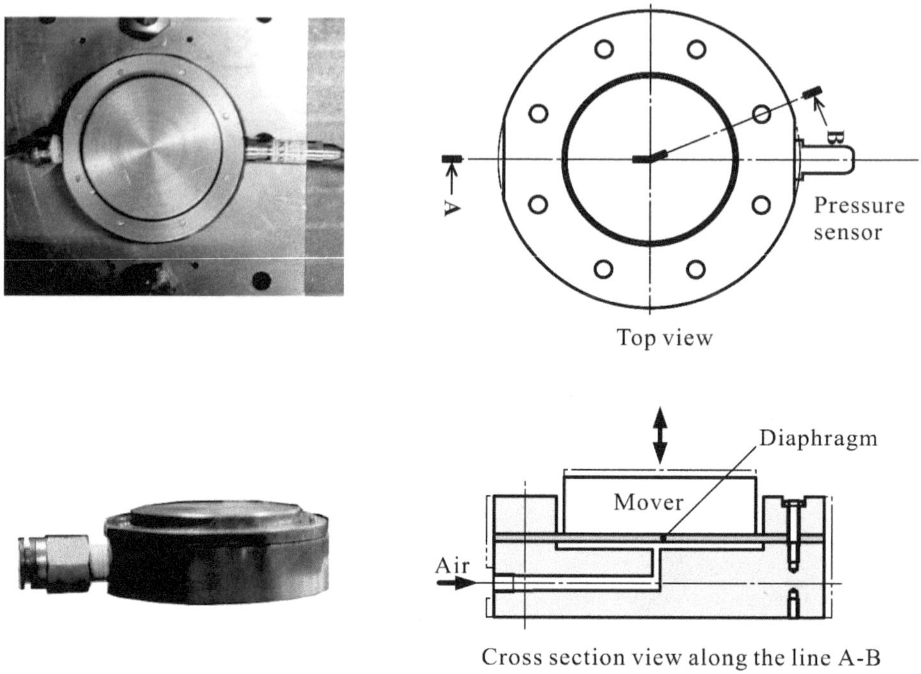

(a) Photograph (b) Schematic drawing

Figure 2. Pneumatic cylinder.

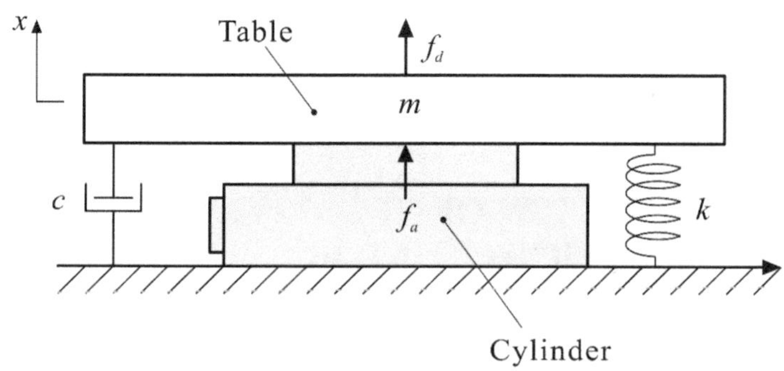

Figure 3. Basic model.

$$m\ddot{x} + c\dot{x} + kx = f_a + f_d, \tag{5}$$

where f_a : force generated by the pneumatic cylinder, and f_d : direct disturbance acting on the table. The force of the actuator is approximately given by

$$F_a(s) = \frac{Ak_i}{Ts+1} I(s), \tag{6}$$

where A: section area of the cylinder, i: coil current of the control valve of the cylinder, and T: time constant depending on the dynamic characteristic of the valve. From Eqs. (5) and (6), we get

$$X(s) = \frac{1}{s^3 + a_2 s^2 + a_1 s + a_0} \{b_0 I(s) + (d_1 s + d_0) F_d(s)\}, \tag{7}$$

where

$$a_2 = \frac{1}{T} + \frac{c}{m}, \qquad a_1 = \frac{k}{m} + \frac{c}{mT}, \qquad a_0 = \frac{k}{mT},$$

$$b_0 = \frac{Ak_i}{mT}, \qquad d_1 = \frac{1}{m}, \qquad d_0 = \frac{1}{mT}.$$

3.2 Control System Design

The displacement of the table is treated as an output signal here. When linear control law is applied, therefore, the control input is generally represented as

$$I(s) = -\frac{h(s)}{g(s)} X(s). \tag{8}$$

When the transfer function of the controller is proper, the polynomials are represented as

$$g(s) = s^n + \sum_{k=0}^{n-1} g_k s^k, \tag{9}$$

$$h(s) = \sum_{k=0}^{n} h_k s^k. \tag{10}$$

Substituting Eq. (8) into Eq. (7) leads to

$$X(s)=\frac{g(s)}{t_o(s)g(s)+b_0h(s)}(d_1s+d_0)F_d(s). \tag{11}$$

The disturbance is assumed to be stepwise so that it can be modeled as

$$F_d(s)=\frac{F_0}{s} \quad (F_0: \text{const}). \tag{12}$$

It is assumed that the controller is selected to stabilize the closed loop system. The steady-state displacement $x(\infty)$ is given by

$$\frac{x(\infty)}{F_0}=\lim_{s\to 0}\frac{(d_1s+d_0)g(s)}{(s^3+a_2s^2+a_1s+a_0)g(s)+b_0h(s)}=\frac{d_0g_0}{a_0g_0+b_0h_0}. \tag{13}$$

For the system to have negative stiffness with a magnitude of k_n, the following equation must be satisfied

$$\frac{d_0g_0}{a_0g_0+b_0h_0}=-\frac{1}{k_n}. \tag{14}$$

For assigning the closed-loop poles arbitrarily, third- or higher-order compensators are necessary. When a third-order compensator is used, the characteristic polynomial of the closed-loop system becomes

$$\begin{aligned} t_c(s)=s^6&+(g_2+a_2)s^5+(g_1+a_2g_2+a_1)s^4\\ &+(g_0+a_2g_1+a_1g_2+a_0+b_0h_3)s^3\\ &+(a_2g_0+a_1g_1+a_0g_2+b_0h_2)s^2\\ &+(a_1g_0+a_0g_1+h_1b_0)s+(a_0g_0+b_0h_0). \end{aligned} \tag{15}$$

To obtain a system with a characteristic equation of the form

$$t_d(s)=\prod_{n=1}^{3}(s^2+2\varsigma_n\omega_ns+\omega_n{}^2), \tag{16}$$

we can match coefficients to obtain g_0, ..., g_2 and h_0, ..., h_3.

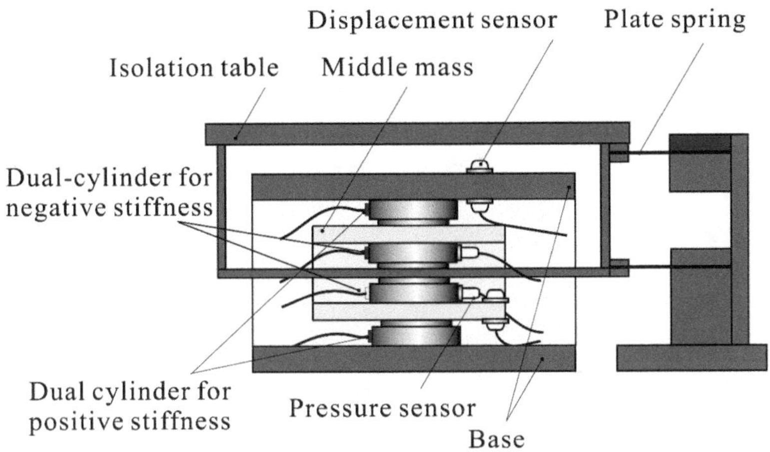

Figure 4. Experimental apparatus.

4. EXPERIMENTAL APPARATUS

Figure 4 shows a schematic drawing of the developed experimental apparatus with four cylinders. Each cylinder has a diaphragm made of rubber with a thickness of 2 mm. Its effective sectional area is 50 cm^2 so that the generated force is approximately 500 N when the gauge pressure of supply air is 0.1 MPa. To reduce the mass of the apparatus, therefore, two cylinders are operated in a differential mode, which are referred to as a dual-cylinder.

One dual-cylinder suspends the middle mass and operates as a suspension with positive stiffness. Another dual-cylinder fixed on the middle mass suspends the isolation table and operates as a suspension with negative stiffness. These cylinders are controlled with flow control valves.

The middle mass and the isolation table are guided to be in translation by plate springs. The relative displacement of the middle mass to the base and that of the isolation table to the middle mass are detected by eddy-current gap sensors with a resolution of 1mm. Designed control algorithms are implemented with a DSP-based digital controller.

5. EXPERIMENTAL RESULTS

Two kinds of experiments are carried out. The first is for estimating positive stiffness and negative stiffness generated by the dual-cylinder fixed on the middle mass. In this experiment, the middle mass is clamped not to move. The other is for estimating the characteristics of the zero-compliance system. Disturbance force is produced by putting weights on the isolation table. In the following graphs, force and displacement in the downward direction are represented to be positive, and vice versa.

First, suspension with prescribed positive stiffness is realized by the dual-cylinder. Its controller has the same form as that for realizing negative stiffness except that stiffness is opposite; Eq.(14) is replaced by

$$\frac{d_0 g_0}{a_0 g_0 + b_0 h_0} = \frac{1}{k_p} \tag{17}$$

Figure 5 shows the measured force-displacement characteristics. The amplitude of negative stiffness is set to be

(a) $k_p = 300$ [kN/m], (b) $k_p = 400$ [kN/m], (c) $k_p = 500$ [kN/m],

in designing the controller. It shows that the characteristics are quite linear. The actual stiffness estimated from the results is

(a) $k_p = 311$ [kN/m], (b) $k_p = 383$ [kN/m], (c) $k_p = 499$ [kN/m]

The differences between the prescribed and experimental values are within 5%.

Next, suspension with prescribed negative stiffness is realized by the dual-cylinder. The amplitude of negative stiffness is set to be

(a) $k_n = 300$ [kN/m], (b) $k_n = 400$ [kN/m], (c) $k_n = 500$ [kN/m].

Figure 6 shows the measured force-displacement characteristics. The estimated amplitude of stiffness is

(a) $k_n = 328$ [kN/m], (b) $k_n = 398$ [kN/m], (c) $k_n = 493$ [kN/m].

The differences between the prescribed and experimental values are of the same order as in the previous experiment. In both the cases, the difference is small when the amplitude of stiffness is large.

In the second experiment, the clamp of the middle mass is released. The two dual-cylinders are designed to have static stiffness of ± 1000 [kN/m]. Figure 7 shows the displacement of the isolation table to the base, that of the isolation table to the middle mass and that of the middle mass to the base. It is observed that the position of the isolation table is maintained at the same position while the position of the middle mass changes proportion to direct disturbance. The estimated stiffness between the isolation table and the base is 8.8×10^5 [kN/m], which is about 90 times the stiffness of each suspension. This result demonstrates that the compliance between the isolation table and the base is made very small by the proposed mechanism.

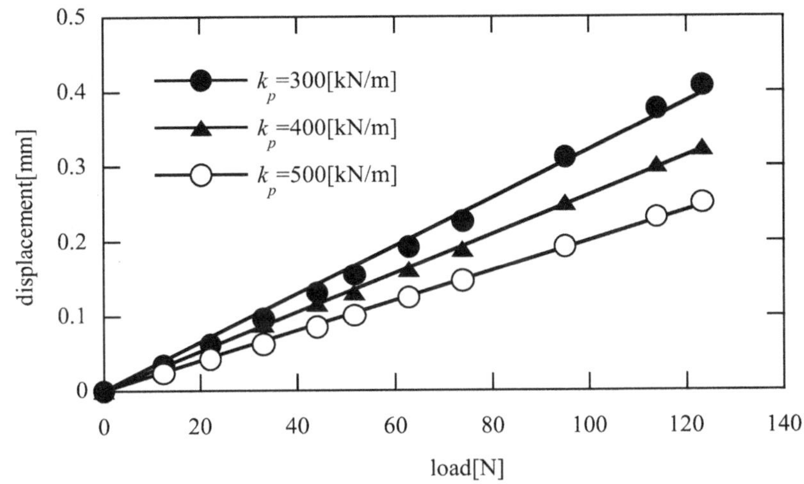

Figure 5. Realization of suspension with positive stiffness.

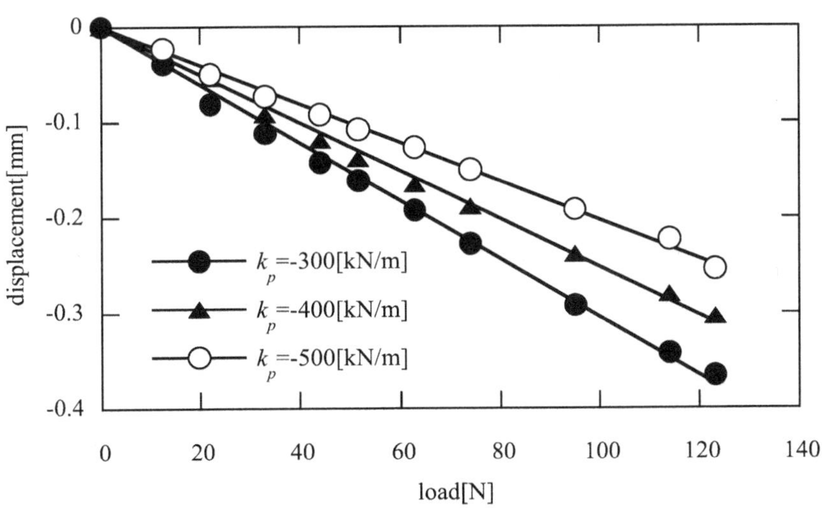

Figure 6. Realization of suspension with negative stiffness.

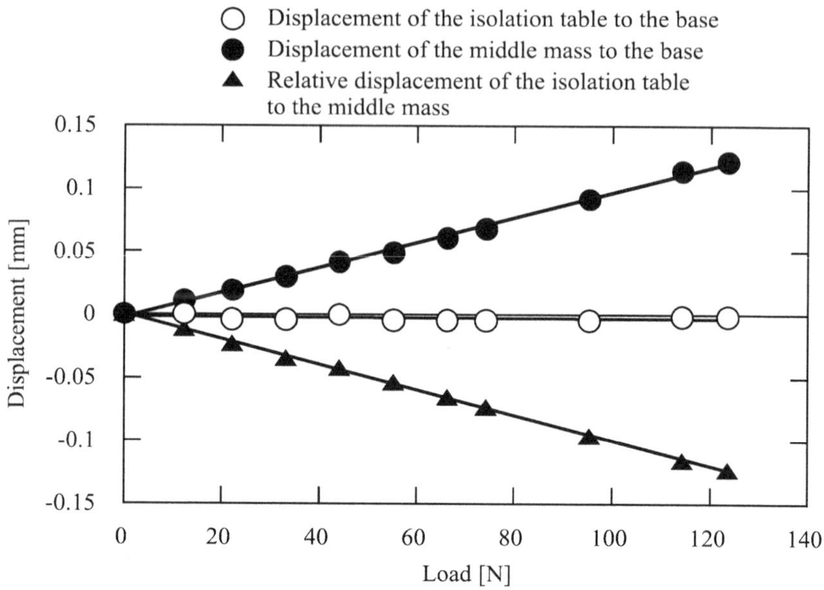

Figure 7. Realization zero-compliance mechanism.

6. CONCLUSIONS

A zero-compliance suspension mechanism with pneumatic actuators was developed. A control method of realizing prescribed negative stiffness by a pneumatic actuator was shown. The fabricated apparatus has a middle mass suspended by a dual-cylinder fixed on the base and an isolation table suspended by a dual cylinder fixed on the middle mass. The former cylinder is controlled to operate as a suspension with positive stiffness while the latter was controlled to operate as a suspension with negative stiffness. It was demonstrated that the equalization of the amplitude of negative and positive stiffness enables the system to have zero-compliance to direct disturbance.

REFERENCES

1. Yoshioka, H., Takahashi, Y., Katayama, K., Imazawa, T. and Murai, N., An Active Microvibration Isolation System for Hi-tech Manufacturing Facilities, *ASME Journal of Vibration and Acoustics*, Vol. 123, 2001, pp. 269–275.
2. Mizuno, T., Vibration Isolation System Using Zero-Power Magnetic Suspension, Preprints of 15th World Congress IFAC, 2002, 955.
3. Mizuno, T., Toumiya, T. and Takasaki, M., Vibration Isolation System Using Negative Stiffness, *JSME International Journal, Series C*, Vol. 46, No. 3, 2003, pp. 807–812.

CONTROL OF STICK-SLIP VIBRATIONS

Marcus Neubauer, Cord-Christian Neuber and Karl Popp
Institute of Mechanics, University of Hannover, Germany

Abstract Many technical applications are subjected to friction induced vibrations. These vibrations are mostly unwanted as they generate noise, increase wear and diminish accuracy. In this paper a feedback control is introduced that is capable to suppress these oscillations. The results are validated experimentally on a squealing brake.

Key words: self excited vibration system, stick-slip vibrations, feedback control, brake squeal.

1. INTRODUCTION

Friction is an all-day-phenomenon that everybody knows. From an engineering point of view two different phenomena can be distinguished. First, the resistance against the start of a relative motion of bodies which is caused by the contact between these bodies. Secondly, the resistance against an existing relative motion of two contacting bodies. In the first case, the phenomenon is called static friction. In the second case, the effort to sustain a sliding motion is called kinetic friction.

The phenomenon of stick-slip occurs if there is an intermittent change between static and kinetic friction. It becomes possible, if the dynamical properties of the regarded system like elasticity, mass distribution and damping allow this kind of motion. Stick-slip vibrations appear in everyday life as well as in engineering applications. Examples are the sound of a violin string, creaking doors, brake groan, chatter of machine tools or vibrations in slides and seals. Mostly, stick-slip has to be avoided, as it creates unwanted noise, diminishes accuracy and increases wear. In nonlinear vibration theory, the stick-slip motion can be treated as a stable limit cycle of a self-excited vibration system.

Different mechanisms to generate friction-induced self-excited vibrations are known. They can be shown by simple models with one or two degrees of freedom. The major physical mechanisms to generate friction induced vibrations are: (i) a friction characteristic decreasing with increasing relative velocity; (ii) periodic fluctuating normal forces in a particular phase relation

H. Ulbrich and W. Günthner (eds), IUTAM Symposium on Vibration Control of Nonlinear Mechanisms and Structures, 223–232.

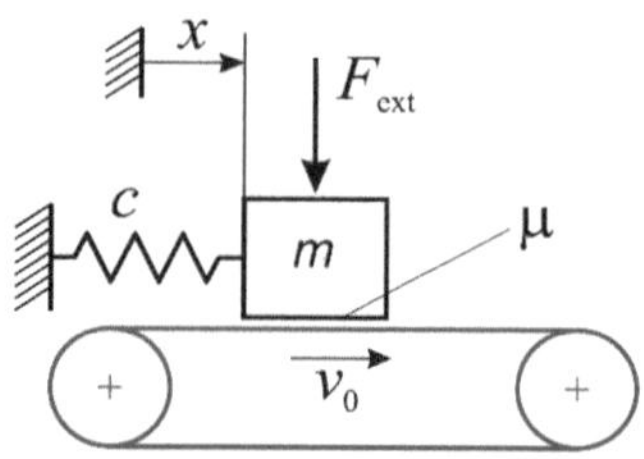

Figure 1. Model of a simple 1 DOF friction oscillator.

to the vibrations in the tangential direction, and (iii) nonconservative restoring forces which depend on the geometrical parameters of the system.

Understanding the mechanisms to generate friction-induced vibrations can help to avoid these unwanted motions. An active suppression of brake squeal has been shown in von Wagner et al. (2004). Based on a multi body system of a floating caliper disc brake an optimal control law has been applied. The control forces are induced by 'smart pads', which incorporate piezoceramics.

In the present paper it is shown how stick-slip motions can be suppressed by applying an active vibration control, which is based on the mechanism of fluctuating normal forces. Choosing a proper phase relation between the vibration amplitude and the normal force results in dissipated energy and, thus, achieves damping. This new concept has successfully been applied to the high frequency vibrations of a squealing brake.

2. ACTIVE CONTROL OF A 1 DOF FRICTION OSCILLATOR

In Popp (2004) a control strategy has been presented to avoid stick-slip motions of a simple 1 DOF friction oscillator as shown in Figure 1. The undamped oscillator is pressed on a base moving with a constant velocity v_0 by an external force F_{ext}. The friction force is given by Coulomb's law

$$F_{\text{R}} = \mu(v_{\text{rel}})F_{\text{N}}, \quad F_{\text{N}} = F_{\text{ext}}, \tag{1}$$

where the normal force F_{N} is equal to the external force F_{ext}. The kinetic coefficient of friction μ is assumed to be linearly decreasing with the negative gradient δ versus the relative velocity v_{rel},

$$\mu(v_{\text{rel}}) = \mu_0 - \delta v_{\text{rel}}, \quad v_{\text{rel}} = v_0 - \dot{x}. \tag{2}$$

The equation of motion around the equilibrium position reads

$$m\ddot{x} - \delta F_{\text{ext}}\dot{x} + cx = 0. \tag{3}$$

The equilibrium state becomes unstable when $\delta > 0$. In this case a limit cycle is reached, which is characterized by a balanced input and output of energy

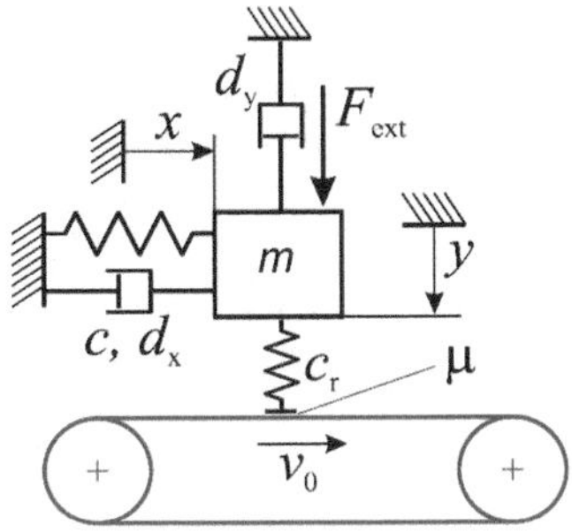

Figure 2. Model of a 2 DOF friction oscillator.

during each period. Assuming a harmonic motion of the block,

$$x(t) = \hat{x} \cos \omega t, \ \omega = \frac{2\pi}{T_s}, \tag{4}$$

and considering a fluctuating normal force which consists of a constant part F_0 and a superimposed sinusoidal part with an amplitude $\hat{F}$ with the same frequency as the block vibration,

$$F_{ext}(t) = F_0 + \hat{F} \cos(\omega t - \varphi), \tag{5}$$

one can show that this results in an input or output of mechanical energy depending on the phase shift φ,

$$\Delta E_{in} = \int_0^{T_s} F_R \dot{x} \, dt = -(\mu_0 - \delta v_0)\pi \sin \varphi \hat{F} \hat{x} + \delta \pi \omega F_0 \hat{x}^2 = E^- + E^+. \tag{6}$$

Expression E^+ gives the energy growth due to the decreasing friction characteristic, and the dynamical part of F_{ext} leads to E^-, which is a dissipated energy if $\sin \varphi > 0$. Optimal damping is achieved if ΔE_{in} becomes a minimum, i.e.

$$\varphi_{opt} = 90°. \tag{7}$$

This control law was realized successfully to avoid stick-slip motions of relatively slow frequencies (see Popp, 2005). But it reaches its limits when rather high frequencies are to be controlled, e.g. as in a squealing brake.

3. ACTIVE CONTROL OF A 2 DOF FRICTION OSCILLATOR

A more detailed model of a friction oscillator is shown in Figure 2. It is the same oscillator as given in Figure 1, but with an additional vertical degree of freedom y. A contact stiffness c_r is placed between the moving base and the oscillator and both degrees of freedom are damped. This oscillator also executes stick-slip motions in x-direction if the equilibrium position is unstable.

The equations of motion of the system read

$$\begin{bmatrix} m & 0 \\ 0 & m \end{bmatrix} \begin{bmatrix} \ddot{x} \\ \ddot{y} \end{bmatrix} + \begin{bmatrix} d_x & 0 \\ 0 & d_y \end{bmatrix} \begin{bmatrix} \dot{x} \\ \dot{y} \end{bmatrix} + \begin{bmatrix} c & 0 \\ 0 & c_r \end{bmatrix} \begin{bmatrix} x \\ y \end{bmatrix} = \begin{bmatrix} F_R \\ F_{ext} \end{bmatrix}, \tag{8}$$

with the friction force F_R

$$F_R(t) = \mu(v_{rel}) F_N(t), \quad F_N(t) = c_r y(t). \tag{9}$$

Note that the vibration in y-direction is decoupled from that in x-direction when no control force is applied, while the vibration in x-direction depends on the y-motion via the friction force F_R. If we apply the same control law for F_{ext} as in the 1 DOF oscillator also for the 2 DOF system, one has to consider the given differences.

Both models differ in the correlation between the external force F_{ext} and the normal force F_N in contact. For the 1 DOF oscillator it follows $F_N = F_{ext}$, cf. eq. (1), while the 2 DOF oscillator forms a standard damped oscillator in y-direction. Therefore the transfer of the external force F_{ext} to the normal force $F_N = c_r y$, cf. eq. (9) is subjected to

- an amplification of the force amplitude, and
- a phase-shift,

which both are functions of the frequency ratio η.

3.1 Force Amplification

Because of the proportional correlation between the normal force $F_N(t)$ and the vertical displacement $y(t)$, cf. eq. (9), the normal force F_N can be expressed by the well known function

$$|F_N(\eta)| = \frac{|F_{ext}|}{\sqrt{\left(1-\eta^2\right)^2 + (2D\eta)^2}}. \tag{10}$$

The parameter η is the ratio of the excitation frequency Ω and the eigenfrequency of the oscillator in y-direction, $\omega_y = \sqrt{\frac{c_r}{m}}$. When applying the control law

$$F_{ext}(t) = F_0 + \hat{F}\cos(\omega_x t - \varphi), \tag{11}$$

which means the frequency of the fluctuating external force is equal to the frequency of the oscillation in x-direction, the ratio η can be expressed as

$$\eta = \frac{\omega_x}{\omega_y} = \sqrt{\frac{c}{c_r}}. \tag{12}$$

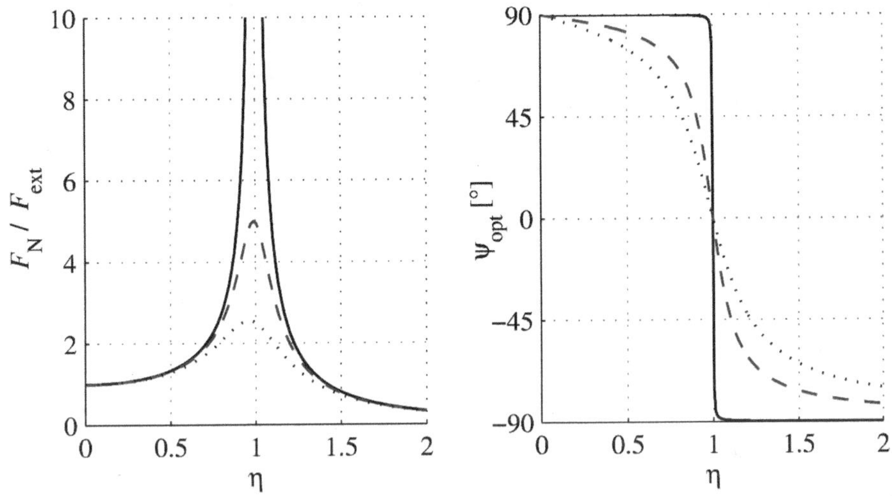

Figure 3. Force amplification and optimal phase shift for $D = 0$ (———), $D = 0.1$ (– – –) and $D = 0.2$ (·······).

The damping ratio D_y of the vertical oscillation is given by

$$D_y = \frac{d_y}{2\sqrt{c_r m}}. \tag{13}$$

The amplification functions for different damping ratios of the system is shown in Figure 3. In the static case ($\eta = 0$) the force amplification is equal to one, which means $F_N = F_{ext}$. For a weak damped system with $D < \sqrt{0.5}$ the amplification grows until reaching a maximum at

$$\eta_{res} = \sqrt{1 - 2D^2}. \tag{14}$$

At this point, the force amplification reads $\frac{F_N}{F_{ext}} = \frac{1}{2D\sqrt{1-D^2}}$. Beyond this frequency the amplification decreases asymptotically towards zero. In the range $0 < \eta < \sqrt{2}\sqrt{1 - 2D^2}$, the ratio $|\frac{F_N}{F_{ext}}|$ is larger than one, so the external force is amplified, while the force is diminished in the range $\eta > \sqrt{2}\sqrt{1 - 2D^2}$.

3.2 Phase Shift

The displacement $y(t)$ does not follow directly the force $F_{ext}(t)$ due to the mass inertia, and the same applies to the normal force in the contact. The corresponding phase shift is given by

$$\tan\varphi = \frac{2D\eta}{1 - \eta^2}. \tag{15}$$

The angle φ is the phase delay between the normal force $F_N(t)$ in the contact and the external force $F_{ext}(t)$. In order to produce a normal force $F_N(t)$ in the friction contact with a phase shift ϕ to the displacement $x(t)$ of the stick-slip motion, the phase shift ψ between the external force $F_{ext}(t)$ and the stick-slip displacement $x(t)$ reads

$$\psi = \phi - \varphi(\eta). \tag{16}$$

As the phase delay between the normal force $F_N(t)$ and the displacement $x(t)$ still has to be tuned to $\phi_{opt} = 90°$, the optimal phase delay between $F_{ext}(t)$ and $x(t)$ results in

$$\psi_{opt} = 90° - \varphi(\eta). \tag{17}$$

According to this relation the optimal phase delay for $\eta \ll 1$ is $\psi_{opt} = +90°$. At $\eta = 1$ the phase shift $\varphi = +90°$ due to the mass inertia, so the optimal phase delay reads $\psi_{opt} = 0$. For $\eta > 1$ the phase shift caused by inertia effects grows towards $\varphi = +180°$, thus the theoretical optimal phase delay ψ_{opt} would be negative, cf. Figure 3. To get a positive phase shift, the displacement of the previous period has to be taken into account. Thus, for $\eta \gg 1$ with $\varphi(\eta) \approx 180°$ one gets $\psi_{opt} = 270°$.

3.3 Control Law

A practical control law will unlikely have the form according to eq. (5). Usually, the amplitude of the controlled force is not constant, but proportional to the measured signal. Also the delay will not be given in an angle but as a time delay,

$$F_{ext}(t) = k\ x(t - \Delta t). \tag{18}$$

The factor k stands for the feedback gain of the control loop. The correlation between the displacement $x(t)$ and the normal force F_N in the contact is given by the effective feedback gain k_{eff},

$$F_N(t) = \underbrace{\frac{k}{\sqrt{\left(1-\eta^2\right)^2 + (2D\eta)^2}}}_{k_{eff}}\ x(t - \Delta t). \tag{19}$$

The time shift is given by the parameter Δt. By transforming the optimal angle ψ_{opt} into a time delay Δt_{opt} is obtained,

$$\Delta t_{opt} = \frac{1}{f_{squeal}} \frac{\psi_{opt}}{360°}, \tag{20}$$

where f_{squeal} is the frequency of the brake-squeal. From eq. (20) the following conclusions can be made:

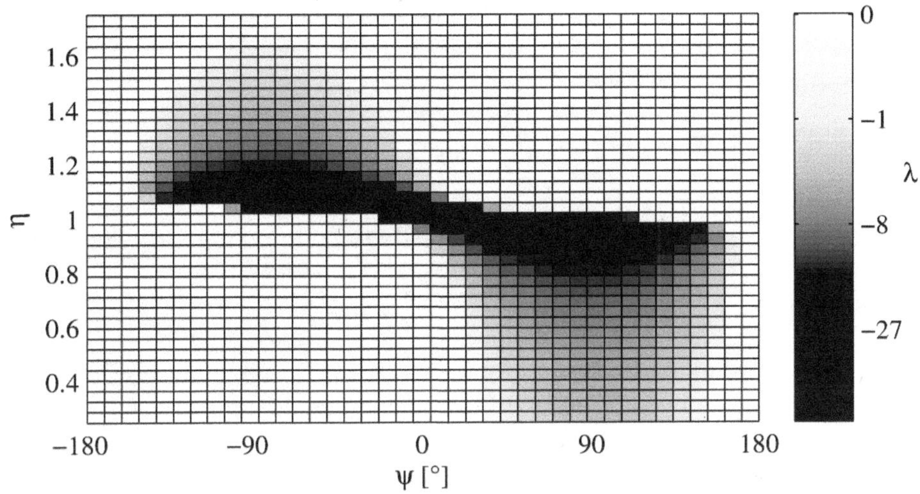

Figure 4. Top Lyapunov exponent versus phase shift ψ and ratio η.

- The squealing frequency f_{squeal} has to be known for adjusting the optimal time delay Δt_{opt}.
- Only one frequency (and certain higher frequencies where the adjusted time delay results in $\psi = \psi_{\text{opt}} + 2\pi n$) can be damped optimally by that time delay.
- Frequencies for which the adjusted time delay results in $\psi \approx -\psi_{\text{opt}}$ are even excited by this control force.

3.4 Numerical Stability Analysis

A mechanical model with a control loop as described in eq. (18) was analyzed numerically using MATLAB/SIMULINK.

If we assume a linearly decreasing friction characteristic, cf. eq. (2), then due to eq. (9) the system becomes nonlinear. Thus, the stability of the system has been classified by the top Lyapunov exponent, which is defined as

$$\lambda_t = \lim_{t\to\infty} \frac{1}{t} \ln \frac{\|\mathbf{z}(t)\|}{\|\mathbf{z}(t_0)\|}, \tag{21}$$

with the state vector $\mathbf{z} = [x\ y\ \dot{x}\ \dot{y}]^{\mathrm{T}}$. A negative value for λ_t indicates an asymptotically stable system, a positive value an unstable system and $\lambda_t = 0$ results for a stable system.

A stability analysis has been performed showing the top Lyapunov exponent λ_t versus the frequency ratio η and the time delay Δt, see Figure 4 for a feedback gain $k = 300$ kN/m and a damping ratio $D_y = 0.001$. The ratio η

has been varied by changing the stiffness c_r and the time delay has been transformed into a phase shift ψ between the measured displacement signal $x(t)$ and the applied external force F_{ext},

$$\psi = 360° \frac{\Delta t}{T_S}, \tag{22}$$

where T_S is the length of one vibration period. It can be seen that optimal damping is achieved when tuning to the optimal phase shift $\psi_{opt}(\eta)$ according to Figure 3. For $\eta < 1$ the optimal phase delay is $\psi_{opt} = 90°$ and $\eta > 1$ results in $\psi_{opt} = -90°$. The effective feedback gain k_{eff} according to eq. (19) results in stronger damping of the system for $\eta \approx 1$, while the damping is diminished for $\eta \gg 1$. Above a certain value of η, the damping effect is not sufficient to brake up the stick-slip limit cycle, and $\lambda_t = 0$ holds.

The control strategy is quite robust to phase angles $\psi = \psi_{opt} \pm \epsilon$ with small deviations ϵ from the optimal phase angle ψ_{opt}, which still result in additional damping of the system. This means that also frequencies $f = f_{squeal} \pm \epsilon$ close to the attuned squealing frequency f_{squeal} are damped.

4. EXPERIMENTAL RESULTS

The control law described above was also verified experimentally on a test rig of an automotive disc brake. The brake shows a strong squealing frequency of $f_{squeal} = 3400$ Hz for a disc rotation of 27 rpm and a brake pressure of 9 bar. The floating caliper brake has been modified in order to incorporate two piezo elements, which are placed between the caliper and the outboard pad. Because of the piezoelectric effect, an applied voltage on the electrodes results in a force acting on the outboard pad in normal direction.

To realize the control strategy, the vibration of the pad in the rotation direction of the disc has been measured by a Laser vibrometer with a sampling rate of 25 kHz. Instead of the displacement $x(t)$ the velocity $v(t)$ has been taken for the control loop, which is possible when the vibration is nearly harmonic. In this case there is a constant phase delay of 90° between the displacement and velocity signal,

$$v(t) = \omega\, x\left(t + \frac{T_S}{4}\right). \tag{23}$$

The phase delay that corresponds to the optimal time delay between the displacement and force signals was found experimentally at $\psi_{opt} = 90°$. The system could be stabilized for amplification factors $6\,\text{N}/\mu\text{m} < k < 8\,\text{N}/\mu\text{m}$. The time signals of the velocity $v(t)$ of the pad and the applied voltage $U(t)$ on the piezo elements are shown in Figure 5 for a gain factor $k = 6.7\,\text{N}/\mu\text{m}$. A constant voltage has been added to the voltage signal in order to keep the overall voltage on the electrodes always positive.

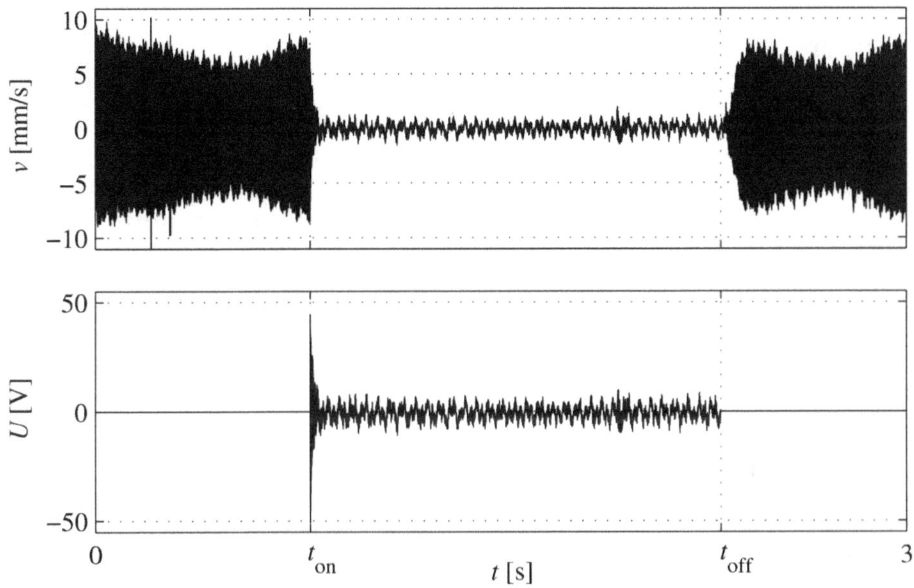

Figure 5. Time signals of the vibration velocity and the control voltage.

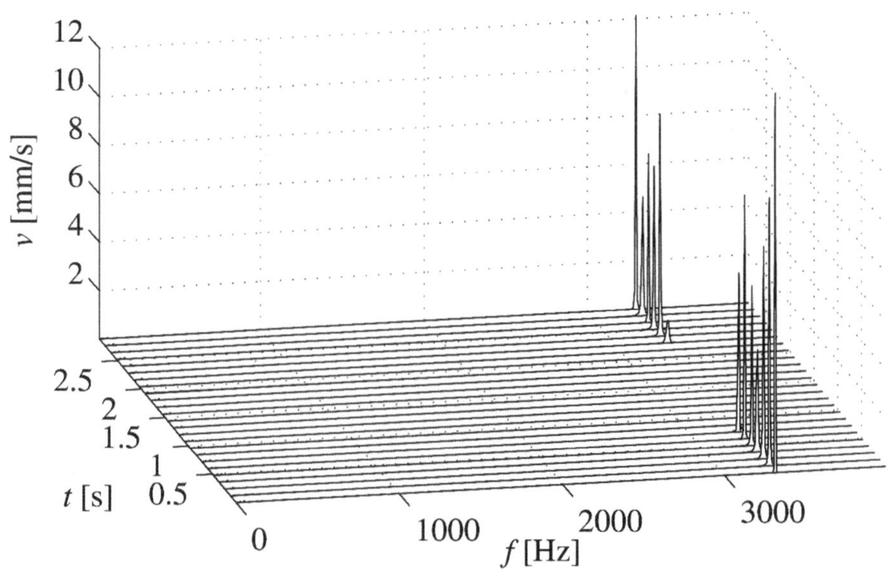

Figure 6. Short time FFT of the velocity signal. Control force activated between $t = 0.8$ and $t = 2.3$ s.

The control force was turned on at $t_{\text{on}} \approx 0.8$ s and turned off at $t_{\text{off}} \approx 2.3$ s. The squealing of the brake was suppressed in less than 0.05 s with a maximum voltage of ± 50 V. The velocity signal and therefore also the voltage signal are reduced to a constant status of $\hat{v} = 1$ mm/s resp. $\hat{U} = 6$ V. While applying the control force the brake was virtually silent. After switching the control off again, the squealing arose within 0.1 s to the original strength and frequency.

A fast Fourier transform of the velocity signal is given in Figure 6. The squealing frequency $f_{\text{squeal}} = 3400$ Hz is nearly totally eliminated for the time the control force was applied. The remaining oscillation is dominated by a frequency of $f = 150$ Hz, which can be regarded as brake groan.

The control strategy is quite robust concerning the feedback gain of the control loop. However, increasing the feedback gain too much results again in a squealing brake because of the excitation in normal direction.

5. SUMMARY

The control strategy to suppress stick-slip motions was extended in order to eliminate high frequencies also. The modified control law was tested numerically with a `MATLAB/SIMULINK` model and verified experimentally on a squealing brake. By applying a voltage signal with a proper time delay and amplification to the measured velocity signal of the brake pad on the electrodes of piezo elements which are placed inside a modified floating caliper brake, the squealing has successfully and reproducibly been eliminated.

REFERENCES

Popp, K. (2005). Modelling and Control of Friction Induced Vibrations. *Mathematical and Computer Modelling of Dynamical Systems*, Vol. 11, to appear.

Popp, K. and Rudolph, M. (2004). Vibration Control to Avoid Stick-Slip Motion. *Journal of Vibration and Control*, Vol. 10, pp. 1585–1600.

von Wagner, U., Hochlenert, D., Jearsiripongkul, T. and Hagedorn, P. (2004). *Active Control of Brake Squeal via 'Smart Pads'*. SAE Technical Papers 2004-01-2773.

ENERGY REGENERATIVE AND ACTIVE CONTROL OF ELECTRO-DYNAMIC VIBRATION DAMPER

Yohji Okada and Keisuke Ozawa
Department of Mechanical Engineering, Ibaraki University, Japan
y.okada@mx.ibaraki.ac.jp

Abstract: Active vibration control is introduced to an electro-dynamic regenerative vibration damper. It is intended to improve vibration reduction capability using the regenerated energy. This idea is applied to a moving mass type vibration absorber and vibration control is performed. In the case of energy regenerative control, a PWM step-up chopper is used to solve the dead zone problem. Active control signal is also produced by the other PWM chopper from the power supply. However, energy regenerative and active control mode can not operate at the same time. A new control law is introduced to switch control mode and to follow the ideal force. The ideal force is calculated using the LQ control theory. Experimental setup is made to confirm the proposed technique and the damping capability is tested.

Key words: regenerative damper, active control, dynamic absorber, digital control.

INTRODUCTION

Vibration control has been widely investigated and reported. Recently active vibration control has been gradually used because of its high damping capability such as high building, automobile suspension, and large space structures (Garcia et al., 1995; Alleyne and Hedrick, 1995). But there exist several difficulties such as high power consumption and complexity of the system even though it has strong ability of vibration reduction. On the contrary, vibration is a kind of energy and the actuator is a kind of energy converter, the vibration energy can be regenerated to electric energy. This idea of utilizing the vibration energy and regenerating it to electric energy can lead to high cost saving and efficiency (Okada and Harada, 1995).

In a previous paper, one of the authors reported the energy regenerative damper using pulse width modulated (PWM) step-up chopper control (Kim and Okada, 2002a). By using the step-up chopper, electric energy can be regen-

H. Ulbrich and W. Günthner (eds), IUTAM Symposium on Vibration Control of Nonlinear Mechanisms and Structures, 233–242.

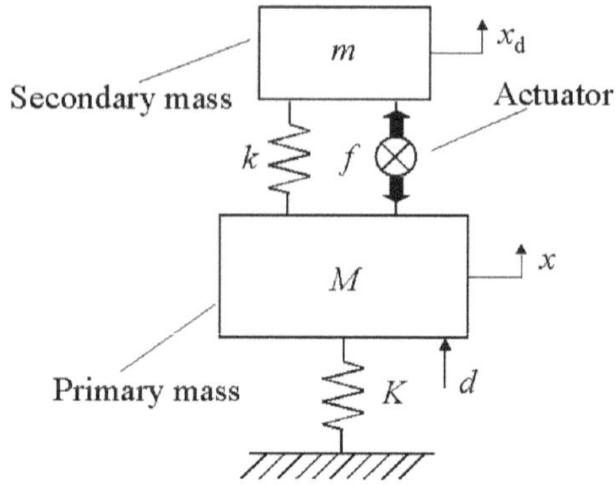

Figure 1. Model of the moving mass type damper system.

erated from the low voltage source to high battery voltage. Hence the damper can regenerate even though the actuator velocity is in the dead zone region. The damping characteristics can be adjusted by changing the PWM duty ratio.

In this paper, active vibration control is introduced to an electro-dynamic regenerative vibration damper. It is intended to improve vibration reduction capability using the regenerated energy. An electro-dynamic actuator is used as an energy regenerative damper and active control actuator. This idea is applied to a moving mass type vibration absorber and vibration control is performed. In the case of energy regenerative control, a PWM step-up chopper is used to solve the dead zone problem. Active control signal is also produced by the other PWM chopper from the power supply. However, energy regenerative and active control mode can not operate at the same time. A new control law is introduced to switch control mode and to follow the ideal force. The ideal force is calculated using the LQ control theory. Experimental setup is made to confirm the proposed technique and the damping capability is tested.

1. MOVING MASS DAMPER

In this paper the proposed active and regenerative damper is applied to a moving mass type vibration damper.

1.1 Modeling of the System

The physical model of the system is shown in Figure 1. The external force d is applied to the main mass M, and the damping force f is applied between the main mass and the damper mass m. Then the equations of motion are written as follows, where the parameters used in the experiment are shown in Table 1.

$$\begin{cases} M\ddot{x} + Kx + f + k(x - x_d) = d \\ m\ddot{x}_d - f + k(x_d - x) = 0 \end{cases} \tag{1}$$

Table 1. Parameters of experimental setup.

Description	Symbol	Value
Primary mass	M	4.85 [kg]
Primary spring constant	K	19809 [N/m]
Primary resonant frequency	ω_n	63.9 [rad/s]
Secondary mass	m	0.8468 [kg]
Secondary spring constant	k	2510.5 [N/m]
Mass ratio	μ	0.1746
Actuator constant	ψ	25 [N/A=Vs/m]

1.2 Damping Actuator

Linear voice coil motor (VCM) is used in this research for the damping actuator. Assuming the actuator is an ideal one, it has the following energy conversion property (Okada and Harada, 1995),

$$f = \psi i, \qquad e = \psi v \tag{2}$$

where f and v are the actuator force and velocity, e and i are the driving voltage and the current, and ψ is the actuator constant, respectively.

2. ACTIVE AND ENERGY REGENERATIVE CONTROL

Single actuator is used as active and regenerative damper. Hence a new control algorithm is introduced to switch the control mode between the active and regenerative one to follow the ideal damping force.

2.1 Ideal Damping Force

The ideal damping force f_d is calculated using the LQ control theory. The performance index is written as

$$J = \int_0^\infty [\ \boldsymbol{x}^T \boldsymbol{Q}\boldsymbol{x} + f_d^2 R\ ]\ dt \tag{3}$$

where the state variable $\boldsymbol{x}$ is composed of the main and submass displacements and their velocities, and $\boldsymbol{Q}$ $[W/m^2]$ and R $[W/N^2]$ are the weights for them.

Solving the Riccati equation we have the following ideal damping force.

$$f_d = -\boldsymbol{K}_f \left[\ x\ \ x_d\ \ \dot{x}\ \ \dot{x}_d\ \right]^T \tag{4}$$

The displacements of the main and sub masses are measured by the laser displacement sensors and the velocities are calculated from the observers.

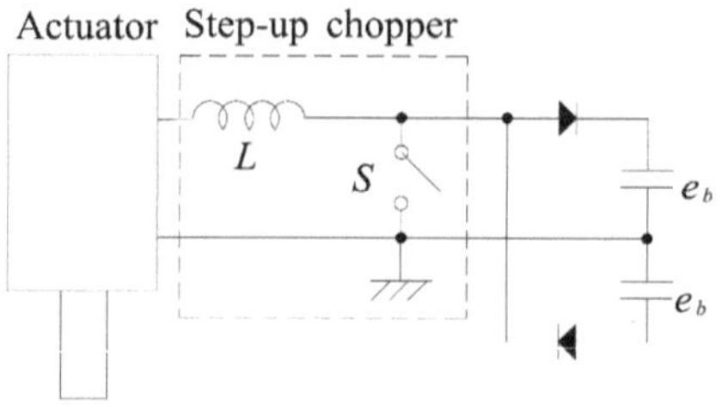

Figure 2. PWM step-up chopper circuit.

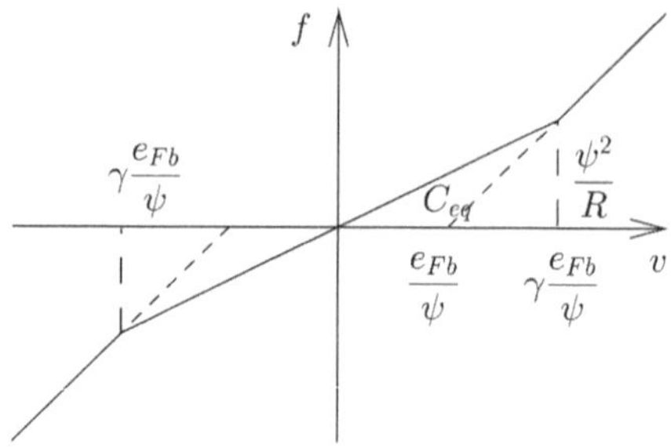

Figure 3. Damping characteristics of PWM step-up chopper.

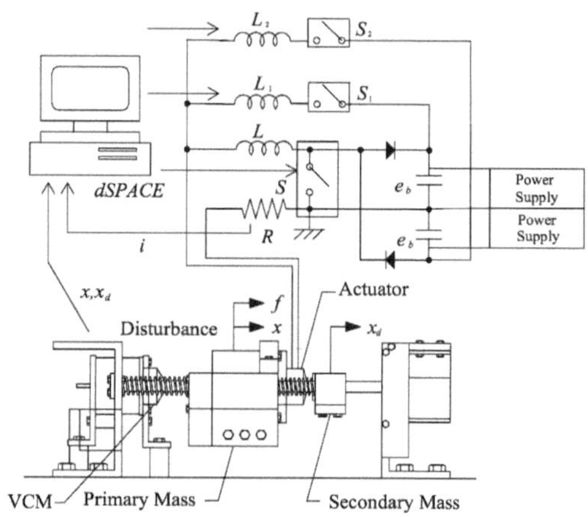

Figure 4. Schematic of control system.

2.2 PWM Stepup Chopper Regenerative Damper

The energy regenerative damper is shown in Figure 2. Without stepup chopper, which is shown in the dashed block, the damper has the undesirable dead-zone as shown by the dotted line in Figure 3.

$$-\frac{e_{Fb}}{\psi} \leq v \leq \frac{e_{Fb}}{\psi} \tag{5}$$

In this region damping force becomes zero and energy is not regenerated. This is improved by introducing the PWM stepup chopper as shown by the solid line in Figure 3 (Kim and Okada, 2002a).

2.3 Incorporating Active Control

The regenerative force is used when the regenerative circuit can control the actuator force close to the ideal force, while the active control is introduced in the other region. For this purpose new control circuit is introduced as shown

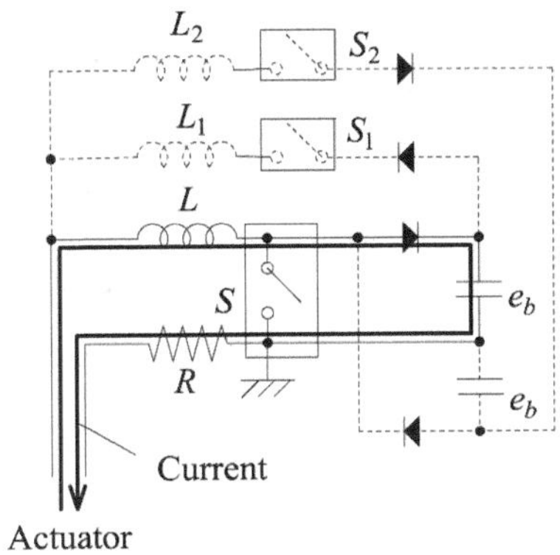

Figure 5. Energy regenerative mode I.

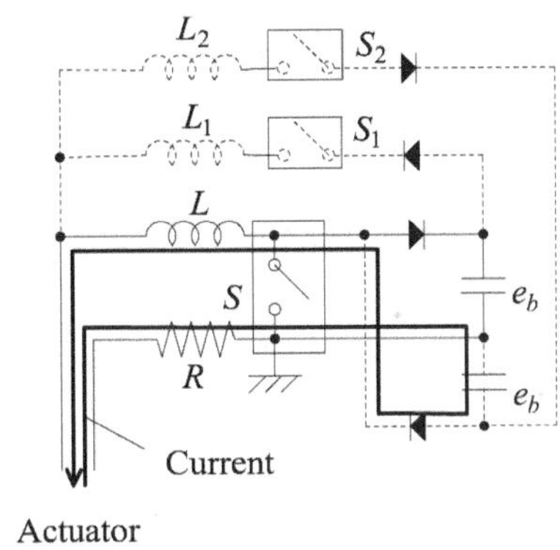

Figure 6. Energy regenerative mode II.

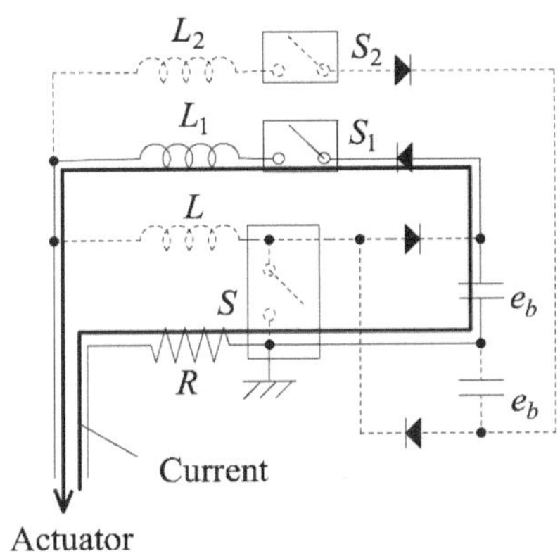

Figure 7. Active control mode I.

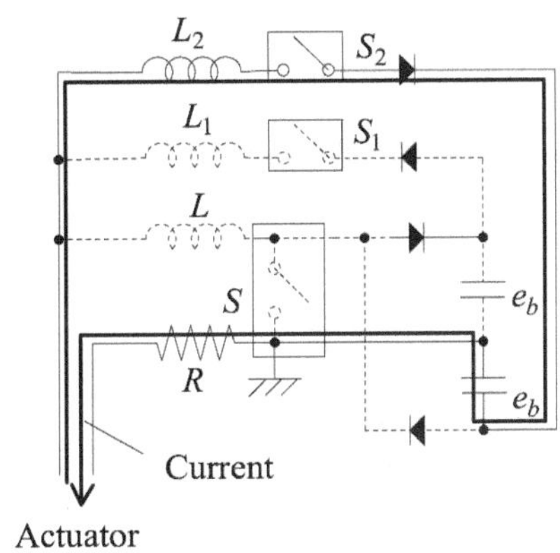

Figure 8. Active control mode II.

in Figure 4. The left side voice coil motor (VCM) is used to excite the main structure, and the bigger VCM is installed between the main mass and the auxiliary mass as the damping actuator. The control is switched as shown in Figures 5–8. All the control modes are controlled with the pulse width modulated (PWM) signals using the analog switches (MAX4601).

The control modes are switched whether the regenerative control is possible or not as:

1) $-\dfrac{\psi^2}{R'}v^2 \leq f_d \cdot v \leq 0$: Energy Regenerative Mode

2) Other cases: Active Control Mode

In case 1), the ideal force and the actuator velocity is out of phase, then the brake force (regenerative force) can produce damping force. According to the velocity direction the actuator current flows as shown in Figure 5 or as in Figure 6. In this case the active control switches S_1 and S_2 are turned OFF, and

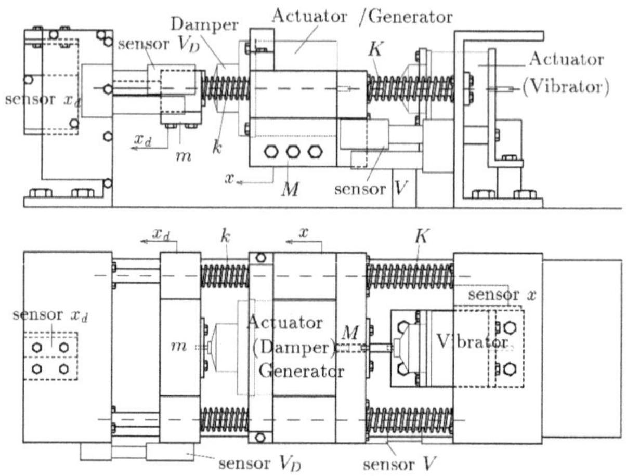

Figure 9. Schematic of experimental setup.

the regenerative current is controlled only by the switch S as

$$|f_d| \geq |f| \quad \rightarrow \quad \text{PWM of switch } S \text{ high}$$
$$|f_d| \leq |f| \quad \rightarrow \quad \text{PWM of switch } S \text{ low}$$

In case 2), the damping force should be produced with the active control. The actuator is controlled as shown in Figure 7 or as in Figure 8 according to the control force direction. Notice that the regenerative switch S should always be OFF and the active control switches are controlled as

$$f_d \geq 0\,,\ f_d \geq f \ \rightarrow \ \text{PWM of } S_1 \text{ high, while } S_2 \text{ OFF}$$
$$f_d \geq 0\,,\ f_d \leq f \ \rightarrow \ \text{PWM of } S_1 \text{ low, while } S_2 \text{ OFF}$$
$$f_d \leq 0\,,\ f_d \leq f \ \rightarrow \ \text{PWM of } S_2 \text{ high, while } S_1 \text{ OFF}$$
$$f_d \leq 0\,,\ f_d \leq f \ \rightarrow \ \text{PWM of } S_2 \text{ low, while } S_1 \text{ OFF}$$

Even all the switches are OFF, the regenerative current will flow when the actuator speed is very high and regenerative voltage is over the battery voltage because the actuator is always connected to the battery through the diodes.

3. EXPERIMENTS

To confirm the proposed active and regenerative control, the experimental setup is made as shown in Figure 9.

3.1 Experimental Setup

The experimental setup and control system are shown in Figures 9 and 4. The two masses are supported with two linear guides horizontally. The damping

Table 2. Experimental conditions.

Power supply [V]	2.4, 6.0, 10.0
Disturbance [N]	1, 5, 10
Switch	PWM driving

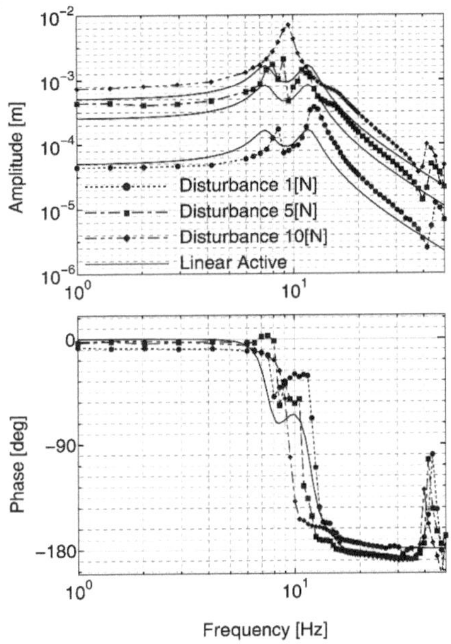

Figure 10. Frequency response: Bat. = 2.4 [V].

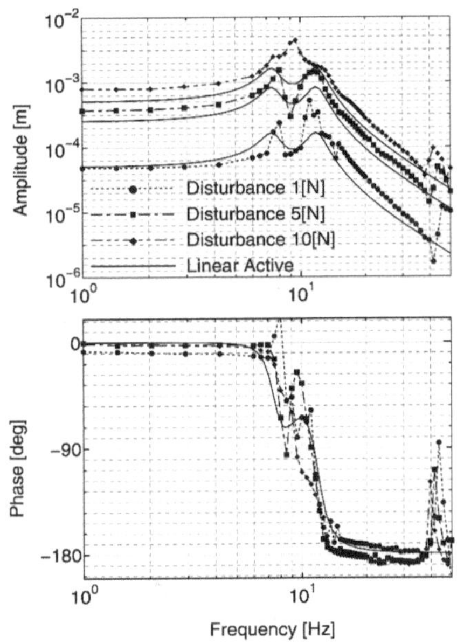

Figure 11. Frequency response: Bat. = 6.0 [V].

factor of the main structure is about $\zeta = 0.1$ ($c = 62$ [Ns/m]). In the left of Figure 9 the auxiliary mass m is connected to the main mass with the spring of k. In the middle a strong actuator is installed; the moving coil is connected to the auxiliary mass and the massive PM part is used as the main mass. They are controlled by a DSP (dSPACE DS-1103). The sinusoidal signal from DSP is added to the disturbance voice coil motor through a power amplifier. The velocity of the damping actuator is measured by the current which flows a small resistance $R = 0.1\Omega$. The displacements of main mass and the auxiliary mass x, x_d are measured by the laser displacement sensors (KEYENCE LB-60). They are put into the DSP through A/D converters. The state feedback signal is calculated in the DSP and the PWM signals from the DSP control three switches S, S_1 and S_2.

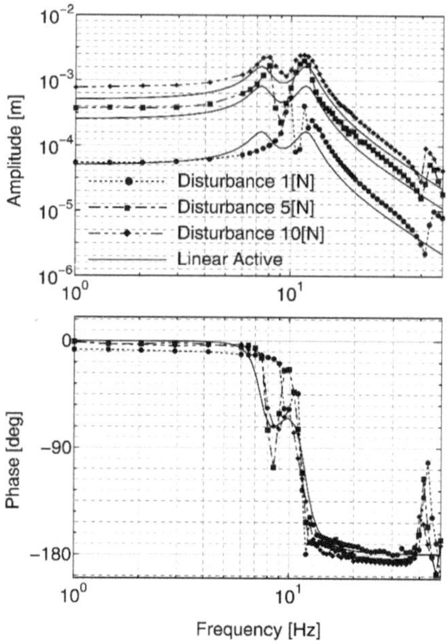

Figure 12. Frequency response: Bat. = 10 [V].

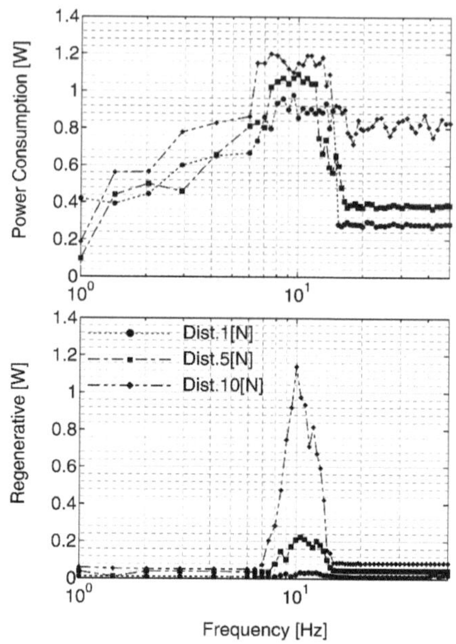

Figure 13. Consumed and regenerated power.

3.2 Results

According to the previous energy regenerative and active control the experiment is performed. The weights for the performance index are determined as

$$Q = \begin{pmatrix} 1 & 0 & 0 & 0 \\ 0 & 0.01 & 0 & 0 \\ 0 & 0 & 0 & 0 \\ 0 & 0 & 0 & 0 \end{pmatrix} [W/m^2], \quad R = 10^{-7}\ [W/N^2] \tag{6}$$

Frequency Response. The frequency responses are measured and the results are shown in Figures 10–12. The responses are compared with the case controlled by the ideal linear damping force f_d. The experimental conditions are shown in Table 2.

A single resonant peak is recognized with the higher disturbance force in Figure 10. This is considered due to the small maximum active force which is not enough for reducing the peak. Increasing the supply voltage to 10 [V] relatively good damping is recognized as shown in Figure 12.

Consumed power and regenerated power are compared in the case of supply voltage 10 [V], as shown in Figure 13. This is the results of the frequency response case shown in Figure 12.

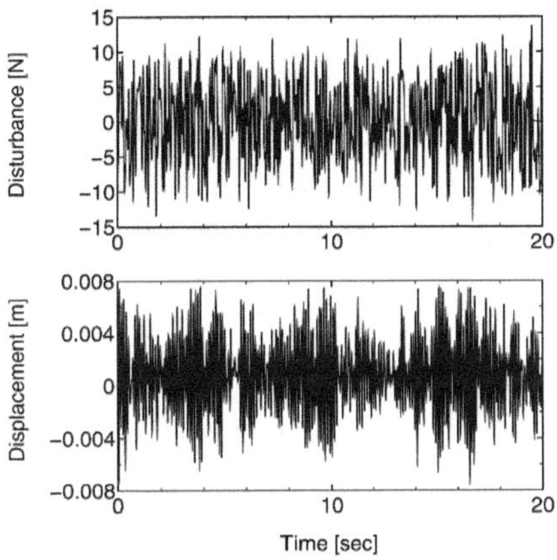

Figure 14. Random excitation (No control).

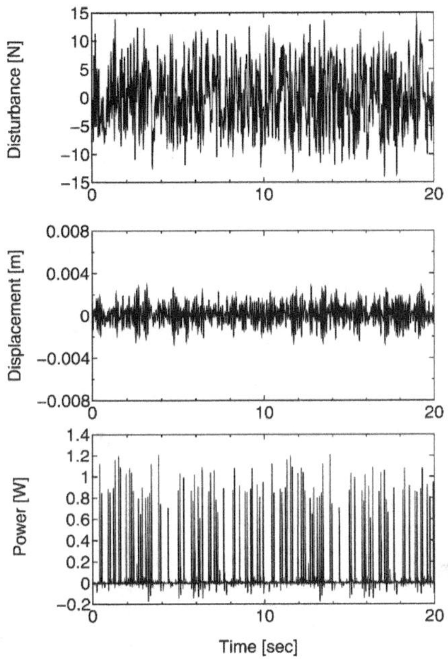

Figure 15. Random excitation (With control).

In all frequency range the power is consumed for active control, while the power is regenerated within the narrow region near resonant peak. This indicates that near the resonant frequency the system can regenerate vibration power. However the active consumed power is much bigger in the other region. This response confirmed that the proposed system can regenerate the vibration power even it is restricted within the narrow frequency region near the resonance and also the active control is operated at the same time.

Response of Random Excitation. A random force excitation is tested and recorded. The mean value of random force is 5 [N], and the supply voltage is 10 [V]. Random signal is supplied from FFT analyzer (MTS SigLab Version3.2.4) with the frequency band of 10 [Hz]. The excited displacement and the power are compared with and without control as shown in Figures 14 and 15. The power response is shown in the bottom of Figure 15, where the positive signal is the consumed power due to the active control while the negative one is the regenerated power.

Comparing the displacement time responses in the middle of Figure 15 with the bottom one in Figure 14, the amplitude is reduced about 1/4 with the proposed control. However the regenerated power is very small and the main is the consumed power by active control as indicated in the bottom of Figure 15.

This might be improved by improving the control circuit using the hardware PWM modulation circuit.

4. SUMMARY

In this paper active control is incorporated to the PWM setup chopper energy regenerative damper. The part of active control energy can be supplied from the regenerated energy. Also regenerative control is planned to reduce the vibration energy. From the experimental results the proposed idea is partly succeeded, but the vibration reduction and the regenerated energy is not enough. This is considered due to the slow PWM duty ratio change using the software change in dSPACE. Further work is continuing to use the hardware PWM modulation circuit to improve the system performance.

REFERENCES

Alleyne, A. and Hedrick, J.K. (1995). Nonlinear Adaptive Control of Active Suspensions. *IEEE Trans. on Control Systems Technology*, Vol. 3, No. 1, pp. 94–101.

Garcia, E., Webb, C.S. and Duke, M.J. (1995). Passive and Active Control of a Complex Flexible Structure Using Reaction Mass Actuators. *ASME Trans. on Vibration and Acoustics*, Vol. 117, No. 1, 116–122.

Kim, S.-S. and Okada, Y. (2002a). Variable Resistance Type Energy Regenerative Damper Using Pulse Width Modulated Step-up Chopper. *ASME Trans. on Vibration and Acoustics*, Vol. 124, No. 1, pp. 110–115.

Kim, S.-S. and Okada, Y. (2002b). Variable Resistance Type Energy Regenerative Suspension. *JSME Trans.*, Series C, Vol. 68, No. 675, pp. 3224–3229 [in Japanese].

Kim, S.-S., Yonemura, J. and Okada, Y. (1997). Regenerative Suspension System with an Optimally Tuned Resonant Circuit. In *Asia-Pacific Vibration Control '97*, pp. 1141–1146.

Okada, Y. and Harada, H. (1995). Active and Regenerative Control of Electrodynamic Vibration Damper. In *Proc. of the 1995 Design Engineering Technical Conf.*, Vol. 3, Part C, ASME DE-Vol. 84-3, pp. 595–602.

Okada, Y. and Harada, H. (1996). Regenerative Control of Active Vibration Damper and Suspension System. In *Proc. of the 35th IEEE Conf. on Decision and Control*, pp. 4715–4720.

INTERNAL VELOCITY FEEDBACK FOR STABILISATION OF INERTIAL ACTUATORS FOR ACTIVE VIBRATION CONTROL

Christoph Paulitsch, Paolo Gardonio and Stephen J. Elliott
Institute of Sound and Vibration Research, University of Southampton, United Kingdom

Abstract: Collocated direct velocity feedback with point force actuators mounted on structures generates active damping and theoretically is unconditionally stable. When inertial actuators are used to generate the control force the actuator becomes unstable even for moderate velocity feedback gains. In this study an internal, relative velocity sensor is used to implement an additional feedback loop in a lightweight, electrodynamic, inertial actuator. It is found that when the internal loop is used in addition to the conventional external velocity feedback loop there is an ideal internal velocity feedback gain for which maximum vibration reduction is achieved through the external velocity feedback loop. Experiments and measurements are compared at the model problem of an inertial actuator attached to a clamped plate.

Key words: active vibration control, electrodynamic actuator, double coil actuator, inertial actuator, velocity feedback, internal damping.

1. INTRODUCTION

In some cases active vibroacoustic control with point force actuators may be more efficient than using piezoelectric patch actuators [1]. In practical applications point forces are normally generated by either reactive or inertial actuators. When inertial actuators are used to implement direct velocity feedback control loops on a structure unconditional stability is no longer guarantees [2]. The actuator may be stabilized by internal feedback of the relative velocity between the moving actuator mass and the structure to be controlled [3]. For too high gains in the internal relative velocity feedback

H. Ulbrich and W. Günthner (eds), IUTAM Symposium on Vibration Control of Nonlinear Mechanisms and Structures, 243–254.

loop the inertial actuator purely acts like an attached point mass. Therefore there is a best gain for which maximum vibration reduction of the controlled structure may be achieved through the external velocity feedback.

This study presents an innovative, lightweight, inertial actuator [4] which has been mounted on a clamped plate in order to implement both relative velocity feedback using an internal velocity sensor and plate velocity feedback using a closely located external sensor. The internal loop, which can also be implemented in a self-sensing way [5], is used to improve the stability of the external loop. The external loop is used in such a way as to reduce plate vibrations at resonance frequencies of the plate by adding active damping. Both simulations and experiments of this model problem are presented.

Section 2 presents a model of the inertial actuator attached to a clamped plate and Section 3 shows velocity feedback results. Sections 3.1 and 3.2 compare experiments and simulations using external or internal velocity feedback respectively. Section 3.3 shows experimental and simulated results if external and internal velocity feedback are implemented simultaneously. They show that there is a best combination of external and internal velocity feedback for maximum vibration reduction for each resonance of the plate.

2. MODEL

The steady state flexural response of a rectangular panel clamped along the four edges which, as shown in Figure 1, is excited by a harmonic primary force, will be used as model problem to illustrate the active damping effects produced by an electrodynamic inertial actuator with internal and external velocity feedback connected to the panel.

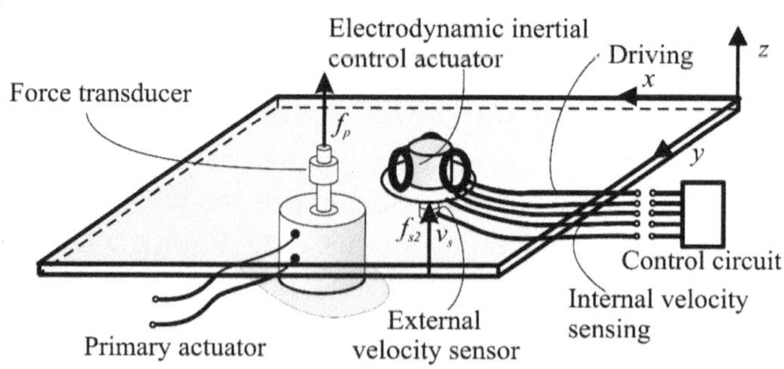

Figure 1. Sketch of the model problem where a test panel is connected to an electrodynamic, inertial actuator with internal velocity sensor which is set to control the vibrations generated by a primary force disturbance using either external, internal or combined velocity feedback.

The inertial actuator has a fundamental resonance frequency of about 13Hz and it is equipped with a double coil system such that the "primary" coil is used to generate the control force and the "secondary" coil is used to detect the relative vibration between the base of the actuator and the inertial mass. It is attached to the aluminium panel test rig of dimensions 0.414m×0.314m×0.001m. The steady state response of the panel has been derived assuming the primary force disturbance to be harmonic, with time dependence of the form $\mathrm{Re}\{\exp(j\omega t)\}$ where ω is the circular frequency and $j=\sqrt{-1}$. The mechanical and electrical functions in the model have therefore been taken to be the real part of anticlockwise rotating complex vectors, i.e. phasors, given in the form $X(\omega)e^{j\omega t}$ where $X(\omega)$ is the phasor at $t=0$. Considering the lumped parameter model in Figure 2 it has been shown [6] that the phasors of the complex velocity at the error sensor, $V_s(\omega)$, and of the moving control actuator mass $V_a(\omega)$ can be expressed in terms of the phasors of the primary, $F_p(\omega)$, and the control actuator force $\mathbf{f_a}=\{F_a(\omega)\ -F_a(\omega)\}^{\mathrm{T}}$, with the following mobility relation

$$\mathbf{v}_s=(\mathbf{I}+\mathbf{Y}_{ss}\mathbf{Z})^{-1}\mathbf{y}_{sp}F_p+(\mathbf{I}+\mathbf{Y}_{ss}\mathbf{Z})^{-1}\mathbf{Y}_{ss}\mathbf{f}_a \tag{1}$$

where the transfer and point mobility matrices are $\mathbf{y}_{sp}=\{Y'_{sp}\quad 0\}^{\mathrm{T}}$, $\mathbf{Y}_{ss}=\left[\{Y'_{ss}\quad 0\}^{\mathrm{T}}\quad \{0\quad Y_m\}^{\mathrm{T}}\right]$ $\mathbf{Y}_{ss}=\left[\{Y'_{ss}\quad 0\}^{\mathrm{T}}\{0\quad Y_m\}^{\mathrm{T}}\right]$, $\mathbf{Z}=[\{1\quad -1\}^{\mathrm{T}}\{-1\quad 1\}^{\mathrm{T}}](K/(j\omega+D))$, $\mathbf{v}_s=\{V_s\ V_a\}^{\mathrm{T}}$ and $\mathbf{I}$ is a 2x2 identity matrix. $Y_m=1/j\omega m$ is the mobility of the moving actuator mass m and K, D are the stiffness and the damping of the actuator suspension respectively. $Y'_{sp}=Y_{sp}/(1+j\omega M_p Y_{pp})$ and $Y'_{ss}=Y_{ss}/(1+j\omega M Y_{ss})$ incorporate the mass of the primary force transducer M_p and the mass of the control actuator housing and moving coil M [6]. The two mobility functions of the plate, $Y_{sp}(\omega)$, $Y_{pp}(\omega)$ and $Y_{ss}(\omega)$, have been derived with the following modal expansions

$$Y_{ij}=j\omega\sum_{n=1}^{N}\frac{\phi(x_i,y_i)\phi(x_j,y_j)}{\rho l_x l_y h\left[\omega_n^2+2j\omega\xi_n\omega_n-\omega^2\right]} \tag{2}$$

where $(i,j)\in\{(s,s),(s,p),(p,p)\}$. In these equations (x_p,y_p) and (x_s,y_s) are the coordinates of the primary and control positions, ρ is the density of the material and, ξ_n is the viscous damping factor of the n-th mode. Finally ω_n and $\phi_n(x,y)$ are respectively the n-th natural frequency and natural mode for a clamped panel [7] where ω_n is adjusted by a factor $\sqrt{K_1/K_2}$ for non-ideally clamped boundary conditions. A vertical actuation Lorentz force

$$F_a = \Psi_p I_c + \Psi_s I_s \tag{3}$$

between the moving mass m and the housing mass M connected to the plate is created by an electric current in the secondary coil I_s and in the primary coil I_c since the coils are immersed in a constant horizontal magnetic field generated by a permanent magnet integrated in the moving mass. Then Ψ_p and Ψ_s are transducer coefficients of the primary and the secondary coil respectively.

For the external velocity feedback loop the plate vibration V_s at the base of the actuator is measured and fed back to the primary coil such that the actuator is driven by a current I_c=-$G_1 V_s$ where G_1 is the external velocity feedback gain. Closed loop stability of external velocity feedback hence is evaluated by investigating the open loop frequency response function (FRF)

$$\frac{V_s}{I_c} = \{1 \quad 0\}\left(\mathbf{I} + \mathbf{Y}_{ss}\mathbf{Z}\right)^{-1} \mathbf{Y}_{ss}\boldsymbol{\psi}_p^{\mathrm{T}} \tag{4}$$

where $\boldsymbol{\psi}_p=\{\Psi_p \ \ {-\Psi_p}\}$. The vibration reduction at the base of the actuator is then indicated by the closed loop FRF

$$\frac{V_s}{F_p} = \{1 \quad 0\}\left(\mathbf{I} + \mathbf{Y}_{ss}\left(\mathbf{Z} + \Psi_p\begin{bmatrix} G_1 & 0 \\ -G_1 & 0 \end{bmatrix}\right)\right)^{-1} \mathbf{y}_{sp} \tag{5}$$

According to Biot-Savart's law [8] the currents in the coils will also generate a time-varying magnetic B field, primarily in vertical direction, in addition to a constant horizontal B field so that a voltage

$$U_s = L_s j\omega I_s + L_{sp} j\omega I_c + \boldsymbol{\psi}_s \mathbf{v}_s \tag{6}$$

will be induced in the secondary coil where L_s is the self-inductance of the secondary coil, L_{sp} is the coupling or mutual inductance between the primary and the secondary coil and $\boldsymbol{\psi}_s = \{\Psi_s \ \ {-\Psi_s}\}$.

As indicated in Figure 2 the primary coil is used to drive the actuator with a current I_c that is controlled by a current command amplifier. The secondary coil circuit is connected to the electrical input impedance of the controller Z_s=-U_s/I_s and it is used to sense the relative velocity. Using Eqs. (1), (3) and (6), with F_p=0, the FRF relative to the internal open loop becomes

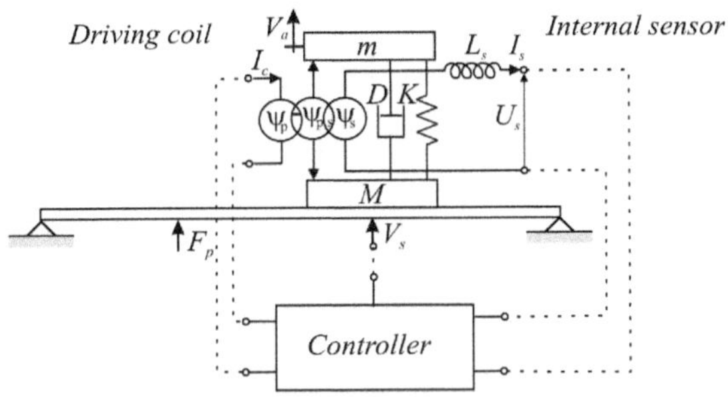

Figure 2. Model of the electrodynamic, inertial actuator connected to the plate.

$$\frac{U_s}{I_c} = L_{sp} j\omega + \boldsymbol{\psi}_s \mathbf{Y}_{ss} \boldsymbol{\psi}_p^{\mathrm{T}} \Big/ \left(1 + \frac{L_s j\omega}{Z_s} + \frac{\boldsymbol{\psi}_s \mathbf{Y}_{ss} \boldsymbol{\psi}_s^{\mathrm{T}}}{Z_s} \right) \tag{7}$$

In order to obtain the wanted relative velocity sensing, the second term in the numerator should be dominant, but the other terms have a disturbing effect especially at higher frequencies as will be illustrated in Section 3. In order to look at the closed loop performance of internal velocity feedback with gain G_2 Eqs. (1), (3), (6) and the feedback current I_c=-G_2U_s are used to obtain the response of the plate at the base of the actuator per unit primary excitation

$$\frac{V_s}{F_p} = \{1 \quad 0\} \frac{\mathbf{y}'_{sp}}{1 + \boldsymbol{\psi}_s \mathbf{Y}'_{ss} \left(G_2 \boldsymbol{\psi}_p^{\mathrm{T}} + \dfrac{\boldsymbol{\psi}_s^{\mathrm{T}}}{Z_s} \right) \Big/ \left(1 + L_{sp} j\omega G_2 + L_s j\omega / Z_s \right)} \tag{8}$$

where $\mathbf{y}'_{sp} = (\mathbf{I} + \mathbf{Y}_{ss}\mathbf{Z})^{-1} \mathbf{y}_{sp}$ and $\mathbf{Y}'_{ss} = (\mathbf{I} + \mathbf{Y}_{ss}\mathbf{Z})^{-1} \mathbf{Y}_{ss}$. Taking into account the effect of internal loop the FRF of the external open loop becomes

$$\frac{V_s}{I_c} = \left(\mathbf{I} + \mathbf{Y}_{ss} \left(\mathbf{Z} + \begin{bmatrix} 1 & -1 \\ -1 & 1 \end{bmatrix} \frac{\Psi_p \Psi_s G_2 Z_s}{Z_s + j\omega \left(L_s + G_2 L_{sp} Z_s \right)} \right) \right)^{-1} \mathbf{Y}_{ss} \boldsymbol{\psi}_p^{\mathrm{T}} \tag{9}$$

Also, the vibration reduction performance is changed to the following expression

$$\frac{V_s}{F_p} = \left(\mathbf{I} + \mathbf{Y}_{ss} \left(\mathbf{Z}' + \begin{bmatrix} 1 & -1 \\ -1 & 1 \end{bmatrix} \frac{\Psi_p \Psi_s G_2 Z_s}{Z_s + j\omega \left(L_s + G_2 L_{sp} Z_s \right)} \right) \right)^{-1} \mathbf{y}_{sp} \tag{10}$$

where $\mathbf{Z}'=\mathbf{Z}+[\{\Psi_p G_1\ -\Psi_p G_1\}^T\{0\ 0\}^T]$. In the following section Eqs. (4) and (5), (7) and (8) and (9) and (10) are investigated for different values of the feedback gains G_1 and G_2.

3. RESULTS

Figure 3 schematically depicts the three circuits that are used to implement (a) external velocity feedback, (b) internal velocity feedback and (c) combined external and internal velocity feedback. A B&K 4375 accelerometer is used together with an integrating B&K 2635 charge amplifier to obtain the plate velocity and the amplification in the external velocity feedback loop. Because of the mutual inductance effect a KEMO VBF8 filter is used to appropriately cut-off the secondary coil voltage signal and to adjust the internal velocity feedback gain. An OPA549 in current command constellation together with a DC300 amplifier drives the control actuator via the primary coil.

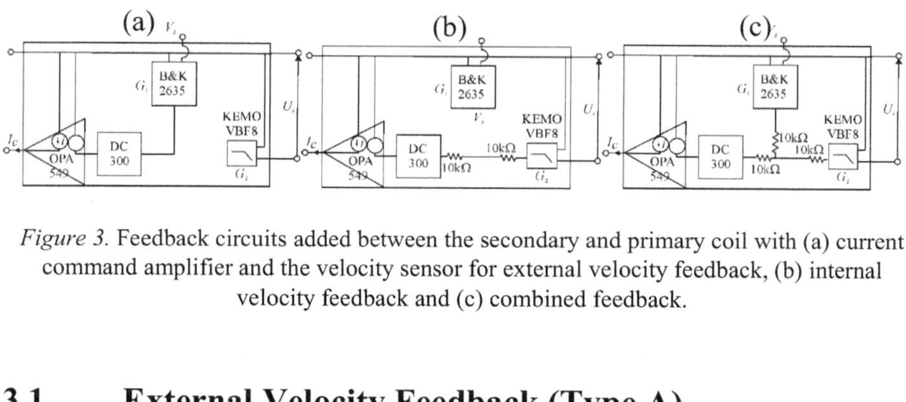

Figure 3. Feedback circuits added between the secondary and primary coil with (a) current command amplifier and the velocity sensor for external velocity feedback, (b) internal velocity feedback and (c) combined feedback.

3.1 External Velocity Feedback (Type A)

Figure 4 shows the Nyquist plot of the simulated (left) and measured (right) open loop FRF between the primary coil current I_c and the plate velocity at the control actuator position V_s. Both measurements and simulations indicate that the locus lies in the positive real half-plane for all frequencies except those around the fundamental resonance of the inertial actuator at about 13Hz. At the resonance frequency of the inertial actuator the locus has the smallest real part. Also, at even lower frequencies the integrator in the charge amplifier distorts measurements. Hence, because of the inertial actuator resonance only medium gains are predicted to guarantee closed loop stability.

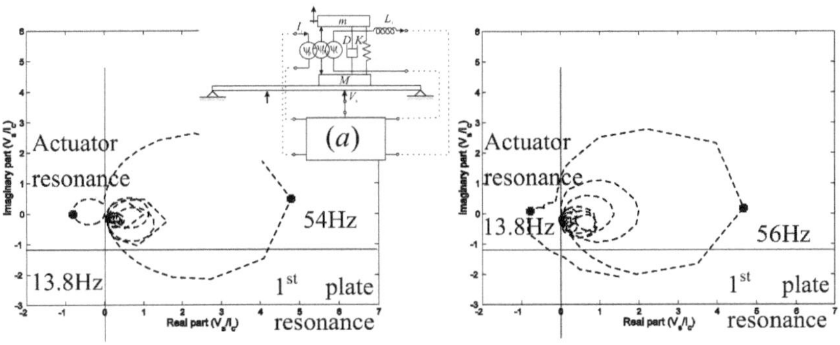

Figure 4. Nyquist plot of the simulated (left) and measured (right) FRF between the primary coil current and the plate velocity at the actuator position.

Figure 5 shows the amplitude of the simulated (left) and measured (right) response of the plate at the base of the actuator per unit excitation force (Eq. (5)) in the external and internal open loop case (solid line) and for external velocity feedback (dashed line) with the gain used to plot Figure 4.

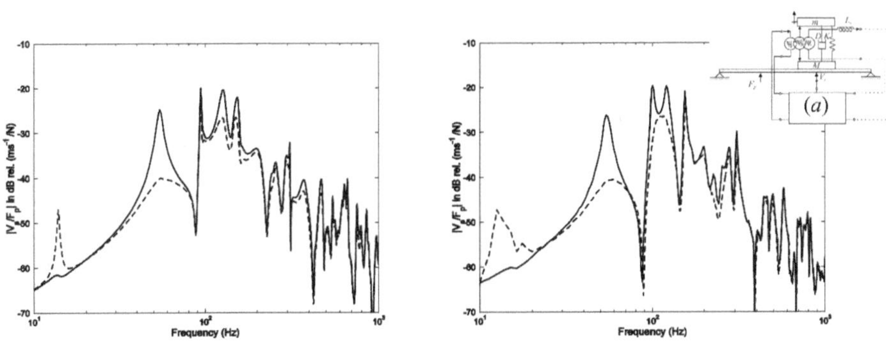

Figure 5. Amplitude of the simulated (left) and measured (right) FRF between the primary disturbance force and the plate velocity at the actuator position in the open loop case (thick solid line) and for external velocity feedback (thick dashed line).

Since I_c and F_a are in phase with V_s at the resonance frequency of the inertial actuator, as shown in Figure 4, vibrations are increased around this frequency. Slightly higher gains would lead to instability. At the resonance frequencies of the plate where F_a is out of phase with V_s vibrations are reduced. Vibration reduction of about 16 dB at the first plate mode and about 5 dB at higher modes are measured and predicted by simulations. There is no spillover at higher frequencies, which indicates that higher order actuator resonance frequencies or non-ideal collocation of the transmitted force and external velocity sensing could be neglected.

3.2 Internal Velocity Feedback (Type B)

Stability of the internal velocity feedback is evaluated by the Nyquist plot of the simulated (left) and measured (right) FRF between the primary coil current and the secondary coil voltage shown in Figure 6. The open loop is shown for high gains with a 240Hz cut-off filter (solid line) and for medium gains with a 700Hz cut-off filter (thick dash-dotted line). The filters are used to cut-off the effect of the coupling between the primary and the secondary actuator coil, which would make the locus encircle the critical point (-1,0j). Both measurements and simulations predict spillover in the closed loop since in both cases the locus enters the circle with radius 1 about the critical point.

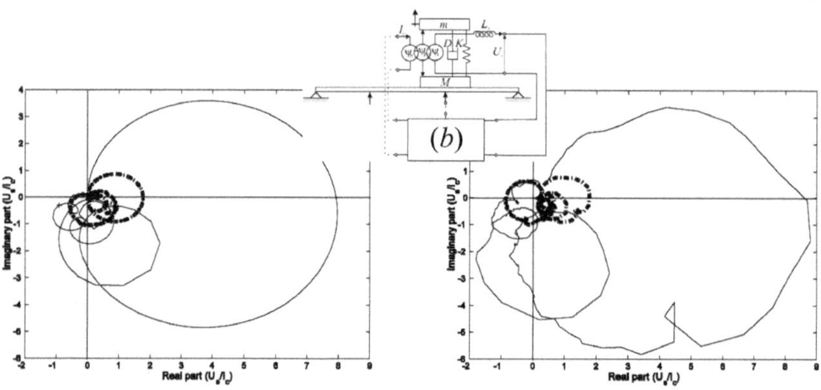

Figure 6. Nyquist plot of the simulated (left) and measured (right) FRF between the primary coil current and the secondary coil voltage for a 240Hz cut-off filter and high gain (solid line) and for a 700Hz filter and medium gain (thick dashed line).

For relatively high gains (thin solid line) a larger number of loops due to plate resonances enter the negative real half-plane, which indicates higher control spillover effects. However, the large loop in the positive real half-plane indicates larger closed loop vibration reduction at the fundamental resonance of the actuator. Fig. 7 shows the amplitude of the simulated (left) and measured (right) FRF between the primary disturbance force and the plate velocity in the open loop case (thick solid line), for medium internal velocity feedback gains (dash-dotted line) and high internal velocity feedback gains (thin solid line). Vibration reduction at the first plate mode are well predicted for medium gains, but worse for high gains because of the filtering effect. As predicted there is significant spillover of high feedback gains at higher resonances and even at the first plate resonance the vibration amplitude seems to increase as the feedback gains are further risen. Vibration reduction of about 8 dB at the first plate resonance and about 5 dB at the second and third resonance for medium gains and about 12 dB at the first plate resonance for large gains are measured and predicted. Further measurements and simulations show that for increasing internal relative velocity feedback gains vibration reduction at the first plate resonance increases, but it levels off or even decreases for increasing gains. For infinite internal velocity feedback gains the actuator mass would act as being attached to the plate so that a new (lower) resonance frequency would replace the first resonance of the plate. Hence, there is a best gain for maximum vibration reduction.

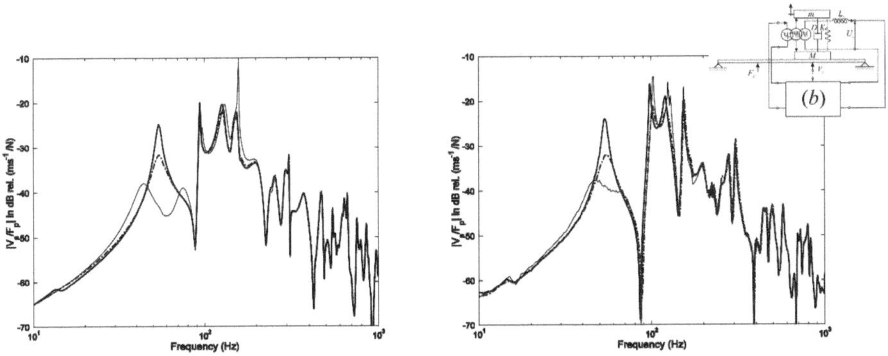

Figure 7. Amplitude of the simulated (left) and measured (right) FRF between the primary disturbance force and the plate velocity in the open loop case (thick solid line), a medium (thick dashed line) and a high internal velocity feedback gain (thick solid line).

3.3 Internal and External Velocity Feedback (Type C)

Combined internal and external velocity feedback is now analysed and discussed. Figure 8 shows the Nyquist plot of the simulated (left) and measured (right) FRF between the primary coil current and the plate velocity at the control actuator position for high internal velocity feedback gains (thin solid line) and for medium internal velocity feedback gains (dash-dotted line) for frequencies up to 160 Hz. Considering equal gain margins, for medium gains the loops associated to plate modes are bigger than in the external feedback case shown in Figure 4 and for measurements they do not significantly move into the left half-plane at higher frequencies. In contrast, simulations show small spillover into the negative real half-plane at higher frequencies because of the use of a different filter supposed to cut-off the non-ideal internal velocity sensing.

The amount of vibration reduction at the first resonance of the plate is indicated by the size of the first loop in the positive real half-plane and it is estimated by the amplitude at the intersection point with the positive real axis δ^2. The amplitude of the intersection point with the negative real axis δ^1

is a measure for stability. The smaller the ratio $\delta_1/\ \delta_2$ the more closed loop vibration reduction is predicted at the first plate mode at equal gain margin. Since this criterion does not consider the smallest distance between the locus and the critical point it does not guarantee robustness.

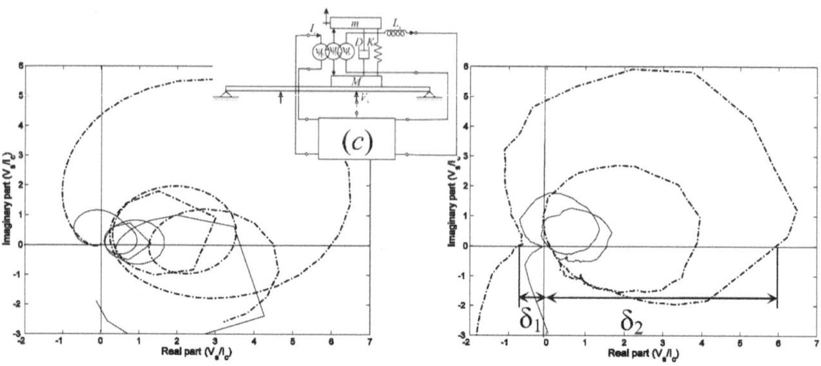

Figure 8. Nyquist plot of the simulated (left) and measured (right) FRF between the primary coil current and the plate velocity at the secondary actuator position between 1Hz and 160Hz for high (thin solid line) and medium (thick dash dotted line) internal velocity feedback gains.

Figure 9 shows the measured (dots) and simulated (continuous lines) ratio δ_1/δ_2 in the case with coupling inductance (solid line) and for ideal internal relative velocity feedback (dashed line). Both measurements and simulations indicate that there is a best gain for which this ratio becomes minimum i.e. maximum vibration reduction is achieved at the first plate resonance for a given gain margin. The non-ideal integrator in the charge amplifier distorts the measurement results. In the case without coupling inductance between the primary and the secondary coils this minimum is smaller, it lies at much higher gains and is much shallower. Figure 10 shows the amplitude of the simulated (left) and measured (right) FRF between the primary disturbance force F_p and the plate velocity V_s in the open loop case (thick solid line) for a high internal / medium external velocity feedback gain (thin solid line) and for a medium internal / high external velocity feedback gain (thick dashed line).

Although these figures show the same discrepancies between predictions and measurements for relatively high internal feedback gains as in Figure 7 the trend for significant spillover at frequencies below the first plate resonance is clearly visible. This trend would even be enhanced for increased external velocity feedback gains. More vibration reduction of up to 21 dB and up to 10 dB at higher plate resonances is possible with only a medium internal and a high external velocity feedback gain. In the case of

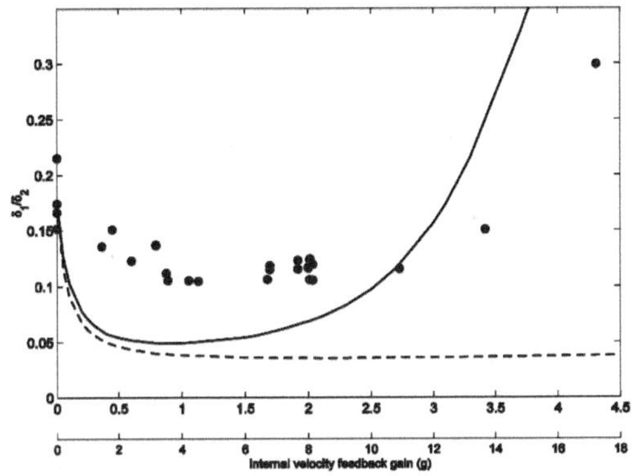

Figure 9. Simulated (lines) and measured (dots) ratio between δ_1 and δ_2 in Figure 8 for different internal velocity feedback gains with secondary coil voltage feedback (solid line and upper x-axis) and relative velocity feedback (dashed line and lower x-axis).

medium internal feedback gain higher external velocity feedback gains cannot be implemented because of stability limits due to the non-ideal integrator of the charge amplifier. Further simulations show that, when the vibration reduction at the first plate resonance only is considered, maximum vibration reduction is achieved at medium internal and large external feedback gains.

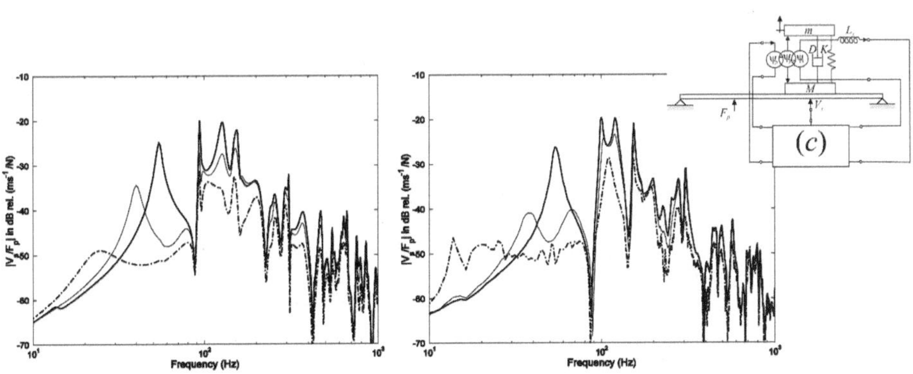

Figure 10. Amplitude of the simulated (left) and measured (right) FRF between the disturbance force and the plate velocity at the secondary actuator position in the open loop case (thick solid line), internal high gain and external velocity feedback (thin solid line) and internal medium gain and external velocity feedback (thick dashed line).

4. CONCLUSIONS

This study investigates the effect of internal relative velocity feedback on an inertial actuator used for active vibration control of a plate. A model shows that internal relative velocity feedback not only influences vibration reduction performance but also enhances stability of external velocity feedback. Measurements and simulations of external velocity feedback using an extra velocity sensor on the plate show that only medium vibration reduction of about 16 dB at the first flexible plate mode and of about 5 dB at higher plate modes can be achieved since the inertial actuator becomes unstable. Internal velocity feedback only achieves about 8 dB at the first flexible plate mode and 4 dB at higher plate modes. There is a best feedback gain for which maximum vibration reduction at a chosen resonance frequency is achieved. In practice vibration reduction is also limited by the coupling effect between the primary and secondary coil used to measure the internal relative velocity. When implementing combined internal and external velocity feedback, measurements and simulations show vibration reduction of about 21 dB at the first flexible plate resonance and about 10 dB at higher plate resonances. Maximum reduction at a chosen resonance can only be achieved when an appropriate relative velocity feedback gain is chosen in order to stabilise the internal actuator resonance. This is explained by a characteristic distortion of the open loop FRF for external velocity feedback due to internal velocity feedback.

REFERENCES

1. Elliott, S.J., Gardonio, P., Sors, T.C. and Brennan, M.J., Active vibroacoustic control with multiple local feedback loops, *J. Acoust. Soc. Am.* **111**(2), 908–915 (2002).
2. Elliott, S.J., Serrand, M. and Gardonio, P., Feedback stability limits for active isolation systems with reactive and inertial actuators, *Journal of Vibration and Acoustics* **123**, 250–261 (2001).
3. Inman, D.J., chapter 20 in: *Mechanics and Control of Large Structures*, edited by J.L. Junkins, American Institute of Aeronautics and Astronautics, 1990, pp. 507–533.
4. Paulitsch, C., Gardonio, P., Elliott, S.J., Sas, P. and Boonen, R., Design of a lightweight, electrodynamic, inertial actuator with integrated velocity sensor for active vibration control of a thin lightly-damped panel, in: *Proceedings of ISMA2004*, 2004, pp. 239–253.
5. Paulitsch, C., Gardonio, P. and Elliott, S.J., Active vibration damping using a self-sensing electrodynamic actuator, in: *Proceedings of SPIE*, Vol. 5386, 2004, pp. 282–293.
6. Paulitsch, C., Gardonio, P. and Elliott, S.J., Active vibration damping using an electrodynamic actuator with internal velocity sensor, in: *Proceedings of SPIE*, Vol. 5760, 2005, pp. 305–316, in print.
7. Gardonio, P. and Brennan, M.J., chapter 9 in: *Advanced Applications in Noise and Vibration*, edited by F. Fahy and J. Walker, E & FN Spon, 2004.
8. Inan, U.S. and Inan, A.S., *Engineering Electromagnetics*, Addison-Wesley, 1998.

A CONTROL CONCEPT FOR PARALLEL KINEMATICS

Alexandra Ratering and Peter Eberhard
Institute B of Mechanics, University of Stuttgart, Germany

Abstract: In machine tools with parallel and hybrid kinematics new problems arise in comparison to standard machines, such as inherently nonlinear dynamics, machine vibrations due to elasticities especially because of the lightweight construction and independently controllable drives being infeasible due to the parallel kinematics. Simulation results with a combination of advanced position control methods such as the presented nonlinear flatness-based controller together with active vibration damping show the potential to significantly improve the machine tool performance with such a control concept.

Key words: parallel kinematics, elastic multibody system, flatness, nonlinear position control, active vibration damping.

1. INTRODUCTION

The development of new machine tools in the last decade, e.g. the parallel and hybrid kinematics, has led to new challenges in the construction and control of such machines. In order to surpass existing machines in speed and accuracy new technologies are used, for example linear drives and lightweight structures. Consequently, new approaches in the control of such machines are needed too, as the dynamics are inherently nonlinear and the movement of the drives can only be controlled independently with difficulty and loss of accuracy. Furthermore, due to the lightweight construction, elasticities become important, so that active vibration damping can further improve the system performance.

The presented system is a high performance machine tool with linear drives and lightweight structures for processing of plate-like workpieces, e.g. milling of large wooden plates. The machine tool has been built at the Institute of Machine Tools headed by Professor Heisel at the University of Stuttgart, whose cooperation within our joint project in the framework fo the DFG SPP 1156 is highly appreciated.

H. Ulbrich and W. Günthner (eds), IUTAM Symposium on Vibration Control of Nonlinear Mechanisms and Structures, 255–265.

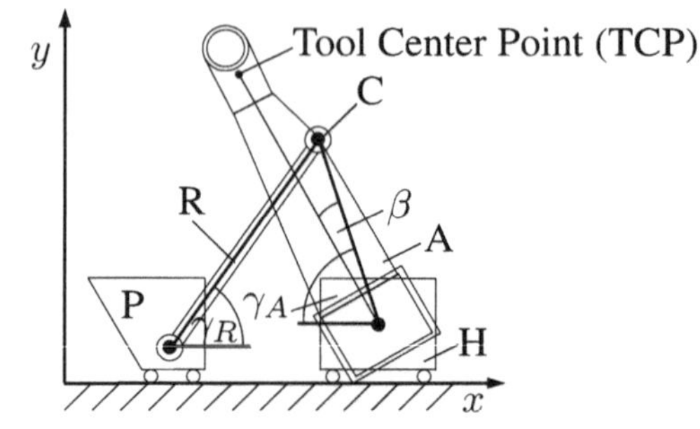

Figure 1. Schematic view of the machine tool.

2. RIGID MACHINE TOOL MODEL

The machine is a hybrid kinematic mechanism, consisting of a scissor-like parallel part for the movement in the xy-plane, see Figure 1. The two supports P and H move along a common feed driven by two linear drives in x-direction. Due to this scissor-like construction the machine tool is called Lambda Kinematics. The processing head attached to the main cantilever can be moved by a serial axle in z-direction in order to move the processing tool into or remove it out of the processing plane.

In a first step, the Lambda Kinematics has been modeled as a rigid multibody system (Schiehlen and Eberhard, 2004) consisting of four rigid bodies – the main cantilever A, the connecting rod R and the two supports P and H – and a closed kinematic loop at point C. Hence the resulting equations of motion for the system with two degrees of freedom are a system of four differential equations of second order with two additional algebraic loop closing conditions

$$\mathbf{M}(\mathbf{q}_t) \cdot \ddot{\mathbf{q}}_t + \mathbf{k}(\mathbf{q}_t, \dot{\mathbf{q}}_t) = \mathbf{Q}^a(\mathbf{q}_t, \dot{\mathbf{q}}_t, t) + \mathbf{H} \cdot \mathbf{u}(t) + \mathbf{G}^T \cdot \boldsymbol{\lambda} \quad (1)$$

$$\mathbf{g}(\mathbf{q}_t) = \mathbf{0}, \quad (2)$$

where $\mathbf{q}_t$ are the joint coordinates of the system, $\mathbf{M}$ is the mass-matrix, $\mathbf{k}$ is the vector of the generalized gyroscopic and centrifugal forces, $\mathbf{Q}^a$ the vector of generalized applied forces, $\boldsymbol{\lambda}$ the vector of Lagrangian multipliers and $\mathbf{H} \cdot \mathbf{u}$ the vector of motor forces with $\mathbf{H}$ being an appropriate Boolean matrix. The constraint equations (2) define implicitly the correlation between the independent minimal coordinates $\mathbf{q}_m$ and the joint coordinates $\mathbf{q}_t$ with

$$\mathbf{q}_m = [x_P \;\; x_H] \quad \text{and} \quad \mathbf{q}_t = [x_P \;\; x_H \;\; \gamma_R \;\; \gamma_A], \quad (3)$$

where x_P and x_H are the positions of the two supports and γ_R and γ_A are the angles between the x-axis of the inertial coordinate system and the rod and cantilever respectively, see Figure 1.

3. NONLINEAR VERSUS LINEAR POSITION CONTROL

The traditionally implemented control system of the real machine consists of two independent standard linear controllers for each support with a PI control loop for velocity and in cascade a P control loop for position. This controller design is equivalent to an incomplete PID controller for position in the sense that no desired velocities are used for comparison. Moreover, no feedforward control was implemented. The linear control design does not take into account that the movement of the tool center point is determined by the concerted motion of the two supports and neglects that the dynamic behavior of the machine tool changes nonlinearly over the workspace. Hence, the machine performance varies over the workspace, in the worst case the two position controllers may even work against each other and the machine tool can become unstable.

An alternative control concept, which takes the nonlinear dynamics and the interconnection of the supports into account, is given by a flatness-based controller design (Ratering and Eberhard, 2004). In a (differentially) flat system all states and inputs can be calculated as an algebraic function of the so-called flat output $\mathbf{y}$ and a finite number of its time derivatives (Rothfuß et al., 1997). For this system one possible flat output is given by the minimal coordinates $\mathbf{y} = \mathbf{q}_m$. Using the angles γ_A and γ_R and the corresponding angular velocities and accelerations, which can be calculated analytically from the explicit constraint equations, one can determine the system state $\mathbf{x}$ and output $\mathbf{u}$ as

$$\mathbf{x} = [x_P \ x_H \ \dot{x}_P \ \dot{x}_H] = [\mathbf{y} \ \dot{\mathbf{y}}], \tag{4}$$

$$\mathbf{u} = [u_1 \ u_2] = \boldsymbol{\Psi}(\mathbf{y}, \dot{\mathbf{y}}, \ddot{\mathbf{y}}). \tag{5}$$

The function $\boldsymbol{\Psi}$ in (5) is the solution of the – for the unknowns $\mathbf{u}$ and $\boldsymbol{\lambda}$ – algebraic system (1), hence it is often called the inverse system. Using $\boldsymbol{\Psi}$ as state feedback and introducing $\mathbf{v}$ as new input

$$\mathbf{u} = \boldsymbol{\Psi}(\mathbf{y}, \dot{\mathbf{y}}, \mathbf{v}) \tag{6}$$

results in a linearized system with two uncoupled linear ODEs of second order, i.e. $\ddot{\mathbf{y}} = \mathbf{v}$. A standard linear PID controller with feedforward can be used for the now linear system with positive definite diagonal matrices to guarantee asymptotic stability. Such a controller design results in a homogeneous machine tool behavior over the whole workspace.

In simulations the nonlinear controller shows a superior performance in comparison to the linear control concept, see Figure 2. Whereas the position errors of the supports for the linear controller are in the range of millimeters (e.g. ± 4 mm in Figure 2), the flatness-based controller achieves position errors in the size of micrometers (e.g. ± 1 μm in Figure 2). The parameters of the linear controller were chosen equal to the parameters implemented in the

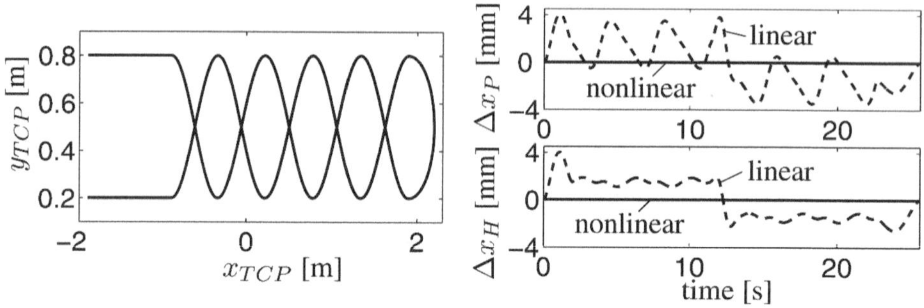

Figure 2. Left: Desired trajectory for the TCP; right: Simulation results for the position errors of the supports P and H with linear (dashed) and nonlinear (solid) controller, assumed disturbance of 30% of the moment of inertia of rod and cantilever.

real machine tool. The simulation results were achieved with a co-simulation between SIMPACK and Simulink.

4. MODEL ENHANCEMENT CONSIDERING ELASTICITIES

The next step in the modeling and analysis of the machine tool is to include elasticities in the multibody system model in order to simulate the vibrational behavior of the Lambda Kinematics. Apart from joint elasticities, there are several bodies which may contribute to vibration problems of the system. Experimental investigations have shown that oscillations in the xy-plane of the machine tool are dominated by a mode shape consisting of a rotation of the cantilever in combination with elastic deformations of the connecting rod, on which large forces act especially in longitudinal direction. An experimental modal analysis showed also that this eigenmode has the lowest resonance frequency and is only lightly damped. Hence, the multibody system model has been enhanced with radial joint elasticities for the three revolute joints and a deformable body for the connecting rod.

Elastic bodies can be incorporated in multibody system models with the "floating frame of reference" formulation (Shabana, 1998). This approximation is used to formulate a finite set of dynamic equations of motion for the multibody system model that contains interconnected deformable bodies. The underlying assumption of this method is that only small and hence linearizable deformations occur in the elastic body, so that the overall body motion can be described by large translations and rotations of a body reference frame superimposed by small linear deformations given by some elastic coordinates. As infinite dimensional models are computational difficult to handle and for complex geometries even harder to derive, classical approximation techniques can be used to introduce these elastic coordinates such as Rayleigh–Ritz methods or condensed finite elements (Schwertassek and Wallrapp, 1999).

4.1 Kinetics of a Deformable Body

The equation of motion for a free deformable body i is

$$\begin{bmatrix} \mathbf{M}_t^i & \mathbf{M}_{tr}^i & \mathbf{M}_{te}^i \\ & \mathbf{M}_r^i & \mathbf{M}_{re}^i \\ sym. & & \mathbf{M}_e^i \end{bmatrix} \cdot \begin{bmatrix} \mathbf{a}^i(t) \\ \boldsymbol{\alpha}^i(t) \\ \ddot{\mathbf{q}}_e^i(t) \end{bmatrix} = \mathbf{h}_\omega^i + \mathbf{h}_a^i + \mathbf{h}_e^i + \mathbf{h}_c^i, \tag{7}$$

where the generalized mass matrix consists of the submatrix for the translation of the reference system $\mathbf{M}_t^i = m^i\mathbf{E}$ with m^i being the total mass of the body, the submatrix for the rotation of the reference system $\mathbf{M}_r^i$ being the deformation dependent moment of inertia and the submatrix for the the elastic deformation $\mathbf{M}_e^i$ consisting of the inertia shape integrals (Shabana, 1998). The submatrices $\mathbf{M}_{tr}^i$, $\mathbf{M}_{et}^i$ and $\mathbf{M}_{er}^i$ represent the coupling between the translational and rotational motion and the elastic deformation, respectively. Apart from $\mathbf{M}_t^i$ and $\mathbf{M}_e^i$ all submatrices are a function of the elastic deformation and hence implicit time dependent. Here, $\mathbf{a}^i(t)$ denotes the translational acceleration and $\boldsymbol{\alpha}^i(t)$ the rotational acceleration of the reference system and $\mathbf{q}_e^i(t)$ the elastic deformation coordinates, $\mathbf{h}_\omega^i$ is the vector of gyroscopic and centrifugal forces. The vector of generalized applied forces $\mathbf{h}_a^i$ is made up of the generalized gravitational, surface and discrete (or point) forces. The vector of generalized internal forces $\mathbf{h}_e^i$ is a function of the elastic body deformation and the material properties. For a linear isotropic material the internal forces can be calculated from the positive definite stiffness and damping matrix $\mathbf{K}_e^i$ and $\mathbf{D}_e^i$

$$\mathbf{h}_e = \begin{bmatrix} \mathbf{0} \\ \mathbf{0} \\ -\mathbf{K}_e^i \cdot \mathbf{q}_e^i - \mathbf{D}_e^i \cdot \dot{\mathbf{q}}_e^i \end{bmatrix}. \tag{8}$$

The vector of constrained forces is

$$\mathbf{h}_c^i = \mathbf{G}^{i\mathrm{T}} \cdot \boldsymbol{\mu} \tag{9}$$

with $\boldsymbol{\mu}$ being the vector of Lagrangian Multipliers and $\mathbf{G}^i$ the constraint Jacobian matrix (Schwertassek and Wallrapp, 1999).

4.2 Modal Condensation

As the adequate representation of the deformations of an elastic body using finite elements may require a large number of nodal coordinates, it is usually necessary to reduce this number of coordinates so that the numerical simulation of the mechanical system can be done in a reasonable computation time. One way to reduce the number of elastic coordinates is to use modal coordinates instead of nodal coordinates (Shabana, 1998). As it can be seen in (7) the kinetics of an elastic body can be partitioned into a part resulting from the motion of the body reference system r and the elastic deformation e

$$\begin{bmatrix} \mathbf{M}_r^i & \mathbf{M}_{re}^i \\ \mathbf{M}_{er}^i & \mathbf{M}_e^i \end{bmatrix} \cdot \begin{bmatrix} \ddot{\mathbf{q}}_r^i \\ \ddot{\mathbf{q}}_e^i \end{bmatrix} = \begin{bmatrix} (\mathbf{h}^i)_r \\ (\mathbf{h}^i)_e \end{bmatrix} + \begin{bmatrix} \mathbf{G}_r^{i\mathrm{T}} \\ \mathbf{G}_e^{i\mathrm{T}} \end{bmatrix} \cdot \boldsymbol{\mu}, \tag{10}$$

where $\mathbf{h}^i$ is the sum of all generalized forces. If the body is assumed to vibrate freely about a reference configuration, eigenvalues and the corresponding eigenvectors of the elastic coordinates can be calculated by solving the standard eigenvalue problem. The eigenvectors are called normal modes or mode shapes. A coordinate transformation from the physical nodal coordinates $\mathbf{q}_e^i$ to the modal elastic coordinates $\mathbf{p}_e^i$ is given by

$$\mathbf{q}_e^i = \bar{\boldsymbol{\Phi}}_m^i \cdot \mathbf{p}_e^i \tag{11}$$

where $\bar{\boldsymbol{\Phi}}_m^i$ is the modal transformation matrix, whose columns are given by the mode shapes. By choosing only a smaller number m of modes shapes than actual degrees of freedom of the finite element model, which are usually in the lower frequency range, a reduced order model can be achieved. High frequency mode shapes typically contribute little to the deformation of the body. In addition to the normal modes so called frequency response modes can be included as well to further improve the accuracy of the elastic body representation taking into account deformations resulting from discrete forces or constraints, which cannot be described accurately with the mode shapes calculated from the free vibration of the elastic body. Details concerning the frequency response modes can be found in Dietz (1999). Using the transformation

$$\begin{bmatrix} \mathbf{q}_r^i \\ \mathbf{q}_e^i \end{bmatrix} = \underbrace{\begin{bmatrix} \mathbf{I} & \mathbf{0} \\ \mathbf{0} & \bar{\boldsymbol{\Phi}} \end{bmatrix}}_{= \boldsymbol{\Phi}_m^i} \cdot \begin{bmatrix} \mathbf{p}_r^i \\ \mathbf{p}_e^i \end{bmatrix} \tag{12}$$

in (10) and premultiplying with $\boldsymbol{\Phi}_m^{i\mathrm{T}}$ results in the equations of motion for a free body in reference and modal coordinates

$$\begin{bmatrix} \bar{\mathbf{M}}_r^i & \bar{\mathbf{M}}_{re}^i \\ \bar{\mathbf{M}}_{er}^i & \bar{\mathbf{M}}_e^i \end{bmatrix} \cdot \begin{bmatrix} \ddot{\mathbf{p}}_r^i \\ \ddot{\mathbf{p}}_e^i \end{bmatrix} = \begin{bmatrix} (\bar{\mathbf{h}}^i)_r \\ (\bar{\mathbf{h}}^i)_e \end{bmatrix} + \begin{bmatrix} \bar{\mathbf{G}}_r^{i\mathrm{T}} \\ \bar{\mathbf{G}}_e^{i\mathrm{T}} \end{bmatrix} \cdot \boldsymbol{\mu}, \tag{13}$$

where

$$\begin{aligned}
\bar{\mathbf{M}}_r^i &= \mathbf{M}_R^i, & \bar{\mathbf{M}}_{re}^i &= \bar{\mathbf{M}}_{er}^{i\mathrm{T}} = \mathbf{m}_{re}^i \cdot \bar{\mathbf{\Phi}}_m^i, \\
\bar{\mathbf{M}}_e^i &= \bar{\mathbf{\Phi}}_m^{i\mathrm{T}} \cdot \mathbf{M}_e^i \cdot \bar{\mathbf{\Phi}}_m^i, & (\bar{\mathbf{h}}^i)_r &= (\mathbf{h}^i)_r, \\
(\bar{\mathbf{h}}^i)_e &= \bar{\mathbf{\Phi}}_m^{i\mathrm{T}} \cdot (\mathbf{h}^i)_e, & \bar{\mathbf{G}}_r^i &= \mathbf{G}_r^i, \\
\bar{\mathbf{G}}_e^i &= \mathbf{G}_e^i \cdot \bar{\mathbf{\Phi}}_m^i. & &
\end{aligned} \tag{14}$$

The eigenvectors are usually normalized such that $\bar{\mathbf{M}}_e^i = \mathbf{E}$.

4.3 Elastic Multibody System Model

The equations of motion for a free cut rigid body have the same form as in (10) or (13) with the elastic coordinates $\mathbf{q}_e^i$, respectively $\mathbf{p}_e^i$, being equal to zero. Taking into account the constraints due to joints in the elastic multibody system, the equations of motion for the whole system can be derived for example in descriptor form as

$$\dot{\mathbf{z}}_I = \mathbf{Z} \cdot \mathbf{z}_{II}, \tag{15}$$

$$\mathbf{M} \cdot \dot{\mathbf{z}}_{II} = \mathbf{h} + \mathbf{G}^{\mathrm{T}} \cdot \boldsymbol{\mu}, \tag{16}$$

$$\mathbf{g}(\mathbf{z}_I, t) = \mathbf{0} \tag{17}$$

with $\mathbf{z}_I$ being the position variables consisting of the position and orientation of the body reference frames and the elastic coordinates of the deformable bodies and $\mathbf{z}_{II}$ being the corresponding velocity variables. Equation (17) is the implicit constraint equation (for holonomic joints) for the position variables with

$$\mathbf{G} = \frac{\partial \mathbf{g}(\mathbf{z}_I, t)}{\partial \mathbf{z}_I} \cdot \mathbf{Z}(\mathbf{z}_I). \tag{18}$$

The descriptor form can also be transformed into a state space representation, if the explicit constraint equations are used, resulting in a state space representation with minimal coordinates and without constraint forces. For details on using joint coordinates and how to handle systems with kinematic loops, see Schwertassek and Wallrapp (1999).

4.4 Simulation Results with Elastic Multibody System

SIMPACK enables the user to incorporate elastic bodies in the multibody system model by using modal coordinates. Results from a finite-element analysis can be imported with a preprocessor, which creates the necessary data for the description of the elastic body as in (13). In the first step only the vibrational behavior of the machine tool in the xy-plane was of interest. Hence, the first six longitudinal and the first six bending modes of a clamped-free beam were

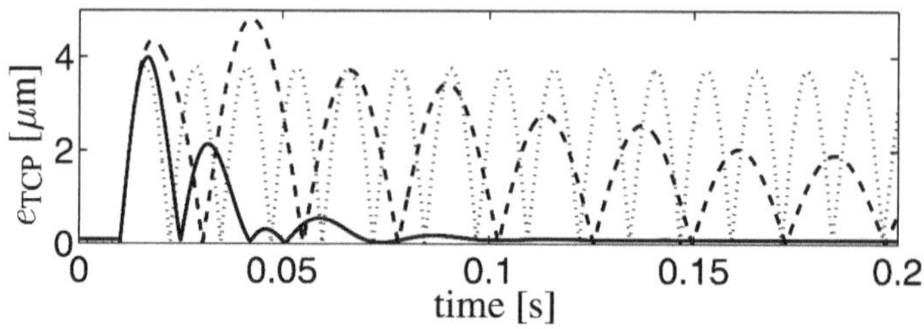

Figure 3. Simulation results for the elastic coordinates of the longitudinal (left) and bending modes (right) with fixed supports and force impulse acting on the TCP in the longitudinal direction of the rod at $t = 0.01$ s.

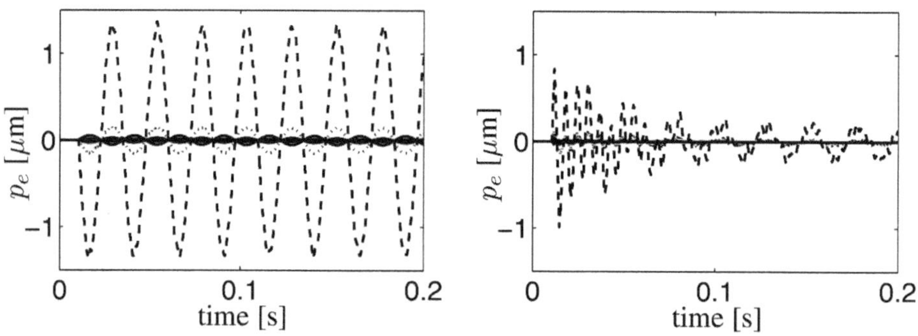

Figure 4. Error of the TCP for the elastic multibody system model with fixed supports (dotted), linear P-PI controller (dashed) and nonlinear flatness-based controller (solid) to an impulse force of 1 kN acting in longitudinal direction of the rod at $t = 0.01$ s.

chosen as mode shapes for the connecting rod. Furthermore, two frequency response modes were taken into account at the free end of the beam to account for deformations resulting from forces due to the spring element at point C, see Figure 1.

Figure 3 shows the simulation results for the elastic coordinates of the mode shapes when a force impulse of 1 kN magnitude acts in the longitudinal direction of the connecting rod at the TCP at $t = 0.01$ s. The position of thc two supports was fixed with the minimal possible distance of 0.54 m between the two, which is not singular but the position most sensitive to vibrations. One can clearly see, that the first longitudinal mode shape (dashed line in the left graphic) has the greatest amplitude and hence influence and that due to the machine setup, now bending and longitudinal mode shapes are not independent any more as for the free vibrating beam. The machine tool vibrations are dominated after the force impulse by the lowest eigenmode with a resonance frequency of about 40 Hz. Figure 4 shows the error of the TCP-position

$$e_{\mathrm{TCP}} = \left\| \begin{bmatrix} x^d_{\mathrm{TCP}} \\ y^d_{\mathrm{TCP}} \end{bmatrix} - \begin{bmatrix} x_{\mathrm{TCP}} \\ y_{\mathrm{TCP}} \end{bmatrix} \right\| \tag{19}$$

for the same impulse force as before acting at $t = 0.01$ s at the TCP. Results are shown with fixed supports (dotted) and with the constant position controlled by the linear P-PI controller (dashed) and the nonlinear flatness-based controller (solid). The linear position controller is able to eventually reduce the vibrations in the machine tool but the resonance frequency is about 50% smaller at this position than for the case of fixed supports. The nonlinear controller is able to reduce the vibrations much faster, whereas the resonance frequency lies at about 36 Hz at this position, being quite close to the case of fixed supports.

5. A SIMPLE MODAL CONTROLLER FOR LOCAL ACTIVE VIBRATION DAMPING

Based on the elastic multibody system model an active vibration control can be developed to further improve the machine tool performance. A local approach is to increase the stiffness and damping of the connecting rod with a simple modal controller (Gawronski, 1998). Based on the previous simulation results, the stiffness and damping of the first longitudinal mode shape were increased. For the controller design a reduced modal model of the elastic deformation of the connecting rod was used

$$\ddot{p}_{e1} + D_{e1}\dot{p}_{e1} + K_{e1}p_{e1} = b_1 F_x \tag{20}$$

with p_{e1} being the modal amplitude of the first longitudinal mode and D_{e1} and K_{e1} being the damping and stiffness factor, respectively. The input force F_x in longitudinal direction can be generated for example with a piezoelectric actor. The factor b_1 depends on the positioning of the actor on the rod. The control law for increasing the stiffness and the damping of p_{e1} then reads to

$$F_x = \frac{1}{b_1}(-D_1\dot{p}_{e1} - K_1 p_{e1}), \tag{21}$$

where K_1 and D_1 have to be chosen in an appropriate way. Because higher modes have been neglected in this approach and all system states are strongly coupled due to the machine structure in the full model, spillover effects occur, when the factors are chosen to high. The machine tool may even become unstable.

Figure 5 shows the simulation results for the error of the TCP with the same trajectory as in Figure 2 (left) and the elastic multibody system model with the nonlinear flatness-based position controller. The position error can be reduced by a factor of two with the additional modal controller for the first longitudinal mode. The dotted line shows the simulation result for the rigid multibody system model for comparison. The right side of Figure 5 shows the TCP error for the impulse force simulation with and without modal controller and nonlinear position controller. Again the amplitude is reduced with the simple modal controller, although it takes longer till the oscillation is damped out. The actuator

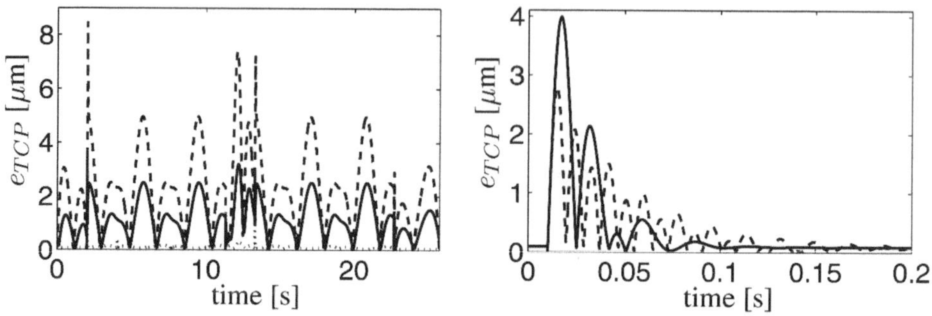

Figure 5. Left: Position error of the TCP with nonlinear, flatness-based position control and rigid multibody system model (dotted), elastic multibody system model (dashed) and elastic multibody system model with modal controller (solid); right: Position error of the TCP for impulse force with (solid) and without (dashed) modal controller and nonlinear position control.

location was chosen in both simulations at 2/3 of the rod length at the node of the second longitudinal mode shape.

6. CONCLUSIONS AND OUTLOOK

Active vibration damping in combination with advanced position control methods is able to improve the machine tool performance significantly. With a simple local approach for a modal controller, the position error could be reduced by a factor of two compared to a system without modal controller. Nevertheless, the success of this method is limited due to the relatively strong coupling between all system states, which leads easily to spillover effects and instability. Especially a modal control approach for the bending modes proved itself to be highly sensitive and hence infeasible. Therefore, alternative approaches have to be tested, including other control concepts than modal control and a global control approach.

Further enhancements of the elastic multibody system have to be incorporated, if tilting of the cantilever is taken into account. Then, bending of the connecting rod in the xz-plane might occur as well. The cantilever itself is planned to be build with a less heavy truss construction in order to further increase the possible machine speed. Here clearly lies the potential to include an active controlled system to damp out system vibrations especially in the z-direction, which strongly affect the processing quality.

The authors propose to use a combination of advanced position control methods, such as the presented flatness-based controller, together with methods of active vibration damping in a hierarchical control system. Such a control concept is able to deal simultaneously with some of the afore mentioned newly arising problems in machine tools with parallel or hybrid kinematics: inherently nonlinear dynamics, machine vibrations and independently controllable drives being infeasible.

REFERENCES

Dietz, S. (1999). *Vibration and Fatigue Analysis of Vehicle Systems Using Component Modes*, Fortschritt-Berichte VDI, Reihe 12, Nr. 401. VDI-Verlag, Düsseldorf.

Gawronski, W.K. (1998). *Dynamics and Control of Structures*. Springer Verlag, Berlin.

Ratering, A. and Eberhard, P. (2004). Flatness-Based Control of a Machine Tool with Lambda Kinematic. In *Proc. of the NOLCOS 2004, 6th IFAC-Symposium on Nonlinear Control Systems*, accepted for publication.

Rothfuß, R., Rudolph, J. and Zeitz, M. (1997). Flachheit: Ein neuer Zugang zur Steuerung und Regelung nichtlinearer Systeme. *Automatisierungstechnik*, 11:517–525.

Schiehlen, W. and Eberhard, P. (2004). *Technische Dynamik*. Teubner, Wiesbaden.

Schwertassek, R. and Wallrapp, O. (1999). *Dynamik flexibler Mehrkörpersysteme*. Vieweg Verlag, Braunschweig/Wiesbaden.

Shabana, A. (1998). *Dynamics of Multibody Systems*. Cambridge University Press, Cambridge.

MOTION PLANNING AND CONTROL OF PARALLEL MECHANISMS THROUGH INVERSE DYNAMICS

M. Necip Sahinkaya and Yanzhi Li
Centre for Power transmission and Motion Control, Department of Mechanical Engineering, University of Bath, UK
ensmns@bath.ac.uk

Abstract: The inverse dynamic analysis of an electro-mechanical system consisting of a three-degree-of-freedom planer parallel mechanism driven by three motors is carried out. A successful linearisation of the system and error dynamics is achieved for each desired motion. The linearised models are used in the adaptive control strategy avoiding the need to perform the computationally intensive inverse dynamic simulation in real-time. It is shown that accurate linear models can be obtained for a given motion, but model parameters change significantly with change of the motion. Significant differences are also observed between the system dynamics and the error dynamics models. The presented simulation results show the effectiveness of the adaptive control strategy.

Key words: inverse dynamics, parallel mechanisms, motion planning, adaptive control.

1. INTRODUCTION

The required performance of advanced machines is constantly increasing in terms of speed and accuracy. In most cases this necessitates accurate dynamic analysis of the whole system including drives and sensors. The design tools should be capable of handling systems, which are inherently multi-physics in nature, i.e. they include elements from various engineering domains. This paper uses a simulated electro-mechanical system consisting of a three dimensional planar parallel mechanism driven by three electrical motors, to discuss a wide range of issues concerning the dynamic analysis and control of non-linear multi-physics systems. Inverse dynamics of planar mechanisms are subject to current studies, for example (Sahinkaya, 2004; Geike and McPhee, 2003). An existing package, Dysim (Sahinkaya, 2004), is used in this study, which utilises a Lagrangian formulation of the equations of motion.

H. Ulbrich and W. Günthner (eds), IUTAM Symposium on Vibration Control of Nonlinear Mechanisms and Structures, 267–276.

It is well known that parallel mechanisms offer many advantages over serial mechanisms, in terms of stiffness, motion accuracy and load to space requirement ratio (Merlet, 2000). However, dynamic and kinematic analyses are more difficult (Merlet, 1996) and classical control techniques for uncoupling the motion either through design or coordinate selection have limited applicability. Even if the system dynamics is accurately modelled, including the coupling between the actuators, the computational requirements prohibit real-time solution of these equations. In the control approach suggested in this paper, inverse dynamic solution of the system equations is completed off-line to tune the feed-forward and feedback adaptive control parameters. Although the inverse dynamic modelling in the paper is handled by the existing package, Dysim (Sahinkaya, 2004), any other inverse dynamic modelling technique can be used. For more complicated path specifications such as following a square path, a piece-wise linearisation approach is used. The demand signal for a point to point motion is pre-shaped to achieve smooth and vibration free motion to simplify the controller's task.

2. INVERSE DYNAMIC ANALYSIS

The simulation package Dysim uses a Lagrangian formulation to develop equations of motion automatically from the kinetic (T) and potential (K) energy expressions and power function P. If the Lagrangian function $\mathcal{L} = T - K$ is specified using N generalised coordinates q_i ($i = 1 \ldots N$), for a system of M ($< N$) degrees of freedom, the N Lagrangian equations of motion are:

$$\frac{d}{dt}\left(\frac{\partial \mathcal{L}}{\partial \dot{q}_i}\right) + \frac{\partial P}{\partial \dot{q}_i} + \sum_{j=1}^{N-M} \lambda_j f_{ji} = Q_i \,, \quad i = 1 \ldots N. \tag{1}$$

The Lagrangian multiplier λ_j corresponding to the jth constraint equation, f_{ji} is the partial derivative of the jth constraint equation with respect to q_i, and Q_i is the generalised input corresponding to q_i. Due to superfluous coordinates, $(N - M)$ constraint equations are needed in the form of

$$\sum_{i=1}^{N} f_{ji}\dot{q}_i + f_{j0} = 0 \,, \quad j = 1 \ldots (N - M). \tag{2}$$

Equation (2) can be differentiated to give $(N - M)$ second order differential equations, which when combined with Eq. (1) gives the following $(2N - M)$ algebraic-differential equations of motion:

$$\begin{bmatrix} \mathbf{A} & \mathbf{F}^T \\ \mathbf{F} & \mathbf{0} \end{bmatrix} \begin{bmatrix} \ddot{\mathbf{q}} \\ \lambda \end{bmatrix} = \begin{bmatrix} \mathbf{D}_1 + \mathbf{Q} \\ \mathbf{D}_2 \end{bmatrix}. \tag{3}$$

The vectors $\mathbf{q}$ and λ contain the generalised coordinates and the Lagrangian multipliers. The matrices $\mathbf{A}$ and $\mathbf{F}$ are functions of $\mathbf{q}$ and t and represent the inertia and Jacobian matrices. The right hand side vectors $\mathbf{D_1}, \mathbf{D_2}$ and the vector of generalised inputs $\mathbf{Q}$ are functions of $\mathbf{q}$, $\dot{\mathbf{q}}$ and t. The above equations can be solved for N generalised coordinates and $(N - M)$ Lagrangian multipliers for given general inputs and initial conditions. This solution is referred to here as *forward dynamic* simulation. The solution exists for non-singular values of the coefficient matrix in Eq. (3). Particular values of $\mathbf{q}$, that make the coefficient matrix singular are termed singularity points; a discussion of these is outside the scope of this paper.

Inverse dynamic analysis involves calculating the required control inputs to achieve a R-degrees-of-freedom desired motion, which is defined by the acceleration-time history of R $(\leq M)$ generalised coordinates. The system is said to be fully actuated if $R = M$, and under-actuated if $R < M$. The number of control inputs must be equal to R. The inverse dynamic solution is achieved in two steps.

1. The rows corresponding to control input coordinates are deleted in Eq. (3), and the terms (i.e. the columns in the coefficient matrix) corresponding to the second derivatives of motion specifying coordinates are moved to the right hand side. This gives $(2N - M - R)$ differential-algebraic equations, which can be solved for the remaining $(N - R)$ generalised coordinates and the $(N - M)$ Lagrangian multipliers.

2. This solution is then inserted in the R unused equations, which were deleted in step 1 above, to solve for the required R control inputs. No integration is necessary.

Although the coefficient matrix for the forward simulation is of structurally full rank, this may not be true for the reduced coefficient matrix in step 1 (Sahinkaya, 2004). The reason for structural non-invertibility is due either to improper choice of control input locations or to mismatch in the order of equations in multi-physics simulations. The latter is relevant to the present paper, where electrical and mechanical components form a multi-physics system.

3. SYSTEM DESCRIPTION AND MODELING

The system simulated in this paper is a three-degree of freedom mechanism moving on a horizontal plane (Gosselin and Angeles, 1988), shown in Figure 1, and driven by three electric motors located on the vertices of the equilateral triangle $A_1A_2A_3$ of dimension $0.3\sqrt{3}$. The platform is connected to ground by three equal legs comprising a crank and a follower link. The manipulator data is given in Table 1. The three motors driving the cranks have the same characteristics: resistance of $R = 0.4\ \Omega$, inductance of $L = 5 \times 10^{-5}$H, torque constant of $K_t = 0.008NmA^{-1}$, and back emf constant of $K_b = 0.008NmA^{-1}$. The power rating of the motors is 50 W.

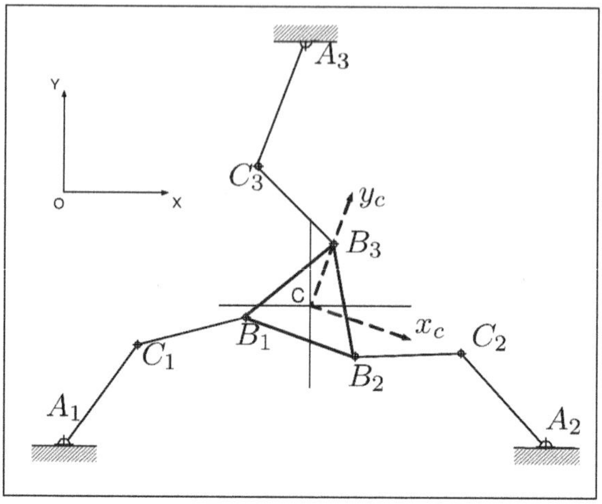

Figure 1. Planar three-degree-of-freedom parallel mechanism.

Table 1. Manipulator data ($i, j = 1..3, i \neq j$).

	A_iC_i	B_iC_i	Platform
Length (m)	0.17	0.15	$0.13\sqrt{3}$
Mass (kg)	0.1442	0.1272	1.331
M.O.I. (10^{-3} kgm^2)	0.3617	0.2513	1.987

An earth-fixed X-Y axis system lying on a horizontal plane is used as shown in Figure 1. The motor angles, θ_1, θ_2 and θ_3, are measured in a counter-clockwise direction from the positive x-axis. The position of the centre of gravity of the base is defined by its Cartesian coordinates x_c, y_c, and its angular position θ_c. The planar mechanism has three degrees of freedom, but the inclusion of electric motors increases the degrees of freedom to six. Electrical charges can be used as generalised coordinates to express the energy and power function for motor circuits. Then the torque input at the three motor angles will be proportional to the current, and control voltages become generalised inputs along the three generalised coordinates representing the electrical drive circuit. Alternatively, the following motor equations of motion can be included:

$$T_i(t) = K_t I_i(t), \quad V_i(t) = L\frac{dI(t)}{dt} + RI_i(t) + K_b\dot{\theta}_i(t). \tag{4}$$

There are many options for the selection of generalised coordinates to represent the parallel mechanism. The motion is defined in terms of the three platform coordinates, but the driving torque generated by the motors provide generalised inputs along the motor angles. Therefore, the selection of at least these six coordinates (with three constraint equations) is desirable for the inverse dynamics solution. The default planar mechanism interface of Dysim uses 3 generalised coordinates for each link (Cartesian coordinates of the centre of gravity and the absolute angle of the body fixed x-axis), which makes a total number of 21 coordinates for the mechanism (or 24 if the motor dynamics are included). Due to its object oriented properties, the Dysim modelling engine can be used in the Matlab/Simulink simulation environment to carry out further analysis and controller design. As mentioned, there are other simulation tools to analyse the inverse dynamics of planar mechanism, and can equally be used for the implementation of the suggested adaptive control strategy in Section 5.

The inclusion of motor dynamics in the Dysim model presents no problem in forward dynamic analysis, but makes the inverse analysis structurally non-invertible due to mismatch of equation orders of the mechanical and electrical components. A segregation method is used to overcome this problem, which involves numerical differentiation of calculated motor torques as explained in Sahinkaya (2004).

4. MOTION PLANNING AND LINEARISATION

The use of inverse dynamic analysis to calculate the required control voltages for the three motors requires the three degrees of freedom motion of the platform to be defined in terms of the acceleration-time history of three independent coordinates. It has been shown elsewhere (Sahinkaya, 2001) that the following third order exponential function can be used to define a vibration free point-to-point motion of a system even with undamped or lightly damped modes. For the x-axis motion:

$$x_c = x_c(0) + \Delta x_c(1 - e^{-(\alpha t)^3}). \tag{5}$$

The function and its derivatives are smooth and defined for $t \geq 0$. The initial and final velocities and accelerations are zero. The speed of the motion can be controlled by a single parameter α, which gives a 99% settling time of $1.66/\alpha$.

In the simulated example, the Cartesian coordinates of the centre of gravity of the platform and its angular position are selected as the motion-defining coordinates. A square path of corner coordinates (25mm,−25mm), (25mm,25mm), (−25mm,25mm) and (−25mm,−25mm) is selected to demonstrate the changing dynamic properties of the system. This path consists of four point-to-point motions between the corner points. Each motion starts and ends at rest and is completed within a specified time before starting the next motion. The platform angle is kept constant at 0 degree for all four motions. The same general function in Eq. (5) is used to define the motion of y_c for all four motions of the square path. The duration of each motion is selected as 0.25 s, which is fast for the given motor specifications.

Table 2. Estimated linearised model parameters.

	R^2	C_1	C_2	C_3	C_0
Motion 1:					
Motor 1	0.924	3.161 (0.069)	31.268 (1.282)	**6.880 (18.228)**	**–29.624 (24.224)**
Motor 2	0.904	2.063 (0.054)	24.662 (1.007)	**–6.042 (14.376)**	**29.633 (22.085)**
Motor 3	0.931	0.106 (0.002)	1.387 (0.038)	4.823 (0.539)	–18.584 (2.120)
Motion 2:					
Motor 1	0.913	0.334 (0.009)	0.754 (0.149)	–15.762 (2.084)	29.265 (3.265)
Motor 2	0.982	1.207 (0.012)	7.982 (0.205)	17.526 (2.874)	–35.393 (4.824)
Motor 3	0.998	0.868 (0.003)	7.226 (0.053)	**–0.823 (0.746)**	**4.633 (2.742)**
Motion 3:					
Motor 1	0.914	1.881 (0.052)	–6.470 (0.941)	–187.401 (12.674)	264.982 (19.558)
Motor 2	0.931	2.894 (0.067)	–9.188 (1.221)	–230.199 (16.556)	453.322 (30.915)
Motor 3	0.995	0.293 (0.001)	1.420 (0.025)	2.963 (0.346)	–11.438 (1.276)
Motion 4:					
Motor 1	0.875	0.312 (0.004)	2.108 (0.082)	21.602 (0.895)	
Motor 2	0.985	1.330 (0.008)	5.121 (0.158)	–42.486 (1.689)	
Motor 3	0.995	1.719 (0.006)	5.681 (0.107)	–33.394 (1.137)	

After calculating the required motor voltages through the inverse dynamic analysis, the calculated voltages are linearised for each motor and for each motion as follows:

$$V_i = C_1\ddot{\theta}_i + C_2\dot{\theta}_i + C_3\theta_i + C_0\,, \quad i = 1\ldots 3. \tag{6}$$

The last two terms are replaced by $C_3(\theta_i - \theta_{i,end})$ for the last motion to ensure that the input voltage is zero when the system reaches its steady state final value as this is a free mass system. A Least Square Estimator (Astrom and Wittenmark, 1990) is used to determine the model parameters. The linearisation is repeated for each motion, and the results are shown in Table 2.

The R^2 column shows the goodness of fit values; the standard deviations of estimated parameters are shown in brackets. The insignificant coefficients are printed in boldface. Figure 2 compares the linearised control voltages with the voltages calculated by inverse dynamic analysis for the four motions.

The high goodness of fit values and the accuracy of the linear models to predict the required control voltages indicate that motion dependent linearised models can be used to represent the system dynamics. This has significant implications in cases where the use of inverse dynamic solutions in real-time is not practical due to computational limitations. The inverse dynamic analysis can be performed off-line and the linearised models can be implemented online for adaptive control of highly non-linear parallel mechanisms. The results in Table 2 show that the linearised model parameters change significantly with the motion. It has also been shown (Sahinkaya and Li, 2005) that the coefficients also change significantly with motion speed in high speed applications.

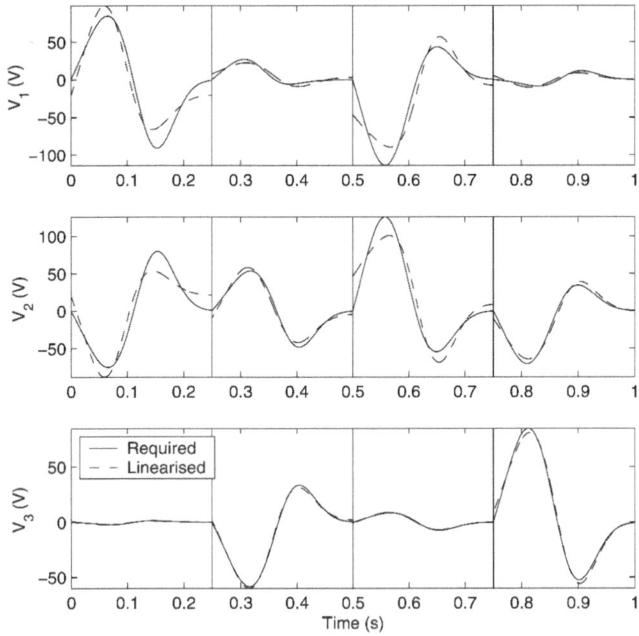

Figure 2. Linearised voltages compared with the calculated required voltages to achieve the desired motion.

5. ADAPTIVE CONTROL

Figure 3 summarises the adaptive control strategy adopted in this paper. All computationally intensive calculations and feedback controller tunings are carried out off-line for each specified motion by utilising inverse dynamic analysis. The real-time controller is linear and includes feed-forward and feedback paths, providing an efficient control system that can be implemented easily in real-time. Figure 4 shows a schematic view of the off-line calculations for a specified desired motion of the platform. The motion in the figures refers to acceleration, velocity and displacement. Since superfluous coordinates are used, the inverse dynamics analysis provides the desired motion in motor coordinates automatically, hence there is no need for coordinate transformations.

Because of the highly non-linear nature of the system dynamics at high speed operation, it is expected that the error dynamics will differ significantly from the motion dynamics. An approach shown in Figure 4 performs the linearisation based on perturbations around the desired motion. Perturbations are added to the calculated motor voltages before they are used as an input to the forward dynamic analysis. A Least Square Estimator (Astrom and Wittenmark, 1990) is used to estimate the coefficient of the linearised error dynamics model

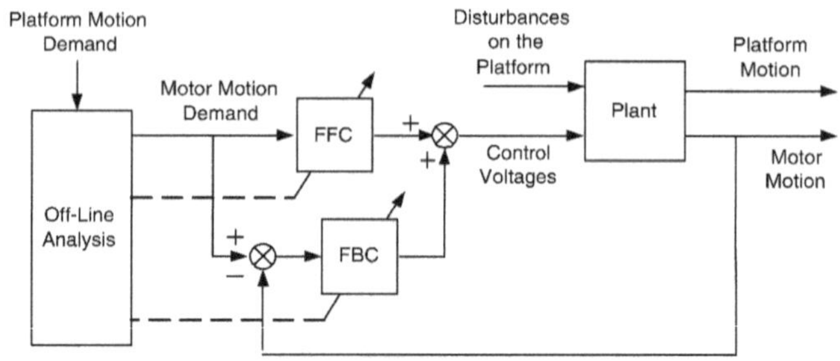

Figure 3. Adaptive control strategy (FFC: feed-forward controller, FBC: feedback controller).

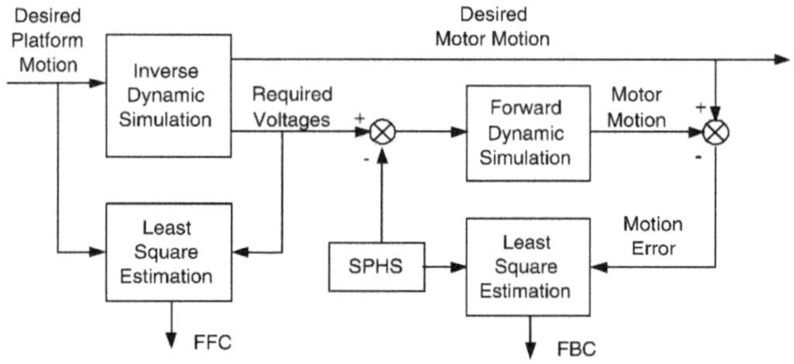

Figure 4. Off-line simulations to tune the adaptive controller.

for the feedback path in the following form:

$$\Delta V_i = C_1 \Delta \ddot{\theta}_i + C_2 \Delta \dot{\theta}_i + C_3 \Delta \theta_i \ , \quad i = 1 \ldots 3. \tag{7}$$

A Schroeder Phased Harmonic Sequence (SPHS) is used as a disturbance signal. This is a periodic low peak factor signal with user specified frequency spectrum characteristics (Schroeder, 1970). The amplitudes of the disturbance is set to 10% of the calculated voltages with a flat spectrum between 4 and 80Hz for each motion of the square path. The estimated parameters of the error dynamic model are shown in Table 3.

The estimated parameters for the linearised error model differ significantly compared with the estimated linearised model parameters, and they also change significantly with motion as expected. This shows the need to perform controller tuning for each motion. To test the performance of the controller, the simulated system shown in Figure 3 was run with disturbance forces and torques acting on the platform. Some results of this simulation in terms of the motion of the platform are shown in Figure 5.

Good tracking and dynamic performance is obtained from the adaptive controller. It is encouraging to obtain such a high accuracy with a simple linear adaptive controller for a high speed motion. Further work is necessary to en-

Table 3. Estimated linearised error model parameters.

	R^2	C_1	C_2	C_3
Motion 1:				
Motor 1	0.984	0.7704 (0.0046)	15.0438 (0.4438)	–50.1286 (4.4385)
Motor 2	0.973	0.5715 (0.0045)	11.8312 (0.4358)	–44.4435 (4.4034)
Motor 3	0.963	0.4945 (0.0045)	10.5797 (0.5502)	–42.2553 (5.6561)
Motion 2:				
Motor 1	0.986	0.7460 (0.0041)	13.8483 (0.4330)	–31.6599 (4.4484)
Motor 2	0.993	0.8151 (0.0032)	15.0993 (0.3015)	–42.4743 (2.9786)
Motor 3	0.998	0.4497 (0.0010)	11.0597 (0.1047)	–12.8660 (1.2549)
Motion 3:				
Motor 1	0.968	0.4990 (0.0043)	9.7395 (0.4617)	–19.8254 (4.9157)
Motor 2	0.986	0.7622 (0.0042)	12.2476 (0.4275)	–34.4538 (4.0778)
Motor 3	0.964	0.6336 (0.0057)	12.7235 (0.4997)	–12.8280 (5.5001)
Motion 4:				
Motor 1	0.633	0.0819 (0.0021)	2.8852 (0.2457)	–12.4249 (7.5468)
Motor 2	0.898	0.3332 (0.0037)	7.9402 (0.4041)	–8.4064 (3.9383)
Motor 3	0.929	0.4966 (0.0045)	11.9194 (0.4930)	–18.6742 (4.8301)

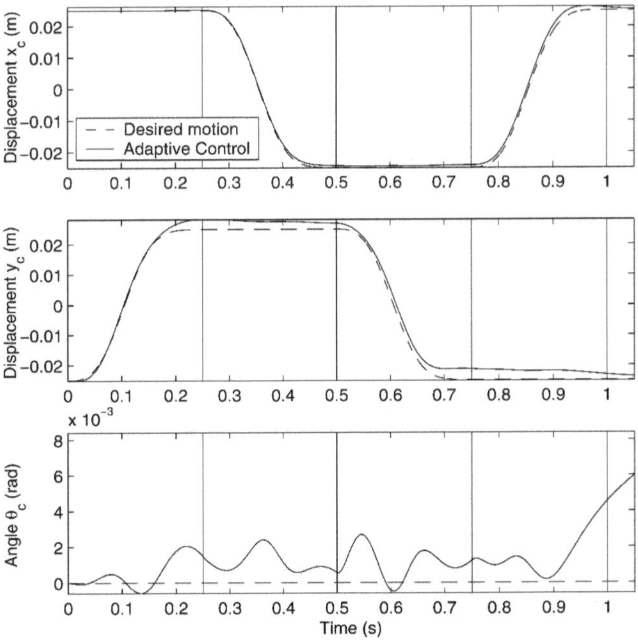

Figure 5. The desired motion of the platform compared with the motion achieved with the adaptive controller.

sure the system stability, as some of the linearised model parameters are negative. However, no problems were encountered during extensive simulations covering a variety of motions at different speeds.

6. CONCLUSIONS

The inverse dynamic analysis of a electro mechanical system is carried out by using Lagrangian dynamics. A three degree of freedom parallel mechanism driven by three electric motors is selected for the simulation study. The trajectory planning for a point-to-point motion of the platform is formulated by an input shaping technique. A square path is planned and the system dynamics are successfully linearised for each section of the path. An adaptive control strategy based on the linearised models are used in order to avoid performing the inverse dynamics simulation in real-time. It is observed that the linearised model parameters change significantly for each motion, and hence an adaptive control strategy is needed for the successful control of high speed parallel mechanisms. The suggested off-line tuning of the control system, and the adaptive control strategy is applied to the simulated example. It is shown that the platform accurately follows the desired path despite the external disturbances on the platform.

REFERENCES

Astrom, K.J. and Wittenmark, B. (1990). *Computer Controlled Systems: Theory and Design*, 2nd edition. Prentice-Hall, Englewood Cliffs, NJ.

Geike, T. and McPhee, J. (2003). Inverse dynamic analysis of parallel manipulators with full mobility. *Mechanism and Machine Theory*, 38(6):549–562.

Gosselin, C. and Angeles, J. (1988). The optimum kinematic design of a planar three degree of freedom parallel manipulator. *Journal of Mechanisms, Transmissions and Automotion in Design*, 110(1):35–41.

Merlet, J.P. (1996). Direct kinematic design of planar manipulators. *Proceedings, IEEE International Conference on Robotics and Automation*, 1–4:3744–3749.

Merlet, J.P. (2000). *Parallel Robots*, 1st edition. Kluwer Academic Publishers, Dordrecht.

Sahinkaya, M.N. and Li, Y. (2005). Inverse dynamic analysis and linearisation of a high speed parallel mechanism. *ASME DETC2005, 29th Mechanisms and Robotics Conference*, Long Beach, CA, USA, submitted.

Sahinkaya, M.N. (2001). Input shaping for vibration free positioning of flexible systems. *Proc. Instn. Mech. Engrs., Part I, Journal of Systems and Control Engineering*, 215(I5):467–481.

Sahinkaya, M.N. (2004). Inverse dynamic analysis of multiphysics systems. *Proc. Instn. Mech. Engrs., Part I, Journal of Systems and Control Engineering*, 218(I1):13–26.

Schroeder, M.R. (1970). Synthesis of low-peak-factor signals and binary sequences with low auto correlation. *IEEE Trans. on Information and Theory*, pp. 85–89.

POWERSAVING CONTROL OF MECHANISMS

Werner Schiehlen and Nils Guse
Institute B of Mechanics, University of Stuttgart, Germany

Abstract: Inverse dynamics control of mechanisms with prescribed motions results in high energy demand which will be reduced by adding springs for local energy storage. Two methods are proposed, adjusting the trajectories of mechanisms to a periodic conservative trajectory as well as finding optimal spring parameters by curve fitting. A control logic is introduced to further reduce the energy cost of transition processes.

Key words: energy consumption, inverse dynamics control, mechanism, shooting method, curve fit.

1. INTRODUCTION

A well established control principle is inverse dynamics which is used to overcome high nonlinearities typical for mechanisms undergoing large displacement motion. Based on an accurate model of the system under consideration, all the nonlinearities are compensated first by control action and, then, the remaining double integrator is controlled by linear feedback. This approach is very attractive to control engineers since a broad variety of design tools can be applied successfully. However, this principle results in high energy demand, see e.g. [6].

Recently, underlying nonlinear conservative vibrations were proposed to generate the required motion of mechanisms coarsely without energy dissipation and to adjust the motion finely by control with strongly reduced energy consumption. Thus, periodic trajectories of an underlying conservative system have to be adapted as closely as possible to the desired trajectories. The fundamentals of energy consumption and energy storage of mechanical systems with rheonomic constraints have been analyzed for a linear harmonic oscillator [5]. In addition it was shown how spring characteristics can be chosen to reduce the energy consumption for arbitrary periodic trajectories of a robot arm, see [4]. Furthermore, a curve fitting method was presented to adjust spring parameters of a mechanical system to an optimal force trajectory computed by inverse dynamics [3].

H. Ulbrich and W. Günthner (eds), IUTAM Symposium on Vibration Control of Nonlinear Mechanisms and Structures, 277–286.

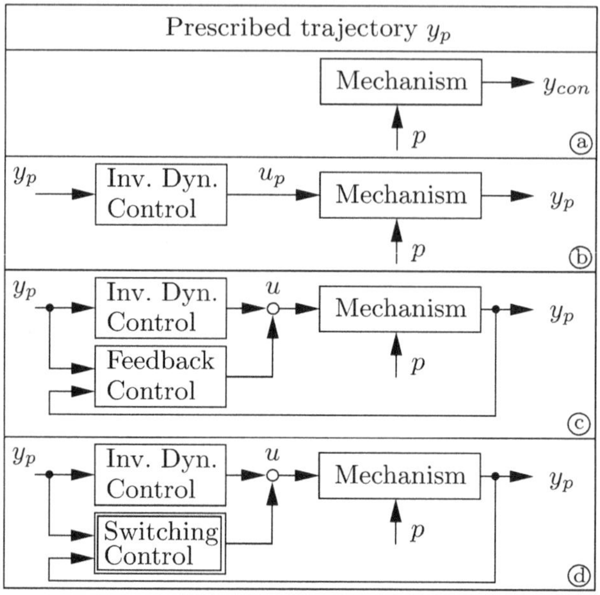

Figure 1. Control strategy.

A similar approach is used for resonant vibration machines, see [2] where, however, the period and desired trajectory are not specified.

In this paper the power aspects of inverse dynamics control systems will be treated for actively controlled mechanisms by trajectory adjustment and curve fitting. To obtain an asymptotically stable limit cycle, the linear control may be supplemented by a switching control.

2. CONSERVATIVE PERIODIC MOTIONS OF MECHANISMS

Nonlinear mechanisms with actuators and springs for energy storage are modelled as multibody systems resulting in equations of motion of the form

$$\mathbf{M}(\mathbf{y}, t)\ddot{\mathbf{y}} + \mathbf{k}(\mathbf{y}, \dot{\mathbf{y}}, t) = \mathbf{q}(\mathbf{y}, \mathbf{p}, t) + \mathbf{C}(\mathbf{y}, t)\,\mathbf{u}(t)\ . \tag{1}$$

The $f \times 1$ vector $\mathbf{y}$ summarizes the f generalized coordinates of the mechanism, $\mathbf{M}$ is its $f \times f$-inertia matrix, the $f \times 1$-vector $\mathbf{k}$ characterizes the Coriolis forces. The $f \times 1$-vector $\mathbf{q}$ of the applied forces depends on the design parameters of the springs represented by the $s \times 1$-vector $\mathbf{p}$ and the $f \times 1$-vector $\mathbf{u}$ of the actuator forces is subject to a $f \times f$ input matrix $\mathbf{C}$. The mechanism is assumed to be completely controllable with one actuator for each generalized coordinate. As a first step, a shooting method will proposed to evaluate the design parameters such that without actuation a conservative periodic vibration $\mathbf{y}_{con}(t) = \mathbf{y}_{con}(t+T)$ is achieved as close as possible to desired prescribed motion $\mathbf{y}_p(t)$, Figure 1a. Alternatively, a curve fitting method can be used to evaluate the design parameters. The motion $\mathbf{y}_{con}$ with period T does not consume any energy. Then, an inverse dynamics control $\mathbf{u}_p(t)$ is added to adjust the motion completely to the desired motion spending some energy, Figure 1b. Moreover, a linear feedback control is used to obtain asymptotic stability approaching the prescribed motion $\mathbf{y}_p(t)$, Figure 1c. Thus, the control law reads as

$$\mathbf{u}(t) = \mathbf{u}_p(t) - \mathbf{D}(\dot{\mathbf{y}} - \dot{\mathbf{y}}_p) \tag{2}$$

with a positive definite $f \times f$ matrix $\mathbf{D}$. The linear feedback is mainly required during transitions, in the steady-state less energy will be consumed. This transition to the desired trajectory will be finished as early as possible by a switching control, Figure 1d, resulting in a true limit cycle.

3. OPEN-LOOP MECHANISMS

Figure 2 shows a mechanism with two degrees of freedom represented by the angles α_1 and α_2. The model consists of two bodies, with mass m_1, m_2 and

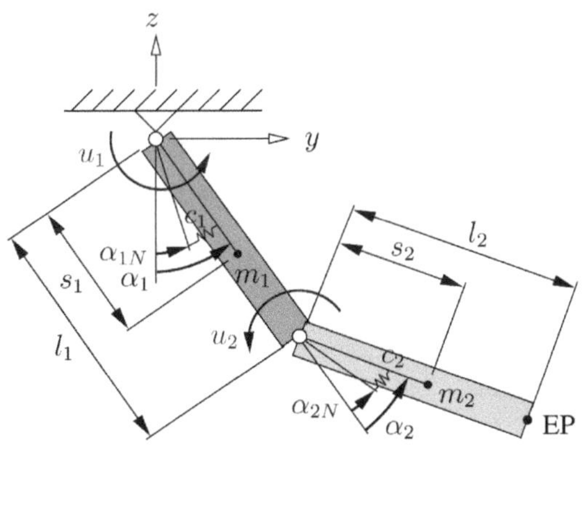

Figure 2. Mechanism with two degrees of freedom.

the inertia J_{11} and J_{21}. Two drive motors generate the torque u_1 and u_2 at the joints of the robot. Two rotational springs (spring stiffness c_1 and c_2, mounting positions α_{1N} and α_{2N}) are added for energy storage. The matrices and vectors of the equations of motion (1) read as

$$\mathbf{M} = \begin{bmatrix} 2m_2 l_1 s_2 \cos\alpha_2 + (l_1^2 + s_2^2)m_2 + s_1^2 m_1 + J_{11} + J_{21} & sym. \\ m_2(s_2^2 + l_1 s_2 \cos\alpha_2) + J_{21} & m_2 s_2^2 + J_{21} \end{bmatrix},$$

$$\mathbf{y} = \begin{bmatrix} \alpha_1 & \alpha_2 \end{bmatrix}^T,$$

$$\mathbf{k} = \begin{bmatrix} -m_2 l_1 s_2 \sin\alpha_2(\dot{\alpha}_2^2 + 2\dot{\alpha}_1\dot{\alpha}_2) \\ m_2 l_1 s_2 \dot{\alpha}_1^2 \sin\alpha_2 \end{bmatrix}, \quad \mathbf{q} = \begin{bmatrix} c_1(\alpha_{1N} - \alpha_1) \\ c_2(\alpha_{2N} - \alpha_2) \end{bmatrix},$$

$$\mathbf{u} = \begin{bmatrix} u_1 \\ u_2 \end{bmatrix}, \quad \mathbf{C} = \mathbf{I} \quad \text{and} \quad \mathbf{p} = \begin{bmatrix} c_1 & c_2 & \alpha_{1N} & \alpha_{2N} \end{bmatrix}^T. \tag{3}$$

The end effector point EP is supposed to follow a given trajectory.

3.1 Conservative Periodic Trajectory

The goal of this method is to adjust the storage springs of the mechanism such that its natural trajectory, without any control action, is close to the desired trajectory. This can be achieved, by solving a boundary value problem. When the system is set to its starting conditions at the starting point of the desired trajectory, it has to return to the endpoint of the desired trajectory. For this boundary value problem, the shooting method is used to find a periodic solution close to the desired trajectory characterized by the design parameters $\mathbf{p}_{opt}$. For details, see [4].

As an example, a harmonic oscillation of the end effector point EP in y-direction is prescribed as

$$y = A\cos\omega t\ , \quad z = -B\ . \tag{4}$$

The results are shown in Figures 3 and 4. There are three cases presented: The optimal conservative periodic solution without control ($\mathbf{p} = \mathbf{p}_{opt}$, $\mathbf{u} = \mathbf{0}$), the inverse dynamics control without springs ($\mathbf{p} = \mathbf{0}$, $\mathbf{u} = \mathbf{u}_{inv}(t)$) and the adjusted periodic solution ($\mathbf{p} = \mathbf{p}_{opt}$, $\mathbf{u} = \mathbf{u}_{opt}(t)$) . The trajectories of both controlled motions are not to be distinguished in Figure 3 (solid line) while the uncontrolled motion shows some differences (dashed line).

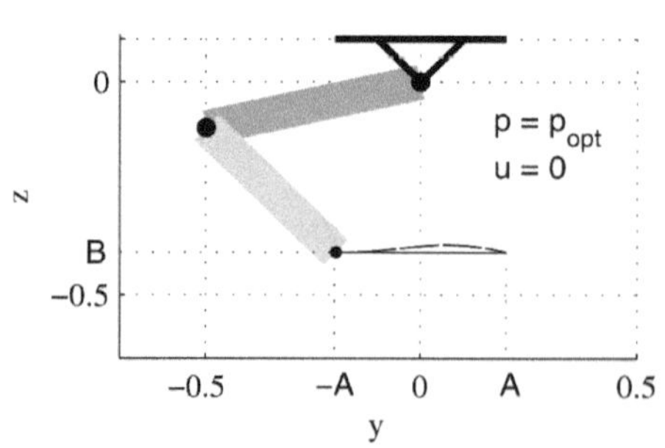

Figure 3. Conservative periodic motion (dashed line) and controlled motion (solid line).

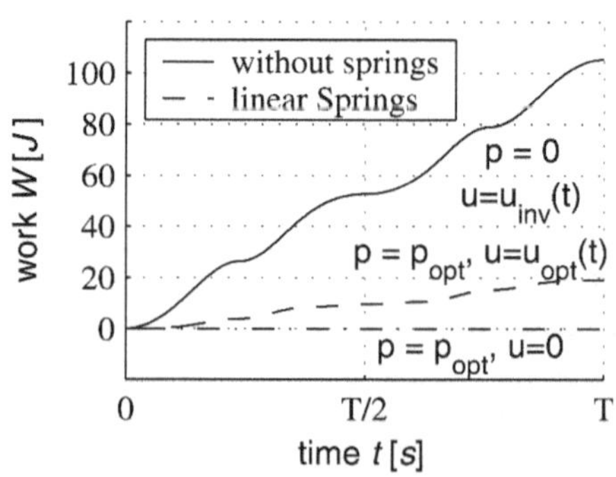

Figure 4. Total work of both motors for one cycle period T.

The energy consumption can be seen in Figure 4. The solid line shows the total work of both motors for one cycle period T without any springs. When springs are added, designed by the shooting method principle, the energy consumption is reduced by more than 80% (dashed line) while the conservative motion does not consume any energy at all.

3.2 Torque Curve Fitting

A more engineering approach is to compute the required control force or torque, respectively, as function of the position, and then to use a polynomial to fit this curve. The area between the linear or cubic fit and the required force represents the work that has to be minimized, see [1] or [3], respectively.

Starting from the trajectory prescribed the required control signal, here the torques of the motor drives, is computed using inverse dynamics. Then, this torque is converted to a function of the position for both bodies. Figure 5a shows the control torque calculated by inverse dynamics for body 1 and angle α_1. The shaded area represents the energy which is required to drive the system. Therefore, this area has to be minimized by curve fitting. In Figure 5b, a linear function is used as a curve fit (dashed line) representing a linear spring where the spring coefficient is the slope of this curve fit and the axis intercept being the spring fastening. Now, the shaded area i.e. the energy is significantly reduced, since the spring generates most of the necessary torque. In Figure 5c a nonlinear spring is used and the energy is even more reduced. It has to be noted that the curve fitting is a minimization task with inequality constraints, since the stiffness coefficient is always positive. The same procedure is performed for body 2, too. Figure 6 shows the work for both motor drives for one cycle. Again, the energy can be reduced significantly by springs designed by the curve fitting method (dashed line). The results are comparable to the previous method, see Figure 4.

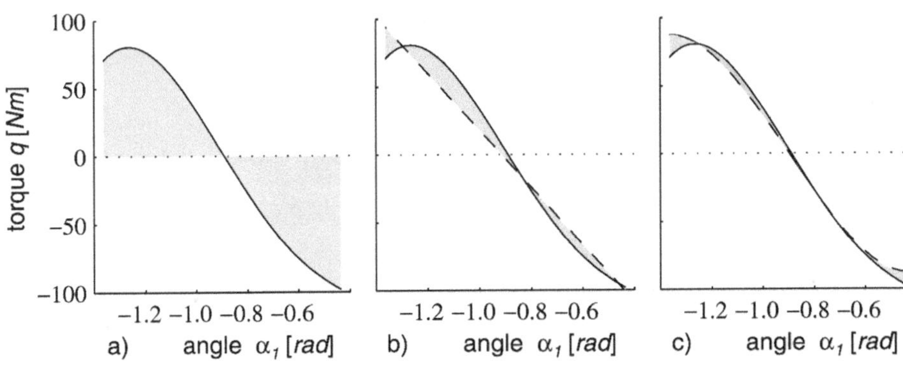

Figure 5. Control torque vs. angle position, a) without springs, b) with linear springs, c) with nonlinear springs. The grey area represents the work used for controlling the system.

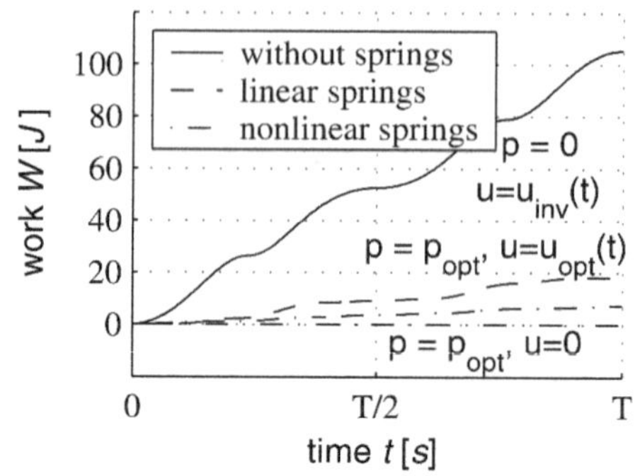

Figure 6. Total work of both motors for one cycle period T.

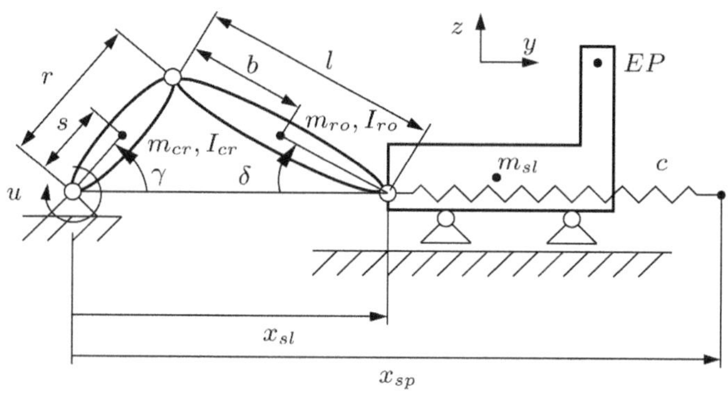

Figure 7. A slider crank mechanism with a spring and a driving torque u.

4. CLOSED-LOOP MECHANISM

Figure 7 shows a slider crank mechanism representing different machine types. The angle γ is chosen as generalized coordinate and describes the only degree of freedom. A torque u is applied to the crank. A spring force with stiffness c and an adjustable mounting position x_{sp} acts on the slider. The scalar quantities of the equation of motion (1) read as

$$
\begin{aligned}
M &= m_{cr}s^2 + m_{ro}(r^2 - 2rb\delta'\cos(\gamma+\delta) + b^2\delta'^2) \\
&\quad + m_{sl}(r\sin\gamma + l\delta'\sin\delta)^2 + J_{cr} + J_{ro}\delta'^2, && (5) \\
y &= \gamma, && (6) \\
k &= \{m_{ro}[br\delta'(1+\delta')\sin(\gamma+\delta) - br\delta''\cos(\gamma+\delta) + b^2\delta'\delta''] \\
&\quad + m_{sl}(r\sin\gamma + l\delta'\sin\delta)[r\cos\gamma + l(\delta''\sin\delta + \delta'^2\cos\delta)] \\
&\quad + J_{ro}\delta'\delta''\}\dot{\gamma}^2, && (7) \\
q &= c(x_{sl} - x_{sp})(r\sin\gamma + l\delta'\sin\delta), \quad C = 1, && (8)
\end{aligned}
$$

$$\delta = \arcsin(\lambda \sin\gamma),\ \ \delta' = \frac{\lambda\cos\gamma}{\cos\delta},\ \ \delta'' = \frac{(\lambda^2-1)\sin\delta}{\cos\delta^3}, \tag{9}$$

$$\lambda = \frac{r}{l},\ \ x_{sl} = r\cos\gamma + l\cos\delta,\ \ \mathbf{p} = \left[\begin{array}{cc} c & x_{sp}\end{array}\right]^T. \tag{10}$$

The end effector point EP is supposed to follow a given time history. It has to be mentioned, that the different types of applications of a slider crank mechanism result in different sizes and different mass and inertia distributions. Here, two basic types will be investigated. First, a mechanism with large mass and inertia of the crank, then a mechanism with a large slider mass is considered.

4.1 Crank Enforced Mechanism

The mass m_{cr} and the inertia J_{cr} of the crank are large compared to the those of the rod and the slider. The desired motion is again the harmonic oscillation (4) with $A = r$, $B = 0$ and a rotating crank. For the optimal design of the springs, using inverse dynamics and curve fitting, an optimal torque function is calculated. This torque function is transformed to a corresponding spring force $F_s(x_{sl})$ of the slider position x_{sl}. This function is plotted in Figure 8 (solid line) for $l = 0.25m$ and an amplitude $r = 0.05m$. As in the previous section, curve fitting is used to design the parameters of the spring such that its force function minimizes energy. Since the spring coefficients can only be positive, the spring force has a negative slope (dashed line) while its amount follows from the preload or x_{sp}, respectively. Therefore, the energy decrease the most, when x_{sp} is negative. Since this might lead to design conflicts it is assumed here that there is some upper bound for x_{sp}. Anyhow, almost 65% of the energy can be saved compared to the system without any springs, see Figure 9 featuring the three cases like in Figure 4.

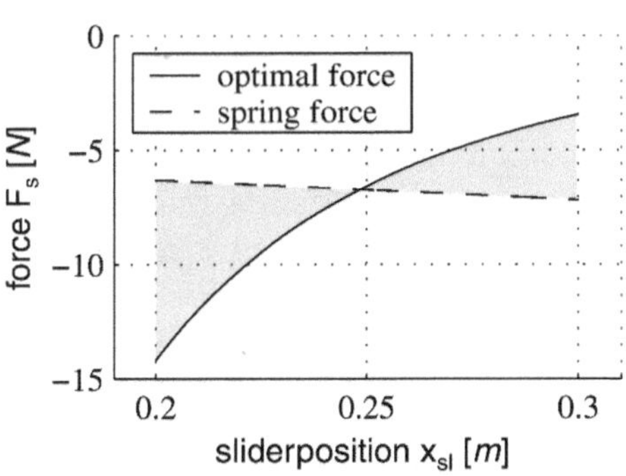

Figure 8. Force function vs. slider position, crank enforced mechanism.

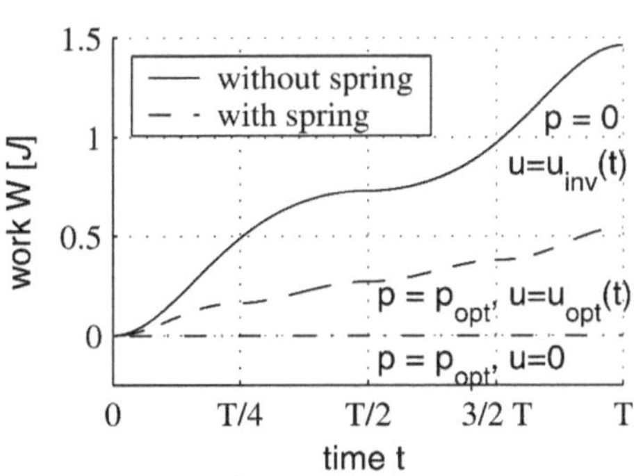

Figure 9. Work of motor for one cycle period T.

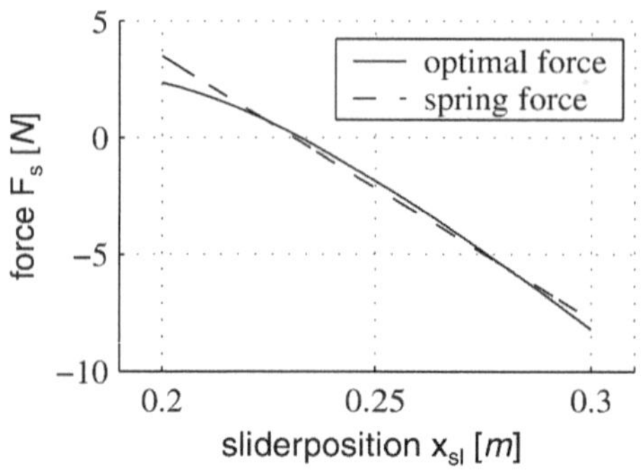

Figure 10. Force function vs. slider position, slider enforced mechanism.

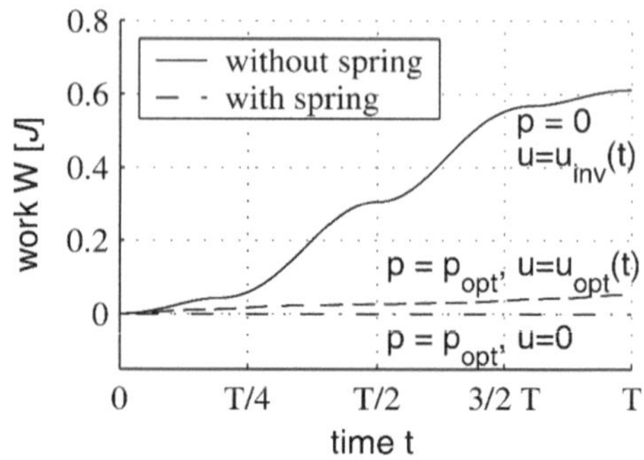

Figure 11. Work of motor for one cycle period T.

4.2 Slider Enforced Mechanism

For slider enforced mechanisms, the design method is the same as for crank enforced mechanisms. The inverse dynamics and the corresponding transformation result in the force-position function shown in Figure 10. In this case it is fairly easy to find an optimal spring coefficient and mounting position such that the curve of the spring force fits the optimal force calculated by inverse dynamics. It is not surprising that also savings of more than 90% can be achieved (see Figure 11).

5. TRANSITION CONTROL

For the transition analysis an oscillator is considered. The linear transition control by damping feedback according to equation (2) is added to the system with inverse dynamics control. For the harmonic motion of an oscillator one gets

$$m\ddot{y} + cy = A(m\omega^2 - c)\cos\omega t - D(\dot{y} - A\omega\sin\omega t). \tag{11}$$

Small damping as well as large damping coefficients result in asymptotic stability but with oscillatory motions with respect to the desired trajectory as shown in Figures 12a,b. Only for a suitable chosen damping a monotonic stability is achieved, see Figure 12c. Therefore, it is proposed to use a switching logic, see Figure 13. The switching logic is turned off when the desired trajectory

$$\frac{y_p^2}{A^2} + \frac{\dot{y}_p^2}{(A\omega)^2} = 1, \tag{12}$$

in the phase space is reached. This logic offers a monotonic behavior of representing a true limit cycle (Figure 12d). As a result the energy consumption

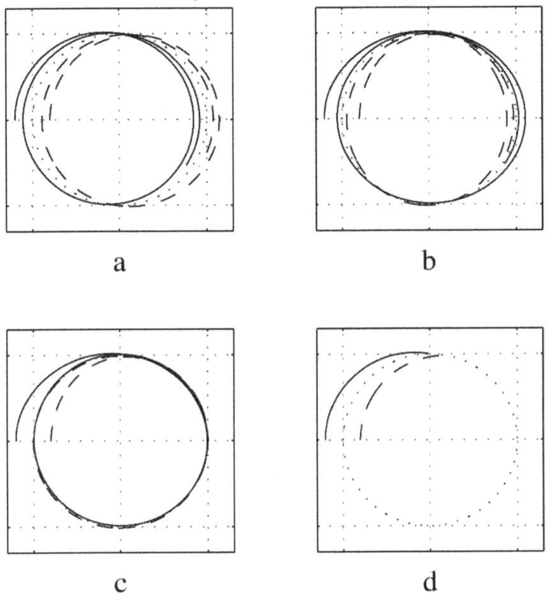

Figure 12. Transition trajectories in phase plane y, $\dot{y}$. Prescribed motion y_p is indicated by the dotted line.

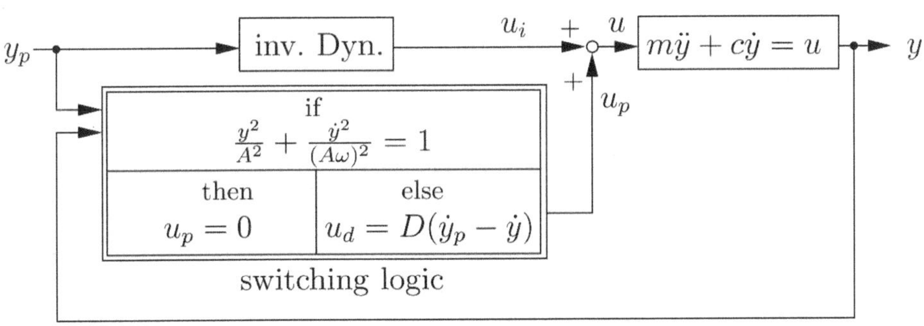

Figure 13. Switching logic of oscillator.

for transitions is reduced during transition. The switching control can be used for the mechanisms discussed in this paper, too.

6. CONCLUSION

It has been shown that adapting the natural motion of a system to a prescribed conservative periodic trajectory offers significant amounts of power savings. A shooting method was applied to adjust the system to the boundary conditions of a conservative periodic vibration close to the desired trajectory. Then, a low energy inverse dynamics control is sufficient to force the system onto the desired trajectory. Alternatively, curve fitting can be used to adjust spring parameters to an optimal force trajectory computed by inverse dynamics. A mechanism with two degrees of freedom and a slider crank mechanism were considered to show the basic concept. For the slider crank mechanism different mass distribution was investigated resulting in different behavior of the mechanism.

Asymptotic stability of the mechanism motion can be achieved by an additional linear control offering damping for transition phases. Due to the fact that the inverse dynamics control changes the dynamical properties of the system vibrations with respect to the prescribed motion occur. It is proposed to use a switching logic to stop the transitional vibrations after crossing the prescribed trajectory. This means minimal cost for the transitions.

REFERENCES

[1] Ackermann, M., Guse, N. and Lakrad, F., Improving Efficiency of Controlled Rheonomic Systems by Curve Fitting, Zwischenbericht ZB-136. Universität Stuttgart, Institut B für Mechanik, 2003.

[2] Babitsky, V.I. and Shipilov, A.V., *Resonant Robotic Systems*, Springer-Verlag, Berlin, 2003.

[3] Guse, N. and Schiehlen, W., Low Energy Control of Periodic Motions in Manufacturing. In *Proc. XXI ICTAM 2004*, Warsaw, Poland, August 15–21, 2004.

[4] Schiehlen, W. and Guse, N., Control of Limit Cycle Oscillations. In *Proceedings of IUTAM Symposium on Chaotic Dynamics and Control of Systems and Processes in Mechanics*, G. Rega and F. Vestroni (eds), Rome, Italy, 8–13 June 2003, Springer, Dordrecht, 2005, pp. 429–440.

[5] Schiehlen, W. and Guse, N., Power Demand of Actively Controlled Multibody Systems. In *ASME Proceedings of DETC2001/VIB-21343*, Pittsburgh, PA, USA, September 9–12, 2001.

[6] Waldron, K.J., Some Thoughts on the Design of Power Systems for Legged Vehicles. In *Advances in Multibody Systems and Mechatronics*, Gerhard-Mercator-Universität, Duisburg, 1999, pp. 389–394.

APPLICATION OF VIBRATION CONTROL IN STEEL INDUSTRIES

Kurt Schlacher[1,2], Gernot Grabmair[1,2] and Johann Holl[1,2]
[1]*Christian Doppler Laboratory for Automatic Control of Mechatronic Systems in Steel Industries;* [2]*Institute of Automatic Control and Control Systems Technology, Johannes Kepler University of Linz, Altenbergerstr. 69, A-4040 Linz, Austria*
kurt.schlacher@jku.at

Abstract: In steel industry, the high quality requirements of rolled products, in particular the restrictive thickness tolerances, mean a challenging task also from a control point of view. One of the key problems in this context are vibrations of the rolling plant. Therefore, this contribution is concerned with the active rejection of two common vibration problems, namely vibrations of so-called wrapper rolls and the chatter phenomenon. A mathematical model based on physical considerations taking into account the essential nonlinearities of the plant provides the basis for the controller design. For vibration rejection two nonlinear control concepts based on energy considerations and the theory of differential flatness are presented.

Key words: steel industry, model-based control, nonlinear control, vibration rejection.

INTRODUCTION

The efficiency of steel rolling mill plants gains more and more in importance mainly due to the considerable demand of high quality steel products. This ambition of high production cycles entails in many cases vibration problems which demand an effective rejection in order to satisfy the restrictive thickness tolerances and surface properties of rolled products.

This contribution is concerned with two applications of vibration rejection in steel rolling mills by nonlinear control techniques. The first section of this paper is devoted to vibrations of so-called wrapper rolls which are typically used for coiling the rolled strip. Since such an arrangement is actuated hydraulically it is often poorly damped. Third octave chatter, which represents the second considered vibration problem, is indicated by unacceptable thickness variations of the rolled product or even damageof the plant due to vertical vibrations of at least one mill stand.

H. Ulbrich and W. Günthner (eds), IUTAM Symposium on Vibration Control of Nonlinear Mechanisms and Structures, 287–296.

1. WRAPPER ROLLS

Wrapper assemblies consist of a lever mechanism actuated by a hydraulic piston. The mechanical part of the wrapper system belongs to the class of Euler-Lagrange (EL) systems with Lagrangian

$$L = \frac{1}{2}\Theta\dot{q}^2 - V\,, \quad V = m_l g l_g \sin(q+\kappa)$$

with the angle of the lever mechanism q, its constant moment of inertia Θ – the moment of inertia of the actuator and its mass is neglected – the mass of the wrapper assembly m_l, the constant of gravity g and the position of the center of gravity at the distance l_g and the angle κ relative to the pivot. The applied method is not restricted to this assumption of a constant Θ. The corresponding Euler–Lagrange equation augmented by a damping term linear in the generalized velocity, or expressed as a Rayleigh dissipation function of the form $R_R = \frac{1}{2}d\dot{q}^2$, $d > 0$, and with the hydraulic force F_h as input reads as

$$\Theta\ddot{q} = -\partial_q V - d\dot{q} + \partial_q x_p(q) F_h - M_l\,, \tag{1}$$

with the piston position $x_p(q)$ and an unknown but constant external load torque M_l. Here and in the following ∂_q indicates the partial derivative with respect to q. Let $\partial_q x_p(q) \neq 0$ be met in the domain under consideration. We assume that each chamber is rigidly connected to a three-land-four-way spool valve. Often, the dynamics of the compensated servo valves are much faster than those of the other parts of the hydraulic adjustment system. Therefore, one can neglect the valve dynamics and consider the valve volume flow q_{vi} as the plant input. This will be assumed in the following. As already shown in Grabmair et al. (2004) the overall system has port Hamiltonian (PH) structure with the fluid masses in both chambers as Casimir functions, i.e., invariants of the system independent of H. In the following the canonical coordinates $x = (q, p, z_{h1}, z_{h2})$, consisting of the wrapper part $x_1 = (q, p, z_{h1})$ and a decoupled part $x_2 = (z_{h2})$, with $p = \Theta\dot{q}$, $F_h = p_{h1}A_1 - p_{h2}A_2$,

$$z_{h1} = F_h + \sum_i (-1)^{i-1} EA_i \ln(V_{hi}V_{pi}^{-1}) \tag{2a}$$

$$z_{h2} = p_{h2}A_2 + EA_2 \ln(V_{h2}V_{p2}^{-1}) \tag{2b}$$

will be used. $E > 0$ indicates the constant hydraulic bulk modulus. The effective piston areas are denoted by A_i, the chamber offset volumes by V_{pi}, the chamber pressures and volumes by p_{hi} and V_{hi}, with $i = 1, 2$ and $V_{h1} = V_{p1} + A_1 x_p(q)$, $V_{h2} = V_{p2} - A_2 x_p(q)$. The PH structure (see Grabmair et al., 2003) is then given by

$$\begin{aligned} \dot{x} &= (J - R)\,\partial_x H^T + Gu \\ y &= G^T \partial_x H^T \end{aligned}$$

with Hamiltonian

$$
\begin{aligned}
H &= H_h + \frac{1}{2}\Theta^{-1}p^2 + V \\
H_h &= \sum_i EV_{hi}\left(\ln(V_{hi}V_{pi}^{-1}) - 1\right) \\
&\quad + V_{p1}(p_{h0} + E)\mathrm{e}^{\frac{z_{h1}+z_{h2}}{A_1 E} - \frac{p_{h0}}{E}} + V_{p2}\,(p_{h0} + E)\,\mathrm{e}^{\frac{z_{h2}}{A_2 E} - \frac{p_{h0}}{E}} \\
&\quad - z_{h1}x_p - (z_{h1} + z_{h2})\,\frac{V_{p1}}{A_1} - z_{h2}\frac{V_{p2}}{A_2}\,,
\end{aligned}
\tag{3}
$$

the input matrix G for the transformed input $u = (u_1, u_2)$

$$
\begin{aligned}
G^T &= \begin{bmatrix} 0 & 0 & 1 & 0 \\ 0 & 0 & 0 & 1 \end{bmatrix}, \quad u_2 = \frac{EA_2}{V_{p2} - A_2 x_p\,(q)} q_{v2} \\
u_1 &= \frac{EA_1}{V_{p1} + A_1 x_p\,(q)} q_{v1} - \frac{EA_2}{V_{p2} - A_2 x_p\,(q)} q_{v2}\,,
\end{aligned}
$$

and the canonical skew symmetric structure matrix J and the positive semidefinite dissipation matrix R with $R_{22} = d$ and zero otherwise. p_{h0} denotes some offset pressure.

Here and in the following, $\bar{x} = x - \breve{x}$ indicates coordinates relative to the equilibrium $\breve{x}$. The simple proportional controller

$$
u_1 = -k_{p1}\,(z_{h1} - z_{h1}\,(\breve{q}))\ , \quad u_2 = -k_{p2}\,(z_{h2} - \breve{z}_{h2})\ , \tag{4}
$$

$k_{pi} > 0$ maintains the plants PH structure, and asymptotically stabilizes the equilibrium $\breve{x} = (\breve{q}, 0, z_{h1}\,(\breve{q})\,, \breve{z}_{h2})$. Nevertheless, this controller has a poor performance if M_l is unknown and further it does not inject additional damping. Additionally, the controller should not rely on the velocity signal because separate velocity sensors are usually not mounted and due to the noisy position signal it cannot be easily used to obtain a good approximation. In order to tackle these problems, one can introduce a dynamical extension of second order by means of a disturbance observer. Towards this end, one considers the mechanical subsystem (1) with fictitious input

$$
u_{obs} = -\partial_q V + \partial_q x_p\,(q)\,F_h
$$

and designs a linear reduced Luenberger observer for (1) extended by $\frac{\mathrm{d}}{\mathrm{d}t}M_l = 0$ with observer state

$$
\hat{x}_{obs} = \begin{bmatrix} \hat{p} \\ \hat{M}_l \end{bmatrix} - \begin{bmatrix} -(\lambda_1 + \lambda_2)\,\Theta - d \\ -\lambda_1\lambda_2\Theta \end{bmatrix} \bar{q}\,.
$$

Then, the observer error dynamics with $e_1 = \hat{p} - p$, $e_2 = \hat{M}_l - M_l$ and eigenvalues λ_1, λ_2 in the left halfplane are given by

$$\begin{bmatrix} \dot{e}_1 \\ \dot{e}_2 \end{bmatrix} = \underbrace{\begin{bmatrix} \lambda_1 + \lambda_2 & -1 \\ \lambda_1 \lambda_2 & 0 \end{bmatrix}}_{A_{obs}} \begin{bmatrix} e_1 \\ e_2 \end{bmatrix} .$$

Since A_{obs} is Hurwitz there exists a positive definite solution P^{-1} of the Lyapunov type equation

$$P^{-1} A_{obs}^T + A_{obs} P^{-1} + 2Q = 0$$

for any positive definite Q. This leads to the decomposition $A_{obs} = (J_o - Q) P$ with $J_o = (A_{obs} + QP) P^{-1}$. This observer is included into an overall desired PH structure (J_a, R_a, H_a) for the extended state $\bar{x}_e = (\bar{q}, p, \bar{z}_{h1}, e_1, e_2, \bar{z}_{h2})$ with some constants $\Gamma_1, \Gamma_2 > 0$, a positive definite augmented Hamiltonian

$$\begin{aligned} H_a &= \frac{p^2}{2\Theta} + V - \frac{\partial_q V(q)\big|_{q=\breve{q}}}{\partial_q x_p(q)\big|_{q=\breve{q}}} \left(x_p - x_p(\breve{q}) \right) \\ &\quad + \Gamma_1 \frac{\bar{z}_{h1}^2}{2} + E \int_{x_p(\breve{q})}^{x_p(\breve{q}+\bar{q})} \left(A_1 \ln \left(\frac{V_{h1}(\breve{x}_p + \tau)}{\breve{V}_{h1}} \right) \right. \\ &\quad \left. + A_2 \ln \left(\frac{\breve{V}_{h2}}{V_{h2}(\breve{x}_p + \tau)} \right) \right) d\tau + \Gamma_2 \frac{\bar{z}_{h2}^2}{2} + \frac{1}{2} e^T P e \end{aligned}$$

and

$$J_a = \begin{bmatrix} 0 & 1 & 0 & 0 & 0 \\ -1 & 0 & \beta + \gamma & 0 & 0 \\ 0 & -\beta - \gamma & 0 & -\delta^T & 0 \\ 0 & 0 & \delta & J_o & 0 \\ 0 & 0 & 0 & 0 & 0 \end{bmatrix}, \quad \tilde{\delta} = \begin{bmatrix} \tilde{\delta}_1 \\ \tilde{\delta}_2 \end{bmatrix},$$

$$R_a = \begin{bmatrix} 0 & 0 & 0 & 0 & 0 \\ 0 & d & -\beta + \gamma & 0 & 0 \\ 0 & -\beta + \gamma & \frac{k_{p1}}{\Gamma_1} & \delta^T & 0 \\ 0 & 0 & \delta & Q & 0 \\ 0 & 0 & 0 & 0 & \frac{k_{p2}}{\Gamma_2} \end{bmatrix}, \quad \delta = P^{-1} \tilde{\delta}$$

with

$$\begin{aligned} \beta &= \frac{\partial_q x_p(q)\big|_{q=\breve{q}+\bar{q}}}{2\Gamma_1}, \quad \gamma = \Theta \frac{k_d}{2} \\ \tilde{\delta}_1 &= -\frac{k_d}{2}, \quad \tilde{\delta}_2 = -\frac{k_p}{2\, \partial_q x_p(q)\big|_{q=\breve{q}+\bar{q}}} . \end{aligned}$$

This choice guarantees local asymptotic stability of the closed loop due to

$$\dot{H}_a = -\partial_{\bar{x}_e} H_a R_a \partial_{\bar{x}_e} H_a^T$$

and LaSalle's invariance principle if $k_{p1}, k_{p2}, k_d > 0$, and if

$$\frac{k_{p1}d}{\Gamma_1} - \left(\frac{\partial_q x_p(q)\big|_{q=\check{q}+\bar{q}}}{2\Gamma_1} - \Theta \frac{k_d}{2} \right)^2 \geq 0$$

is fulfilled. These conditions are obtained by the decomposition R_a into the sum of positive semidefinite matrices. The control law is given by

$$\begin{aligned} u_1 &= -k_{p1}\left(z_{h1} - \frac{1}{\partial_q x_p(q)} \left(\hat{\bar{u}}_{obs} + \partial_q V \right) \right) \\ &\quad -k_{p1} E A_1 \ln \left(V_{h1} V_{h2}^{-\frac{A_2}{A_1}} \right) \Bigg|_{q=\check{q}} - k_d \Theta \hat{p} \\ u_2 &= -k_{p2} \bar{z}_{h2} \end{aligned}$$

with the observer equations and the transformations eq. (2a). The tuning parameters of the controller are the proportional gains k_{p1}, k_{p2}, the damping gain k_d and the eigenvalues of the observer. The performance of the energy-based dynamic controller with damping compared to the pure static controller is demonstrated by the simulations in Figure 1. At the scaled time value 0.5 an external load torque is applied, which is removed at 1.5. At the time values 0.1 and 0.8 the tracking signal is changed.

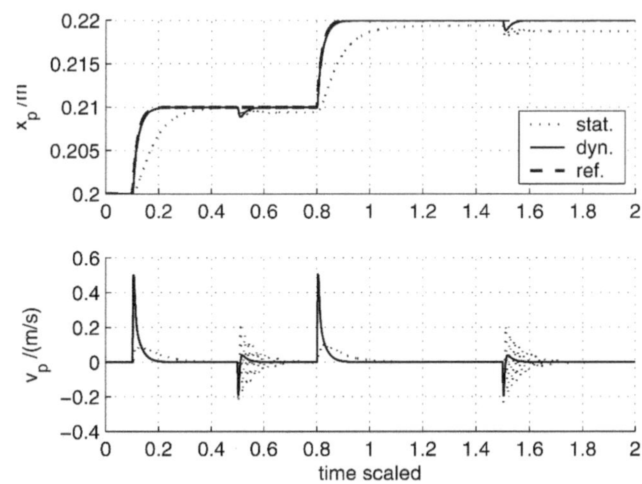

Figure 1. Wrapper position (x_p) and velocity (v_p) responses with static and dynamic controller.

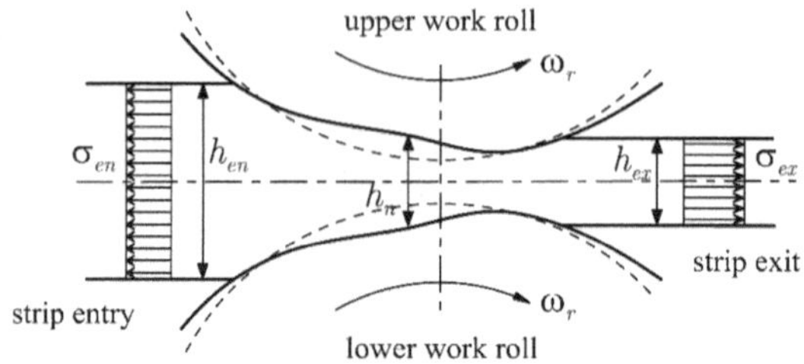

Figure 2. Deformation of the strip in the roll-gap.

2. THIRD OCTAVE CHATTER

Particularly, in thin product cold rolling mill plants, where the material deformation takes place under a considerable tension, so-called third octave chatter vibrations are observed frequently. These vibrations are characterized by a nearly exponential increase of the vibration amplitude and a frequency of about 100–150 Hz. This phenomenon would inevitably lead to breaks of the rolled material or damages of the rolling mill plant since it can be explained by an unstable operating point of the mechanical arrangement. This fact as well as two active rejection approaches are illustrated by means of a mill plant model.

2.1 Mathematical Model of a Rolling Mill Plant

The process of rolling typically takes place in several successive deformation steps under the effect of an adequate tension to influence the material properties and surface quality systematically. Therefore, the mathematical model of such a rolling mill plant contains the individual mill stands including the deformation process and the material to roll, also called strip. The mill stand model we use here, is based on a finite element calculation of the mill housing where also all rolls and roll chocks are considered as separate rigid bodies coupled elastically. Moreover, friction effects between the roll chocks and the mill housing are taken into account by Coulomb, viscous and sticking friction. The material deformation of the strip is effectively accomplished by the work rolls in the so-called roll-gap. Under the action of the roll force F_r and the entry as well as the exit strip tensions σ_{en}, σ_{ex} the strip, which enters the roll-gap with the entry thickness h_{en}, is deformed elasto-plastically in order to achieve the exit thickness h_{ex} (see Figure 2). This process is mathematically described by a so-called roll-gap or roll force model as one can find in the literature, see, e.g., Bland et al. (1951). Especially for thin strip rolling the assumption of a circular work roll deformation is not valid anymore. Therefore, in Fuchshumer et al. (2004) a non-circular roll/strip contact arc is considered and a Ritz approximation of the radial and circumferential displacement fields is proposed. In general, the dynamics of the strip deformation are sufficiently fast such that

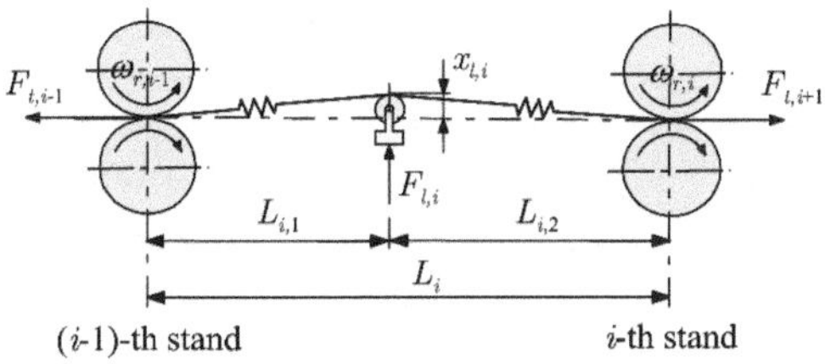

Figure 3. On the coupling of adjacent mill stands and the arrangement of a looper.

it is appropriate to set up a steady-state roll-gap model. The obtained model follows as an implicit algebraic equation of the form

$$f_{RGM}\left(F_r, h_{en}, h_{ex}, \sigma_{en}, \sigma_{ex}\right) = 0\,. \tag{5}$$

The strip element between two adjacent mill stands couples the individual deformation processes on the one hand by the strip thickness and on the other hand by the strip tension. It is assumed that the material behavior of the strip element is linear-elastic with Young's modulus E_s and furthermore, the eigenfrequencies of the distributed parameter system are much higher than the dynamics of the considered system. Thus, the tension relation of the i-th strip element reads as

$$\frac{\mathrm{d}}{\mathrm{d}t}\sigma_{s,i} = \frac{E_{s,i}}{L_{s,i}}\left(v_{en,i} - v_{ex,i-1}\right)\,, \tag{6}$$

where $L_{s,i}$ is the length of the strip element and $v_{ex,i-1}$, $v_{en,i}$ are the velocity of the strip at the exit of the $(i-1)$-th stand as well as at the entry of the i-th stand, respectively, see Figure 3. The interaction of the roll force, the strip tensions, and the strip thicknesses, described by (5) and (6), allow an explanation of third octave chatter occurrence by an instability of the nonlinear mechanical interconnected rolling mill stand. The occurrence of the instability clearly depends on the operation point of the mill plant. For a more detailed discussion of this observation, the reader is referred to, e.g., Holl et al. (2004b).

In Figure 3, additionally, a so-called looper system is considered which is acting on the strip in order to adjust the length of the strip between adjacent mill stands with high dynamics. For modelling it is assumed that the looper is permanently in contact with the strip. The mathematical model of the looper arrangement as well as of the main drives follows as an Euler–Lagrange system and from the strip elongation x_s we derive

$$\begin{aligned}\dot{x}_{s,i} &= \frac{h_{n,i}}{h_{en,i}}R_i\omega_i - \frac{h_{n,i-1}}{h_{ex,i-1}}R_{i-1}\omega_{i-1} \\ &\quad + \left(\frac{1}{\sqrt{L_{i,1}^2 + x_{l,i}^2}} + \frac{1}{\sqrt{L_{i,2}^2 + x_{l,i}^2}}\right)x_{l,i}v_{l,i}\,,\end{aligned} \tag{7}$$

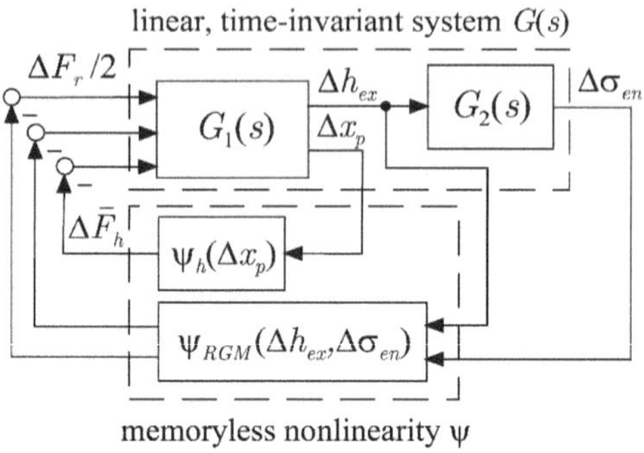

Figure 4. On the stability investigation of the mill plant.

where x_l, v_l denote the displacement and the velocity of the looper, and R, ω the radius and the angular velocity of the work rolls.

2.2 Nonlinear Control

In this contribution it is the intention to discuss two independent approaches for active chatter rejection. The first concept means an add-on system for existing plant and uses the hydraulic adjustment system, which effects primarily the thickness and the quality of the rolled product, to inject additional damping into the mechanical system by nonlinear control and therefore, to stabilize the mill plant system. In combination with the static position control law (4) this can be obtained by an additional flow to the cylinder chamber indicated by index 1 in the form

$$q_{v1,DI} = -\alpha V_{h1} a_p \,, \tag{8}$$

where a_p denotes the piston acceleration relative to the cylinder and α is the tuning parameter of the controller (Holl et al., 2004). In order to show the stability of the closed loop system, which contains the finite element mill stand model including the roll chocks under the assumption of pure viscous friction, consider a change of coordinates such that the operating point is shifted to the origin. In these coordinates, denoted by Δ, the mill stand system including the control law (8) can be represented as a feedback connection of a linear dynamic system and a memoryless nonlinear function as depicted in Figure 4. Here, the transfer function $G_1(s)$ contains the mill stand model and $G_2(s)$ the linear elastic strip element (6). The static nonlinear functions

$$\begin{aligned}
\psi_h &= \Delta F_h - \Delta z_{h1} = \Delta \bar{F}_h = \sum\nolimits_i (-1)^{i-1} EA_i \ln\left(V_{hi}\left(\Delta x_p\right) V_{pi}^{-1}\right) , \\
\psi_{RGM} &= \left[-\frac{1}{2}\hat{f}_{RGM}\left(\Delta h_{ex}, \Delta\sigma_{en}\right), -\frac{1}{2}\hat{f}_{RGM}\left(\Delta h_{ex}, \Delta\sigma_{en}\right)\right]^T
\end{aligned}$$

take into account the nonlinearities of the hydraulic control system, cf. (2a) and (Kugi, 2001), as well as of the roll-gap model (5) written in an explicit form under the assumption of constant entry strip thickness and exit strip tension, $F_r = \hat{f}_{RGM}(\Delta h_{ex}, \Delta\sigma_{en})$. In this representation the proof of stability is a direct consequence of the circle criterion (Vidyasagar, 1993; Khalil, 2002).

In the second approach we overcome the instability caused by the tension feedback loop by means of a strip tension control. For this purpose we suppose to have a looper system at our disposal and the force acting on the looper as well as the torque of the main mill drives are used as control inputs. With regard to a flatness-based controller design the quest for a flat output can be accomplished by the fact that a system that has the property that it is state feedback equivalent to a linear time-invariant one (see, e.g., Isidori, 1995), implies the property of differential flatness and the associated output y is also a flat output (Fliess et al., 1995; Rudolph, 2003). Considering the mathematical model of a coupled mill plant system it can be shown that for two adjacent mill stands the particular output

$$y = \left(\frac{E_{s,i}}{L_{s,i}} x_{s,i},\ \omega_{r,i},\ x_{s,i} - \sqrt{L_{i,1}^2 + x_{l,i}^2} - \sqrt{L_{i,2}^2 + x_{l,i}^2} \right) \tag{9}$$

represents a flat output (Holl et al., 2004a). It has to be mentioned that the flat output has the physical meaning of the strip tension, the angular velocity of the i-th stand, and the strip elongation according to the main drives. Moreover, it is not difficult to verify that this flatness-based approach is also appropriate for a plant with more than two mill stands. The concept of differential flatness allows in addition a systematic approach for solving the trajectory tracking problem. For the asymptotic stabilization of the trajectory tracking error $e_i = y_i - y_{i,d}$, $i = 1, \ldots, m$, the dynamics of the tracking error are adjusted by m linear, time-invariant differential equations. It is worth mentioning that the torque required for the strip deformation is known only insufficiently with respect to an exact input-to-state linearization. Therefore, to obtain stationary accuracy for the flat output an integral term is added to the control law.

Finally, in Figure 5 the concepts of active chatter rejection are illustrated in simulation. In order to demonstrate the energy-based controller at time $t = 0.5$ s a disturbance in the strip entry thickness of 10 μm (e.g., due to a welding seam) is assumed and additionally at time $t = 0.5$ s a step in the reference piston position signal of 50 μm is applied to the hydraulic adjustment system. The performance of the disturbance rejection by means of the flatness-based control is nearly the same as for the energy-based approach. Moreover, a change of the operating point can be applied by means of a reference trajectory, see also Figure 5.

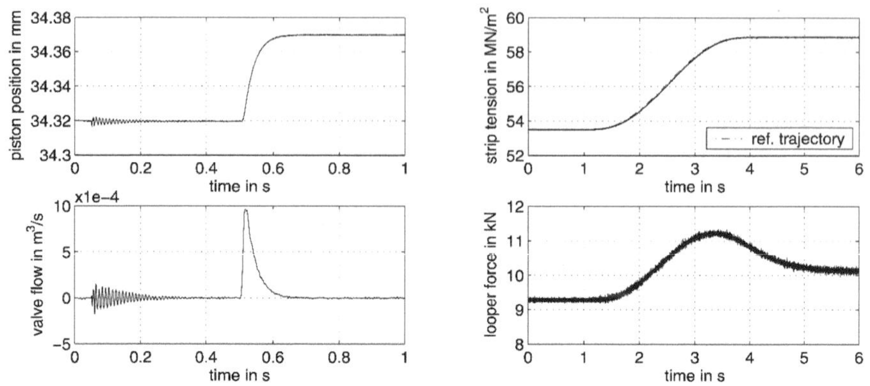

Figure 5. Simulation of active chatter rejection by means of the hydraulic actuator (left) and the looper system (right).

3. CONCLUSIONS

This contribution deals with two applications of active vibration rejection by model-based nonlinear control in steel rolling industry. Considerations of a mechanical Euler–Lagrange system driven by a hydraulic actuator motivate an energy-based nonlinear controller design for vibration rejection. Moreover, for the control of a multi-stand rolling mill plant by specifying desired operational sequences a flatness-based approach is discussed. Finally, it has to be mentioned that the presented methods are protected by patent.

REFERENCES

Bland, D.R., Ford, H. and Ellis, F. (1951). Cold rolling with strip tension. part I. *Journal of Iron and Steel Institute*, pp. 57–72.

Fliess, M., Levine, J., Martin, P. and Rouchon, P. (1995). Flatness and defect of non-linear systems: Introductory theory and examples. *Internat. J. Control*, 61:1327–1361.

Fuchshumer, S., Schlacher, K. and Keintzel, G. (2004). A novel non-circular arc rollgap model, designed from the control point of view. In *Proc. MMM04*, Nancy, France, to appear.

Grabmair, G., Schlacher, K. and Keintzel, G. (2004). Energy-based control of hydraulic actuators in steel rolling mills. In *Proc. MMM04*, Nancy, France, to appear.

Grabmair, G., Schlacher, K. and Kugi, A. (2003). Geometric energy based analysis and controller design of hydraulic actuators applied in rolling mills. In *Proc. ECC2003*.

Holl, J., Schlacher, K. and Keintzel, G. (2004a). Active rejection of chatter-vibrations in steel rolling mills. In *Proc. Third European Conference on Structural Control*, Vienna, Austria, to appear.

Holl, J., Schlacher, K. and Keintzel, G. (2004b). Chatter rejection by nonlinear control in steel rolling mills. In *Proc. MMM04*, Nancy, France, to appear.

Isidori, A. (1995). *Nonlinear Control Systems*, 3rd edition, Springer.

Khalil, H. (2002). *Nonlinear Systems*, 3rd edition. Prentice Hall.

Kugi, A. (2001). *Nonlinear Control Based on Physical Models*. Lecture Notes in Control and Information Sciences 260. Springer.

Rudolph, J. (2003). *Flatness Based Control of Distributed Parameter Systems*. Shaker.

Vidyasagar, M. (1993). *Nonlinear Systems Analysis*, 2nd edition. Prentice Hall.

ACTIVE CONTROL FOR LIGHTWEIGHT ISOLATION SYSTEMS

Kazuto Seto, Masahiko Naruke, Taichi Watanabe and Hiroko Morino
Department of mechanical Engineering, College of Science and Technology, Nihon University, 1-8 kanda Surugadai, Chiyoda-ku, Tokyo, Japan

Abstract: The isolation performance of a scaled three-dimensional active isolation system with a low stiffness isolation table assumed as a large-size, lightweight isolation system is reported. In a large-size but lightweight isolation table, there are multi vibration modes in a lower frequency region in the operational frequency. Thus, a 5-DOF reduced order model of control using a reduced order physical modeling method is applied for effectively controlling multi mode vibration in the vertical direction. In the horizontal direction, a rigid body model is created. For controller design, a 2-DOF controller combining a feedback controller with a feedforward controller is designed.

Key words: active isolation, vibration control, lightweight table, modelling, reduced order model, LQ control, feedforward control.

1. INTRODUCTION

Recently, ultra precision machining apparatus, like semiconductor manufacturing apparatus, have tended to require super precision processing. To help achieve this, environmental vibrations around such apparatus must be significantly reduced to obtain highly accurate processing. There have been many studies [1, 2] carried out on active isolation systems to reduce the effects of vibrations on apparatus. However, these studies have dealt exclusively with isolation systems with rigid bodies. In the future, expansion of the available fields and enlargement of processing machines sizes are expected; therefore, isolation systems with rigid bodies will not be realistic, because a lot of energy is required to support such an isolation system.

This paper focuses on the basic research needed to realize active isolation systems that are large size yet lightweight [3, 4]. As the tables of isolation systems become larger in size and lighter in weight, they become flexible and the elastic modes on the isolation table appear at a lower frequency region. Therefore, it is necessary to control the elastic vibrations of the table. On

H. Ulbrich and W. Günthner (eds), IUTAM Symposium on Vibration Control of Nonlinear Mechanisms and Structures, 297–307.

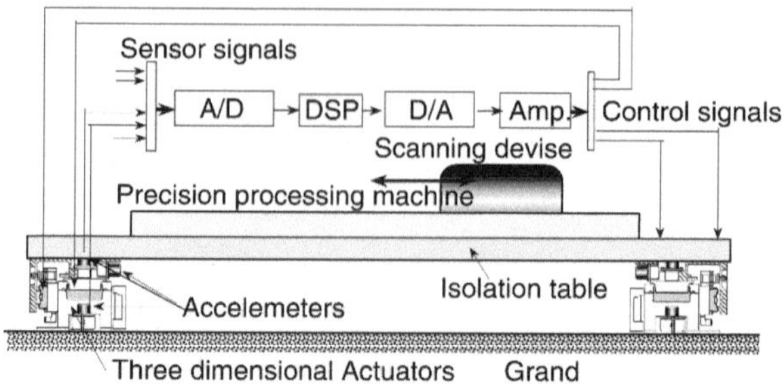

Figure 1. Planned large size of isolation system.

one hand, a lightweight isolation table is susceptible to vibrations of a loaded structure; therefore, it is necessary to consider the dynamic characteristics of a loaded structure in order to prevent any performance decrement of an isolation system caused by vibrations of the loaded structure.

In this paper, a scaled isolation system with a lightweight low stiffness table is studied. In the vertical direction, a 5-DOF lumped mass model to control the first fifth vibration modes is created by using a reduced order physical modelling method [5]. In the horizontal direction, a rigid body model is created. By combining these models, a 3-dimensional model of an active isolation system is constructed. On consideration of the isolation table with a load structure, isolation performance improvements are studied in the horizontal direction. The control system uses a 2-DOF control [6, 7] which includes a feedback controller based on LQ control and a feedforward controller for vibration isolation. The effectiveness of the vibration control system for controlling a scaled isolation system is investigated through numerical calculations and experimentation.

2. ACTIVE ISOLATION SYSTEM WITH LOW STIFFNESS TABLE

2.1 Constriction of Its Active Isolation System

A planned active isolation system with a large size isolation table is shown in Figure 1. In this case, a length of 20 m is required to load a long precision processing machine with a scanning device. The system needs to control the elastic modes on the isolation table appearing at a lower frequency region, along with a directly loaded disturbance due to the scanning device. Since the scanning device moves quickly to work processing other precision devices, it is necessary to isolate any influence of the moving mass. Such problems will be happen in reality in the future.

To resolve the above-mentioned examples, a scaled isolation system with a lightweight table is constructed. The experimental setup and the main as-

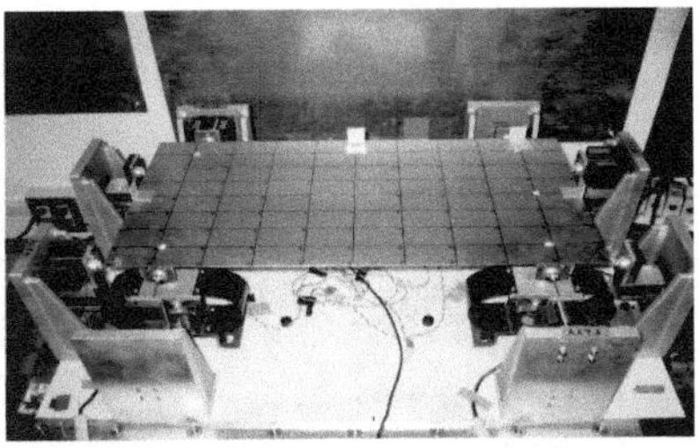

Figure 2. Outside view of active isolation system.

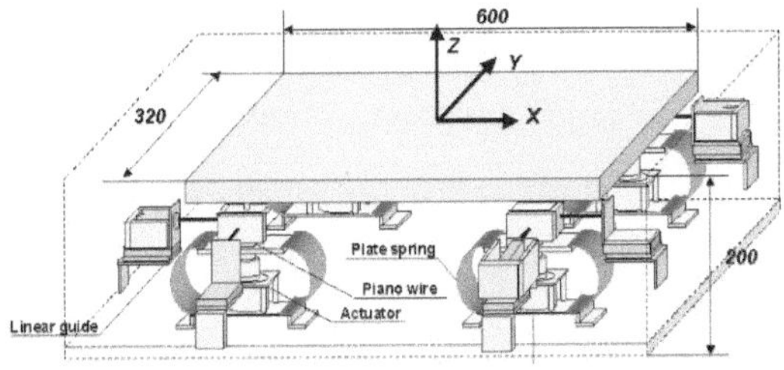

Figure 3. Dimensions of main assembly.

sembly are shown in Figures 2 and 3, respectively. This 3-dimensional active isolation system is used as the control object in this research paper. The isolation table uses an aluminum plate 0.6 m in length, 0.32 m in width and 0.002 m in thickness. The isolation table is supported at twelve points (four vertical and eight horizontal) by voice-coil motor actuators, magnet coil actuators and linear guides as shown in Figure 3. Since they are joined with piano wire, the table and actuator can be moved in both vertical and horizontal directions without any interaction. To measure the absolute displacement from the ground and mass points, laser sensors are located at the ground and the modeling point on the table.

2.2 Vibration Mode Shapes of the Low Stiffness Table

For constructing the lightweight isolation table, it is important to examine the effect of the vibration of the loaded structure on isolation performance. Vibration mode shapes of the isolation table in the vertical direction is shown in Figure 4 and in the horizontal direction in Figure 5. In this research, we consider these modes to be controlled.

3. MODELING APPROACH

3.1 Reduced Order Modeling Method

In this research, the reduced order physical modelling method is applied. The key points of the modelling procedure are:

(1) Analyse the vibration mode shapes of the flexible structures.

(2) Determine the strategically selected mass points on the vibration modes.

(3) Make a temporal modal matrix constricted by modal data obtained at modelling points.

(4) Determine the values of masses and spring constants using modified modal matrix.

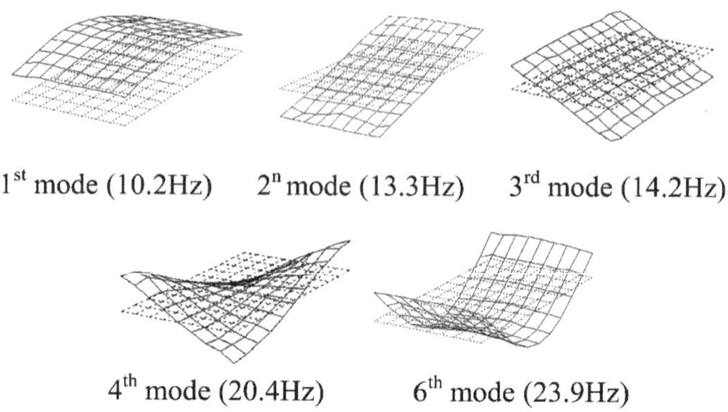

Figure 4. Vibration mode shapes in the vertical direction.

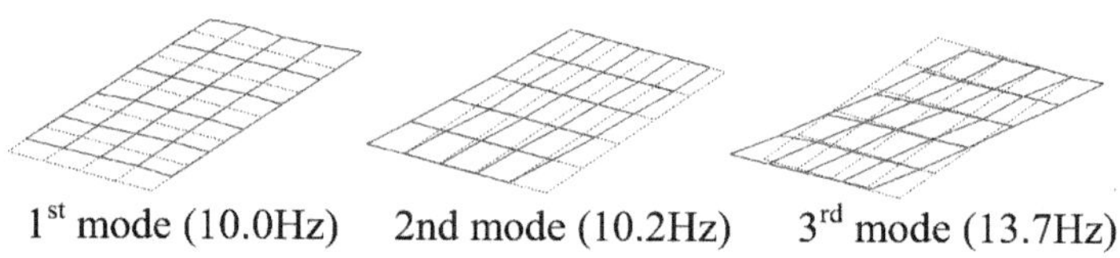

Figure 5. Vibration mode shapes in the horizontal direction.

Normalizing the modal matrix Φ yields the following set of relationships for the mass $\boldsymbol{M}$ and stiffness matrixes $\boldsymbol{K}$ of the physical domain:

$$\boldsymbol{M} = (\boldsymbol{\Phi\Phi}^T)^{-1}, \tag{1}$$

$$\boldsymbol{K} = (\Phi^T)^{-1}\Omega^2\Phi^{-1}, \tag{2}$$

where Ω^2 is a diagonal matrix given by the square of the natural frequencies of each mode. However, the modal matrix directly obtained from the FEM modal analysis data or an experimental modal analysis is not guaranteed to satisfy Equation (1), where both sides should be a diagonal matrix. Therefore, Φ means a temporal modal matrix for making a lumped parameter model. Substituting Φ into Equation (1) results in a matrix, $\Phi\Phi^T$, whose off-diagonal positions usually contain non-zero values defined as an error vector ε. By defining the modification vector of Φ as $\delta\Phi$, which is assumed to change the error vector from its initial value to 0, the correcting procedure then start and the new modal matrix is calculated by $\Phi + \delta\Phi \rightarrow \Phi$. Iterative calculation is usually necessary in this procedure so that converges to 0. After a certain iteration so that a preset accuracy has been attained, a final modified modal matrix can be obtained that will satisfy the diagonal conditions of Equation (1). Consequently, $\boldsymbol{M}$ and $\boldsymbol{K}$ can be obtained.

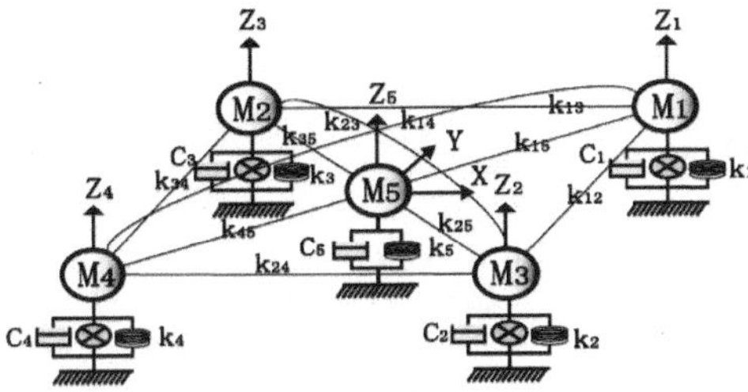

$M_1 = 0.537$ kg	$C_5 = 2$ Ns / m	$k_{25} = -2.069$ kN / m
$M_2 = 0.562$ kg	$k_1 = 3.521$ kN / m	$k_3 = 3.841$ kN / m
$M_3 = 0.534$ kg	$k_{12} = -0.647$ kN / m	$k_{34} = -0.588$ kN / m
$M_4 = 0.577$ kg	$k_{13} = -0.877$ kN / m	$k_{35} = -2.234$ kN / m
$M_5 = 0.464$ kg	$k_{14} = 1.630$ kN / m	$k_4 = 3.708$ kN / m
$C_1 = 6$ Ns / m	$k_{15} = -2.145$ kN / m	$k_{45} = -2.234$ kN / m
$C_2 = 7$ Ns / m	$k_2 = 4.087$ kN / m	$k_5 = -1.698$ kN / m
$C_3 = 7$ Ns / m	$k_{23} = 1.788$ kN / m	
$C_4 = 7$ Ns / m	$k_{24} = -0.884$ kN / m	

Figure 6. Vertical direction model.

3.2 Modeling in the Vertical Direction

In the vertical direction, the isolation system essentially has a flexible table with a finite number of vibration modes. To express the first to fifth modes of the table, reduced order physical modelling method is applied, so that the temporal modal matrix takes the following form:

$$\Phi = \begin{bmatrix} \phi_{11} & \cdots & \phi_{15} \\ \vdots & \ddots & \vdots \\ \phi_{51} & \cdots & \phi_{55} \end{bmatrix}. \tag{3}$$

Mass points selected to equal numbers of controlling modes are placed on the table by this method. The centre point corresponds to the maximum amplitude point in the 5th mode, and the fixed points of four actuators in the vertical direction are selected as modelling points. The temporal modal matrix is constructed on the modelling points based on the vibration mode shapes shown in Figure 4. Using the above indicated modelling procedure, a 5-DOF reduced order model is obtained as shown in Figure 6. In this figure, five masses are combined with springs and dampers. These parameters are also indicated in Figure 6. Then, after obtaining the final modified modal matrix, the parameters of the masses and springs are determined by Equations (1) and (2). Since the parameters of the dampers have not yet been obtained at this stage, they are assumed from the peaks of the measured frequency responses.

3.3 Modeling in the Horizontal Direction

In the horizontal direction, the model is created as a rigid body, because the vibration modes of the table are located at a frequency range that is high enough to be negligible when considering frequency. A rigid body model and each of the physical parameters are shown in Figure 7.

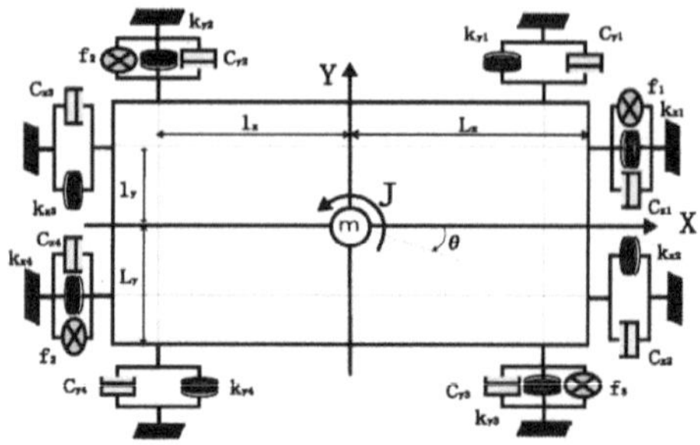

$m = 2.480$ kg $\quad ,J = 0.1$ kgm^2
$L_x = 0.3$ m $\quad ,I_x = 0.24$ m^4
$L_y = 0.16\ m \quad ,I_x = 0.12$ m^4
$k_{x1} = k_{x2} = k_{x3} = k_{x4} = 2.448$ kN / m
$k_{y1} = k_{y2} = k_{y3} = k_{y4} = 2.547$ kN / m
$C_{y1} = C_{y2} = C_{y3} = C_{y4} = 5.5$ Ns/ m
$C_{y1} = C_{y2} = C_{y3} = C_{y4} = 7.0$ Ns / m

Figure 7. Horizontal direction model.

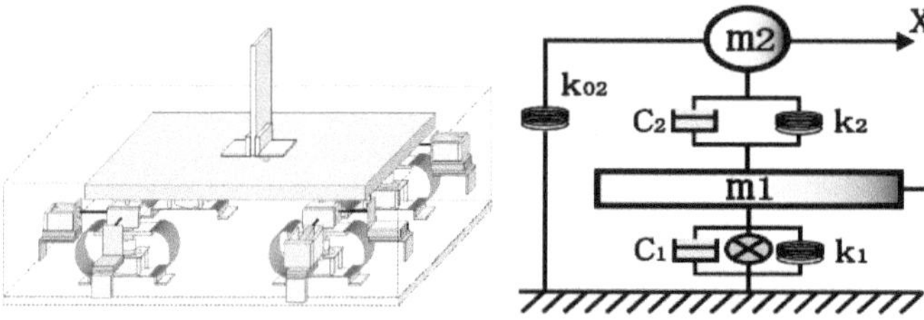

$m_1 = 1.495$ kg
$m_2 = 0.0084$ kg
$C_1 = 14$ Ns / m
$C_2 = 0.012$ Ns / m
$k_1 = 5.728$ kN / m
$k_2 = 0.031$ kN / m
$k_{02} = 0.047$ kN / m

Figure 8. 2-DOF model combined table with loaded structure.

3.4 Modeling of the Table with Loaded Structure

Vibration modes of a loaded structure often exist in a low frequency range; thus, it is necessary to consider the influence of a loaded structure on the isolation performance of the table. Therefore, a thin aluminium plate, 0.25 m in length, 0.05 m in width and 0.001 m in thickness, is installed at the center of the table as the loaded structure. In this research, the 1st mode of the loaded structure is considered for control. Because the influence of the loaded structure strongly acts in the horizontal direction, a 2-DOF model which combines a 1-DOF model of the table with a 1-DOF model of the loaded structure is expressed in the X-direction as shown in Figure 8.

3.5 Verification of the Constructed Model

To verify the validity of these models, simulated frequency responses are compared with the experimental frequency ones. Figure 9. shows the transfer function from exciting the mass point 1 to measuring the mass point 1 in the vertical direction. Figure 10 shows the transfer function from exciting the table to measuring the table in the X-direction. Figure 11 shows the transfer function from exciting the table to measuring the loaded structure in the X-direction. These figures show that the simulation results agreed well with the experimental ones. It is confirmed that both of the constructed models well explain the dynamic behaviour of the table.

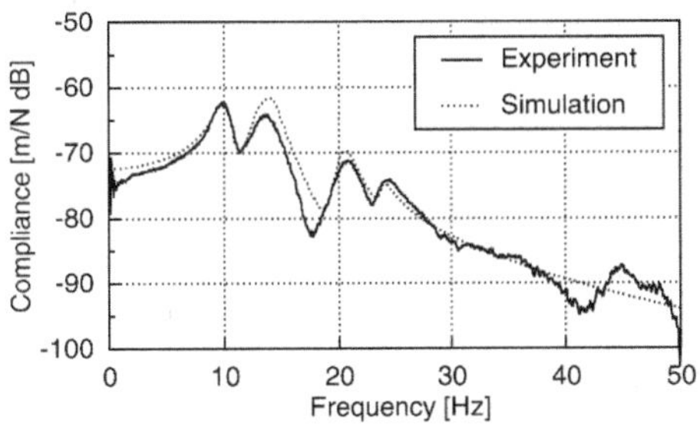

Figure 9. Validity of the vertical model.

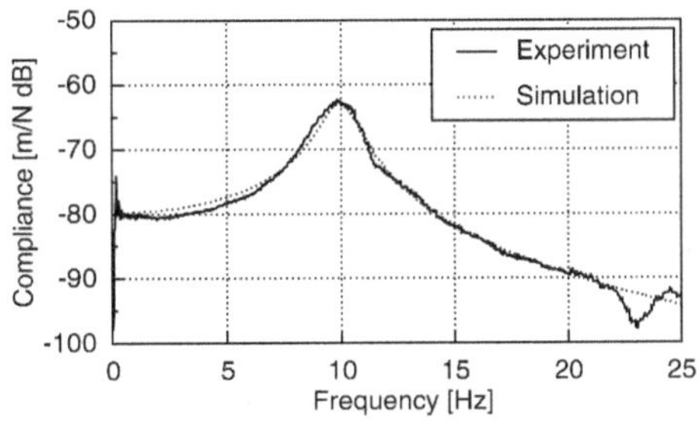

Figure 10. Validity of the horizontal model.

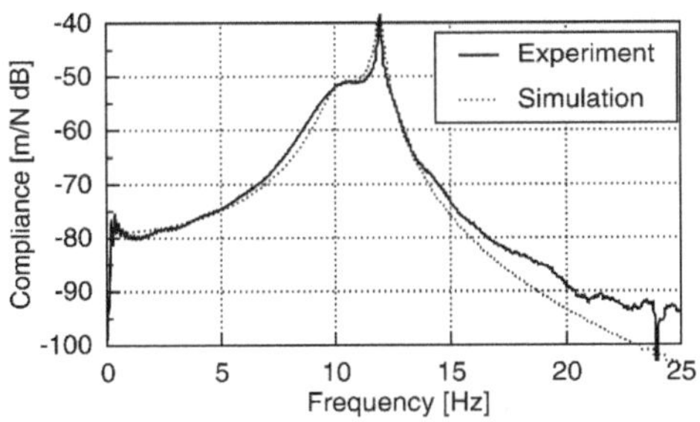

Figure 11. Validity in case of loading structure.

4. CONTROL SYSTEM DESIGN

4.1 State Equation of Controlled Objects

For considering the dynamics of actuators with inductance L, resistance R, coefficients of driving and inverse electromotive force K_c and K_a, a state equation of the controlled object in the vertical direction can be expressed as

$$\dot{\boldsymbol{X}}_v = \boldsymbol{A}_v\boldsymbol{X}_v + \boldsymbol{B}_v\boldsymbol{u}_v, \tag{4}$$

$$\boldsymbol{Y}_v = \boldsymbol{C}_v\boldsymbol{X}_v, \tag{5}$$

where the state vector $\boldsymbol{X}_v$, control vector $\boldsymbol{u}_v$, coefficient matrixes $\boldsymbol{A}_v$, $\boldsymbol{B}_v$ and $\boldsymbol{C}_v$ are as follows:

$$\boldsymbol{X}_v = [\dot{z}_1 \ \ \dot{z}_2 \ \ \dot{z}_3 \ \ \dot{z}_4 \ \ \dot{z}_5 \ \ z_1 \ \ z_2 \ \ z_3 \ \ z_4 \ \ z_5 \ \ I_{v1} \ \ I_{v2} \ \ I_{v3} \ \ I_{v4}]^T,$$

$$\boldsymbol{v}_v = [u_{v1} \ \ u_{v2} \ \ u_{v3} \ \ u_{v4}]^T,$$

$$\boldsymbol{A}_v = \begin{bmatrix} -\boldsymbol{M}^{-1}\boldsymbol{C}(5\times 5) & -\boldsymbol{M}^{-1}\boldsymbol{K}(5\times 5) & \boldsymbol{A}_{13}(5\times 4) \\ \boldsymbol{I}(5\times 5) & \boldsymbol{0}(5\times 5) & \boldsymbol{0}(5\times 5) \\ \boldsymbol{A}_{31}(4\times 5) & \boldsymbol{0}(4\times 5) & \boldsymbol{0}(4\times 4) \end{bmatrix},$$

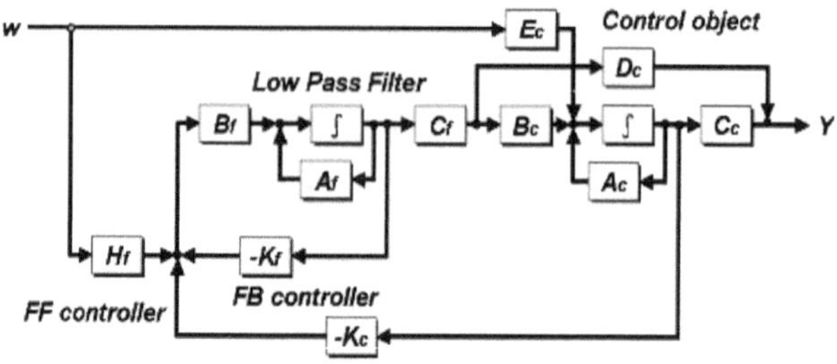

Figure 12. Block diagram of 2-DOF control system.

$$\boldsymbol{B}_v = \begin{bmatrix} \mathbf{0}(10 \times 4) \\ \mathrm{diag}\left[\frac{1}{L_{v1}} \ \frac{1}{L_{v2}} \ \frac{1}{L_{v3}} \ \frac{1}{L_{v4}}\right] \end{bmatrix}.$$

In the horizontal direction, the state equation is also obtained with subscript h using it in the same way.

$$\dot{\boldsymbol{X}}_h = \boldsymbol{A}_h \boldsymbol{X}_h + \boldsymbol{B}_h \boldsymbol{u}_h, \tag{6}$$

$$\boldsymbol{Y}_h = \boldsymbol{C}_h \boldsymbol{X}_h. \tag{7}$$

Variables shown in the following state vector are based on Figure 7.

4.2 2-DOF Control System Including Low Pass Filter

A model of a 3-dimensional active isolation system is constructed by combining these models. In this research, LQ control theory is used to control the multi modes vibration of the table. Moreover, feedforward control is applied to cancel ground disturbances. However, there is neglected higher order modes that can invite unstable spillover; to suppress this, a low pass filter is inserted in the control system. The block diagram of a 2-DOF control system combined with LQ based state feedback control with feedforward control is shown in Figure 12.

In this figure, subscripts f and c indicate matrixes in respect to the low pass filter and the control object, respectively. Moreover, the matrixes $\boldsymbol{W}$, $\boldsymbol{H}_f$ and $\boldsymbol{K}_c$ show a disturbance, feedforward gain and feedback gain, respectively. These gain matrixes are designed using the following expanded argument system:

$$\dot{\boldsymbol{X}} = \boldsymbol{A}\boldsymbol{X} + \boldsymbol{B}\boldsymbol{u}, \tag{8}$$

$$\boldsymbol{Y} = \boldsymbol{C}\boldsymbol{X}, \tag{9}$$

where the state vector and coefficient matrixes are follows:

$$\boldsymbol{X} = \{\boldsymbol{X}_{vf} \ \ \boldsymbol{X}_v \ \ \boldsymbol{X}_{hf} \ \ \boldsymbol{X}_h\}^T,$$

$$A = \begin{bmatrix} \boldsymbol{A}_{vf} & \mathbf{0} & \mathbf{0} & \mathbf{0} \\ \boldsymbol{B}_{vf}\boldsymbol{C}_{vf} & \boldsymbol{A}_{v} & \mathbf{0} & \mathbf{0} \\ \mathbf{0} & \mathbf{0} & \boldsymbol{A}_{hf} & \mathbf{0} \\ \mathbf{0} & \mathbf{0} & \boldsymbol{B}_{h}\boldsymbol{C}_{hf} & \boldsymbol{A}_{h} \end{bmatrix},$$

$$\boldsymbol{B} = \begin{bmatrix} \boldsymbol{B}_{vf} & \mathbf{0} \\ \boldsymbol{B}_{v} & \mathbf{0} \\ \mathbf{0} & \boldsymbol{B}_{hf} \\ \mathbf{0} & \boldsymbol{B}_{h} \end{bmatrix}, \quad \boldsymbol{C} = \begin{bmatrix} \boldsymbol{C}_{vf} & \mathbf{0} & \mathbf{0} & \mathbf{0} \\ \mathbf{0} & \boldsymbol{C}_{v} & \mathbf{0} & \mathbf{0} \\ \mathbf{0} & \mathbf{0} & \boldsymbol{C}_{hf} & \mathbf{0} \\ \mathbf{0} & \mathbf{0} & \mathbf{0} & \boldsymbol{C}_{h} \end{bmatrix}.$$

5. EXPERIMENT

5.1 Experimental Setup

The experimental setup is used for an active isolation system; its dimensions are shown in Figures 2 and 3. Four actuators of a voice-coil type linear motor with 10 N force capacity are used in the vertical direction. Four magnet coil type linear actuators composed of a pair of permanent magnets and a moving coil are used in the x- and y-horizontal directions, each direction having two sets. To measure the absolute displacements of the five mass points and the ground disturbance, six laser sensors are employed for the fundamental research. These could be replaced by accelerometers.

Displacement signals measured by the lasers are input to the controller through an A/D converter. The velocities of the control object are obtained by differentiating the displacement signals. DSP is used as the controller. The actuators are driven by amplified output signals from D/A converter generated control forces. The shaking table generates ground disturbances as an option.

5.2 Performance of LQ Based State Feedback Control

An isolation system must control not only vibrations but also direct disturbances to the isolation table. The initial vibration control results using LQ based state feedback as the isolation table is excited by impulse signal, are shown in Figure 13, which also shows compared frequency responses without and with LQ based state feedback control, impulsively excited at mass point 1 in the vertical direction. Figure 14 shows the same compared frequency responses in the X-direction. Since the resonance peaks are well suppressed, the effectiveness of the vibration control is demonstrated.

5.3 Performance of 2-DOF Control

In order to evaluate the effectiveness of the isolation of the vibrations, transmissibility of vibration from the ground disturbance to the measurements of

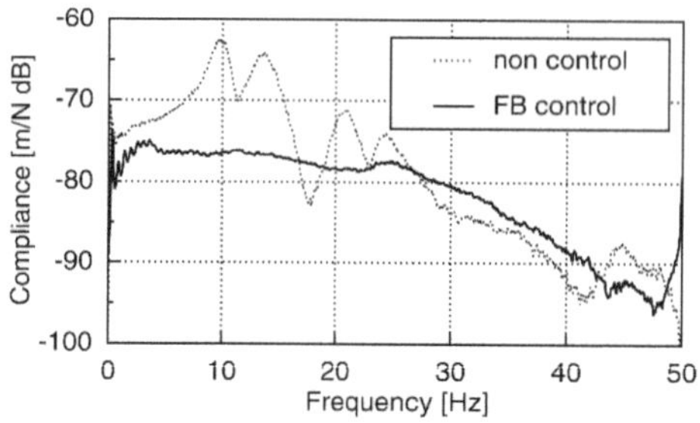

Figure 13. Impulse responses (vertical).

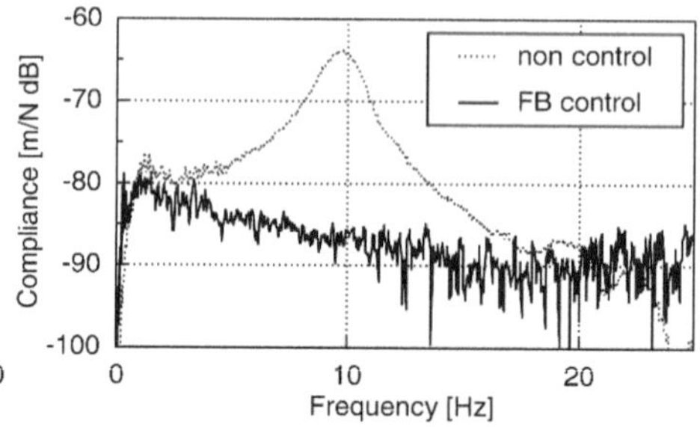

Figure 14. Impulse responses (horizontal).

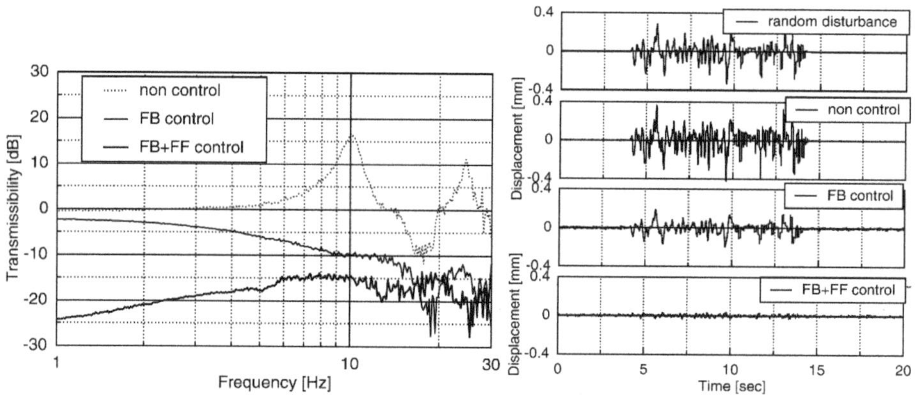

Figure 15. Transmissibility of vibration (vertical).

the table are obtained under a sinusoidal sweep. Time responses are obtained by inputting the white noise from the shaker. Figure 15 shows the responses in the vertical direction. With state feedback control, the resonance peaks of the table are well suppressed, but they are not so isolated, while the 2-DOF control system effectively achieves excellent vibration isolation.

5.4 Performance of the Isolation System to Loaded Structure

Finally, the isolation performance of the table for the loading structure is examined. Figure 16 shows the transmissibility of vibrations from the ground disturbance to the mass point of the loaded structure without and, by way of contrast, with the model that includes the dynamic characteristics of the loaded structure. Though the loaded structure resonates, the vibrations of the table and the loaded structure are well suppressed using the model, which includes the dynamic characteristics of the loaded structure.

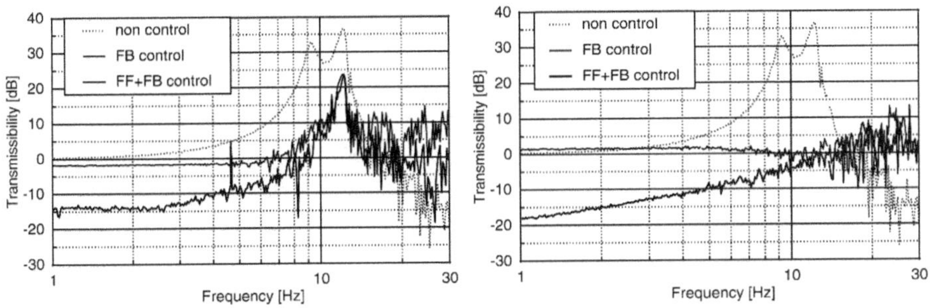

Figure 16. Without and with consideration of dynamics of loaded structure.

6. CONCLUSIONS

In this paper, a scaled isolation system with a lightweight low stiffness table was studied. In the vertical direction, a 5-DOF lumped mass model to control the first fifth vibration modes was created by using the reduced order physical modelling method; while in the horizontal direction, a rigid body model was created. The 3-dimensional active isolation system combining these models has effectively acted to obtain good isolation performance. The control system with a 2-DOF control system, which includes a feedback controller, based on LQ control and a feedforward controller was effective for this isolation system. When the isolation system has a moving or vibrated load structure, it is important to consider the dynamics in the modelling stage in order to obtain very high isolation performance.

REFERENCES

[1] Tagawa, Y., Seino, Y., Yasuda, M. and Fujita, T., "Active 6-DOF Micro vibration Control System Using Air Actuators", in *JSME, 2nd International Conference on Motion and Vibration Control (MOVIC)*, Vol. 1, 1994, pp. 13–18.

[2] Cui, W., Nonami, K. and Kanemitsu, Y., "Isolation Performance of Hybrid Isolation System by H Control and Disturbance Cancellation Control", in *JSME, 3rd International Conference on MOVIC*, Vol. 3, 1996, pp. 7–12.

[3] Shono, T., Nagasaka, N., Watanabe, T. and Seto, K., "Multi-Mode Control for an Isolation System with an Elastic Table", In *6th International Conference on MOVIC*, Vol. 1, 2002.

[4] Naruke, M., Nishi, M., Watanabe, T. and Seto, K., "Three-Dimensional Active Control for Lightweight Isolation Table Taken into Account of Vibration in Loaded Structure", in *7th International Conference on MOVIC*, CD-Rom, 2004.

[5] Seto, K. and Mitsuda, S., "New Method for Making a Reduced Order Model of Flexible Structures Using Unobservability and Uncontrollability and Its Application in Vibration Control", in *1st International Conference on MOVIC*, 1992, pp. 152–158.

[6] Ikeda, M. and Yasuda, M., "Double-Active Control of Micro Vibration Isolation Systems to Improve Performances", *The Japan Society of Mechanical Engineers, C*, 63-562, 1993, 1694–1701.

[7] Miura, T., Ikeda, M. and Yasuda, M., "Robust Position and Attitude Control of an Active Vibration Isolation", in *5th International Conference on MOVIC*, Vol. 2, 2000, pp. 469–474.

DYNAMICS AND CONTROL OF TENSEGRITY SYSTEMS

Robert Skelton
Mechanical and Aerospace Engineering, UCSD, USA
bobskelton@ucsd.edu

Abstract: Rather than the traditional vector differential equation, this paper introduces rigid body dynamics in a new form, as a matrix differential equation. For a system of n_b rigid bodies, the forces are characterized in terms of network theory, and the kinematics are characterized in terms of a directed graph of the connections of all members. The dynamics are characterized by a second order differential equation in a $3 \times 2n_b$ *configuration matrix*. The first contribution of the paper is the dynamic model of a broad class of systems of rigid bodies, characterized in a compact form, requiring no inversion of a variable mass matrix. The second contribution is the derivation of all equilibria as linear in the control variable. The third contribution is the derivation of a linear model of the system of rigid bodies. One significance of these equations is the exact characterization of the statics and dynamics of all class 1 tensegrity structures, where rigid bar lengths are constant and the string force densities are control variables. The form of the equations allow much easier integration of structure and control design since the control variables appear linearly. This is a significant help to the control design tasks.

Key words: tensegrity, rigid body dynamics.

1. INTRODUCTION

This paper introduces rigid body dynamics in a new form, as a matrix differential equation, rather than the traditional vector equation. This paper describes the dynamics and the static equilibria of a set of discontinuous rigid bodies, connected via a continuous set of strings to stabilize the system. In our theory, the "strings" are "springs" which can take compression or tension. However, in the special application of greatest interest, the "strings" can only take tension. All equilibria of such bar and string connections are described, and the dynamics of such systems are described in a new form, a second order differential equation of a $3 \times 2n_b$ matrix, called the *configuration matrix*. By parametrizing the configuration in terms of the components of vectors, the usual

H. Ulbrich and W. Günthner (eds), IUTAM Symposium on Vibration Control of Nonlinear Mechanisms and Structures, 309–318.

nonlinearities of angles, angular velocities and coordinate transformations are avoided. Indeed, there are no trigonometric functions in this formulation. We seek simplicity in the analytical form of the dynamics, for ease in designing control laws later. Among all the available equations for a system of rigid bodies, these equations produce the simplest form. Our model of dynamics is in a matrix form, opening new control research challenges to develop control theories for matrix models of the dynamics.

In the 60s and 70s, a variety of Newtonian and energy approaches (Hamilton and Lagrange) were introduced and traded for numerical efficiencies. NASA had great interest in building accurate deployable spacecraft simulations composed of a large number of rigid bodies in a topological tree. The typical form of these equations was vector-second order. For a large class of problems it is reasonable to assume that the rigid bodies are rod-shaped and have negligible inertia about their longitudinal axes. We will make this assumption.

1.1 Tensegrity Systems

Class k tensegrity systems are defined by the number (k) of rigid bodies that connect to each other (with frictionless ball joints) at a specific point (node). This paper entertains only class 1 tensegrity systems, so no rigid bodies are in contact, and the system is stabilized only by the presence of tensile members connecting the rigid bodies. In the steady state, such a system has only axially-loaded members, since the rigid bodies do not touch each other and the strings connected to the rigid bodies cannot apply torques at the site of the attachment. These features not only simplify the equations of motion, but the resulting models will be much more accurate than models of rigid bodies that are subject to bending moments. That is, the internal stresses in the rigid bodies have a specific direction.

Tensegrity systems have been around for over 50 years as an artform, with some architectural appeal, but analytical tools to design engineering structures from tensegrity concepts are still inadequate. The primary motivation for this paper is to provide a convenient analytical tool to describe both the statics and dynamics of class 1 tensegrity systems.

1.2 Notation

DEFINITION 1 *The set of vectors* $\underline{e}_i$, $i = 1, 2, 3$, *form a* dextral set, *if the dot products satisfy* $\underline{e}_i \cdot \underline{e}_j = \delta_{ij}$ *(where* δ_{ij} *is a Kronecker delta), and the cross products satisfy* $\underline{e}_i \times \underline{e}_j = \underline{e}_k$, *where the indices* i, j, k *form the cyclic permutations,* $i, j, k = 1, 2, 3$ *or* $2, 3, 1$, *or* $3, 1, 2$.

DEFINITION 2 *Let* $\underline{e}_i$, $i = 1, 2, 3$ *define a* dextral set *of unit vectors fixed in an inertial frame, and define the* vectrix $\underline{E}$ *by* $\underline{E} = [\ \underline{e}_1 \quad \underline{e}_2 \quad \underline{e}_3\ ]$.

The item we call $\underline{r}$ is a Gibbs vector. The items we call r^X and r^E are vectors in the linear vector spaces of linear algebra, where we use the notation, $r^X \in \mathbb{R}^3$ and $r^E \in \mathbb{R}^3$ to denote that the items r^X and r^E live in a real three-dimensional space. However, the items r^X and r^E tell us nothing unless we have previously specified the frames of reference $\underline{X}$ and $\underline{E}$ for these quantities. If we must assign a "dimension" to these quantities $\underline{X}$ and $\underline{E}$, then we must say they are 3×1 arrays, composed of the three elements $\underline{e}_i$, $i = 1, 2, 3$. However, these arrays contain quantities we call Gibbs vectors $\underline{e}_i$. So the 3×1 item $\underline{E}$ is not a *vector* in either the sense of Gibbs, nor in the sense of linear algebra. For these reasons Peter Hughes makes the logical choice to call the quantity $\underline{E}$ a *vectrix*.

Unlike many problems in aerospace, where multiple coordinate frames are utilized, this paper uses only one coordinate frame to describe all vectors. Since we always use the same frame of reference, the inertial frame, described by the vectrix $\underline{E}$, we will not complicate the notation of vectors with different superscripts, as would be required above to distinguish between components of a vector represented in different frames. Hence, we use the notation for the vector $\underline{n}_i$, as follows

$$\underline{n}_i = \underline{E} n_i^E, \qquad n_i^E = n_i, \tag{1}$$

where $n_i^T = [\, n_{i_1}, \quad n_{i_2}, \quad n_{i_3} \,]$ describes the components of the vector n_i in coordinates $\underline{E}$, where we have dropped the superscript E that would be used in the more complete and more general notation above (n_i^E), and we will write only n_i, hereafter, instead of n_i^E.

We generate a diagonal $n \times n$ matrix from an n-dimensional vector $v^T = [\, v_1 \quad v_2 \quad v_3 \quad v_4 \quad \ldots \,]$, by denoting the *hat* operator by

$$\hat{v} = \operatorname{diag}\left[\, v_1 \quad v_2 \quad v_3 \quad v_4 \quad \ldots \,\right]. \tag{2}$$

We generate a 3×3 matrix $\tilde{v}$ from the 3-dimensional vector $v^T = [\, v_1 \quad v_2 \quad v_3 \,]$ by the *tilde* operator as follows

$$\tilde{v} = \begin{bmatrix} 0 & -v_3 & v_2 \\ v_3 & 0 & -v_1 \\ -v_2 & v_1 & 0 \end{bmatrix}. \tag{3}$$

We often use the fact that for any two n-dimensional vectors $v \in \mathbb{R}^n$, and $x \in \mathbb{R}^n$,

$$\hat{v} x = \hat{x} v. \tag{4}$$

2. DESCRIPTION OF A NETWORK OF BARS/STRINGS

We now show how to organize the equations for n_b rigid bars. We will show below how to describe all dynamics in the $\underline{E}$ frame, after the usual definition of dot and cross products. The 3×1 matrix b_i represents the components of vector $\underline{b}_i$ with respect to the fixed frame $\underline{E}$. That is,

$$\underline{b}_i = \sum_{j=1}^{3} \underline{e}_j b_{i_j} = \begin{bmatrix} \underline{e}_1 & \underline{e}_2 & \underline{e}_3 \end{bmatrix} \begin{bmatrix} b_{i_1} \\ b_{i_2} \\ b_{i_3} \end{bmatrix} = \underline{E} b_i . \tag{5}$$

LEMMA 3 *Let for some chosen inertial reference frame $\underline{E}$,*

$$\underline{b}_i = \underline{E} b_i, \qquad \underline{f}_i = \underline{E} f_i, \qquad \underline{n}_i = \underline{E} n_i .$$

Then the cross product is given by

$$\underline{b}_i \times \underline{f}_{i+n_b} = (\underline{E} b_i) \times (\underline{E} f_{i+n_b}) = \underline{E} \tilde{b}_i f_{i+n_b}$$

and the dot product is given by

$$\underline{b}_i \cdot \underline{f}_{i+n_b} = (\underline{E} b_i) \cdot (\underline{E} f_{i+n_b}) = b_i^T \underline{E}^T \cdot \underline{E} f_{i+n_b} = b_i^T f_{i+n_b}$$

where the dot product $\underline{E}^T \cdot \underline{E} = I$ since $\underline{e}_i, i = 1, 2, 3$ form a dextral set of unit vectors.

Let a structural system be composed of n_b bars and n_s strings. The definitions below will later allow us to describe the connections between the rigid members and the strings.

DEFINITION 4 *A* node *(the ith node $\underline{n}_i$) of a structural system is a point in space at which members of the structure are connected. The coordinates of this point in the $\underline{E}$ frame are $n_i \in \mathbb{R}^3$, as in (1).*

DEFINITION 5 *A* string *(the ith string) is characterized by these properties:*

- *A massless structural member connecting two nodes.*
- *A vector connecting these two nodes is $\underline{s}_i$. The direction of $\underline{s}_i$ is arbitrarily assigned.*
- *The string provides a force to resist lengthening it beyond its rest-length, but provides no force to resist shortening the string below its rest-length.*
- *A* string *has no bending stiffness.*

DEFINITION 6 *A* bar *(the ith bar of a n_b bar system) is characterized by these properties:*

- *A structural member connecting two nodes $\underline{n}_i$ and $\underline{n}_{i+n_b}$.*

- *The vector along the bar connecting nodes* $\underline{n}_i$ *and* $\underline{n}_{i+n_b}$ *is* $\underline{b}_i = \underline{n}_{i+n_b} - \underline{n}_i, i = 1, 2, \ldots n_b$.

- *The bar* $\underline{b}_i$ *has* length $\|b_i\| = \mathcal{L}_i = \sqrt{b_i^T b_i}$.

DEFINITION 7 *The vector* $\underline{r}_i$ *locates the mass center of bar* $\underline{b}_i$, *and* $\underline{r}_i = \underline{E} r_i$.

DEFINITION 8 *The vector* $\underline{t}_i$ *represents the force exerted on a node by string* $\underline{s}_i$, *where the direction of* $\underline{t}_i$ *is defined to be parallel to string vector* $\underline{s}_i$. *That is,* $\underline{t}_i = \gamma_i \underline{s}_i$ *and hence,* $t_i = \gamma_i s_i$ *for some positive scalar* γ_i.

DEFINITION 9 *The* force density γ_i *in string* s_i *is defined by* $\gamma_i = \frac{\|t_i\|}{\|s_i\|}$.

DEFINITION 10 $\underline{f}_i$ *represents the net sum of vector forces external to bar* $\underline{b}_i$ *terminating at node* $\underline{n}_i$. *The net sum of vector forces acting at the other end of bar* $\underline{b}_i$ *is* $\underline{f}_{i+n_b}$.

From these definitions, define matrices, $F \in \mathbb{R}^{3\times 2n_b}$, $N \in \mathbb{R}^{3\times 2n_b}$, $T \in \mathbb{R}^{3\times n_s}$, $S \in \mathbb{R}^{3\times n_s}$, $B \in \mathbb{R}^{3\times n_b}$, $\Gamma \in \mathbb{R}^{n_s\times n_s}$, as follows

$$F = \begin{bmatrix} F_1 & F_2 \end{bmatrix} = \begin{bmatrix} f_1 & f_2 & \ldots & f_{n_b} & | & f_{n_b+1} & \ldots & f_{2n_b} \end{bmatrix} \tag{6}$$

$$N = \begin{bmatrix} N_1 & N_2 \end{bmatrix} = \begin{bmatrix} n_1 & n_2 & \ldots & n_{n_b} & | & n_{n_b+1} & \ldots & n_{2n_b} \end{bmatrix} \tag{7}$$

$$T = \begin{bmatrix} t_1 & t_2 & \ldots & t_{n_s} \end{bmatrix} \tag{8}$$

$$S = \begin{bmatrix} s_1 & s_2 & \ldots & s_{n_s} \end{bmatrix} \tag{9}$$

$$B = \begin{bmatrix} b_1 & b_2 & \ldots & b_{n_b} \end{bmatrix} \tag{10}$$

$$R = \begin{bmatrix} r_1 & r_2 & \ldots & r_{n_b} \end{bmatrix} \tag{11}$$

$$\hat{\gamma} = \Gamma = \text{diag}\begin{bmatrix} \gamma_1 & \ldots & \gamma_{n_s} \end{bmatrix}, \tag{12}$$

where $\hat{\gamma}$ represents the diagonalizing operation on the vector $\gamma \in \mathbb{R}^{n_s}$. It follows from (7), (10), and Definition 6 that

$$B = N_2 - N_1 = N \begin{bmatrix} -I \\ I \end{bmatrix}, \tag{13}$$

and the locations of the mass centers of all bars are described by

$$R = N_1 + \frac{1}{2} B. \tag{14}$$

It follows from Definition 9 and (8), (9) that

$$T = S\Gamma. \tag{15}$$

2.1 Angular Momentum

LEMMA 11 *Assume that the mass of the bar is uniformly distributed only along its length, and that its length is much longer than its diameter. Then the angular momentum of the bar b_i about the center of mass of bar b_i, expressed in the $\underline{E}$ frame, is*

$$h_i = \frac{m_i}{12}\tilde{b}_i\dot{b}_i \tag{16}$$

3. DYNAMICS OF A RIGID BAR

For a single bar, with bar vector $\underline{b}$, at nodes $\underline{n}_1$ and $\underline{n}_2$, these forces are applied $\underline{f}_1$ and $\underline{f}_2$.

LEMMA 12 *The translation of the mass center of bar $\underline{b}$, located at position $\underline{r}$ obeys*

$$m\ddot{\underline{r}} = \underline{f}_1 + \underline{f}_2 \tag{17}$$

or, in the $\underline{E}$ frame of reference,

$$m\ddot{r} = f_1 + f_2 \tag{18}$$

LEMMA 13 *The rotation of bar b_i about it mass center obeys*

$$\frac{m}{12}\tilde{b}\ddot{b} = \frac{1}{2}\tilde{b}(f_2 - f_1). \tag{19}$$

3.1 Constrained Dynamics

We now wish to develop the dynamics constrained for constant bar lengths. We add a non-working constraint torque τ to get

$$\frac{m}{12}\tilde{b}\ddot{b} = \frac{1}{2}\tilde{b}(f_2 - f_1) + \tau \tag{20}$$

$$\kappa = b^T b - \mathcal{L}^2 = 0, \tag{21}$$

where the added constraint is $\kappa = 0$, and τ is the non-working torque associated with this constraint. The torque τ does no work in the presence of any feasible perturbation of the generalized coordinate b. Hence, $\tau^T \delta b = 0$. The constraint must also hold in the presence of a feasible perturbation. Hence, $d\kappa = (\frac{\partial\kappa}{\partial b})^T \delta b = 0$. Thus,

$$\begin{bmatrix} \tau^T \\ (\frac{\partial\kappa}{\partial b})^T \end{bmatrix} \delta b = 0, \tag{22}$$

requiring that the matrix coefficient of δb must have deficient rank. Thus, $\tau = (\frac{\partial \kappa}{\partial b})\zeta$, for some ζ (called a Lagrange multiplier). Furthermore, $\frac{\partial \kappa}{\partial b} = 2b$. Hence, the constrained dynamic system obeys,

$$\frac{m}{12}\tilde{b}\ddot{b} = \frac{1}{2}\tilde{b}(f_2 - f_1) + b\zeta \tag{23}$$

$$\kappa = b^T b - \mathcal{L}^2 = 0. \tag{24}$$

where we have absorbed some constants into the scalar ζ. Note that the constraint holds over time, hence $\kappa = \dot{\kappa} = \ddot{\kappa} = 0$. Differentiating the constraint yields,

$$\dot{b}^T b + b^T \dot{b} = 0 = 2b^T \dot{b}. \tag{25}$$

Differentiating (25) yields

$$\dot{b}^T \dot{b} + b^T \ddot{b} = 0,$$

or,

$$b^T \ddot{b} = -\dot{b}^T \dot{b}. \tag{26}$$

The conclusion thus far is that constant length rigid bar rotations obey, for some scalar ζ,

$$\begin{bmatrix} \tilde{b} \\ b^T \end{bmatrix} \ddot{b} = \begin{bmatrix} \tilde{b}(f_2 - f_1)\frac{6}{m} \\ -\dot{b}^T \dot{b} \end{bmatrix} + \begin{bmatrix} b \\ 0 \end{bmatrix} \zeta. \tag{27}$$

The following identity will be useful, and is proved by substitution.

LEMMA 14 *For any skew-symmetric matrix $\tilde{b}$, the following is true.*

$$\tilde{b}^2 = bb^T - b^T b I.$$

The properties of the Moore–Penrose inverse are well known, and are used to obtain the following.

LEMMA 15 *The unique Moore-Penrose inverse of* $\begin{bmatrix} \tilde{b} \\ b^T \end{bmatrix}$ *is given by*

$$\begin{bmatrix} \tilde{b} \\ b^T \end{bmatrix}^+ = \begin{bmatrix} -\tilde{b} & b \end{bmatrix} \mathcal{L}^{-2}$$

LEMMA 16 *The solution of (27) for $\ddot{b}$ has the unique solution*

$$\ddot{b} = \frac{6}{m}(f_2 - f_1) - b\left(\frac{\dot{b}^T \dot{b}}{\mathcal{L}^2} + \frac{6}{m\mathcal{L}^2} b^T (f_2 - f_1)\right) \tag{28}$$

The results of this section applies for any number of bars. The next section will write the matrix construction for the general case.

3.2 An n_b-Bar System

It follows clearly that these equations apply to any number of bars, so that the following is true, where θ_i is the ith element of the vector $\theta \in \mathbb{R}^{n_b}$.

THEOREM 17 *Consider an n_b-bar system with constant length bar vectors $b_i, i = 1, 2, \ldots, n_b$, and matrices defined by,*

$$R = N_1 + \frac{1}{2}B \tag{29}$$

$$B = \begin{bmatrix} b_1 & b_2 \ldots & b_{n_b} \end{bmatrix} = N_2 - N_1, \quad N = \begin{bmatrix} N_1 & N_2 \end{bmatrix} \tag{30}$$

$$N_1 = \begin{bmatrix} n_1 & n_2 \ldots & n_{n_b} \end{bmatrix}, \quad N_2 = \begin{bmatrix} n_{n_b+1} & \ldots & n_{2n_b} \end{bmatrix} \tag{31}$$

$$F = \begin{bmatrix} F_1 & F_2 \end{bmatrix}, \qquad F_1 = \begin{bmatrix} f_1 \ldots & f_{n_b} \end{bmatrix} \tag{32}$$

$$F_2 = \begin{bmatrix} f_{n_b+1} \ldots & f_{2n_b} \end{bmatrix} \tag{33}$$

$$\mathcal{Q} = \begin{bmatrix} B & R \end{bmatrix} \tag{34}$$

$$K_0 = \begin{bmatrix} I \\ 0 \end{bmatrix} \hat{\theta} \begin{bmatrix} I & 0 \end{bmatrix} \tag{35}$$

$$\theta_i = b_i^T (f_{n_b+i} - f_i)/2\mathcal{L}_i^2 + m_i \|\dot{b}_i\|^2/12\mathcal{L}_i^2 \tag{36}$$

$$\Phi = \begin{bmatrix} -\frac{1}{2}I & I \\ \frac{1}{2}I & I \end{bmatrix} \tag{37}$$

$$M = \text{diag}\begin{bmatrix} \ldots & m_i & \ldots \end{bmatrix} \tag{38}$$

$$\mathcal{M} = \begin{bmatrix} \frac{1}{12}M & 0 \\ 0 & M \end{bmatrix}. \tag{39}$$

Then the rigid body dynamics are given by

$$\ddot{\mathcal{Q}}\mathcal{M} + \mathcal{Q}K_0 = F\Phi, \qquad \mathcal{Q}\Phi^T = N, \tag{40}$$

where the coordinate transformation from coordinates N to coordinates $\mathcal{Q}$ is provided by the invertible matrix Φ^T.

For a given square matrix J, define $\lfloor J \rfloor = \text{diag}\begin{bmatrix} \ldots & J_{ii} & \ldots \end{bmatrix}$. Then, it may be shown that,

COROLLARY 18

$$\hat{\theta} = \frac{1}{2}\hat{\mathcal{L}}^{-2}\lfloor B^T(F_2 - F_1) + \frac{1}{6}\dot{B}^T\dot{B}M\rfloor \tag{41}$$

$$= \frac{1}{2}\hat{\mathcal{L}}^{-2}\lfloor\begin{bmatrix} I & 0 \end{bmatrix}\mathcal{Q}^T(F_2 - F_1) + \frac{1}{6}\begin{bmatrix} I & 0 \end{bmatrix}\dot{\mathcal{Q}}^T M\rfloor \tag{42}$$

$$\mathcal{L} = \begin{bmatrix} L_1 & L_2 & \cdots & L_{n_b} \end{bmatrix}^T. \tag{43}$$

4. CHARACTERIZING BAR/STRING CONNECTIONS

DEFINITION 19 *Define the "string connectivity matrix" C by*

$$C_{ij} = \begin{cases} 1 & \textit{if string vector } s_i \textit{ terminates on node } n_j. \\ -1 & \textit{if string vector } s_i \textit{ eminates from node } n_j. \\ 0 & \textit{if string vector } s_i \textit{ does not connect with node } n_j. \end{cases} \tag{44}$$

$$C = \begin{bmatrix} C_1 & C_2 \end{bmatrix}, \quad C_1 \in \mathbb{R}^{n_s \times n_b}, \quad C_2 \in \mathbb{R}^{n_s \times n_b}$$

DEFINITION 20 *Define the "disturbance connectivity matrix" D by*

$$D_{ij} = \begin{cases} 1 & \textit{if disturbance vector } w_i \textit{ connects to node } n_j. \\ 0 & \textit{if disturbance vector } w_i \textit{ does not connect with node } n_j. \end{cases} \tag{45}$$

$$D = \begin{bmatrix} D_1 & D_2 \end{bmatrix}, \quad D_1 \in \mathbb{R}^{n_w \times n_b}, \quad D_2 \in \mathbb{R}^{n_w \times n_b}$$

For n_w disturbance vectors applied at nodes selected by the matrix D,

$$W = \begin{bmatrix} w_1 & w_2 \cdots & w_{n_w} \end{bmatrix} \tag{46}$$

$$D = \begin{bmatrix} D_1 & D_2 \end{bmatrix}, \quad D_1 \in \mathbb{R}^{n_w \times n_b}, \quad D_2 \in \mathbb{R}^{n_w \times n_b}. \tag{47}$$

THEOREM 21 *Let any connection of rigid bars and elastic strings be described by the string connectivity matrix C and the disturbance connectivity matrix D in Definitions (44-45), and let the arbitrary convention be established that vectors entering a node have a positive sign. Then, the sum of all forces entering the nodes of a class 1 tensegrity structure may be computed by,*

$$F = -(TC + WD), \tag{48}$$

and the string vectors s_i, $i = 1, \ldots, n_s$ are linearly related to the nodal vectors n_j, $j = 1, \ldots, n_{2n_b}$ by

$$S = NC^T.$$

LEMMA 22

$$F = -\mathcal{Q}\Phi^T C^T \Gamma C - WD, \qquad \Phi^T = \begin{bmatrix} -\frac{1}{2}I & \frac{1}{2}I \\ I & I \end{bmatrix}. \tag{49}$$

THEOREM 23 *The dynamics of all Class 1 tensegrity systems with rigid, fixed length bars are described by*

$$\ddot{\mathcal{Q}}\mathcal{M} + \mathcal{Q}\mathcal{K} + WD\Phi = 0, \tag{50}$$

$$\mathcal{Q} = \left[\begin{array}{cc} B & R \end{array} \right], \tag{51}$$

$$\mathcal{M} = \left[\begin{array}{cc} \frac{1}{12}M & 0 \\ 0 & M \end{array} \right], \tag{52}$$

$$\mathcal{K} = \left[\begin{array}{cc} \hat{\theta} & 0 \\ 0 & 0 \end{array} \right] + \Phi^T C^T \Gamma C \Phi, \tag{53}$$

$$\Phi^T = \left[\begin{array}{cc} -\frac{1}{2}I & \frac{1}{2}I \\ I & I \end{array} \right] \tag{54}$$

$$\begin{aligned} \hat{\theta} \; &= \; \frac{1}{12}\hat{\mathcal{L}}^{-2} \Bigg[6 \left[\begin{array}{cc} I & 0 \end{array} \right] \mathcal{Q}^T (\mathcal{Q}\Phi^T C^T \Gamma C + W D) \left[\begin{array}{c} I \\ -I \end{array} \right] \\ &+ \left[\begin{array}{cc} I & 0 \end{array} \right] \dot{\mathcal{Q}}^T \dot{\mathcal{Q}} \left[\begin{array}{c} I \\ 0 \end{array} \right] M \Bigg]. \end{aligned} \tag{55}$$

From (55), the ith element of the diagonal matrix $\hat{\theta}$ is given by

$$\theta_i = \frac{1}{2}\mathcal{L}_i^{-2} b_i^T (\mathcal{Q}\Phi^T C^T \hat{C}_{\Delta_i} \gamma + W D_{\Delta_i}) + \frac{m_i}{12\mathcal{L}_i^2} ||\dot{b}_i||^2, \tag{56}$$

where the following definitions characterize the ith columns of the matrices $C_1 - C_2$ and $D_1 - D_2$.

$$C_{\Delta_i} = i^{th} col(C_1 - C_2) \tag{57}$$

$$C_{+_i} = i^{th} col(C_1 + C_2) \tag{58}$$

$$D_{\Delta_i} = i^{th} col(D_1 - D_2) \tag{59}$$

$$D_{+_i} = i^{th} col(D_1 + D_2) \tag{60}$$

5. CONCLUSIONS

The dynamics of a system of n_b rigid bodies, connected by tensile elements, has been derived from a network point of view. The resulting equations are in matrix form, rather than the traditional vector form. The nonlinear equations are given in the form of a second order differential equation of a $3 \times 2n_b$ *configuration matrix*. These equations contain no trigonometric nonlinearities, and require no inversion of a mass matrix containing configuration variables.

These equations open the door a bit wider for feedback control design and structure design as integrated activities, since all freedom in the desired equilibria can be utilized in the control problem. The result should be more efficient controlled structures. This efficiency will be demonstrated in a future paper.

NEW RESULTS OF THE DEVELOPMENT AND APPLICATION OF ROBUST OBSERVERS

Dirk Söffker
Chair of Dynamics and Control, University Duisburg-Essen, Germany
soeffker@uni-duisburg.de

Abstract: This contribution deals with a model-based approach for fault detection and vibration control of elastic mechanical structures. Especially in the case of lightweighted structures, online and in-situ fault diagnosis becomes important due to internal delamination processes of modern CFK-materials. Additionally faults like cracks or impacts acting on a flexible structure has to be observed. Structural-Health-Monitoring (SHM) concepts deal with different tools to realize safe adaptronic structures, realizing fault monitoring and repair mechanism in combination with intelligent monitoring of the hazard rate of the complete structural integrity.

Core of the contribution are new results of the development and application of the robust PI-Observer. Due to the fact that the lifetime of modern actuators (piezoelectric or shape-memory-alloy based) strongly depends on the use-intensity safety aspects become more and more relevant. Here briefly the SRCE-concept will be declared.

Key words: elastic structure, diagnosis, control, monitoring, SHM, adaptronic, PI-Observer.

INTRODUCTION

The control of mechanical structures has been the focus of several scientific and industrial efforts of the last decade in the area of robotics, vehicle- and rotor dynamics, cf. [12]. Actual works are focused on flexible (space) structures concerning aspects of practicability/realizability and precision [3]. In all these mentioned areas, the classical fields of mechanics/dynamics and control theory as well as data processing are connected. Advanced control approaches usually need some knowledge about the system to be controlled. Smart structures integrate sensors, actuators and algorithms for control and diagnosis to respond to the changing environment and to compensate faults within the structure. A review of the state of the art of smart structures and integrated systems is given in [17]. For practical applications, smart structures should become intelligent

H. Ulbrich and W. Günthner (eds), IUTAM Symposium on Vibration Control of Nonlinear Mechanisms and Structures, 319–330.

structures with self-diagnosis properties. Signal based methods [15] do not need a model of the structure while model-based methods [8, 5, 11] need an analytical model of the structure or/and the fault in order to detect and localize or isolate a fault avoiding failures of the considered system. Robust observers based on model-less disturbance observations are used and enhanced in the last years. Advanced simulation results are given in [18, 19, 20], experimental results of the PI-Observer for practical applications are given in [1], new detailed analytical considerations when measurement noise and model uncertainties are present are given in [10]. In recent years, a number of control approaches have been developed for disturbance rejection and accomodation. Davison [4] introduced a dynamical compensation of disturbances in output variables using an additional equation for the disturbance. Müller presented a disturbance rejection approach [13] to compensate the effect of disturbances on defined output variables using the PI-Observer. Friction forces are estimated and compensated using the above mentioned approach to realize exact position control of robots. In [2] the PIO is used for observer-based control which is less sensitive to parameter variation of a system, here unknown inputs are not estimated. The observer is applied in combination with a nonlinear PD controller to reject the estimated disturbances [6]. An H_∞ approach is used in [16] for disturbance attenuation using high gain observers to minimizes the norm of a specific transfer function.

In [21] two dynamic compensation approaches are proposed extending the Davison approach [4] using the estimated states of the PI-Observer. The disturbance compensation approach introduced by Müller [13] is extended and improved by the work of Krajcin [10].

The common strategy to design a fault tolerant, robust and reliable system consists of a structural reliability analysis of a planned system's function design. If faults still occur, with which the cause cannot be eliminated by a redesign, a Fault Detection and Isolation (FDI) module may be added to the system [14]. If a possible fault is affecting the safety of the system or the user's health Limp Home Mode (LHM) functionality may be added to the system. Its function is to supply a minimal functionality of the system in the case of faults and subsystem failures. In [23] a new design strategy is proposed which combines methods and procedures of structural reliability analysis, FDI (including observer-based techniques) and LHM techniques to achieve a system with optimal functional and spatial integrated components. One core element is the Safety and Reliability Control Engineering Concept (SRCE) [22]. The concept enables a used, stressed technical system to run up to a specific lifetime in compliance with a defined failure probability by controlling the increase of the failure probability. This ideas are graphically repeated within the last section.

1. DESIGN OF A ROBUST PI-OBSERVER

The design of observers should be also valid if measurement noise and model uncertainties are present. It may be necessary to consider the uncertainties in the design approach realizing robust approaches. Assume the system with additive unknown inputs $n(t)$, additive measurement noise $d(t)$ and additive model uncertainties $h(t)$

$$\dot{x}(t) = Ax(t) + Bu(t) + Nn(t) + Hh(t) , \quad (1)$$

$$y(t) = Cx(t) + d(t) . \quad (2)$$

The matrix H describes the input of the model uncertainties $h(t)$.

The error dynamics becomes

$$\begin{bmatrix} \dot{e}(t) \\ \dot{f}_e(t) \end{bmatrix} = \begin{bmatrix} A - L_1 C & N \\ -L_2 C & 0 \end{bmatrix} \begin{bmatrix} e(t) \\ f_e(t) \end{bmatrix} + \begin{bmatrix} L_1 d(t) - Hh(t) \\ L_2 d(t) - \dot{n}(t) \end{bmatrix} . \quad (3)$$

Assuming the extended system as observable and assuming that the feedback matrices L_1, L_2 are calculated solving the Riccati equation

$$A_e P + P A_e^T + Q - P C_e^T R^{-1} C_e P = 0 , \quad (4)$$

the observer feedback matrix L is denoted by

$$L = \begin{bmatrix} L_1 \\ L_2 \end{bmatrix} = \begin{bmatrix} P_{11} & P_{12} \\ P_{21} & P_{22} \end{bmatrix} \begin{bmatrix} C^T \\ 0 \end{bmatrix} R^{-1} = \begin{bmatrix} P_{11} C^T R^{-1} \\ P_{21} C^T R^{-1} \end{bmatrix} . \quad (5)$$

The matrix L_2 depends on $P_{21} = P_{12}^T$. If the extended system is observable, the gains $\|L_2\|$ increases by increasing q. To achieve the decoupling from from the disturbance inputs to the states

$$\|[Is - (A_e - LC_e)]^{-1} N_e\|_\infty < \epsilon . \quad (6)$$

$\epsilon \to 0$ is required, so the weighting parameter has to be $q \to \infty$ and in practical applications the parameter should be $q \gg 1$ which yields from

$$q \gg 1 \quad \text{to} \quad \|L_2\| \gg \|L_1\| . \quad (7)$$

In frequency domain the error dynamics is described by

$$e(s) = -[Is - (A - L_1 C)]^{-1} (Hh(s) + L_1 d(s) + N f_e(s)) \quad (8)$$

$$f_e(s) = -[Is - L_2 C G(s) N]^{-1} (L_2 C G(s) L_1 d(s) + L_2 C G(s) H h(s) + L_2 d(s) - s\, n(s)) , \quad (9)$$

with $G(s) = (Is - (A - L_1 C))^{-1}$. From this equation it can be seen that four parts for the estimation error exist. To minimize the transfer function from

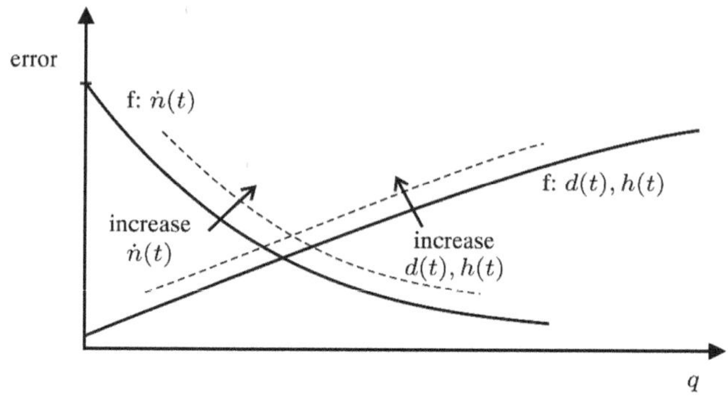

Figure 1. Schematic behavior of the estimation error.

$\|sn(s)\|$ to $\|f_e(s)\|$ described by $[Is - L_2C(Is - (A - L_1C))^{-1}N]^{-1}$ it is necessary to chose $q \gg 1$ if the changes of the unknown input $(\dot{n}(t), sn(s))$ are large. This leads to $\|L_2\| \gg \|L_1\|$. Unfortunately, in this case, (as $q \gg 1$, $|L_2\| \gg \|L_1\|$) the other additive effects will increase, so that the resulting error $f_e(t)$ also increases. Details of the approach are given in [10]. The feedback matrices have to be designed by the LTR method, but the gain parameter q cannot be arbitrary increased. The estimation error depending on the LTR design parameter q is illustrated qualitatively in Figure 1. The curve $f : \dot{n}(t)$ in Figure 1 denotes the error caused by the derivative of the unknown inputs and the curve $f : d(t), h(t)$ denotes the error caused by the uncertainties. The optimal parameter q depends on the qualities of the model and the measurement and on the derivative of the unknown input. As result in the best case the PI-Observer estimates with weak conditions regarding the observability of the inputs from the measurements the external inputs and internal states (used for diagnosis purposes or for control).

2. DESIGN OF AN ADVANCED PIO-BASED DISTURBANCE COMPENSATION AND SYSTEM CONTROL APPROACH

The goal is to control the system given by (A, B, C) and to compensate the influence of unknown inputs $n(t)$ to the selected coordinates $z(t)$. The feedback can be calculated as

$$u(t) = -K_s\hat{x}(t) - K_{v1}\hat{n}(t) - K_{v2}\dot{\hat{n}}(t) + Vw(t)\,, \tag{10}$$

where $K_s\hat{x}(t)$ realizes the classical state feedback, $K_{v1}\hat{n}(t)$ and $K_{v2}\dot{\hat{n}}(t)$ are the disturbance compensation parts and $Vw(t)$ realizes the reference signal. The unknown inputs $\hat{n}(t)$ and $\dot{\hat{n}}(t)$ can be estimated using the PI-Observer. Assuming successful estimation of the PI-Observer without loss of generality, $\hat{x}(t)$ is set to $x(t)$, $\hat{n}(t)$ is set to $n(t)$ and $\dot{\hat{n}}(t)$ is set to $\dot{n}(t)$. The state feedback matrix K_s is calculated by classical approaches. The controlled system is given by

$$\dot{x}(t) = (A - BK_s)x(t) + (N - BK_{v1})n(t) - BK_{v2}\dot{n}(t) \; . \tag{11}$$

Using the substitution

$$v(t) = x(t) + BK_{v2}n(t) \; , \tag{12}$$

the system description yields to

$$\begin{aligned} \dot{v}(t) &= (A - BK_s)v(t) \\ &\quad +[(A - BK_s)BK_{v2} - (N - BK_{v1})]n(t) \; , \qquad (13) \\ z(t) &= F(v(t) - BK_{v2}n(t)) \; . \qquad (14) \end{aligned}$$

The transfer function between the disturbances $n(t)$ and the output $z(t)$ to be controlled results to

$$\begin{aligned} G_{nz}(s) &= F[[Is - (A - BK_s)]^{-1}[(N - BK_{v1}) \\ &\quad -(A - BK_s)BK_{v2}] - BK_{v2}] \; . \end{aligned} \tag{15}$$

The goal is to minimize the transfer function from $n(t)$ to $z(t)$. Considering the stationary case ($\dot{v}(t) = 0$), the transfer function $G_{nz}(s)$ is zero if

$$[F(A - BK_s)^{-1}B]K_{v1} = F(A - BK_s)^{-1}N \; . \tag{16}$$

To decouple the effect of a disturbance to an output $z(t)$ in the stationary case, the feedback matrix K_{v1} has to fulfill Eqn. (16). This is given by Müller in [13] to compensate the effect of disturbances on interesting coordinate $z(t)$, where the condition is formulated for a more generale case. The necessary condition for the compensation is

$$\text{rank}\begin{bmatrix} A & B \\ F & 0 \end{bmatrix} = \text{rank}\begin{bmatrix} A & B & N \\ F & 0 & 0 \end{bmatrix} . \tag{17}$$

If it is desired to compensate the effect of the disturbances in all the states $z(t) = x(t)$, the requirement

$$BK_{v1} = N \tag{18}$$

can be derived from Eqn. (16). The related matching condition

$$\text{rank } B = \text{rank } [B \;\; N] \tag{19}$$

has to be fulfilled, than the disturbances can only be rejected in all the states if they act in the system inputs but this is a rare case in practical applications.

The approach used by Müller contains a vibration control $K_s\hat{x}(t)$, the part $K_{v1}\hat{n}(t)$ accomodate the effect of a disturbance on defined outputs $z(t)$ only in the stationary case. The approach developed by Krajcin and Söffker [9] with the additional feedback $K_{v2}\dot{\hat{n}}(t)$ improves the dynamical behavior of the disturbance accommodation, which is compared to [21] a new feature. The feedback $K_{v1}\hat{n}(t)$ is necessary for the stationary case and has to be calculated using Eqn. (16). The feedback term K_{v2} should be calculated minimizing $||G_{nz}(s)||_\infty$ and has no effect on the stationary case as it can be seen from Eqns. (14) and (16).

3. EXPERIMENTAL RESULTS

Simulation and experimental results of previous development stages of the PIO are related to different applications within the field of machine dynamics, robotics, etc. [1, 7, 18, 20]. Using the above mentioned PIO-based approaches the following new and improved results are obtained.

3.1 One Side Clamped Elastic Beam

The one side clamped elastic beam is shown in Figure 2(a). The scheme of the test rig is presented in Figure 2(b). An elastic beam clamped on one side gets in contact with the contact surface after a short excursion of the beam (dashed position in Figure 2(b)). The impact contact force is measured by quartz force sensors. The displacement of the beam is measured at two points with non-contacting optical measurement systems. There are also 3 strain gages mounted on the beam to measure the strain. The contact force is estimated using the displacement measurements and (or) the strains, the measurement of the contact force is only used for validation. The beam is modeled using five equal finite beam elements. Two displacement measurements are taken in node 3, 4 or 5 of the finite element mesh. The contact point is at the end of the beam and is realized by a steel tip. Additionally, the beam can be excited by a modal hammer which measures the force. The beam gets in contact with the contact

(a) Figure of test rig (SRS, U DuE)

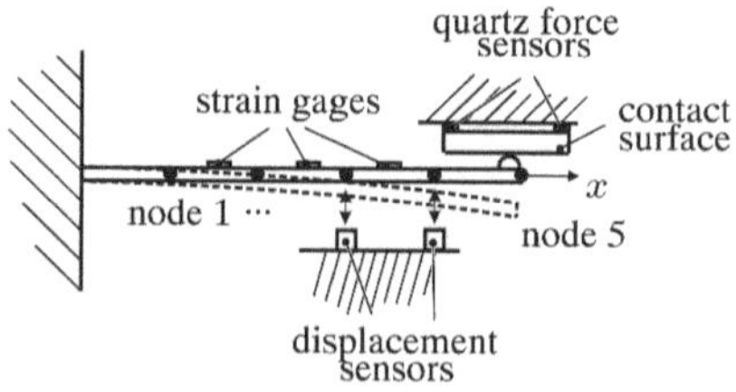

(b) Sketch of the test rig

Figure 2. One side clamped elastic beam.

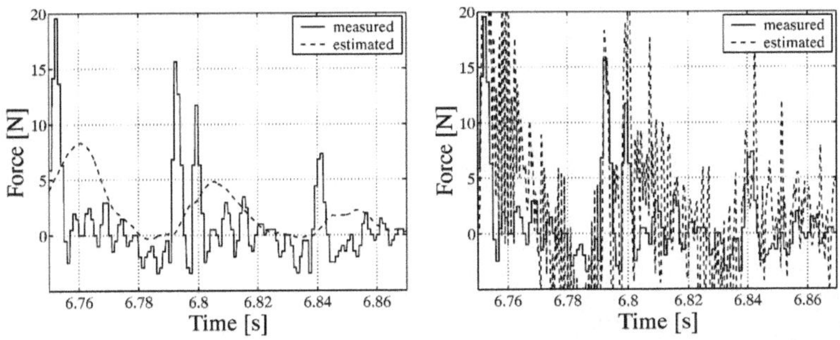

(a) Measurement of displacements in node 4 and 5, $q = 10^3$, too small

(b) Measurement of displacements in node 4 and 5, $q = 10^8$, too high

Figure 3. Contact in node 5.

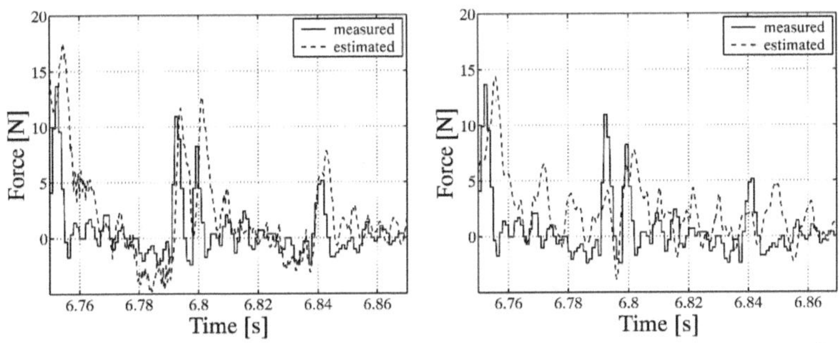

(a) Measurement of displacements in node 4 and 5, $q = 10^6$

(b) Strain gage measurements, $q = 10^6$

Figure 4. Contact in node 5.

surface after a short excursion of the beam (dashed position in Figure 2(b)). The displacement is measured in node 4 and 5. In Figures 3(a) and 3(b) the measured and the estimated force are compared. In Figure 3(a) the design parameter $q = 10^3$ is too small, the observer cannot follow the fast dynamic. In Figure 3(b) the design parameter $q = 10^8$ is too high, the error caused by the measurement noise and the model uncertainties prevails. Choosing $q = 10^6$ the PI-Observers can estimate the contact force very good as can be seen from Figure 4(a). The contacts are very fast, up to 5 ms for a contact. In Figure 4(b) the displacements in node 4 and 5 are calculated by the strain gage measurements. The force can also be estimated for measurements in other nodes and for contacts in other nodes. For slower contacts the estimation results get better.

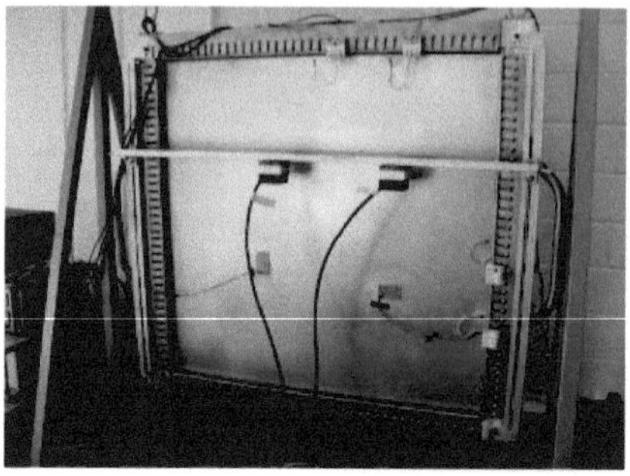

(a) Figure of the test rig (SRS, U DuE)

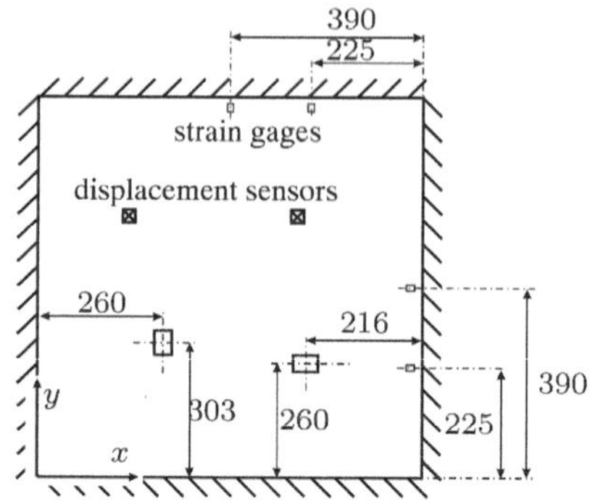

(b) Sketch of the test rig

Figure 5. All side clamped plate.

3.2 All Side Clamped Elastic Plate

The experimental setup of the elastic plate is given in Figure 5 (dimensions in mm). The plate measures 780 × 780 mm and has the thickness $d_p = 0.7$ mm. Two piezo actuators (PZT patches) are bonded on the plate as shown. The displacements are measured at two points with non-contacting optical measurement systems and the strains are measured at the illustrated positions in length direction of the drawn rectangles (see Figure 5(b)). The plate is excited by a hammer, which directly measures the contact force. The plate is modeled using 64 equal plate elements (with different boundary conditions). The first eigenfrequency of the plate e.g. varies temperature-dependent between 9 Hz to 18 Hz. The moments/forces of the PZT patches acts on the plate structure. For the model used for control, the plate is divided into 169 equal plate elements. The mesh of the finite elements is chosen such that each corner of the PZT patches is approximately on one finite element node. Applying voltage to the PZT patches they expand and induces forces parallel to the plate surface. In Figure 6(a) the measured and the estimated contact force are compared, details are given in [10]. The estimation is done using the displacement measurement and the strain gage measurement. The contact duration is up to 0.2 s, the estimation is satisfactory. Faster contacts as shown in Figure 6(b) cannot be estimated satisfactory. The contact can be detected but after the contact the estimated force oscillate with the same amplitude as the contact itself. The model of the plate seems not to be sufficient.

In Figure 7 simulation results are given. Here, parallel to the first experimental realization the disturbance compensation part is not applied to the plate. The PI-Observer is – due to practival limitations regarding the power supply for control – only used to estimate the states for the state feedback. The unknown inputs are assumed as to be present in the inputs of the system (N=B). There-

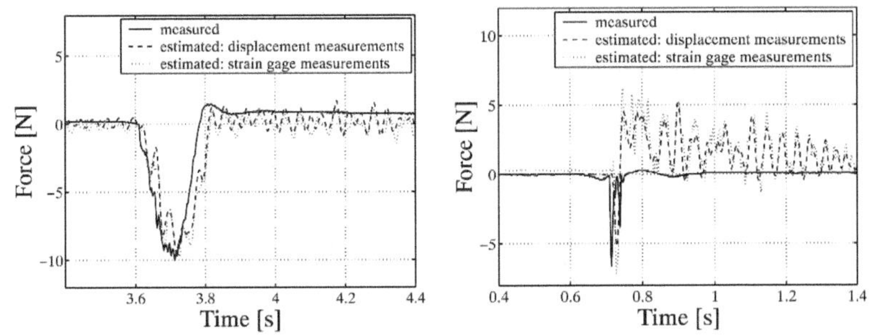

(a) Measurement position 1, contact 1, $q = 10^6$
(b) Measurement position 1, contact 1, $q = 10^6$

Figure 6. Estimation of a contact force applied to the plate.

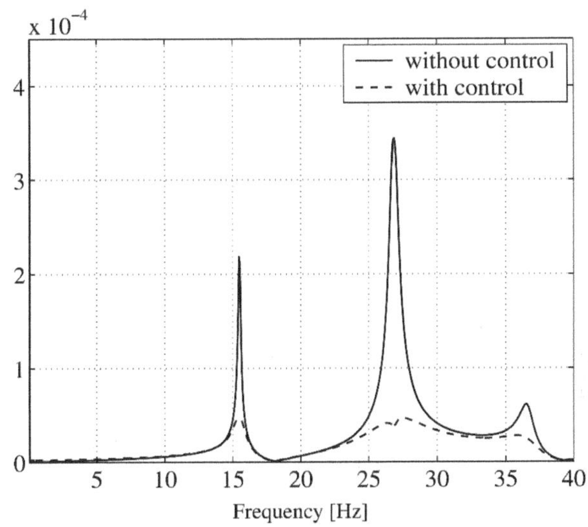

Figure 7. Vibration control of the plate, Simulation.

fore, the states can be estimated sufficiently, even if the PZT patch models are assumed as linear approximations of the real behavior. The control is tuned to reduce the first two eigenmodes. The results show significant reduction of the step response which also includes noise reduction.

4. INTEGRATING DIAGNOSIS AND CONTROL INTO RELIABILITY-BASED CONCEPTS REALIZING SAFE STRUCTURES

The recent developed supervision concept [23] integrates LHM and FDI, so the system design also includes beside control approaches diagnosis and supervision properties, cf. Figure 8a. The whole new system also includes sensory and

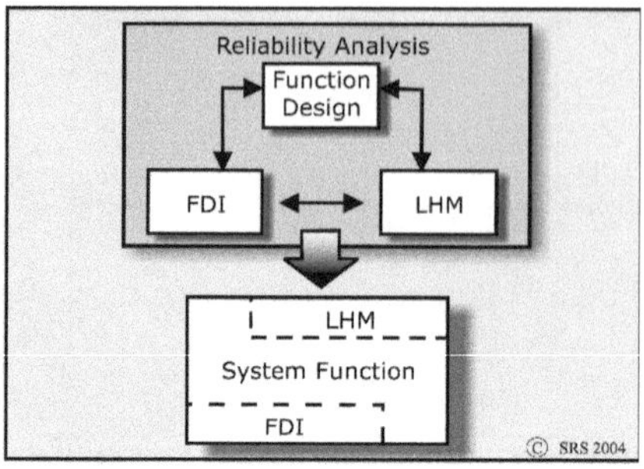

(a) Integration of system, LHM, and FDI

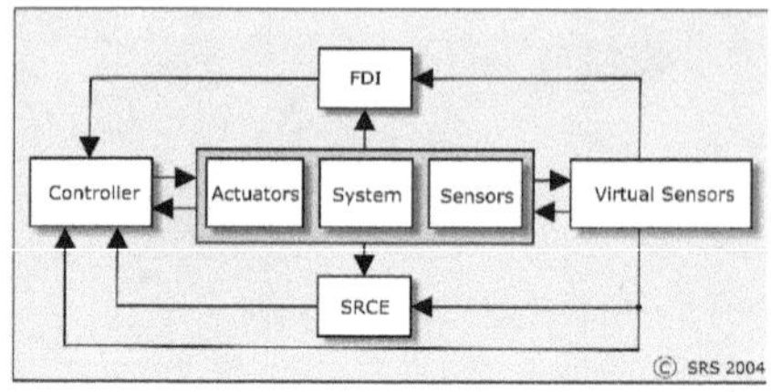

(b) The new integrated system

Figure 8. New system functionalities integrating diagnosis and supervision modules.

actuation devices as well as computing modules, cf. Figure 8b and allows supervision of the system functionalities and dynamical changes within (the soft) parts of the system. Core of the approach is the SRCE-concept [22]. Here in contrast to usual assumptions in reliability engineering the life cycle/the life time is assumed as depending on the load history and the actual load, so appears as a variable to be supervised and/or controlled. One important aspect is – that dependent on the system – diagnosis approaches like the PIO allow an inside view into the systems behavior or the faults (like crack depths) etc. and are used to determine the loss of lifetime and/or reliability. Details are given in the cited literature of the author.

5. CONCLUSIONS AND OUTLOOK

The development of new elastic mechanical structures especially in the context of multifunctional structures like adaptronics systems leads to new functionalities, but also to new dependabilities. Therefore the combination of robust observer, this means model-based techniques like the proposed PI-Observer and the strong integration of the systems hardware with the systems software (PIO, Monitoring techniques) may lead to highly integrated and self-monitoring systems. The proposed PI-Observer techniques allows the estimation of internal and external unknown effects and is now theoretical and experimental validated. The key advantage of the proposed approach is, this only a few measurements are necessary to estimated system states and additive unknown effects, representing internal or external forces or moments acting to the structures. The contribution gives the main results developed at the Chair of Dynamics and Control improving former results from a theoretical point of view. Simulation and experimental results show the applicability of the technique for diagnosis and control of elastic mechanical structures. The illustrated extended control approach allow advanced dynamic feedback and will also give new control qualities. The SRCE-concept introduced and developed with the Chair of Dynamics and Control allows the monitoring of the ageing process of new sensory and actuation devices and therefore ensures safe operation. The next step will be the full hardware integration of the techniques within one structure. Other work will care about the integration of acceleration measurements within this new structural health monitoring (SHM) concept for adaptronic structures.

REFERENCES

[1] Abicht, C., Bormann, J., Müller, P.C., Söffker, D. and Ulbrich, H. (2001). Model-Based Estimation of Impact Forces Affecting Elastic Structures: Simulation and Experiment. In *18th ASME-DECT Conference on Mechanical Vibration and Noise*, Pittsburgh PA, USA, 6 pp.

[2] Beale, S. and Shafai, B. (1989). Robust Control Design with a Proportional Integral Observer. *Int. J. Control*, Vol. 50, pp. 97–111.

[3] Cordes, J. (1992). *Robuste Regelung eines elastischen Teleskoproboterarmes*. Fortschritte der Robotik, Bd. 12, Vieweg Verlag, Braunschweig, Germany.

[4] Davison, E.J. (1972). The Output Control of Linear Time-Invariant Multivariable Systems with Unmeasurable Arbitrary Disturbances. *IEEE Trans. AC*, Vol. 17, pp. 621–630.

[5] Frank, P.M. and Wünnenberg, F. (1989). Robust Fault Diagnosis Using Unknown Input Observer Scheme. In: *Fault Diagnosis in Dynamic Systems; Theory and Application*, Patton, R.J., Frank, P.M. and Clark, R.N. (eds.), Prentice Hall, Engelwood Cliffs.

[6] Gao, Z., Hu, S. and Jiang, F. (2001). A Novel Motion Control Design Approach Based on Active Disturbance Rejection. In: *Proc. 40th IEEE Conference on Decision and Control*, Orlando, FL, pp. 4877–4882.

[7] Kashi, K. and Söffker, D. (2004). Model-Based Estimation of a Force and Displacement of a Hydraulic Cylinder. In: *Proc. 7th International Symposium on Advanced Vehicle Control KIVI-NIRIA*, Arnheim, the Netherlands.

[8] Krajcin, I. and Söffker, D. (2003). Modified PIO Design of Robust Unknown Input Estimation. In: *Proc. 19th ASME-DECT Biennial Conference on Mechanical Vibration and Noise, Symposium on Systems with Frictions and Impacts: Modeling, Stability and Control*, Chicago, IL, USA, 6 pp.

[9] Krajcin, I. and Söffker, D. (2005). Advanced Model-Based Compensation Control Using Proportional-Integral Observer. In: *Proc. of DECT05, 20th Biennial Conference on Mech. Vibration and Noise,* Long Beach (CA), September 24–28, 2005, accepted.

[10] Krajcin, I. and Söffker, D. (2005). Diagnosis and Control of 3D Elastic Mechanical Structures. In: *Proc. 12th SPIE Symposium on Smart Structures and Materials*, March 6–10, 2005, San Diego, CA, USA.

[11] Kranock, S.J. and Peterson, L.D. (1998). Real-Time Structural Health Monitoring Using Model-Based Observers. In: *Proc. SPIE Conference on Mathematics and Control in Smart Structures*, San Diego, CA, USA, pp. 711–722.

[12] Marghitu, D.B. and Diaconescu, C.I. (1999). Control Techniques for Impacting Flexible systems. *Arch. Appl. Mech.*, Vol. 69, pp. 555–568.

[13] Müller, P.C. and Lückel, J. (1977). Zur Theorie der Störgrössenaufschaltung in linearen Mehrgrössenregelsystemen. *Regelungstechnik*, Vol. 25, pp. 54–59.

[14] Patton, R.J., Frank, P.M. and Clark, R.N. (2000). *Issues in Fault Detection for Dynamic Systems*. Springer, London.

[15] Pedemonte, P., Staszewski, W.J., Aymerich, F., Found, M.S. and Priolo, P. (2001). Signal Processing for Passive Impact Damage Detection in Composite Structures. In: *Proc. SPIE Smart Structures and Materials*, pp. 169–178.

[16] Petersen, I.R. and Hollot, C.V. (1988). High-Gain Observer Approach to Disturbance Attenuation Using Measurement Feedback. *Int. J. Control*, Vol. 48, pp. 2453–2464.

[17] Shopra, I. (2002). Review of State of Art of Smart Structures and Integrated Systems. *AIAA J.*, Vol. 40, No. 11, pp. 2145–2187.

[18] Söffker, D., Bajkowski, J. and Müller, P.C. (1993). Detection of Cracks in Turbo Rotors – A New Observer Based Method. *ASME Journal of Dynamic Systems, Measurements, and Control*, Vol. 3, pp. 518–524.

[19] Söffker, D., Yu, T.J. and Müller, P.C. (1995). State Estimation of Dynamical Systems with Nonlinearities by Using Proportional-Integral Observer. *Int. J. Sys. Sc.*, Vol. 26, pp. 1571–1582.

[20] Söffker, D. (1995). A New Model-Based Tool for Failure Detection and Isolation in Machine- and Rotordynamics. ASME DE-Vol. 83-2, pp. 233–242.

[21] Söffker, D. (1995). Elastic Robot Arms with Varying Length – Part II: Robust Control of Elastic Vibrations. In: *Vibration of Nonlinear, Random, and Time-Varying Systems – Time Varying Systems and Structures*, Sinha, S.C., Cusumano, J.P. and Pfeifer, F. (eds), ASME DE-Vol. 84-1, pp. 109–119.

[22] Söffker, D. (2000). Zur Online-Bestimmung von Zuverlässigkeits- und Nutzungskenngrössen innerhalb des SRCE-Konzeptes. *at-Automatisierungstechnik*, Vol. 48, pp. 383–391.

[23] Söffker, D., Kashi, K. and Wolters, K. (2005). Integration of Reliability Concepts, Diagnosis, and Control to Realize Safe Systems. In: *Proc. of ESREL 2005*, Poland, August 2005, accepted.

BIFURCATIONS CAUSED BY SAMPLING EFFECTS IN ROBOTIC FORCE CONTROL

Gábor Stépan and László L. Kovács
Department of Applied Mechanics, Budapest University of Technology and Economics, H-1521 Budapest, Hungary
stepan@mm.bme.hu, kovacs@mm.bme.hu

József Kövecses
Department of Mechanical Engineering, McGill University, Montreal, QC, Canada, H2A 2K6
jozsef.kovecses@mcgill.ca

Abstract: The empirically developed force control in cases of the robotic polishing and the rehabilitation robots serve as a motivation for the study of the peculiar dynamic behaviour of digital force control. The effect of the sampling times of the digital controllers are studied analytically, and the corresponding stability charts are presented for different gain and mechanical parameters describing also the different sampling frequencies at the force sensors and in the digital control loop. The types of bifurcations are also identified at the stability limits. As one of the practical conclusions, the negative role of differential gain is explained in digital force control.

Key words: force controlled robots, differential gain, delayed oscillator.

INTRODUCTION

When a robot has to interact with the environment, the control of the contact force between its actuator and the workpiece is often required. Force control tries to maintain prescribed contact force. Apart from the disturbing effects of the so-called unmodelled high-frequency dynamics of these systems, basic textbooks often call the attention to the destabilizing digital effects, like sampling (Slotine and Li, 1991). At the stability limits presented in the parameter space of the sampling time, control gains and further mechanical parameters, several kinds of bifurcations occur, showing a large variety of nonlinear dynamic behaviour. These analytical results have a central role in understanding

H. Ulbrich and W. Günthner (eds), IUTAM Symposium on Vibration Control of Nonlinear Mechanisms and Structures, 331–341.

the technical phenomena and in forming our physical sense needed during the design of force controlled systems.

The present study has been motivated by two laboratory projects. One is the Rehabilitation Robotics (REHAROB IST-1999-13109) project (Arz et al., 2003) that uses force control during the teaching-in phase of the antispastic physiotherapy of patients suffering from the spastic hemiparesis of the upper limbs (Kovács and Stépán, 2003). The other project is the hybrid position/force controlled Newcastle robot designed for turbine blade polishing (Stépán and Haller, 1995). In both practical applications, peculiar dynamical behaviour and unexpected vibrations occurred referring to several bifurcation phenomena during the experiments.

The present study first describes the basic problems of applied force control, then briefly reviews the motivation of this study. The analytical study of the simplified 1 degree-of-freedom (DoF) mechanical models with different sampling effects at the force sensor and in the control loop is explained in details. The results are presented in the form of stability charts referring also to the types of bifurcations in the system. In the concluding section, qualitative explanation is given why differential gains are avoided in practical force control applications like rehabilitation robotics or robotic polishing.

1. APPLIED FORCE CONTROL

Several excellent books (like Gorinevsky et al., 1997; Siciliano and Villani, 1999; Natale, 2003) have been published recently on force control of robot manipulators, showing the great demand for understanding and synthesizing experiences in this field. These books investigate models of several degrees of freedom in cases of continuous-time force control. The experiments validating the proposed control algorithms, however, are exclusively carried out using digitally controlled experimental testbeds. Only brief qualitative discussions and over-simplified analytical studies are presented regarding the dynamics of the digitally controlled counterparts of these systems. The major simplification in the analytical studies is, that the analyzed mathematical models are scalar and of first order only, i.e., the inertial forces are often neglected compared to the viscous damping ones. In the meantime, the persistent investigation of the delayed oscillators (Stépán, 1989) call the attention to the difficulty of the mathematical problem here, also having important physical consequences, explaining unexpected vibration phenomena.

The reassuring statement, that the digital controller with increasing sampling frequency should approach the characteristics of the corresponding continuous controller, can be posed in a confusing way, too: any finite sampling time results in substantially smaller gain parameters than the corresponding analogue controllers may have. Apart from these general (and not contradictory) statements, it is much less studied that the decrease of sampling frequency does not decrease maximum gains uniformly at the limit of stability, and a large sampling time may result in better stability properties in oscillatory systems than small ones have.

It depends on the control parameters and mechanical properties of the system whether a control algorithm with a certain sampling frequency can be considered continuous, or the digital effects have to be taken into account. For example, a digital force control algorithm with high sampling frequency can be considered "continuous" if the end effector of the robot comes to contact with a soft environment where the effective stiffness of the system is very low and the effective (or modal) mass is high. However, if the environment is very stiff, or there are high-frequency vibration modes with low modal damping, then the effects of the discrete-time nature of the controller will have significant influence on the dynamic behaviour of the force controlled system even at high sampling rate. This significant influence means, for example, that the maximum stable proportional gains are severely limited in these cases.

2. MOTIVATION

The stability properties of the 3 degree-of-freedom (DoF) hybrid position/force controlled Newcastle robot was analysed in detail both theoretically and experimentally (Stépán and Haller, 1995). The robot was used and designed for turbine blade polishing, and it maintained constant contact force between the polisher and the blade. The bending stiffness of the polished blade was strongly determined by the principal directions of the matrix of area moment of inertia of the wing-shaped cross section of the blade. Depending on the direction of the force control that was normal to the blade surface, the robot lost stability and started self-excited oscillations with relatively low frequencies in the range of a couple of Hertz. Figure 1 shows the 2 DoF mechanical model of the robot in the force controlled direction representing the elasticity of the force sensor and also that of the contacted environment, which is the turbine blade in this case.

The experimentally confirmed stability chart shows the stable regions in the plane of the sampling time τ of the digital controller and the proportional gain P for the identified mechanical parameters not presented here. There are two important conclusions of this series of experiments. First, the largest gains can be found not at the lowest realisable sampling times. Second, any attempt to apply the derivative of the force error multiplied with a differential gain D in the digital control resulted in further reduction of the stable domains.

The other example for stability problems with force control comes from the so called RehaRob project (Arz et al., 2003). During the teaching-in phase, the force is controlled between the patient's arm (attached to the orthosis in

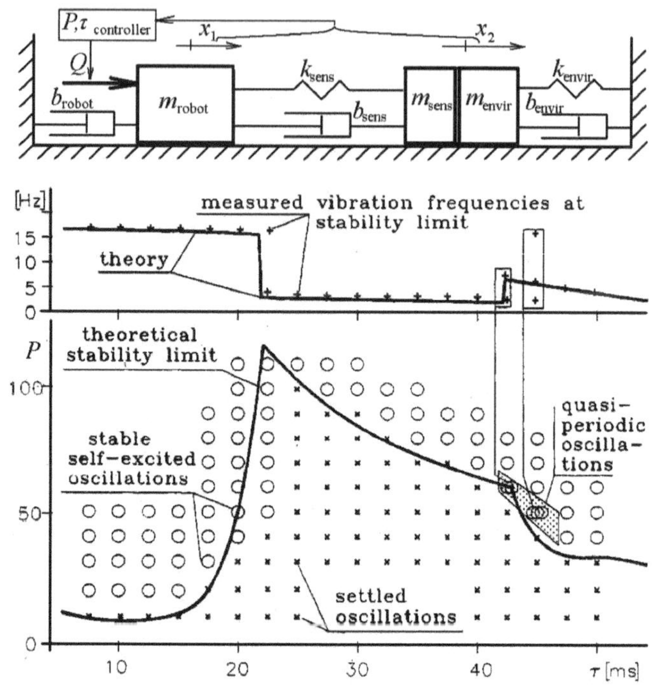

Figure 1. Mechanical model and stability chart for force control in turbine blade polishing.

Figure 2) and the robotic arm. Figure 2 shows also the corresponding mechanical model and stability chart. The parameter values are not listed here, but the corresponding stability chart shows the stable control parameter region for the real mechanical parameter values. Since the differential gains caused stability problems, again, only proportional gain P and integral gain I were used. Actually, the integral terms did not improve the system behaviour much. In this case, we also experienced the improved stability for large sampling times, but those sampling time values were unreasonably high, so we did not study them further.

All these experiences, and similar reports in the literature directed our investigation to the mechanical root of the problem. In the subsequent sections, we study the simplest possible 1 DoF force control model in case of a PD controller with respect to the force error in the presence of the digital effects, namely for finite sampling times.

3. MECHANICAL MODEL

Consider the mechanical model shown in Figure 3. This is a 1 DoF model that can give a good approximation for the behavior of a robotic arm with force control in one direction. The equivalent mass m and equivalent stiffness k represent

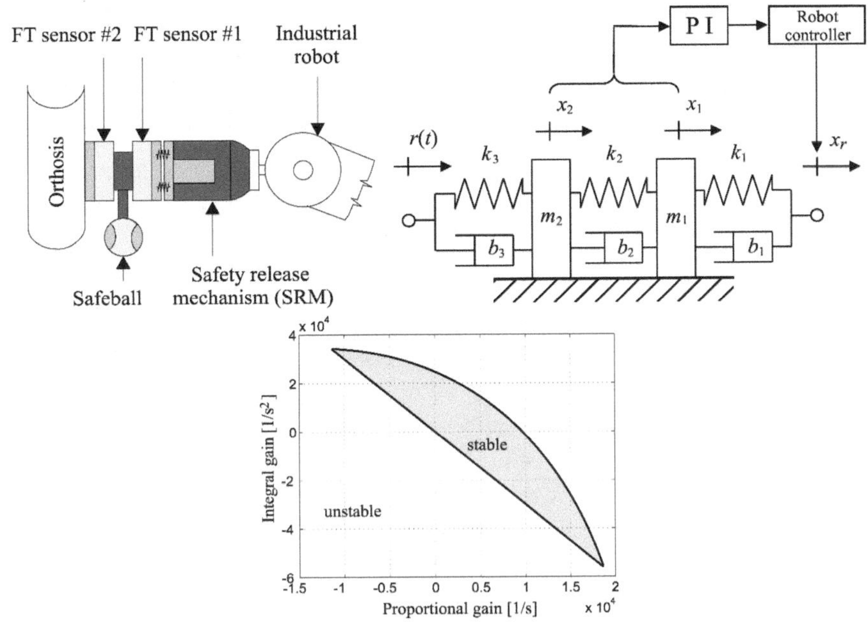

Figure 2. Mechanical model and stability chart for force control applied in rehabilitation robotics.

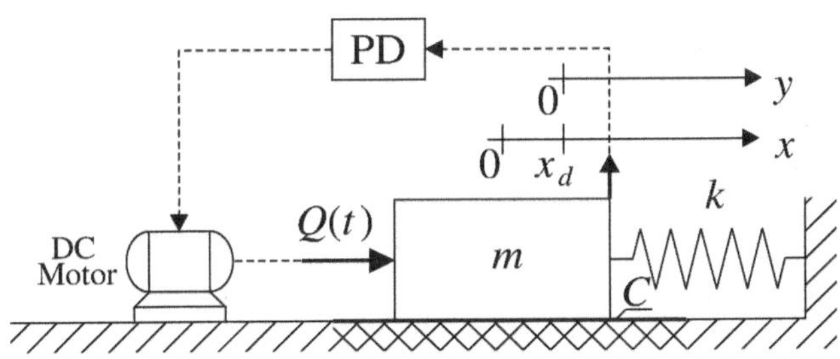

Figure 3. Mechanical model of unidirectional PD force control.

the inertia and stiffness of the robot and the environment in the force controlled direction. These parameters can either be identified experimentally or calculated using the constraint Jacobian representing the force controlled direction, and the mass and stiffness matrices of the robot (Kövecses et al., 2003). The generalized force Q represents the effects of the joint drives. Similar models are frequently used in other papers and books to analyze force control (Stépán, 2001; Craig, 1986; Gorinevsky et al., 1997; Quian and Schutter, 1992).

In the model presented in Figure 3, the notation x_d refers to the position that corresponds to the desired constant force $F_d = kx_d$, while the coordinate $y = x - x_d$ measures the deformation of the spring relative to this desired equilibrium position. The spring is used to represent the interaction force with the environment. Using the force error signal detected via the spring deformation, the simple PD controller determines the control signal for the DC motor that provides the control force (or torque) Q at the joint drive of the robot. The equation of motion of the above mechanical model can be written as

$$\begin{aligned} m\ddot{x}(t) &= -kx(t) - C\,\mathrm{sgn}\dot{x}(t) + Q(t) \\ Q(t) &= F_m(t) - P(F_m(t) - F_d) - D\dot{F}_m(t), \end{aligned} \tag{1}$$

where P and D are the proportional and the differential gains of the PD controller, respectively. In addition, $F_m(t) = kx(t)$ denotes the time-dependent measured force. If there is no dry friction considered in the model, i.e. $C \equiv 0$, the trivial solution $x(t) \equiv x_d$ satisfies equation (1). In this case, the system can reach the desired equilibrium position x_d without a steady state error in principle. Dry friction results in non-zero steady force error, and the higher the proportional gain is, the less this steady force error is (Craig, 1986). This is one of the main reasons why we are interested in applying as large proportional gains as possible, while in the meantime, we often run into stability problems this way. In other words, the steady-state force error cannot be simply eliminated by increasing the proportional gain without the risk of losing stability.

In case of analogue control, however, there is no stability problem to occur in the above simple 1 DoF model. Introducing a perturbation around the desired equilibrium position as

$$x(t) = x_d + y(t)\,, \tag{2}$$

the equation of motion and its characteristic polynomial $p(\lambda)$ obtained by the exponential trial solution $y(t) = c\exp(\lambda t)$, $c \in \mathbb{R}$, $\lambda \in \mathbb{C}$ or by Laplace transformation has the form

$$m\ddot{y}(t) + Dk\dot{y}(t) + Pky(t) = 0\,, \tag{3}$$

$$p(\lambda) = \lambda^2 + D\omega_\mathrm{n}^2\lambda + P\omega_\mathrm{n}^2 = 0, \tag{4}$$

where $\omega_\mathrm{n} = \sqrt{k/m}$ is the natural angular frequency of the uncontrolled mechanical system. The Routh–Hurwitz criterion yields that the solution $y(t) \equiv 0$ corresponding to the desired contact force $F_m(t) \equiv F_d$, is asymptotically stable for any control gains $P > 0,\ D > 0$. There is no upper limit for the gains, which, according to Craig (1986), also means that the steady state force error could be eliminated in principle.

However, stable and accurate digital force control requires the analysis of the more refined discrete-time dynamics of the system. As we show in the subsequent sections, this will explain the unexpected bifurcation phenomena for certain proportional gains and sampling frequencies, as well as the undesired destabilising effect of the differential gain.

4. DIGITAL CONTROL MODEL

To model the digital computer controlled system, we consider a zero-order-hold (ZOH). The force sensor is sampled with the frequency $1/\tau$, while the digital processor sets the control output at the time instants $t_n = n\Delta t$, $n = 0, 1, 2, \ldots$, where Δt is the sampling time of the digital control that is considered to be a large integer multiple of the sampling time of the force sensor. Thus, the measured force is available at every sampling instants of the controller, and the time derivative of the measured force can be estimated by finite differences of the measured force values in practice.

The conventional form of the digital control force for a PD controller would be

$$Q(t) = (1 - P)ky((n-1)\Delta t) - Dk\dot{y}((n-1)\Delta t) + kx_d\,, \qquad t \in [n\Delta t, (n+1)\Delta t) \tag{5}$$

and the corresponding equation of motion has the form

$$\ddot{y}(t) + \omega_{\mathrm{n}}^2 y(t) = (1 - P)\omega_{\mathrm{n}}^2 y((n-1)\Delta t) - D\omega_{\mathrm{n}}^2 \dot{y}((n-1)\Delta t), \qquad t \in [n\Delta t, (n+1)\Delta t) \tag{6}$$

To reduce the number of parameters, introduce the dimensionless time $T = \omega_n t$, and the notation for the derivatives as

$$\frac{d}{dT}(\,) = (\,)' \quad \text{and} \quad \frac{d}{dt}(\,) = \omega_{\mathrm{n}} \frac{d}{dT}(\,). \tag{7}$$

Then the equation of motion is simplified to

$$y''(T) + y(T) = (1 - P)y_{n-1} - D\omega_{\mathrm{n}} y'_{n-1}, \quad T \in [T_n, T_{n+1}) \tag{8}$$

To realize digital force control in the presence of the force error derivative, some kind of digital approximation is needed in practice for the time derivative. The simplest possibility is to use a finite difference approximation in the form

$$\dot{F}_m(n\Delta t) \cong \frac{1}{\tau}\left(F_m((n-1)\Delta t) - F_m\left((n-1)\Delta t - \tau\right)\right). \tag{9}$$

Now, according to (6), (8) and (7), the equation of motion of the system in the dimensionless time domain can be written in the form

$$y''(T) + y(T) = (1 - P)y_{n-1} - \frac{D\omega_n}{(1-q)\Delta T}\left(y_{n-1} - y_{n-2+q}\right)\,, \qquad T \in [T_n, T_{n+1}) \tag{10}$$

where

$$q = \frac{\Delta t - \tau}{\Delta t}, \qquad \Delta T = \omega_n \Delta t \quad \text{and} \quad y_{n-2+q} = y\left((n-2+q)\Delta T\right) . \quad (11)$$

The new parameter q can be interpreted as a sampling quotient relating the sampling time of the force sensor to the sampling time of the controller. Comparing the resulting equation with (8), we can see that the differential term here is divided by the dimensionless sampling time ΔT. Therefore, we can foresee that for high ratios of the mechanical system natural frequency f_n and the control system sampling frequency f_s, i.e., for $f_n/f_s = \Delta T/(2\pi)$, the control will be increasingly similar to the simple proportional controller case when $d = D\omega_n = 0$.

5. STABILITY ANALYSIS AND BIFURCATIONS

The procedure of the stability analysis of the above equation of motion (10) is based on the construction of a discrete map using the piecewise analytic solution of the non-homogeneous equation of motion for each sampling interval, where the non-homogeneous term is piecewise constant. We need to calculate the position of the robot not only at the sampling instants $T_n = n\Delta T$ of the controller, but also at every $T_{n-1+q} = (n-1+q)\Delta T$ instants in order to obtain the measured contact force for the finite difference approximation of the contact force derivative. The lengthy algebraic manipulation results in a simple discrete map in the form

$$\mathbf{z}_{n+1} = \mathbf{W}\mathbf{z}_n , \quad (12)$$

where the 7 dimensional discrete state vector is chosen naturally as

$$\mathbf{z}_n = \left[y_n\,,\ y'_n\,,\ y_{n-2+q}\,,\ y_{n-1}\,,\ y'_{n-1}\,,\ y_{n-3+q}\,,\ y_{n-2}\right]^T . \quad (13)$$

Using the common robotic notations $\cos(\Delta T) = c_{\Delta T}$ and $\sin(\Delta T) = s_{\Delta T}$, the transition matrix $\mathbf{W}$ of the above mapping can be written in the form

$$\mathbf{W} = \begin{pmatrix} c_{\Delta T} & s_{\Delta T} & (1-c_{\Delta T})\tilde{D} & (1-c_{\Delta T})\tilde{P} & 0 & 0 & 0 \\ -s_{\Delta T} & c_{\Delta T} & s_{\Delta T}\tilde{D} & s_{\Delta T}\tilde{P} & 0 & 0 & 0 \\ 0 & 0 & 0 & c_{q\Delta T} & s_{q\Delta T} & (1-c_{q\Delta T})\tilde{D} & (1-c_{q\Delta T})\tilde{P} \\ 1 & 0 & 0 & 0 & 0 & 0 & 0 \\ 0 & 1 & 0 & 0 & 0 & 0 & 0 \\ 0 & 0 & 1 & 0 & 0 & 0 & 0 \\ 0 & 0 & 0 & 1 & 0 & 0 & 0 \end{pmatrix} , \quad (14)$$

where

$$\tilde{D} = \frac{D\omega_n}{(1-q)\Delta T} \quad \text{and} \quad \tilde{P} = 1 - P - \frac{D\omega_n}{(1-q)\Delta T} . \quad (15)$$

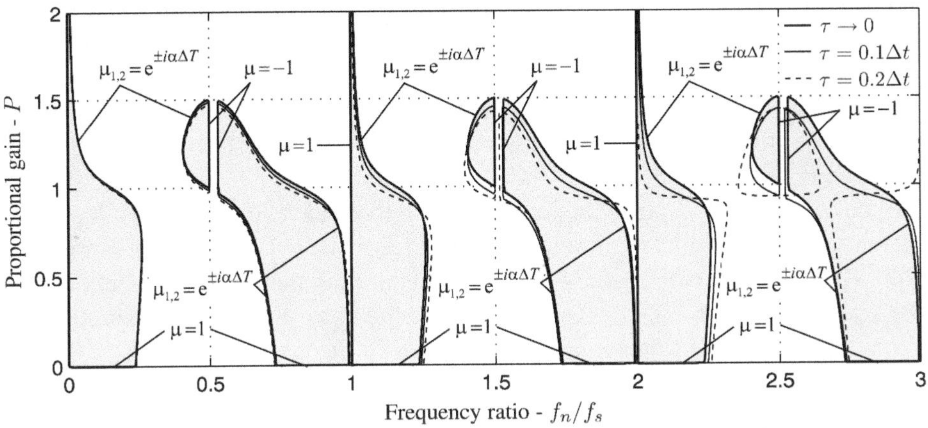

Figure 4. The effect of the finite difference approximation, $d = 0.1$.

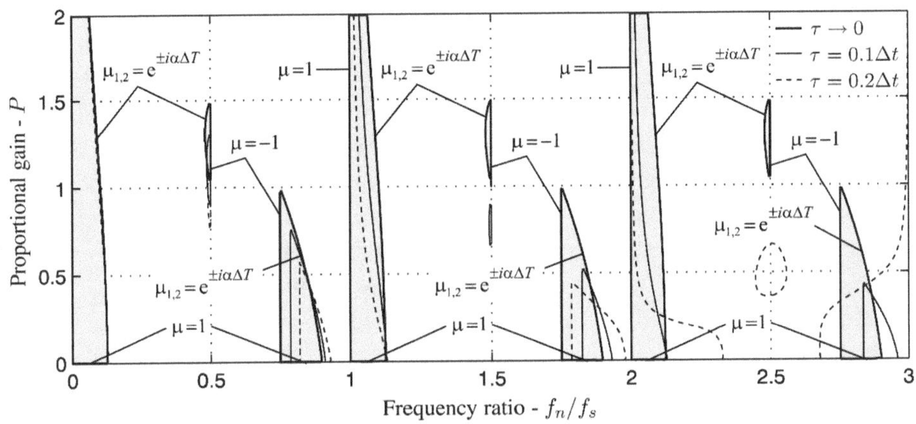

Figure 5. The effect of the finite difference approximation, $d = 1$.

The exponential stability of the digital force control is equivalent to the stability of the above discrete mapping, i.e., to the convergence of this vector geometric series. Consequently, the stability of the system can be investigated by checking the 7 eigenvalues $\mu_{1,\dots,7}$ of the transition matrix $\mathbf{W}$, whether these eigenvalues are located within the open unit disc of the complex plane. With the help of standard numerical methods, we can check this condition of stability.

The corresponding $f_n/f_s - P$ stability charts are calculated for small ($d = 0.1$) and high ($d = 1$) dimensionless differential gains and presented in Figures 4 and 5, respectively. These figures show the deviations from the ideal charts constructed for zero force sensor sampling time $\tau = 0$ as the finite difference approximation of the derivative of the measured force shows up with

$\tau > 0$. The shaded domain bounded by the thick solid lines refer to the case when the time derivative of the measured force is considered as a continuous input signal of the controller, i.e., the sampling time of the force/torque sensor is negligible compared to the sampling time of the digital force control loop. In this case, the shaded stable domains also illustrate the periodic nature of the $f_n/f_s - P$ stability charts.

The thin solid line presents the deviations from these charts, when the finite difference approximation is characterized by $\tau = 0.1\Delta t$, i.e. the sampling frequency of the force sensor is ten times higher than the sampling frequency of the force controller. The dashed line shows the case when the measured force is sampled only five times in a sample period of the controller, i.e. $\tau = 0.2\Delta t$.

Figure 4 shows that the finite difference approximation of the measured force causes that the periodic nature of the stability chart disappears with the increase of the frequency ratio. Moreover, the shape of the stable domain of control parameters will converge to the shape of the stability chart obtained for zero differential gain (see Stépán, 2001). Thus, for sufficiently high frequency ratios, the PD digital force control with finite difference approximation works as a pure proportional control.

Figure 5 presents the case when the differential gain is large. In this case, the above described convergence of the stable domains is less apparent. In the meantime, the deviations of the stability boundaries that correspond to the finite difference approximation at the differential part of the digital force controller are better illustrated. For low frequency ratios, there is not much change in the stability domain due to the finite difference approximation.

The comparison of the charts in Figures 4 and 5 clearly show, that the increase of the differential gain causes loss of most of the stable regions.

The charts also represent those critical eigenvalues μ of the transition matrix $\mathbf{W}$ that are of modulus 1 at the limit of stability. This clearly shows that either secondary Hopf (in other words, Neimark-Sacker) bifurcations or period doubling (in other words, flip) bifurcations can occur, corresponding to the complex conjugate pairs of eigenvalues $\mu_{1,2}$, or to the $\mu = -1$ eigenvalue, respectively. This means, that the largest frequency of the arising self-excited vibrations cannot be greater than the half of the sampling frequency of the system.

6. CONCLUSION

The application of differential terms in the digital force control results in the reduction of most of the regions of stability in the parameter space. This effect is the same if a finite difference approximation is used for the force derivative term in the control loop. The finite difference approximation provides a simple kind of filtering of the force derivative signal. For stiff mechanical systems having large natural frequency f_n, the force control stability properties converge to that of the simple proportional controller.

The stability charts of digital force control show an intricate structure, and several bifurcations can occur at the limits of stability either with or without differential gains in it. The digital effects need special attention during the design of force control in case of stiff mechanical systems with low internal viscous damping to be contacted.

ACKNOWLEDGMENTS

This research was supported by the Hungarian Scientific Research Foundation grant No. T043368, the Natural Sciences and Engineering Research Council of Canada grant No. 206256, and the Bilateral Hungarian-Canadian Society and Technology Program grant No. TéT-BILAT CAN-1/03.

REFERENCES

Arz, G., Tóth, A., Fazekas, G., Bratanov, D. and Zlatov, N. (2003). Three-dimensional anti-spastic physiotherapy with the industrial robots of "Reharob". In *Proceedings of The Eight International Conference on Rehabilitation Robotics (ICORR 2003)*, Republic of Korea, pp. 215–218.

Craig, J.J. (1986). *Introduction to Robotics Mechanics and Control*. Addison-Wesley, Reading, MA.

Gorinevsky, D.M., Formalsky, A.M. and Schneider, A.Yu. (1997). *Force Control of Robotics Systems*. CRC Press, Boca Raton, FL.

Kovács, L.L. and Stépán, G. (2003). Dynamics of digital force control applied in rehabilitation robotics. *Meccanica*, 38(2):213–226.

Kövecses, J., Piedboeuf, J.C. and Lange, C. (2003). Dynamics modeling and simulation of constrained robotic systems. *IEEE/ASME Transactions on Mechatronics*, 8(2):165–177.

Natale, C. (2003). *Interaction Control of Robot Manipulators*. Springer-Verlag, Berlin.

Quian, H.P. and Schutter, J. De (1992). The role of damping and low pass filtering in the stability of discrete time implemented robot force control. In *Proceedings of the 1992 IEEE International Conference on Robotics and Automation*, Nice, France, p. 1368–1373.

Siciliano, B. and Villani, L. (1999). *Robot Force Control*. Kluwer Academic Publishers, Dordrecht.

Slotine, J.J.E. and Li, W. (1991). *Applied Nonlinear Control*. Prentice-Hall, Englewood Cliffs, NJ.

Stépán, G. (1989). *Retarded Dynamical Systems*. Longman, London.

Stépán, G. (2001). Vibrations of machines subjected to digital force control. *International Journal of Solids and Structures*, 38:2149–2159.

Stépán, G. and Haller, G. (1995). Quasiperiodic oscillations in robot dynamics. *Nonlinear Dynamics*, 8:513–528.

VIBRATION CONTROL OF ELASTIC JOINT ROBOTS BY INVERSE DYNAMICS MODELS

Michael Thümmel, Martin Otter and Johann Bals
DLR, Institute of Robotics and Mechatronics, Germany

Abstract: In this article tracking control of nonlinear plants using a two degree of freedom controller structure is considered. The work described herein focuses on the design of feedforward controllers based on automatic generation of inverse plant models. The method is applied to the problem of vibration control of elastic joint robots and is demonstrated with simulations and experimental results.

Key words: inverse dynamics model, elastic joint robot, two degree of freedom controller, Modelica, Dymola.

1. INTRODUCTION

This article addresses model based control of mechanical systems, in particular robots. When the tracking problem of such a plant is considered, the controller often includes some kind of non-linear inverse dynamics of the system. Depending on the controller, the inverse dynamics may be used in the feedback part, e.g. in feedback linearization, or in the feedforward part, e.g. in output regulation. For complex systems with fast dynamics like robots, computation of the inverse dynamics involves a large computational effort. In particular use of the inverse dynamics in the fast feedback loop often fails due to real-time requirements. Therefore, considering practical relevant robots, the inverse dynamics is used in the feedforward controller. Here sampling times can be larger and there are no hard real-time constraints.

In the following section a general framework is discussed to automatically derive a non-linear inverse model from a given plant model. This inverse model is used in the feedforward path of an appropriate two degree of freedom controller which is introduced in Section 3. Section 4

H. Ulbrich and W. Günthner (eds), IUTAM Symposium on Vibration Control of Nonlinear Mechanisms and Structures, 343–353.

highlights practical aspects of inverse model generation. In Section 5, the explained technique is applied to a robot with elastic joints. Section 6 presents experimental results from our laboratory robot illustrating the reduction of vibrations at the tool. Section 7 concludes the paper.

2. DERIVATION OF INVERSE MODELS

The problems related to computation of non-linear inverse models shall be illustrated by calculation of the inverse dynamics of a simple robot having one elastic joint as shown below.

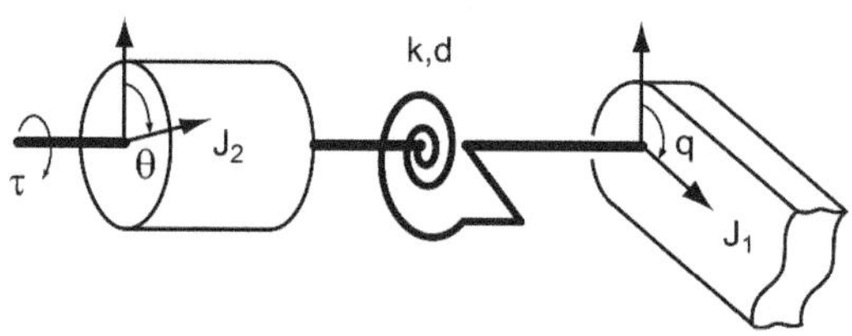

Figure 1. Robot with one elastic joint.

The equations of motion are given by

$$\begin{aligned} 0 &= J_1\ddot{q} + k(q-\theta) + d\cdot(\dot{q}-\dot{\theta}) + g(q) \\ \tau &= J_2\ddot{\theta} - k(q-\theta) - d\cdot(\dot{q}-\dot{\theta}) \end{aligned} \tag{1}$$

with joint position q(t), motor position $\theta(\tau)$, motor torque $\tau(\tau)$, link inertia J_1, motor inertia J_2, a strictly increasing stiffness characteristics $k(.)\in C^1$ which is symmetric with respect to the origin, damping constant d and torque due to gravity g(q). To obtain the inverse dynamics, τ must be computed from a given q(t) and higher derivatives of q(t). In a first step, θ is calculated from the upper part of Eq. (1). It can be shown that this differential equation with state θ, which is known as the "internal dynamics" of the system, is stable and therefore the solution $\theta(t)$ can be further utilized. Note, since k(.) is a non-linear function, the solution $\theta(t)$ is usually calculated by numerical integration. In order to compute τ from the lower part of Eq. (1), the second derivative of θ must be calculated by differentiation of the first equation. This results in a new algebraic equation.

$$\ddot{\theta} = \frac{1}{d}\left(J_1 q^{(3)} + \frac{\partial k}{\partial(q-\theta)}\cdot(\dot{q}-\dot{\theta}) + \frac{\partial g}{\partial q}\dot{q} \right) + \ddot{q} \tag{2}$$

As illustrated with this simple example, the derivation of the inverse dynamics model may require to differentiate a subset of the equations, select appropriate states and solve the resulting system of differential and algebraic equations numerically. Further, derivatives of the desired trajectory must be provided up to a certain order. Below it will be shown how these tasks can be handled automatically using algorithms for solution of differential algebraic equations.

These algorithms start with a general model representation described by a set of differential and algebraic equations (= DAE)

$$\mathbf{0} = \mathbf{f}(\dot{\mathbf{x}}, \mathbf{x}, \mathbf{y}, \mathbf{u}) \tag{3}$$

where $\mathbf{x}$(t) are variables that appear differentiated in the DAE, $\mathbf{y}$(t) are algebraic and $\mathbf{u}$(t) are known input functions of time t. In order to solve Eq. (3) it is, at least numerically, possible to perform an index reduction step and transform the system to state space form

$$\begin{bmatrix}\dot{\mathbf{x}}_1 & \mathbf{x}_2 & \mathbf{y} & \mathbf{w}\end{bmatrix}^T = \mathbf{f}_S(\mathbf{x}_1, \mathbf{u}) \tag{4}$$

Here $\mathbf{x}_1$ and $\mathbf{x}_2$ form the vector $\mathbf{x}$ such that the subset vector $\mathbf{x}_1$ is the *state vector* and contains the independent variables of $\mathbf{x}$. For example in multi-body systems, $\mathbf{x}_1$ are the minimal coordinates and $\mathbf{x}_2$ are the coordinates that are constrained. The vector $\mathbf{w}$ contains higher order derivatives of $\dot{\mathbf{x}}_1$, $\mathbf{x}_2$ and $\mathbf{y}$ that emerge when differentiating equations of $\mathbf{f}$(.). They are treated as algebraic variables. The equations to be differentiated can be determined with the algorithm of Pantelides (1988). The selection of the state variables $\mathbf{x}_1$ can be performed with the "dummy derivative method" of Mattsson and Söderlind (1993). Both algorithms are, for example, available in the Modelica (Modelica Association 2005) simulation environment Dymola (Dynasim 2005). Computation of $\mathbf{f}_S$(.) might include solution of linear and/or non-linear algebraic systems of equations.

In a plant model for controller design and simulation, the control variables are chosen as inputs and the variables that are required to track as outputs. In robotics this kind of model is known as forward dynamics.

Equivalently an *inverse* model of the plant can be constructed by exchanging the meaning of variables: A subset of the control variables, $\mathbf{u}_{inv}$, with dimension n_{inv}, is no longer treated as known but as *unknown*, and n_{inv} previously unknown tracking variables are treated as *known* inputs. This kind of model is named inverse dynamics in robotics. It can still be written in form of Eq. (3) and handled with the same methods as any other DAE. Some more practical aspects of derivation of inverse models from plant models can be found in Section 4.

Use of the inverse model in the controller as described in the next section is only possible if its DAE has a *unique solution* and if it is *stable*. Therefore the internal dynamics of the plant associated with the output to be tracked has to be stable. For linear systems this means, that the plant may not have unstable transmission zeros. For a general DAE no stability proof exists, but for certain classes of DAEs, including models of robots with elastic joints, it is possible to prove the stability of the inverse model. Since the internal dynamics is usually non autonomous, only boundedness of the internal states can be guaranteed. This can e.g. be analyzed using the notion of input-to-state stability from Sontag (1989). For the one link robot of Eq. (1), after a coordinate transformation $(\gamma,\delta)=(q-q_0,q-q_0-\theta)$ with q_0 such that $g(\gamma=0)=0$, one can show that the integral of $k(\delta)$ is an ISS-Lyapunov function for the inverse model. Therefore θ is bounded, provided $q(t) \in C^2$.

In case it is not possible to prove stability, simulations can be performed to check whether the inverse model is stable in the desired operation region. An alternative is to linearize the plant model around several stationary operating points and check whether the transmission zeros are stable. Of course, none of these checks can guarantee that the inverse DAE is stable.

If the inverse model is unstable, *approximate* inversion might be used for the controller. Therefore, since the internal dynamics is associated with the output, often an output redefinition approach is used. For linear single-input/single-output systems this can e.g. be achieved by removing unstable zeros before inverting the plant. For a non-linear DAE, one might choose other outputs that are similar to the original outputs, but don't lead to instability. Alternatively, the DAE might be modified before inversion, e.g., by introducing additional damping terms.

Since the transformation from Eq. (3) to Eq. (4) might differentiate a subset of the model equations, the known inputs of the model may be differentiated too. Therefore, the derivatives of these inputs must exist and must be provided up to a certain order. These derivatives can be computed, e.g., if the inputs are available as functions that can be differentiated sufficiently often. Alternatively, a desired reference model might be used that, in combination with the inverse DAE, results in a DAE that does not require derivatives of inputs. The reference model could, e.g., be selected as a filter such that a combination of the filter states yields the needed derivatives.

3. CONTROLLER USING AN INVERSE MODEL

The question arises how to use an inverse DAE model in a controller. Kreisselmeier (1999) has proposed and analyzed a controller structure with two structural degrees of freedom for linear, single-input/single-output systems. A generalization of this controller structure to *nonlinear* multi-input/multi-output systems is shown in Figure 2.

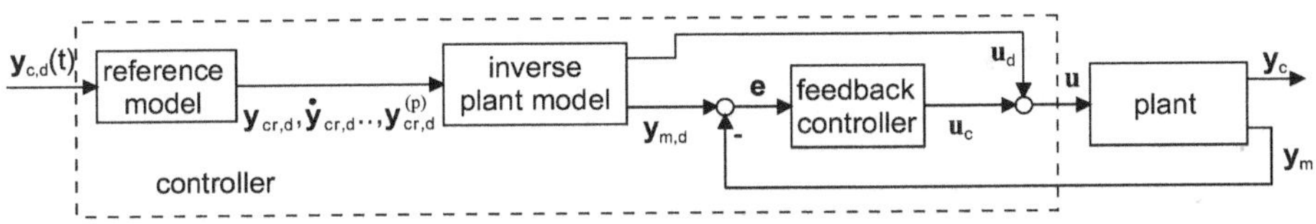

Figure 2. Controller with two structural degrees of freedom.

In Figure 2 the multi-input/multi-output plant has inputs $\mathbf{u}$, measured signals $\mathbf{y}_m$ and outputs $\mathbf{y}_c$ that are primarily controlled. The number of inputs must be identical to the number of controlled variables: $\dim(\mathbf{y}_c) = \dim(\mathbf{u})$. In this case the inverse plant model with (known) inputs $\mathbf{y}_c$ and unknown outputs $\mathbf{u}$ is used in the feedforward path of the controller to compute the desired actuator inputs $\mathbf{u}_d$ to the plant.

A "reference model" is used to provide a smooth trajectory $\mathbf{y}_{cr,d}$ and its derivatives, which are the inputs of the inverse plant model. The inverse plant model computes the desired measurement signals $\mathbf{y}_{m,d}$ and the desired plant inputs $\mathbf{u}_d$. For stabilization and robustness purposes a correction is added that is computed by a feedback controller that has the error between the measured signals and the desired measured signals as input. The feedback controller might be a simple state controller.

The *feedback controller* must be able to stabilize the system around the desired trajectory. Assuming the inverse plant model computes the same value for $\mathbf{y}_{m,d}$ as the measured quantity $\mathbf{y}_m$, then control error $\mathbf{e}$ is zero. Further, the feedback controller starts from a stationary state which is selected such that $\mathbf{u}_c=0$. Then $\mathbf{u}$ equals $\mathbf{u}_d$ which is computed in the inverse plant model. If the inverse model of the plant is ideal and starts at the same initial values as the plant, the plant will produce trajectories $\mathbf{y}_m(t)=\mathbf{y}_{m,d}(t)$ (because the inverse plant model was constructed in this way). Then again the preliminary assumption of zero control error $\mathbf{e}$ is fulfilled and the controller has no effect.

In case a control error occurs, the controller has to stabilize the system and cope with the imprecise inverse plant model. The proposed controller structure is advantageous because design of the feedback controller is decoupled from the feedforward controller that is responsible for tracking the given reference input.

Note, that due to the particular controller structure, the *reference model* determines essentially the input/output behaviour $\mathbf{y}_{c,d} \rightarrow \mathbf{y}_c$ of the overall system: According to the analysis above, the "feedback controller" in Figure 2 has no effect for an ideal inverse plant model. In this case, the overall system consists of a series connection of reference model, inverse plant model and plant model. Using the same assumptions as above, the output of the plant model is identical to the input of the inverse plant model. As a result, the reference model can be interpreted as the "desired transfer function" of the overall system. Since the inverse plant model is in the feedforward path, the calculation of $\mathbf{u}_d$ and of $\mathbf{y}_{m,d}$ might be performed *offline*, so that hard real-time requirements for the solution of the inverse plant model are not present.

4. CONSTRUCTING INVERSE MODELS USING MODELICA AND DYMOLA

Deriving a non-linear inverse model manually can be a challenging task. For complex models this is not practical. Our approach is to use the Modelica modeling language (Modelica Association 2005) to define the model and the Modelica simulation environment Dymola (Dynasim 2005) to derive the inverse model automatically and perform simulations. The object-oriented modeling language Modelica is designed to allow convenient, component-oriented modeling of complex physical systems, e.g., systems containing mechanical, electrical, electronic, hydraulic, thermal, control, electric power or process-oriented subcomponents. The free Modelica language and libraries are available, ready-to-use and have already been utilized in demanding industrial applications. In Figure 3, some example models of available Modelica libraries useful for robot simulation are shown.

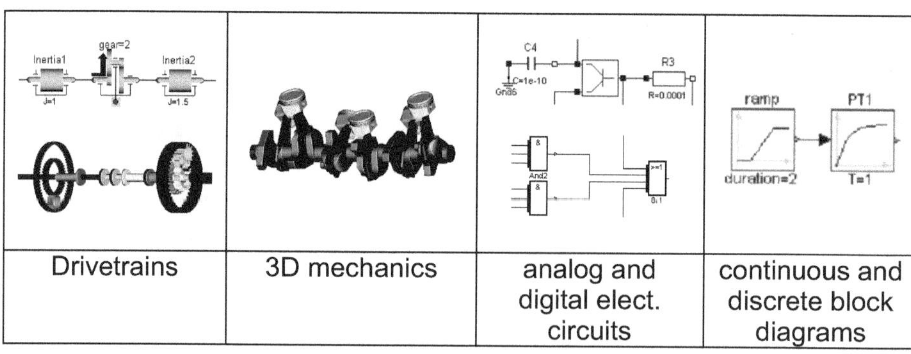

Figure 3. Modelica component libraries.

With the Modelica simulation environment Dymola, the practical derivation of inverse models is straightforward, even for complex systems:

1. Define the plant model and include input and output signals of the plant over which the inversion shall take place.

2. If necessary, provide a reference model or input filter of appropriate relative degree. The relative degree may be known from physical knowledge of the plant dynamics, or can be automatically derived as described below.
3. Connect the components using a twoInputs block as shown in Figure 4. The twoInputs block is necessary because in block diagrams it is not allowed to connect two output signals with each other.

On the left side of Figure 4 the plant model with one input and one output is present.

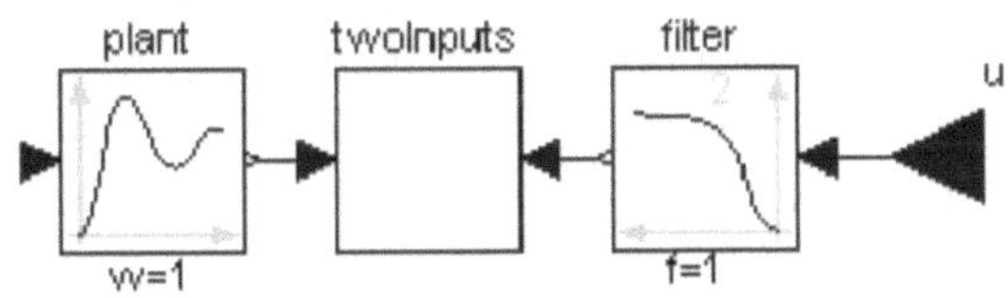

Figure 4. Definition of inverse model with Modelica.

On the right side, a filter is used as reference model. If the filter order is too low the DAE is not causal and Dymola prints an error message of the following form:

Error: The model requires derivatives of some inputs as listed below:

Order of derivative	Input
4	u1
2	u2
3	u3

Error: Failed to reduce the DAE index

In the second column the Modelica names of the input signals are listed that need to be differentiated according to the differentiation order of the first column. The numbers in the first column are therefore the minimum order of the corresponding filters. The higher the required filter order, the more problems usually occur when applying the inverse model in a control system. In such cases, one might simplify the plant model.

5. ELASTIC JOINT ROBOTS

The computation of the inverse dynamics for rigid robots is a standard task. Using the well-known equations of motion for tree-structured rigid multi-body systems

$$\mathbf{M}(\mathbf{q})\ddot{\mathbf{q}} + \mathbf{C}(\mathbf{q},\dot{\mathbf{q}}) + \mathbf{g}(\mathbf{q}) = \boldsymbol{\tau} \tag{5}$$

and given the desired joint positions $\mathbf{q}(t)$ and their derivatives with respect to time up to a order of two, the generalized forces exerted by the drives $\boldsymbol{\tau}$ can be calculated algebraically. De Luca (1996) showed that, when more realistic models including drive dynamics and elasticity in the joints are taken into account the computation becomes more complex. Derivatives of the desired trajectory $\mathbf{q}(t)$ must be provided up to a certain order, because the equations of motion must be differentiated several times in order to calculate $\boldsymbol{\tau}$. The necessary number of derivatives of $\mathbf{q}(t)$ depends on the robot kinematics and cannot be given easily in closed form. An upper bound for an robot having N joints is 2(N+1) without damping in the joints, and 2+N, if damping is present in all joints. A first application of the proposed controller using an inverse dynamics model for elastic joint robots can be found in Thümmel (2001).

The framework presented in the previous sections is now applied to the DLR lightweight robot LBR2 shown in Figure 5. This manipulator arm of approximately 1 m length is able to handle a payload of 8 kg with a total weight as low as 17 kg. It is a chain-structured mechanism with seven revolute joints. The motors are mounted in the joints, their rotors' axes of rotation coincide with the joint axes, as depicted in the right part of Figure 5. The drive trains are known to be elastic due to harmonic drive reduction gears and torque sensors. They are modeled like in the example of Section 2, with non-linear joint stiffness that are roughly 10^4 Nm/rad and damping coefficients that are roughly 15 Nms/rad. A more detailed description of the LBR2 is given in Albu-Schäffer (2002). In Höpler (2004) the inverse dynamics problem for this robot was solved using spatial operator algebra and the reasons for the necessity of differentiation of the equations of motion were given as well.

Figure 5. DLR light weight robot LBR 2.
Left part: complete arm, right part: joint with rotor and drive.

Starting from a detailed Modelica model of the LBR 2, the inverse dynamics model of the robot was automatically derived, as described before. It was found, that this model requires derivatives of the equations of motion up to an order of 4, what is undesirable for use in a controller. This higher order results from gyroscopic effects introduced by the motors which are mounted on the robot arms. It is well known, that these effects decrease with increasing gear ratio. Taking into account the high gear ratios of the LBR2, it seems justified to neglect gyroscopic effects *of the motors*. Therefore, in a simplified model the motors are assumed to be not moving with the bodies to which they are attached but mounted on the ground. This significantly simplifies the model and reduces computation time since now the equations of motion only need to be differentiated once. A comparison of the models can be found in Table 1.

Table 1. Comparision of required differentiability.

Type of model	Detailed model	Simplified model
Required order of derivatives of the equations of motion	4	1
Required order of derivatives of q(t)	6	3

In order to verify the assumption that gyroscopic effects of the motors can be neglected, simulations have been performed to check the errors introduced by the simplification of the model. In the example shown in Figure 6 the desired trajectory for all joints was a step input from one to two radian filtered with a 8th order filter with cut-off frequency of 5Hz.

As can be seen, the errors with the simple model are roughly one percent of the overall torque and much smaller than the errors of a rigid model where the elasticity in the joints is neglected. Therefore the simple model is a good approximation for control purposes.

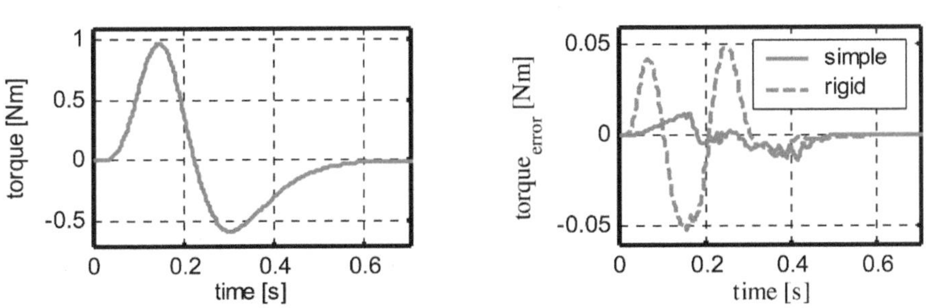

Figure 6. Torques computed with inverse dynamics models.
Left part: torque for detailed model. Right Part: torque errors with respect to detailed model.

6. EXPERIMENTAL RESULTS

In order to check the performance of the proposed controller, tests with our laboratory robot have been carried out using the controller structure shown in Figure 2. The feedback controller was chosen to be a PD controller for every joint and a low-pass filter was used as a reference model.

On the left hand side of Figure 7 the tool vibrations after a robot motion are shown. The motion was executed two times: one time without feedforward controller (PD) and one time with the inverse dynamics model in the feedforward part (PD+inv.). As can be seen the inverse dynamics model reduces vibrations for positioning tasks.

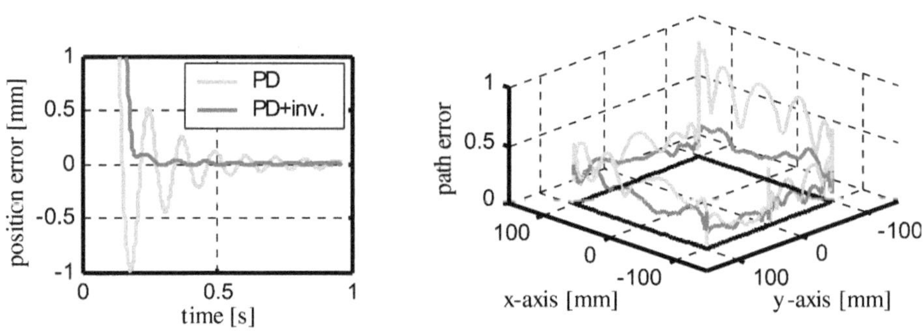

Figure 7. Comparison of control errors. Bright line: controller without feedforward. Dark line: controller using inverse dynamics model. Left part: positioning experiment. Right part: tracking experiment.

On the right hand side of Figure 7 the results of a trajectory tracking experiment are shown. The desired trajectory was chosen to be a square with start point [-100, 0, 0] mm traversed in clockwise direction as indicated by the black line. Again this movement was executed with and without the inverse dynamics model in the feedforward part and measured with an optical coordinate measurement system. The figure shows normalized absolute deviations from the desired path displayed on the z axis. As can be seen also the tracking error can be reduced significantly when using the inverse dynamics model.

7. CONCLUSION

Vibration control of elastic joint robots with automatically generated inverse dynamics models in the feedforward controller was considered. It has been shown, that inverse models of general plants can be derived automatically using existing algorithms for DAE systems. The method was illustrated by simulations with a model of the LBR2, and a simplified model for use in a controller was derived. As shown in experiments, the proposed controller can reduce robot vibrations in positioning tasks as well as tracking errors significantly.

ACKNOWLEDGEMENT

This work was in parts supported by "BMBF Verbundprojekt PAPAS" (Plug And Play Antriebs- und Steuerungskonzepte für die Produktion von Morgen) under contract number 02PH2060.

REFERENCES

Albu-Schäffer, A. (2002). Regelung von Robotern mit elastischen Gelenken am Beispiel der DLR-Leichtbauarme. *Ph.D. Dissertation*, Technische Universität München.

De Luca, A. and Tomei, P. (1996). Elastic Joints. In: *Theory of Robot Control*, de Wit, C.C., Siciliano, B. and Bastin, G. (eds), Springer Verlag, Berlin, pp. 179–217.

Dynasim, 2005, Dymola – Users Manual; http://www.dynasim.com.

Höpler, R. and Thümmel, M. (2004). Symbolic Computation of the Inverse Dynamics of Elastic Joint Robots. In: *Proc. of the 2004 IEEE Intern. Conference on Robotics and Automation*, pp. 4314–4319.

Kreisselmeier, G. (1999). Struktur mit zwei Freiheitsgraden. *Automatisierungstechnik* **6**:266–269.

Mattsson, S.E. and Söderlind, G. (1993). Index Reduction in Differential-Algebraic Equations Using Dummy Derivatives. *SIAM Journal of Scientific and Statistical Computing* **14**:677–692.

Modelica Association (2005). Modelica™ – A Unified Object-Oriented Language for Physical Systems Modeling. Tutorial,Version 1.4; http://www.modelica.org.

Pantelides, C.C. (1988). The Consistent Initialization of Differential-Algebraic Systems. *SIAM Journal of Scientific and Statistical Computing* **9**:213–231.

Sontag, E.D. (1989). Smooth Stabilization Implies Coprime Factorization. *IEEE Trans. Aut. Control* **34**:435–443.

Thümmel, M., Otter, M. and Bals, J. (2001). Control of Robots with Elastic Joints based on Automatic Generation of Inverse Dynamics Models. In: *Proc. of 2001 IEEE/RSJ International Conference on Intelligent Robots and Systems*, pp. 925–930.

MOTION CONTROL DESIGN OF A MOLDED PANTOGRAPH MECHANISM WITH LARGE-DEFLECTIVE HINGES

Thomas Thümmel and Robert Huber
Institute for Applied Mechanics, Technical University of Munich, Germany

Mikio Horie and Chikara Ishikawa
Precision and Intelligence Laboratory, Tokyo Institute of Technology, Japan

Abstract: The paper deals with a material coherent miniature pantograph mechanism. It consists of large-deflective hinges and links made from Polypropylene by injection molding. Experiments and simulations help to verify and improve an elastic Multi-Body-System model, especially different input motions are applied by a shaker in horizontal direction. Laser sensors provide the signals for parameter identification in time and frequency domain. The verified MBS model allows to design the motion control for a picking and placing task with 150 ms cycle time. An improved trajectory could reduce the dynamic displacement error of the tool point E below 0.1 mm within the placing dwell.

Key words: motion control, pantograph, multi-body-system, modelling, vibration.

1. INTRODUCTION

Along with device miniaturization such as portable electric devices (Figure 2a) and computer accessories, minute devices at a size of less than some mm squares are currently bound or assembled on electric circuit boards. However, most of the minute devices above are often assembled by meter-ordered huge surface mounting systems, e.g. 2 m (width) $\times$ 4 m (length) $\times$ 2 m (height).

Therefore a new miniature surface mount system composed of parallel arrangement pantograph mechanisms and a 3-DOF parallel mechanism shown in Figure 1 was proposed in order to use this new surface mount system in the one-room factory [1–3]. In this contribution we consider only one separated plane manipulator. The molded miniature pantograph mechanism consists of large-deflective hinges and links (material coherent, Polypropylene) and two linear actuators (Figure 2b) to realize input x and y.

H. Ulbrich and W. Günthner (eds), IUTAM Symposium on Vibration Control of Nonlinear Mechanisms and Structures, 355–364.

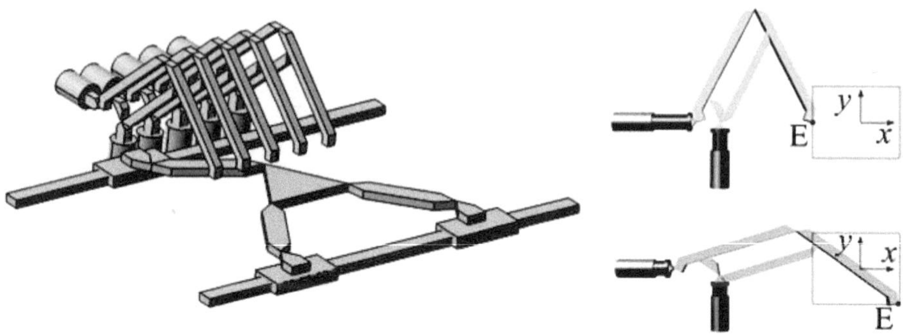

Figure 1. The new surface mount system and one pantograph in two positions.

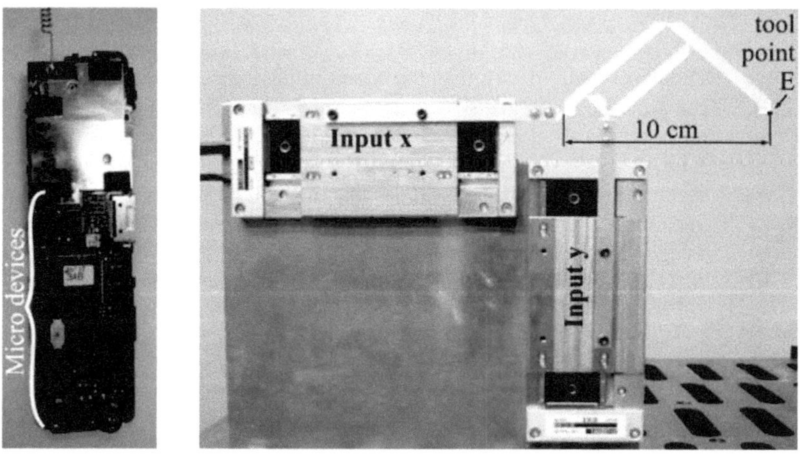

Figure 2. (a) Substrate in the cellular phone. (b) Pantograph and actuators.

The manipulator can work with a speed up to 20 cycles per second in a steady state operation mode. Horie et al. [1] arranged several pantograph mechanisms whose tool point had a $\pm 11\ \mu$m static positioning repeatability during picking and placing works, i.e. the point E of the output drew a U-shape trajectory in the working plane (length 50 mm, height 40 mm).

In this paper we lay emphasis on the input motion design in respect of low dynamic positioning errors at the tool point E.

2. MODELLING AND SIMULATION

The motion control design is based on elastic Multi-Body-System (MBS) modelling and simulation. These investigations apply experimental modal analysis and measurements by laser sensors to verify the elastic MBS model.

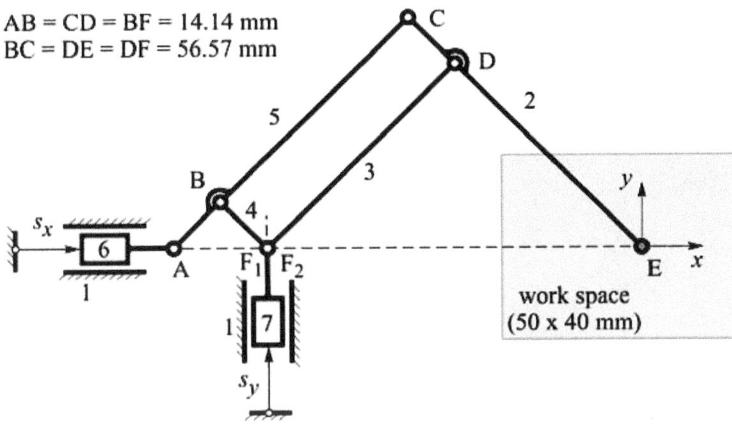

Figure 3. Reference model – Pantograph as classical linkage mechanism.

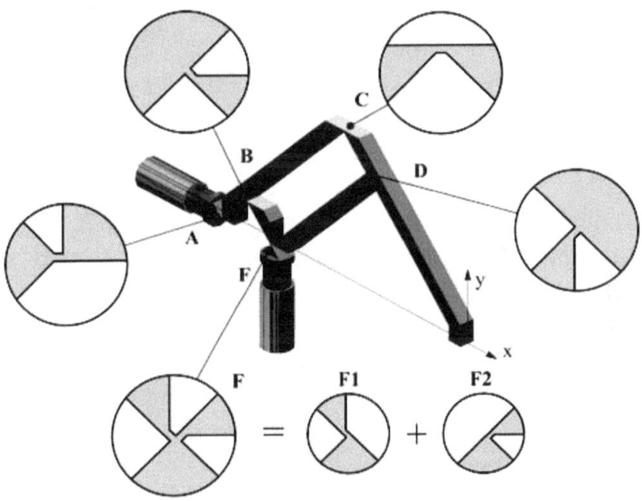

Figure 4. The large-deflective hinges of the molded pantograph.

2.1 Two Different Models

As a first step, a *reference model* of the pantograph is assumed. The classical plane linkage mechanism includes rigid bodies, revolute joints and sliding pairs, shown in Figure 3. The reference motion is defined as follows:

$$x = -4s_x \quad \text{and} \quad y = 5s_y. \tag{1}$$

The plane work space of the tool point E is enclosed by the rectangle of 50 $\times$ 40 mm. The above mentioned ± 11 μm static positioning repeatability was measured by Horie et al. [1] with respect of the motion of this reference model.

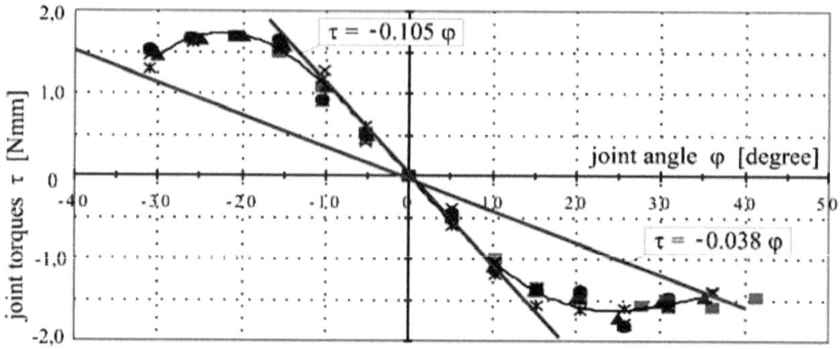

Figure 5. Rotational spring characteristics of large deflective hinges.

The thickness of the links with a square cross section is about 5 mm. The large-deflective hinges A, B, C, D and F consist of the same material as the links, i.e. Polypropylene (PP) as shown in Figure 4. The hinges (length 180 μm and width 200 μm) are modelled by FEM to analyze the restoring torque of rotational elastic deformation. The restoring torque acting at every hinge has similar nonlinear characteristics, although the geometry of each type of hinge is different. The stiffness coefficient (see Figure 5) in the range of ± 10 degree is approximated by 0.105 Nmm/degree or 6.0 Nmm/rad.

The torsional springs can reduce the necessary input forces of the pantograph because of storing energy [2]. The damping characteristics of the large-deflective hinges will increase the restoring torque, assuming a viscous damping coefficient of 0.027 Nmms available from experiments.

The second model is based on the flexible MBS theory. The *MBS model* includes discrete rotational springs and dampers in the joints. The elastic beams, especially the beams 2, 3 and 5 are defined as shape functions (see Figure 6).

Experimental modal analysis and calculations by FEM provides the mode shapes and eigenfrequencies of the pantograph mechanism. Figure 7 illustrates the interesting bending modes of the pantograph. The deformation of the small beam 4 was neglected.

The properties of the links are given by the material parameters of Polypropylene (PP) or have been estimated by adapting simulation results, e.g. Poisson rate 0.4, modulus of elasticity 1240 MPa and density of 827 kg/m^3.

2.2 Parameter Identification and Validation

Successful motion design based on modelling and simulation requires experimental validation of parameters, sometimes the most expensive task.

Horie et al. [1] verified the static behavior of the pantograph and Ishii et al. [2] investigated dynamic properties. This paper presents additional experiments (Figure 9) about eigenfrequencies and damping in frequency domain.

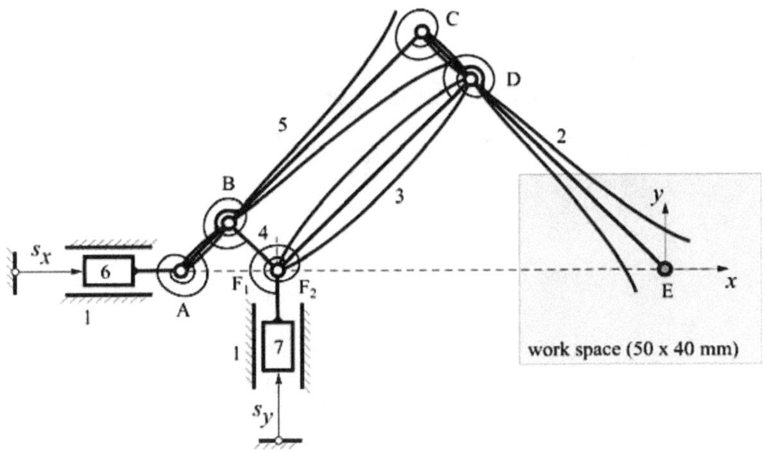

Figure 6. MBS model including rotational springs and elastic beams.

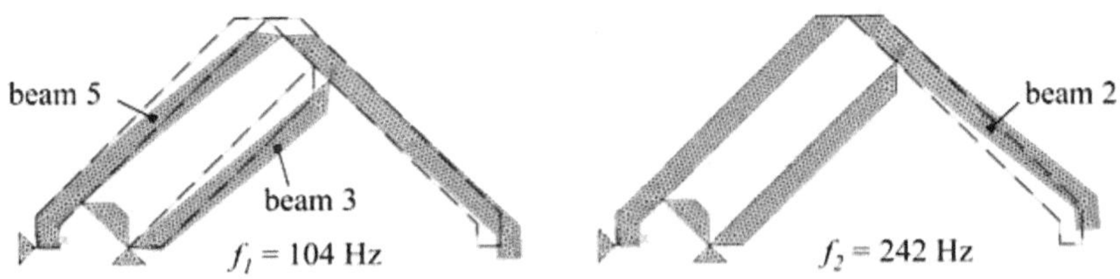

Figure 7. Two bending mode shapes of the pantograph mechanism.

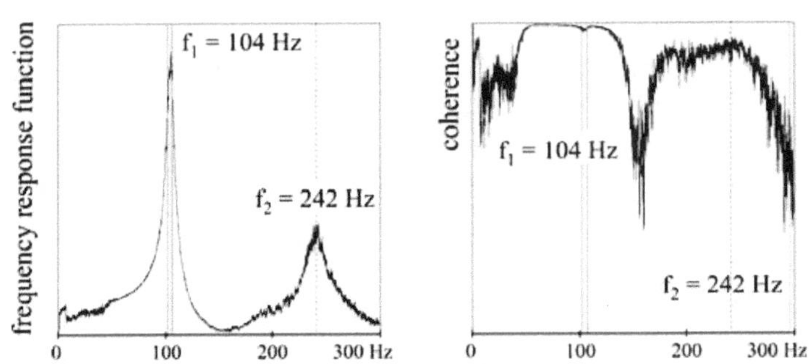

Figure 8. Frequency response function and eigenfrequencies of the pantograph.

The eigenfrequencies could be estimated by measuring the frequency response function (FRF), shown in Figure 8, i.e. $f_1 = 103.6$ Hz and $f_2 = 242$ Hz. A shaker excites the pantograph at the x-input. Two laser triangulation sensors measure the shaker motion and the tool point output motion. From this experiment the damping ratio is estimated as 2.9% for f_1 and 3.1% for f_2.

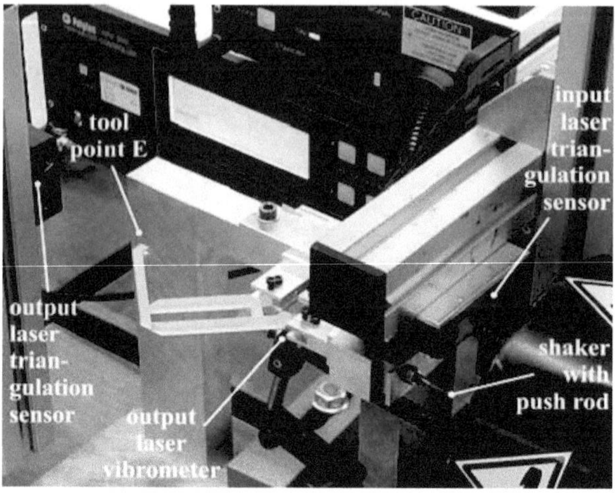

Figure 9. Experimental setup including pantograph, shaker and sensors.

Table 1. Eigenfrequencies in different pantograph positions.

	frequency	pos E1	pos E0	pos E3	pos E5
experiment	f_1 [Hz]	101	104	105	106
simulation	f_1 [Hz]	101.2	104.8	104.9	110.4
experiment	f_2 [Hz]	240	242	242	251
simulation	f_2 [Hz]	245.4	240.6	240.7	253.6

Of course the dispersion of every pantograph specimen should be considered, the eigenfrequency could vary ±1.5%. On the other hand eigenfrequency will change in different pantograph positions. That is why FRFs are measured in the pantograph position for tool point E1, E3, E5 (compare the positions in Figure 11). The tool point E0 corresponds to the static equilibrium position.

The simulation indicates the same eigenfrequencies in positions at a vertical line, e.g. E0 and E3, E1 and E2 or E4 and E2 (results in Table 1). The simulation is done with MBS approach 'SIMPACK'. To describe the elastic behavior the MBS codes must be combined with finite element method (FEM) codes. This was realized by the tool 'FEMBS'.

Another test validates the upper results in position E0. The impact response is measured in both x and z direction and deliver the eigenfrequencies $f_{1x} =$ 105 Hz and $f_{1z} = 83$ Hz and damping ratios $D_x = 2.6\%$ and $D_z = 5.9\%$. The MBS model verified by horizontal test motion in Section 3.1 will use the damping ratios $D_1 = 4.04\%$ and $D_2 = 4.95\%$.

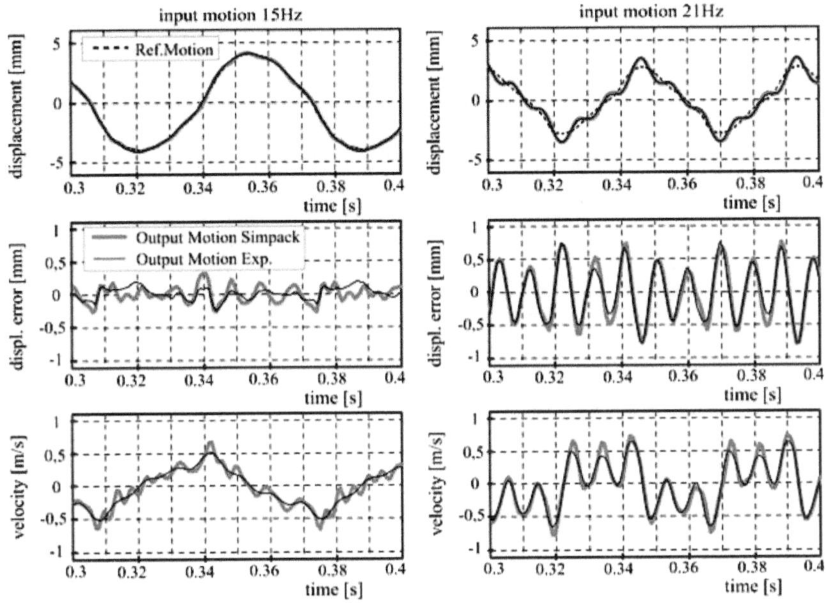

Figure 10. Dynamic displacement error for horizontal test motions.

3. MOTION DESIGN AND CONTROL

3.1 Horizontal Test Motion

Now the dynamic displacement error of the tool point E should be investigated. A special experimental setup was chosen to get a high accuracy (see Figure 9). The shaker could apply sin-like or rectangle-like motion in x-input direction with more than 250 Hz.

The shaker input motion is generated by function generator using a rectangle signal and different frequencies, e.g. 15 and 21 Hz shown in Figure 10. The frequency of 21 Hz match a higher excitation orders, especially the 5th E0.

The test and verification in time domain compares the real output motion to the reference motion calculated with the reference model ($x = -4s_x$). The difference of both defines the dynamic displacement error. Figure 10 includes an additional curve simulated by MBS model using the measured input motion.

The lowest diagram in Figure 10 compares two output point velocities: the first measured by laser vibrometer and the second simulated by SIMPACK. An iterative tuning between experiment and simulation improves the confidence of parameter estimation.

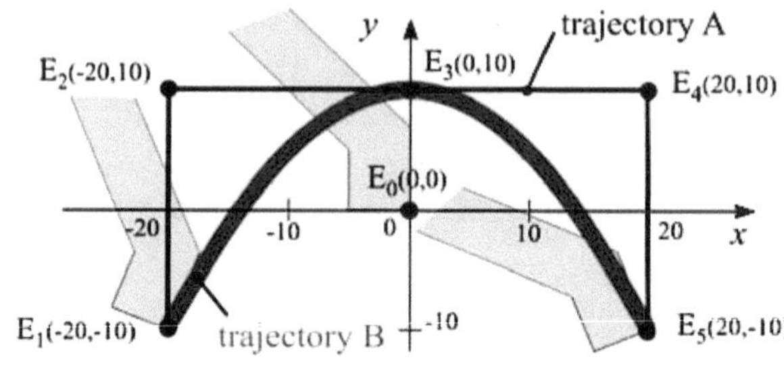

Figure 11. Surface mount trajectories of the tool point E.

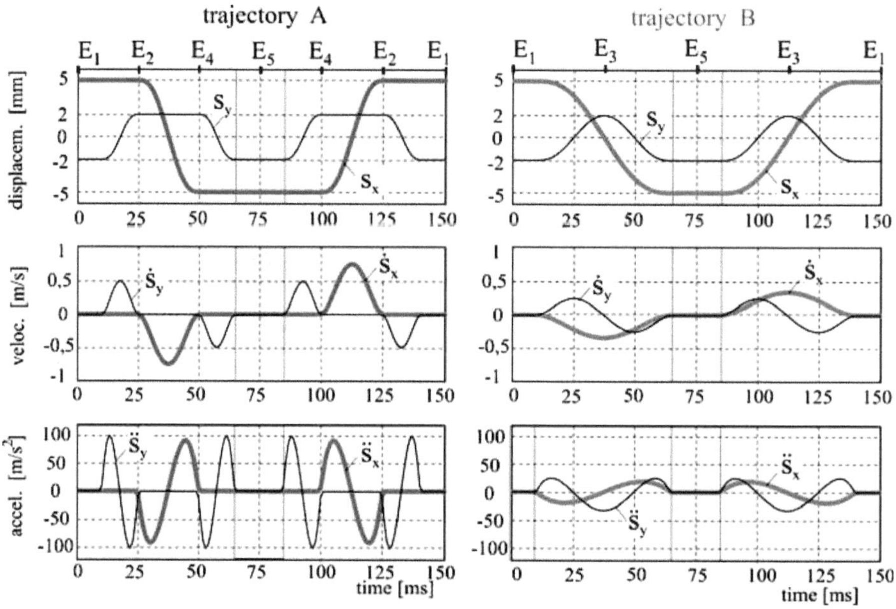

Figure 12. Input functions for both trajectories.

3.2 2D Surface Mount Trajectory

Figure 11 shows two chosen trajectories of the tool point E for the surface mount job. Trajectory A is one of the simplest ones and trajectory B is an improved version of trajectory A in order to reduce the dynamic displacement error of output point E.

Only Ishii et al. [2] studied a U-trajectory with the same cycle time of 150 ms. But now, our trajectory includes two additional dwells of about 20 ms for picking up in point E1 and placing in point E5, instead of the academic trajectory in the earlier paper. More especially, the vibrations in the placing

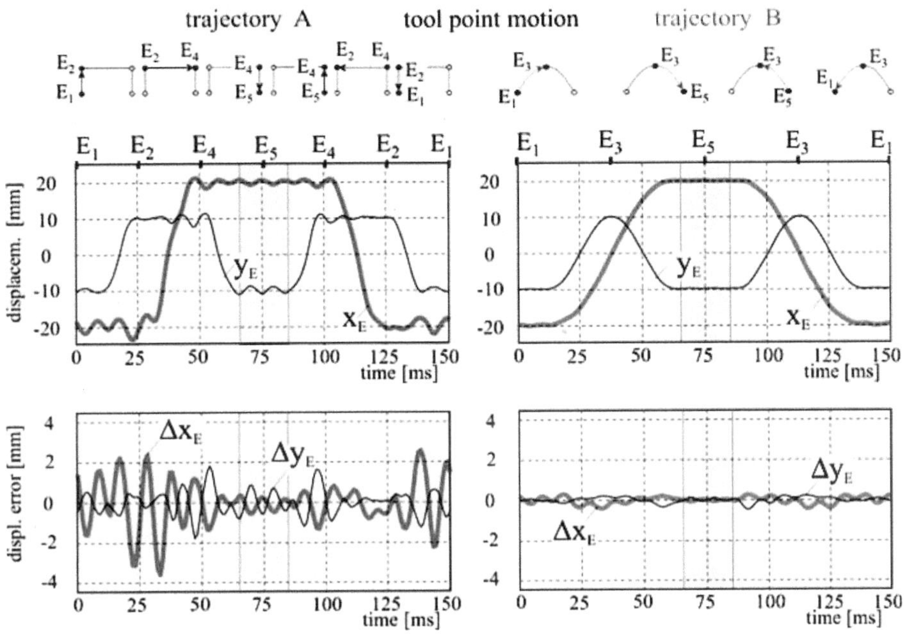

Figure 13. Dynamic displacement errors for both trajectories.

point E5 should be limited by a smoother motion. The upper positions enable free optimization potential.

Trajectory B satisfies the task without corners against trajectory A. Nevertheless, input functions s_x and s_y in Figure 12 for both trajectories include connections by a 5th order polynomial in time domain to ensure no jumps in velocity and acceleration. The rigid linkage model would deliver the reference without dynamic displacement error and vibrations, it is the reference trajectory.

These input functions used can be designed in many ways however the comparison of trajectory A and B is a sufficiently simple one to investigate the dynamic characteristics of the displacement error of the tool point E.

Figure 13 shows the dynamic displacement error of the output point E in comparison to the reference trajectory. The frequency of the displacement error in the x direction of trajectory A and B is approximately 100 Hz depending on the first natural frequency of the pantograph mechanism. Comparing the displacement error of trajectory A and B, a successful reducing of vibrations is evident, while changing from trajectory A to B, because of lower acceleration peaks. Trajectory B could reduce the dynamic displacement error of the tool point E to less than 0.1 mm within the placing dwell near point E5 and less than 0.2 mm within the picking up dwell near point E1.

4. SUMMARY AND REMARKS

Modelling and simulation of the pantograph mechanism by MBS approach SIMPACK allows to predict the dynamic displacement error for every required motion task.

Practical application and realization of the input motion by special step actuators (see Figure 2) is attended by new problems. The actuator approximates the designed input motion by steps or ramps in position, velocity or acceleration level stepwise. The exciting frequencies will differ from the expected ones. This is why the special actuator and servo controller should be modelled to calculate the dynamic output displacement error. A real practical application would require special experimental validation once more.

REFERENCES

[1] Horie, M., Uchida, T. and Kamiya, D., A Pantograph Mechanism with Large-Deflective Hinges for Miniature Surface Mount Systems. In: *Proceedings of ROMANSY 13 – Theory and Practice of Robots and Manipulators in Zakopane 2000*, Springer-Verlag, Vienna, 2002, pp. 93–102.

[2] Ishii, Y., Thümmel, T. and Horie, M., Modelling and Simulation of a Molded Pantograph Mechanism with Large-Deflective Hinges and Links. In: *Proceedings of the 11th World Congress in Mechanism and Machine Science*, Tianjin, China, April 1–4, 2004, Vol. 3, pp. 1437–1441.

[3] Horie, M., Palmtop Surface Mount System Composed of Molded Pantograph Mechanisms with Large-Deflective Hinges and Links Using in a One-Room Factory. In: *Proceedings of the 4th International Workshop on Microfactories (IWMF2004)*, Shanghai, China, October 15–17, 2004, Vol. 1, pp. 98–103.

STRUCTURAL CONTROL ENERGY EFFICIENCY BASED ON ELASTIC DISPLACEMENT

Kevin K. F. Wong
Department of Civil and Environmental Engineering, University of Utah,
122 South Central Campus Drive, EMRO 210, Salt Lake City, UT 84112, USA

Abstract: A modified optimal linear control algorithm is proposed to demonstrate the effectiveness of the active control system in reducing inelastic structural dynamic responses using elastic displacement feedback. Effectiveness of the proposed algorithm is then demonstrated by studying the control energy of the controlled structure. Numerical simulation using various single degree of freedom systems is performed and results show that there always exists a point where the control energy dissipation is maximal, and this optimal point is dependent on the structural characteristics as well as the earthquake excitation.

Key words: optimal linear control, force analogy method, control energy, plastic energy.

1. INTRODUCTION

Structural control is a desirable method of protecting structures from the seismic hazard point of view because it provides controlling ability to the structure while maintaining the structural dynamic properties. Among different types of structural control, active control has attracted major attentions from different researchers. Several active control algorithms have been proposed (Soong, 1990), and optimal linear control and instantaneously optimal control algorithms are among the most widely studied (Yang, 1984; Yang et al., 1991, 1992a). However, these active control algorithms are quite limited in seismic applications because they are applied to structures responding in the elastic domain. Even though active control is used, the structure may still respond inelastically under major earthquake excitations. Only limited research has been done in active control of inelastic structures,

H. Ulbrich and W. Günthner (eds), IUTAM Symposium on Vibration Control of Nonlinear Mechanisms and Structures, 365–374.

and Reinhorn et al. (1987) and Yang et al. (1992b, 1992c) were among the first to develop numerical control algorithms. However, no control algorithm for inelastic structures has been theoretically derived to date.

The objective of this research is to propose a computational algorithm for maximizing the control effectiveness using optimal linear control based on elastic displacement, which represents the portion of total displacement that is recoverable after structural vibration. The cost function is chosen to be composed of only the elastic states of the structure and the control system. Efficiency of control system is then derived based on maximizing control energy dissipation, and the optimal control parameter is studied for various earthquake ground motions on different single degree of freedom systems.

2. FORCE ANALOGY METHOD

The procedure of force analogy method has been discussed in Wong and Yang (1999). The important equations for an n degrees of freedom system with m plastic hinges in a moment-resisting frame are here summarized.

The total displacement of the structure is equal to elastic displacement plus inelastic displacement, i.e.

$$\mathbf{x}(t) = \mathbf{x}'(t) + \mathbf{x}''(t) \tag{1}$$

where $\mathbf{x}(t)$ is the total displacement vector, $\mathbf{x}'(t)$ is the elastic displacement vector, and $\mathbf{x}''(t)$ is the inelastic displacement vector.

Elastic moment at the plastic hinge locations (PHLs) of the moment-resisting frame is related to the elastic displacement, and the relationship is:

$$\mathbf{m}'(t) = \overline{\mathbf{K}}^T \mathbf{x}'(t) \tag{2}$$

where $\mathbf{m}'(t)$ is the elastic moment vector and $\overline{\mathbf{K}}$ is the $n \times m$ matrix that relates plastic rotation with the restoring forces at the structural degrees of freedom (DOFs). The governing equations of force analogy method are:

$$\mathbf{m}(t) + \mathbf{K}'\boldsymbol{\theta}(t) = \overline{\mathbf{K}}^T \mathbf{x}(t) \tag{3}$$

$$\mathbf{x}''(t) = \mathbf{K}^{-1}\mathbf{K}'\boldsymbol{\theta}(t) \tag{4}$$

where $\mathbf{m}(t)$ is the total moment vector, $\boldsymbol{\theta}(t)$ is the plastic rotation vector, $\mathbf{K}$ is the $n \times n$ global stiffness matrix, and $\mathbf{K}'$ is the $m \times m$ matrix that relates the plastic rotation vector with the restoring moments at the PHLs.

3. STATE SPACE RESPONSE

The dynamic equilibrium equation of a controlled structure with *n* DOFs and *p* control actuators can be written as

$$\mathbf{M}\ddot{\mathbf{x}}(t) + \mathbf{C}\dot{\mathbf{x}}(t) + \mathbf{K}\mathbf{x}'(t) = -\mathbf{M}\ddot{\mathbf{g}}(t) + \mathbf{D}\mathbf{f}(t) \tag{5}$$

where $\mathbf{M}$ and $\mathbf{C}$ are the $n \times n$ mass and damping matrices, respectively; $\dot{\mathbf{x}}(t)$, and $\ddot{\mathbf{x}}(t)$ are the $n \times 1$ velocity and acceleration vectors, respectively; $\mathbf{f}(t)$ is the $p \times 1$ active control force vector; $\mathbf{D}$ is the $n \times p$ control force distribution matrix; and $\ddot{\mathbf{g}}(t)$ is the earthquake ground acceleration vector corresponding to each DOF. Now define state variables as

$$\mathbf{z}(t) = \begin{Bmatrix} \mathbf{x}(t) \\ \dot{\mathbf{x}}(t) \end{Bmatrix}, \quad \mathbf{z}'(t) = \begin{Bmatrix} \mathbf{x}'(t) \\ \dot{\mathbf{x}}(t) \end{Bmatrix}, \quad \mathbf{z}''(t) = \begin{Bmatrix} \mathbf{x}''(t) \\ \mathbf{0} \end{Bmatrix} \tag{6}$$

This gives an equation similar to Eq. (1) as

$$\mathbf{z}(t) = \mathbf{z}'(t) + \mathbf{z}''(t) \tag{7}$$

where $\mathbf{z}(t)$ represents the total states of the structure, $\mathbf{z}'(t)$ represents the elastic states of the structure, and $\mathbf{z}''(t)$ represents the inelastic states of the structure that contains only the inelastic displacement $\mathbf{x}''(t)$.

After defining the elastic states $\mathbf{z}'(t)$ given in Eq. (6), Eq. (5) can be represented in a simple state space form as

$$\dot{\mathbf{z}}(t) = \mathbf{A}\mathbf{z}'(t) + \mathbf{B}\mathbf{f}(t) + \mathbf{H}\mathbf{a}(t) \tag{8}$$

where $\mathbf{a}(t)$ is the 6×1 ground acceleration vector in 6 directions, both translations and rotations; $\mathbf{A}$ is the $2n \times 2n$ continuous time state transition matrix; $\mathbf{B}$ is the $2n \times p$ control force transition matrix; and $\mathbf{H}$ is the $2n \times 6$ ground acceleration transition matrix. These matrices are

$$\mathbf{A} = \begin{bmatrix} \mathbf{0} & \mathbf{I} \\ -\mathbf{M}^{-1}\mathbf{K} & -\mathbf{M}^{-1}\mathbf{C} \end{bmatrix}, \quad \mathbf{B} = \begin{bmatrix} \mathbf{0} \\ \mathbf{M}^{-1}\mathbf{D} \end{bmatrix}, \quad \mathbf{H} = \begin{bmatrix} \mathbf{0} \\ -\mathbf{1} \end{bmatrix} \tag{9}$$

where $\mathbf{0}$ is the zero matrix, $\mathbf{I}$ is the identity matrix, and $-\mathbf{1}$ is the matrix that contains only 0's and –1's and relates each DOF with the corresponding direction of motion. Solving for the elastic states $\mathbf{z}'(t)$ in Eq. (7) and substituting the result into Eq. (8) gives

$$\dot{\mathbf{z}}(t) = \mathbf{A}\mathbf{z}(t) + \mathbf{B}\mathbf{f}(t) + \mathbf{H}\mathbf{a}(t) - \mathbf{A}\mathbf{z}''(t) \tag{10}$$

The solution of Eq. (10) can be obtained as follows:

$$\mathbf{z}(t+\Delta t) = \mathbf{e}^{\mathbf{A}\Delta t}\mathbf{z}(t) + \mathbf{e}^{\mathbf{A}t}\int_t^{t+\Delta t} \mathbf{e}^{-\mathbf{A}s}(\mathbf{B}\mathbf{f}(s) + \mathbf{H}\mathbf{a}(s) - \mathbf{A}\mathbf{z}''(s))ds \tag{11}$$

where Δt is the integration time step. Let the time steps be $t_{k+1} = t + \Delta t$ and $t_k = t$. Assuming that the control force $\mathbf{f}(s)$ and inelastic states $\mathbf{z}''(s)$ are uniform over the duration $[t_k, t_{k+1})$ and that the ground motion $\mathbf{a}(s)$ as pulses for the purpose of integration, then integrating Eq. (11) gives

$$\mathbf{z}_{k+1} = \mathbf{F}_s\mathbf{z}_k + \mathbf{G}\mathbf{f}_k + \mathbf{H}_d\mathbf{a}_k - \mathbf{A}_d\mathbf{z}''_k \tag{12}$$

where

$$\begin{aligned} \mathbf{F}_s &= \mathbf{e}^{\mathbf{A}\Delta t}, \quad \mathbf{G} = \mathbf{A}^{-1}\left(\mathbf{e}^{\mathbf{A}\Delta t} - \mathbf{I}\right)\mathbf{B} \\ \mathbf{H}_d &= \mathbf{e}^{\mathbf{A}\Delta t}\mathbf{H}\Delta t, \quad \mathbf{A}_d = \mathbf{A}^{-1}\left(\mathbf{e}^{\mathbf{A}\Delta t} - \mathbf{I}\right)\mathbf{A} \end{aligned} \tag{13}$$

and $\mathbf{z}_k$, $\mathbf{f}_k$, $\mathbf{a}_k$, and $\mathbf{z}''_k$ are discretized forms of $\mathbf{z}(t)$, $\mathbf{f}(t)$, $\mathbf{a}(t)$, and $\mathbf{z}''(t)$, respectively. In Eq. (13), the matrix $\mathbf{A}_d$ can be simplified as follows:

$$\begin{aligned} \mathbf{A}_d &= \mathbf{A}^{-1}\left(\mathbf{e}^{\mathbf{A}\Delta t} - \mathbf{I}\right)\mathbf{A} = \mathbf{A}^{-1}\left(\mathbf{A}\Delta t + \frac{(\mathbf{A}\Delta t)^2}{2!} + \frac{(\mathbf{A}\Delta t)^3}{3!} + \ldots\right)\mathbf{A} \\ &= \mathbf{A}\Delta t + \frac{(\mathbf{A}\Delta t)^2}{2!} + \frac{(\mathbf{A}\Delta t)^3}{3!} + \ldots = \mathbf{e}^{\mathbf{A}\Delta t} - \mathbf{I} \\ &= \mathbf{F}_s - \mathbf{I} \end{aligned} \tag{14}$$

Substituting this result into Eq. (12) gives

$$\mathbf{z}_{k+1} = \mathbf{F}_s\mathbf{z}'_k + \mathbf{G}\mathbf{f}_k + \mathbf{H}_d\mathbf{a}_k + \mathbf{z}''_k \tag{15}$$

4. OPTIMAL LINEAR CONTROL

The control force at time step k (i.e., $\mathbf{f}_k$) in Eq. (15) is regulated by the structural responses when feedback control is used. Since inelastic displacement represents the permanent deformation of the structure that cannot be changed once it happens, the primary objective of applying control force is therefore to reduce the elastic responses. To determine this control force that gives an optimal linear control, define the cost function J to be

$$J = \frac{1}{2}\sum_{k=0}^{N}\left(\mathbf{z}_k'^T \mathbf{Q}\mathbf{z}_k' + \mathbf{f}_k^T \mathbf{R}\mathbf{f}_k\right) \tag{16}$$

where **Q** and **R** are weighting matrices that gives a representation of energy in the structure and N is the last time step of the analysis. Minimizing the cost function J in Eq. (16) in optimal linear control has been discussed in many literatures (Hart and Wong, 2000), and after the minimization process is performed the control force is found to be:

$$\mathbf{f}_k = \left(\mathbf{I} + \mathbf{G}\mathbf{R}^{-1}\mathbf{G}^T\mathbf{P}\right)^{-1}\mathbf{F}_s\mathbf{z}_k' \tag{17}$$

where **P** is the steady state Riccati matrix and is given as

$$\mathbf{P} = \mathbf{Q} + \mathbf{F}_s^T\mathbf{P}\left(\mathbf{I} + \mathbf{G}\mathbf{R}^{-1}\mathbf{G}^T\mathbf{P}\right)^{-1}\mathbf{F}_s \tag{18}$$

5. ENERGY BALANCE

Once the response time histories are known, the variation of energy with time can be calculated. The energy balance equation can be derived based on Eq. (5). Define the absolute acceleration $\ddot{\mathbf{y}}(t)$ to be $\ddot{\mathbf{y}}(t) = \ddot{\mathbf{x}}(t) + \ddot{\mathbf{g}}(t)$. Then it follows from Eq. (5) that

$$\mathbf{M}\ddot{\mathbf{y}}(t) + \mathbf{C}\dot{\mathbf{x}}(t) + \mathbf{K}\mathbf{x}'(t) = \mathbf{D}\mathbf{f}(t) \tag{19}$$

Energy responses are obtained by integrating both sides of Eq. (19) over the path of structural response from time zero to time t_k, where t_k represents the kth time step of interest. Doing so gives

$$\int_{t=0}^{t=t_k} \ddot{\mathbf{y}}^T\mathbf{M}d\mathbf{x} + \int_{t=0}^{t=t_k} \dot{\mathbf{x}}^T\mathbf{C}d\mathbf{x} + \int_{t=0}^{t=t_k} \mathbf{x}'^T\mathbf{K}d\mathbf{x} = \int_{t=0}^{t=t_k} \mathbf{f}^T\mathbf{D}^T d\mathbf{x} \tag{20}$$

Since $d\mathbf{x}(t) = d\mathbf{y}(t) - d\mathbf{g}(t)$ where $\mathbf{y}(t)$ is the absolute displacement vector and $\mathbf{g}(t)$ is the ground displacement vector, Eq. (20) becomes

$$\int_{t=0}^{t=t_k} \ddot{\mathbf{y}}^T \mathbf{M} d\mathbf{y} + \int_{t=0}^{t=t_k} \dot{\mathbf{x}}^T \mathbf{C} d\mathbf{x} + \int_{t=0}^{t=t_k} \mathbf{x}'^T \mathbf{K} d\mathbf{x} - \int_{t=0}^{t=t_k} \mathbf{f}^T \mathbf{D}^T d\mathbf{x} = \int_{t=0}^{t=t_k} \ddot{\mathbf{y}}^T \mathbf{M} d\mathbf{g} \tag{21}$$

In addition, since $d\mathbf{x}(t) = d\mathbf{x}'(t) + d\mathbf{x}''(t)$, Eq. (21) becomes

$$\int_{t=0}^{t=t_k} \ddot{\mathbf{y}}^T \mathbf{M} d\mathbf{y} + \int_{t=0}^{t=t_k} \dot{\mathbf{x}}^T \mathbf{C} d\mathbf{x} + \int_{t=0}^{t=t_k} \mathbf{x}'^T \mathbf{K} d\mathbf{x}' + \int_{t=0}^{t=t_k} \mathbf{x}'^T \mathbf{K} d\mathbf{x}'' - \int_{t=0}^{t=t_k} \mathbf{f}^T \mathbf{D}^T d\mathbf{x} = \int_{t=0}^{t=t_k} \ddot{\mathbf{y}}^T \mathbf{M} d\mathbf{g} \tag{22}$$

Note the first three terms on the left hand side of Eq. (22) represent the kinetic energy (*KE*), damping energy (*DE*), and strain energy (*SE*) of the system, respectively. The fourth term represents the plastic energy (*PE*), and the fifth term represents the control energy (*CE*), and these two energy terms are the main focus of this research. Finally, the term on the right hand side represents the input energy (*IE*) due to earthquake excitation. In summary, Eq. (22) can be written as

$$KE + DE + SE + PE + CE = IE \tag{23}$$

6. NUMERICAL SIMULATIONS

Consider the single degree of freedom (SDOF) system shown in Figure 1. Assume the mass of the structure is $m = 175{,}100$ kg and critical damping ratio is $\zeta = 5\%$. Let the weighting matrices of the control system be:

$$\mathbf{Q} = \begin{bmatrix} \mathbf{K} & 0 \\ 0 & \mathbf{M} \end{bmatrix}, \quad \mathbf{R} = 10^{-r} \tag{24}$$

where r is the control parameter used to study the effectiveness of the control system. The plastic hinges of the structure is assumed to yield at a horizontal displacement of 7.6 cm, and the post yield stiffness is assumed to be 10% of the initial stiffness.

Figure 2 shows the maximum control energy and plastic energy responses for various r values and natural period of vibrations subjected to the recorded 1995 Kobe earthquake ground motion at KJM station, and Figure 3 shows the same responses but subjected to the 1994 Northridge earthquake at Sylmar. From these figures, it is noted that applying small maximum control force (i.e., small r values) results in practically no displacement response reduction. But when large maximum control force is applied (i.e., large r values), significant displacement response reduction occurs. This observation is consistent with the results obtained in other research works and it is intuitively correct.

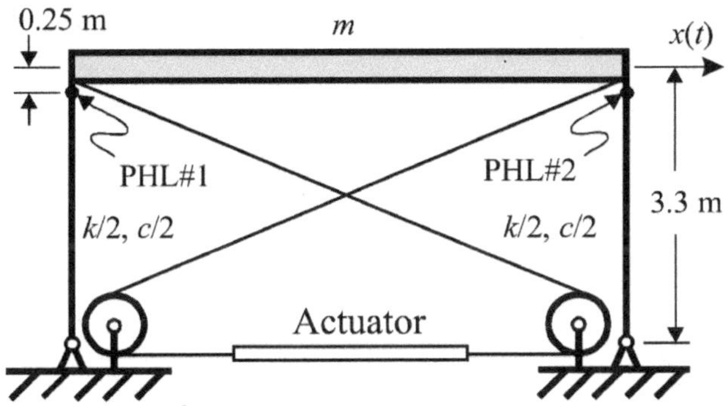

Figure 1. Single degree of freedom system.

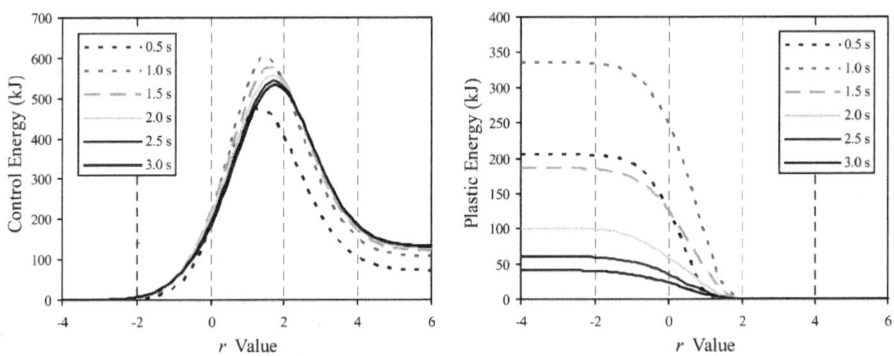

Figure 2. Maximum responses due to 1995 Kobe earthquake.

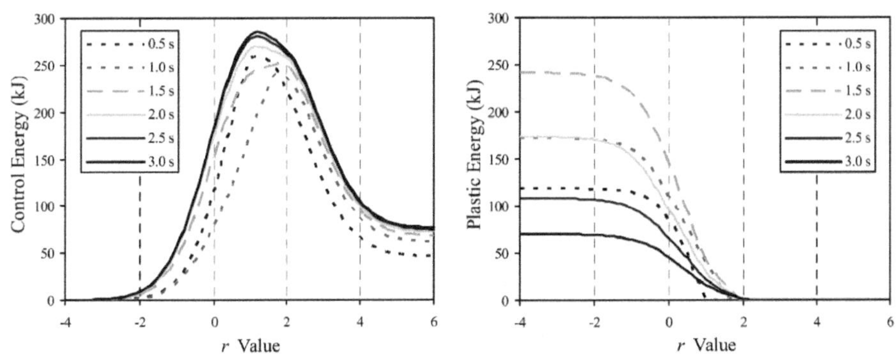

Figure 3. Maximum responses due to 1994 Northridge earthquake.

Now consider the control energy response. Figure 4 summarizes the control energy dissipation of the structure due to another six earthquake ground motions. In addition, Table 1 summarizes the r values at which maximum control energy occurs and the corresponding plastic energy dissipations for various natural periods of vibration. Based on this figure and the control energy figures in Figs. 2 and 3, a common trend is observed:

- At small r values, the displacement is large and the control force is small. This results in small control energy dissipation.

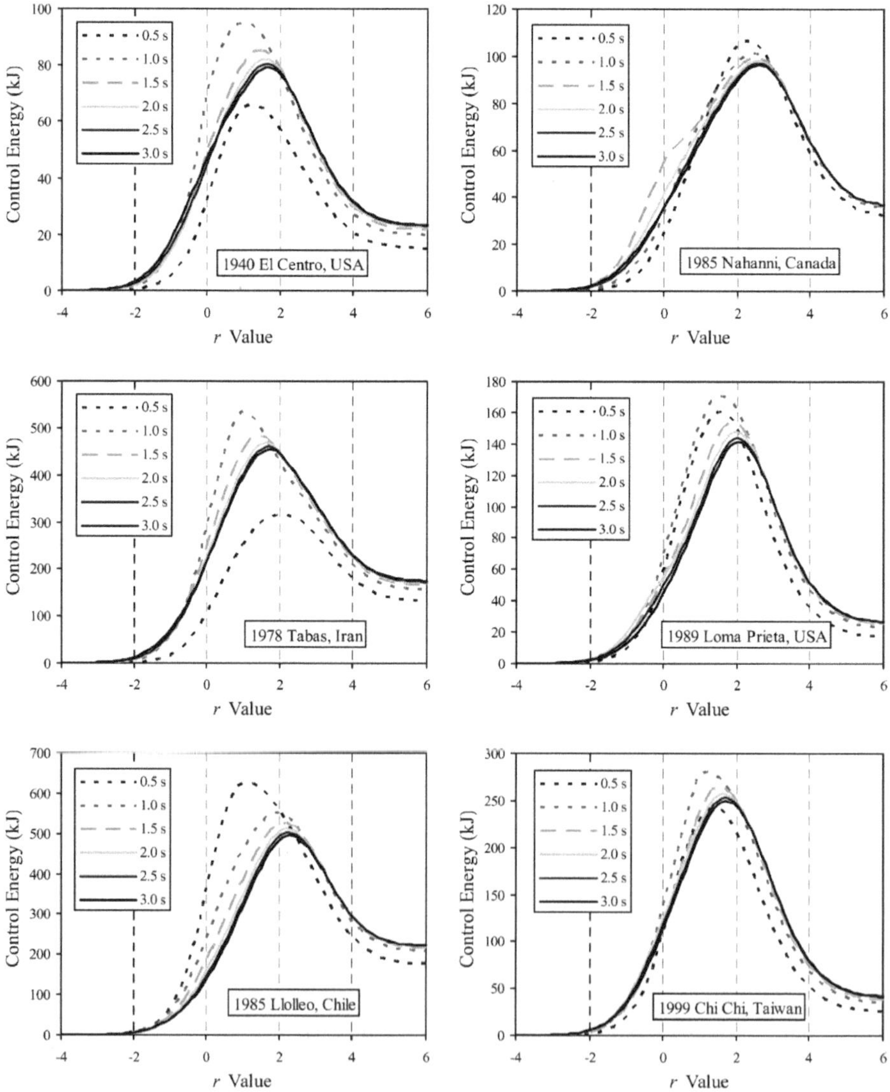

Figure 4. Maximum control energy responses for various earthquake motions.

Table 1. r values corresponding to maximum CE for various earthquake motions.

Earthquake	r value	CE (kJ)	PE (kJ)	r value	CE (kJ)	PE (kJ)	r value	CE (kJ)	PE (kJ)
	$T_n = 0.5$ s			$T_n = 1.5$ s			$T_n = 2.5$ s		
Kobe	1.3	475	0	1.6	577	6.3	1.7	543	1.0
Northridge	1.1	259	0	1.9	253	1.7	1.2	280	15.9
El Centro	1.3	66	0	1.5	85	0	1.6	80	0
Tabas	2.1	317	0	1.4	484	3.4	1.7	461	1.5
Llolleo	1.1	628	0	2.2	526	0	2.3	502	0
Nahanni	2.3	107	0	2.5	99	0	2.6	97	0
Loma Prieta	1.6	160	0	1.9	154	0	2	144	0
Chi Chi	1.4	242	0	1.5	264	0	1.7	253	0
Mean	**1.53**			**1.81**			**1.85**		
	$T_n = 1.0$ s			$T_n = 2.0$ s			$T_n = 3.0$ s		
Kobe	1.5	606	18.3	1.7	557	1.8	1.8	532	0.2
Northridge	1.7	240	0	1.2	270	22.0	1.2	285	11.9
El Centro	1	95	0	1.6	82	0	1.7	79	0
Tabas	1	534	7.8	1.6	469	2.1	1.8	454	0.8
Llolleo	2	551	0	2.2	512	0	2.3	496	0
Nahanni	2.5	101	0	2.6	98	0	2.6	96	0
Loma Prieta	1.6	171	0	2	148	0	2	141	0
Chi Chi	1.2	280	0	1.6	257	0	1.7	250	0
Mean	**1.56**			**1.81**			**1.89**		

- At large r values, the control force is large and the displacement is small. This also results in small control energy dissipation but no plastic energy dissipation (i.e., the structure remains in elastic domain).
- At moderate r values, both control force and displacement are at moderate levels. This results in large control energy dissipation and very small plastic energy dissipation.

Based on the above three points, it can be seen that the control system is ineffective when small control force is applied (i.e., small r values) since the plastic energy dissipation in the structure remains practically the same as the uncontrolled structure. On the other hand, the control system is too costly when large control force is applied (i.e., large r values) since less control force can be applied to achieve the same result of having no damage in the structure. This suggests that the most effective control system is achieved at the point where largest control energy dissipation occurs.

Observations based on Table 1 suggest that the optimal r values corresponding to maximum control energy dissipation occurs within the range $1.0 \le r \le 3.0$. Note that this point does not necessarily correspond to having no plastic energy dissipation, but large control energy dissipation can certainly reduce the plastic energy dissipation and thereby reducing damages in the structure.

7. CONCLUSION

A modified optimal linear control algorithm based on elastic displacement is proposed to control the inelastic structural responses. This algorithm uses elastic displacement in the cost function, and it reflects the portion of total displacement that needs to be controlled. Force analogy method is employed to identify both the elastic and inelastic portions of the total displacement. Effectiveness of the proposed control algorithm is then studied by considering the control energy dissipation.

Based on the study of SDOF system subjected to different earthquake ground motions, it is demonstrated that the proposed control algorithm is very effective in reducing the structural responses. It is also observed that both control force and displacement are moderate at moderate r values, and this combination results in large control energy dissipation. Optimal r value is then studied for the SDOF system, and it is found that optimal the r value based on the control parameters used in Eq. (24) is within the range $1.0 \le r \le 3.0$.

REFERENCES

Hart, G.C. and Wong K. (2000). *Structural Dynamics for Structural Engineers*, John Wiley and Sons, New York, pp. 556–568.

Reinhorn, A.M., Manolis, G.D. and Wen, C.Y. (1987). Active control of inelastic structures, *Journal of Engineering Mechanics ASCE* **113**:315–333.

Soong, T.T. (1990). *Active Structural Control, Theory & Practice*, Longman Scientific and Technical, U.K.

Wong, K.K.F. and Yang, R. (1999). Inelastic dynamic response of structures using force analogy method, *Journal of Engineering Mechanics ASCE* **125**:1190–1199.

Yang, J.N. (1984). Application of optimal control theory to civil engineering structures, *Journal of Engineering Mechanics ASCE* **101**:819–838.

Yang, J.N., Li, Z. and Liu, S.C. (1991). Instantaneous optimal control with velocity and acceleration feedbacks, *Journal of Probabilistic Engineering Mechanics* **16**:204–221.

Yang, J.N., Li, Z. and Liu, S.C. (1992a). Stable controllers for instantaneous optimal control, *Journal of Engineering Mechanics ASCE* **118**:1612–1630.

Yang, J.N., Li, Z., Danielians, A. and Liu, S.C. (1992b). Aseismic hybrid control of non-linear and hysteretic systems I, *Journal of Engineering Mechanics ASCE* **118**:1423–1440.

Yang, J.N., Li., Z., Danielians, A. and Liu, S.C. (1992c). Aseismic hybrid control of non-linear and hysteretic systems II, *Journal of Engineering Mechanics ASCE* **118**:1441–1456.

STABILITY ANALYSIS OF ROLL GRINDING SYSTEM WITH DOUBLE TIME DELAY EFFECTS

L. Yuan[1], E. Keskinen[1,2] and V.M. Järvenpää[1]
[1]*Machine Design, Tampere University of Technology, Finland;* [2]*Current address: Institut für Technische Mechanik, Universität Karlsruhe, Germany*

Abstract: The grinding system of paper machine rolls is studied. The MDOF model with double time delays is used to describe the dynamical behavior of the system. By analyzing the simplified linear and non-linear models the stability characteristics of the system has been investigated by different stability methods. Numerical simulation results are used to illustrate stability behaviors of the models.

Key words: grinding, time delay, stability analysis, regenerative chatter, perturbation, limit cycle.

1. INTRODUCTION

The roll grinding process studied in this work is used in paper manufacturing industry in processing steel roll surfaces. To ensure the quality requirements of the paper manufacturing process the roll surfaces should not have significant irregularities. The finishing grinding of the surfaces is applied either for new rolls manufactured or as a regular maintenance of rolls.

The grinding, turning and milling processes are machining operations in which material is removed from a work piece to produce a part. There exist many researches and studies about the stability of the dynamics related to these processes. For grinding problems the classical theory of chatter growth by Thompson [1] considered the double delay effects. Moon [2] and others explored the dynamical phenomena of several manufacturing processes. Pratt and Nayfeh [4] considered the chatter control and the stability analysis of a cantilever boring bar under regenerative cutting conditions. Landers et al. [10] present the control of machining, Altintas [6] described the chatter vibration in cutting with delay and Olgac and Huang [11] studied about the active vibration absorption and delayed feedback tuning.

The instability of the grinding process is related to the regenerative chatter theory with double time delay effects. Regenerative chatter results from the

H. Ulbrich and W. Günthner (eds), IUTAM Symposium on Vibration Control of Nonlinear Mechanisms and Structures, 375–387.

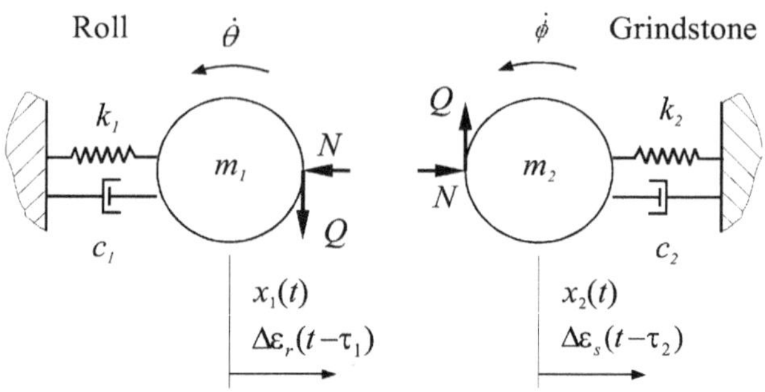

Figure 1. The roll grinding model.

unstable interaction in the machine tool-work piece structures and may cause excessive force, tool wear, tool failure and scrap parts of unacceptable surface finish.

A typical roll-grinding machine consists of a work piece and a tool, in this case a metal roll and a grindstone, respectively. The first chatter source is on the roll. During machining the overlap of the grinding path is needed to ensure that all surface of the roll will be ground. However, this overlap causes that all surface error shapes on the roll surface from the previous round will carry on and then become one excitation source. The resulting phenomena can be expressed by a time delay term. The second chatter source is on the grindstone side. The surface of the grindstone drops chips from its surface because of high friction and that makes the original round shape become irregular. Thus another error shape from the grindstone is an excitation source too and then second delay term is introduced to the model.

2. MODELING OF ROLL GRINDING PROCESS

Dynamics of roll grinding systems will be studied in order to detect the main characteristics of this system and to state the key parameters which influence the stability. The time delay effects from both roll and grindstone will be considered in a multi-degree-of-freedom (MDOF) model presented in Figure 1. The model consists of the time delay equations of the roll and the grindstone and also the speed controllers of their drives. Based on the above descriptions about the physical background of the system, a spring and mass model of it with double regenerative chatter can be illustrated as in Figure 2. Accordingly a mathematical model can be established as MDOF equations with double delay terms.

The total penetration without time delays can be expressed as

$$\varepsilon(t) = \varepsilon_r(t) + \varepsilon_s(t) = \varepsilon_{\text{nom}} + \Delta\varepsilon(t), \qquad (2.1)$$

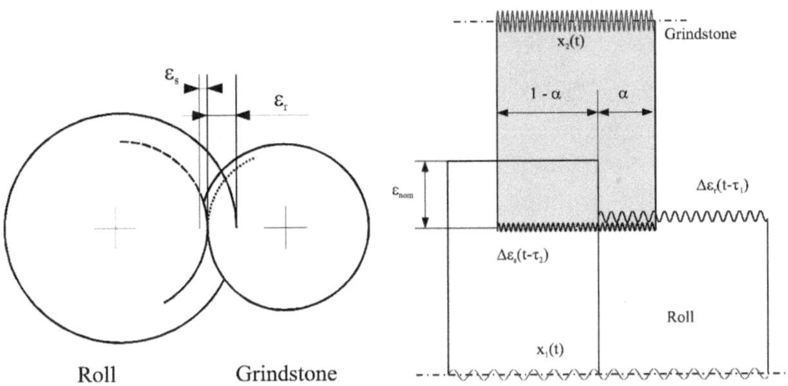

Figure 2. The grinding contact.

where ε_s and ε_r are penetrations on the roll and grindstone surfaces, respectively, and the nominal cutting feed ε_{nom} is a pre-set constant value x_f. The dynamic variation is defined based on the relative motion of the roll and the grindstone

$$\Delta\varepsilon(t) = x_1(t) - x_2(t). \tag{2.2}$$

Due to the normal force equilibrium in the contact the penetrations on the surfaces are

$$\varepsilon_r(t) = \gamma\varepsilon(t) = \gamma\varepsilon_{\text{nom}} + \Delta\varepsilon_r(t) = \gamma\varepsilon_{\text{nom}} + \gamma\Delta\varepsilon(t), \tag{2.3}$$

$$\varepsilon_s(t) = (1-\gamma)\varepsilon(t) = (1-\gamma)\varepsilon_{\text{nom}} + \gamma\Delta\varepsilon_s(t) = (1-\gamma)\varepsilon_{\text{nom}} + \gamma\Delta\varepsilon(t), \tag{2.4}$$

where $\gamma = k_s/(k_r + k_s)$ is the cut ratio and k_r and k_s are the cutting stiffness of the roll and the grindstone, respectively . The total penetration with delays τ_1 and τ_2 (see Figure 2).

$$\begin{aligned} &\varepsilon(t) - \Delta\varepsilon_r(t-\tau_1) - \Delta\varepsilon_s(t-\tau_2) \\ &\quad = \varepsilon_{\text{nom}} + \Delta\varepsilon(t) - \gamma\Delta\varepsilon(t-\tau_1) - (1-\gamma)\Delta\varepsilon(t-\tau_2). \end{aligned} \tag{2.5}$$

The normal force

$$\begin{aligned} N &= k_r\gamma\varepsilon(t, \tau_1, \tau_2, \alpha) \qquad (2.6) \\ &= k_r\gamma\{(1-\alpha)\varepsilon_{\text{nom}} + \Delta\varepsilon(t) - \alpha\gamma\Delta\varepsilon(t-\tau_1) - (1-\gamma)\Delta\varepsilon(t-\tau_2)\}, \end{aligned}$$

where α is the overlap ratio. The governing equations of motion for the workpiece-tool system can be written as

$$m_1\ddot{x}_1 + c_1\dot{x}_1 + k_1x_1 = -N(t; \tau_1, \tau_2, \alpha), \tag{2.7}$$

$$m_2\ddot{x}_2 + c_2\dot{x}_2 + k_2x_2 = -N(t;\tau_1,\tau_2,\alpha), \tag{2.8}$$

$$J\ddot{\theta} = -r_1N(t,\tau_1,\tau_2,\alpha) + K_{p1}\dot{\theta}_{d1}t - K_{p1}\theta + K_{D1}\dot{\theta}_{d1} - K_{D1}\dot{\theta}, \tag{2.9}$$

$$I\ddot{\phi} = -r_2N(t,\tau_1,\tau_2,\alpha) + K_{p2}\dot{\theta}_{d2}t - K_{p2}\theta + K_{D2}\dot{\theta}_{d2} - K_{D2}\dot{\theta}, \tag{2.10}$$

where r_1 and r_2 are radiuses and θ and ϕ are the rotations of the roll and the grindstone, respectively. In (2.9) and (2.10) the remaining terms are related to PD-controllers. The grinding force is assumed to be a non-linear function of the sliding speeds in the contact

$$k_r = \frac{w}{k_w} \cdot \frac{|\dot{\theta}|r_1}{|\dot{\phi}r_2 + \dot{\theta}r_1|}, \tag{2.11}$$

where w is the width of the contact and k_w the wear coefficient.

3. METHODS AND ANALYSIS

The responses in time domain can be solved by using numerical time integration methods and these results can be used to illustrate the stability behavior of the MDOF model with certain system parameters. However, when the general solution of the stability is considered the analysis becomes more difficult due to the double time delays, the non-linear contact stiffness and the multiple degrees of freedom. To overcome these problems the stability characteristics of this double delay system have been investigated by simplified approaches. MDOF model (2.7) and (2.8) can be reduced to two independent equations by considering the roll and the grindstone degrees of freedoms separately. Here (2.7) is mainly considered. After this, different stability analysis methods were applied to obtain more comprehensive view to the system characteristics. The following studies were carried out:

1. The classical stability analysis of metal cutting process by Thompson to determine stability boundaries of the contact stiffness.
2. A single degree of freedom analysis in frequency domain to illustrate frequency response of the delay system.
3. Non-linear limit cycle analysis.
4. Finally, discussion of the applicability of Lyapunov method for the full MDOF grinding equations.

3.1 Classical Stability Chart Analysis

Only the motion of the roll is considered. The delay terms are considered independently and linear analysis is used. Assuming $x_f = 0$ and using x to take the place of x_1, m for m_1, c for c_1, k for k_1, (2.7) then becomes

$$m\ddot{x} + c\dot{x} + (k + k_r\gamma)x - B\gamma k_r x(t-\tau) = 0, \tag{3.1}$$

with the cases $\tau = \tau_1$, $B = (1-\alpha)$ and $\tau = \tau_2$, $B = (1-\gamma)$. The characteristic equation is

$$m\lambda^2 Ae^{\lambda t} + c\lambda Ae^{\lambda t} + (k + k_r\gamma)Ae^{\lambda t} - B\gamma k_r e^{-\lambda t} Ae^{\lambda t} = 0 \tag{3.2}$$

and by substituting $\lambda = i\omega$ the real and imaginary parts becomes

$$\begin{cases} -m\omega^2 + k + \gamma k_r - B\gamma k_r \cos\omega\tau = 0, \\ c\omega + B\gamma k_r \sin\omega\tau = 0. \end{cases} \tag{3.3}$$

The solution takes the form

$$k_r(\omega, \tau) = (m\omega^2 - k)/(\gamma - B\gamma \cos\omega\tau(\omega)), \tag{3.4}$$

$$\tau(\omega) = \frac{2j\pi}{\omega} - \frac{1}{\omega}\arccos\left(\frac{1 \pm A\sqrt{B^2 - 1 + A^2B^2}}{B(1 + A^2)}\right), \tag{3.5}$$

where $A = -(m\omega^2 - k)/(c\omega)$. The stability chart [12] can be plotted according to (3.4) and (3.5).

3.2 Frequency Response Analysis

Frequency response analysis is related to steady state responses of the system. It will not consider the stability of the system, but it illustrates the response amplitudes to harmonic excitation when the system is stable. Again like in (3.1), (2.7) is considered by now adding harmonic excitation force

$$m\ddot{x} + cx + (k + k_r\gamma)x = \alpha\gamma^2 k_r x(t - \tau_1) + \gamma(1 - \gamma)k_r x(t - \tau_2) + Ye^{i\Omega t}. \tag{3.6}$$

By introducing $x = Xe^{i\Omega t}$ the steady state solution takes the form

$$X = \left\{-\Omega^2 m + i\Omega c + (k + k_r\gamma) - \alpha\gamma^2 k_r e^{-i\Omega\tau_1} - \gamma(1 - \gamma)k_r e^{-i\Omega\tau_2}\right\}^{-1} Y. \tag{3.7}$$

3.3 Non-Linear Limit Cycle Analysis by Perturbation Method

A two-degree of freedom model of (2.7) and (2.9) is considered with the simplified non-linear term (2.11) by setting rotation speed of the grindstone as constant.

$$m\ddot{x} + c\dot{x} + kx = -wp\dot{\theta}(x - \alpha\gamma x(t - \tau_1) - (1 - \gamma)x(t - \tau_2) - \beta x_f), \tag{3.8}$$

$$\begin{aligned} J\ddot{\theta} &= wpr\dot{\theta}(x - \alpha\gamma x(t - \tau_1) - (1 - \gamma)x(t - \tau_2) - \beta x_f) \\ &\quad + K_p\dot{\theta}_d t - K_p\theta + K_D\dot{\theta}_d - K_D\dot{\theta}, \end{aligned} \tag{3.9}$$

where $p = \gamma r_1/(k_w r_2\dot{\phi})$, $\beta = 1 - \alpha$ and $r = r_1$.

Due to the presence of the non-linear contact stiffness, the perturbation analysis can be performed using the method of multiple scales [3]. In this method, a fast time-scale $T_0 = t$ and a slow time-scale $T_1 = \varepsilon^2 t$. A second-order expansion for (2.1) and (2.2) is in the form

$$x = \varepsilon x_0(T_0, T_1) + \varepsilon^2 x_1(T_0, T_1) + \varepsilon^3 x_2(T_0, T_1) + O(\varepsilon^4), \tag{3.10}$$

$$\theta = \varepsilon\theta_0(T_0, T_1) + \varepsilon^2\theta_1(T_0, T_1) + \varepsilon^3\theta_2(T_0, T_1) + O(\varepsilon^4). \tag{3.11}$$

The effective width of contact w is perturbed by letting $w = w_c + \varepsilon^2 w_1$, where w_c is the critical cutting width, which can obtain from the stability chart by (3.4) and (3.5). Substituting (3.10) and (3.11) into (3.8) to (3.9), and equating coefficients of power of ε

Order ε^1:

$$m\frac{\partial^2 x_0}{\partial T_0^2} + c\frac{\partial x_0}{\partial T_0} + kx_0 = w_c p\beta x_f \frac{\partial\theta_0}{\partial T_0}, \tag{3.12}$$

$$J\frac{\partial^2\theta_0}{\partial T_0^2} = w_c pr\beta x_f \frac{\partial\theta_0}{\partial T_0} - K_p\theta_0 - K_D\frac{\partial\theta_0}{\partial T_0}. \tag{3.13}$$

Order ε^2:

$$m\frac{\partial^2 x_1}{\partial T_0^2} + c\frac{\partial x_1}{\partial T_0} + kx_1$$
$$= -w_c p\left(\frac{\partial\theta_0}{\partial T_0}(x_0 - \alpha\gamma x_{0\tau 1} - (1-\gamma)x_{0\tau 2} - \beta x_f\frac{\partial\theta_1}{\partial T_0}\right), \tag{3.14}$$

$$J\frac{\partial^2\theta_1}{\partial T_0^2} = w_c pr\left(\frac{\partial\theta_0}{\partial T_0}(x_0 - \alpha\gamma x_{0\tau 1} - (1-\gamma)x_{0\tau 2}) - \beta x_f\frac{\partial\theta_1}{\partial T_0}\right)$$
$$-K_p\theta_1 - K_D\frac{\partial\theta_1}{\partial T_0}. \tag{3.15}$$

Order ε^3:

$$m\left(\frac{\partial^2 x^2}{\partial T_0^2} + 2\frac{\partial^2 x_0}{\partial T_0\partial T_1}\right) + c\left(\frac{\partial x_2}{\partial T_0} + 2\frac{\partial x_0}{\partial T_1}\right) + kx_2$$
$$= \frac{\partial\theta_1}{\partial T_0}(x_0 - \alpha\gamma x_{0\tau 1} - (1-\gamma)x_{0\tau 2})$$
$$-w_c p\left(\frac{\partial\theta_0}{\partial T_0}(x_1 - \alpha\gamma x_{1\tau 1} - (1-\gamma)x)_{1\tau 2} - \beta x_f\left(\frac{\partial\theta_2}{\partial T_0} + 2\frac{\partial\theta_0}{\partial T_1}\right)\right)$$
$$+w_1 p\beta x_f\frac{\partial\theta_0}{\partial T_0}, \tag{3.16}$$

$$
J\left(\frac{\partial^2\theta_2}{\partial T_0^2}+2\frac{\partial^2\theta_0}{\partial T_0\partial T_1}\right)
$$

$$
=w_c pr\left(\frac{\partial\theta_0}{\partial T_0}(x_1-\alpha\gamma x_{1\tau 1}-(1-\gamma)x_{1\tau 2})-\beta x_f\left(\frac{\partial\theta_2}{\partial T_0}+2\frac{\partial\theta_0}{\partial T_1}\right)\right)
$$

$$
+\frac{\partial\theta_1}{\partial T_0}(x_0-\alpha\gamma x_{0\tau 1}-(1-\gamma)x_{0\tau 2}
$$

$$
-w_1 pr\beta x_f\frac{\partial\theta_0}{\partial T_0}-K_p\theta_2-K_D\left(\frac{\partial\theta_2}{\partial T_0}+2\frac{\partial\theta_0}{\partial T_1}\right). \tag{3.17}
$$

The solution of (3.12) and (3.13) can be expressed as

$$
\begin{aligned}
x_0 &= A_1(T_1)e^{i\omega_c T_0}+\bar{A}_1(T_1)e^{-i\omega_c T_0},\\
x_{0\tau 1} &= A_1(T_1)e^{i\omega_c(T_0-\tau_1)}+\bar{A}_1(T_1)e^{-i\omega_c(T_0-\tau_1)},\\
x_{0\tau 2} &= A_1(T_1)e^{i\omega_c(T_0-\tau_2)}+\bar{A}_1(T_1)e^{-i\omega_c(T_0-\tau_2)},
\end{aligned} \tag{3.18}
$$

$$
\theta_0=A_2(T_1)e^{i\omega_c T_0}+\bar{A}_2(T_1)e^{-i\omega_c T_0}, \tag{3.19}
$$

where ω_c is the chatter frequency at the Hopf bifurcation. Also let

$$
x_1=P_1e^{2i\omega_c T_0}+P_2e^{-2i\omega_c T_0}+P_3, \tag{3.20}
$$

$$
\theta_1=R_1e^{2i\omega_c T_0}+R_2e^{-2i\omega_c T_0}+R_3. \tag{3.21}
$$

Substitute (3.18) to (3.21) into (3.14) and (3.15) leads

$$
\begin{aligned}
&-4m\omega_c^2(P_1e^{2i\omega_c T_0}+P_2e^{-2i\omega_c T_0}+2ci\omega_c(P_1e^{2i\omega_c T_0}-P_2e^{-2i\omega_c T_0})\\
&+k(P_1e^{2i\omega_c T_0}+P_2e^{-2i\omega_c T_0}+P_3)\\
&=-w_c pi\omega_c(A_2e^{i\omega_c T_0}-\bar{A}_2e^{-i\omega_c T_0})(A_1e^{i\omega_c T_0}+\bar{A}_1e^{-i\omega_c T_0}\\
&-A_1\alpha\gamma e^{i\omega_c(T_0-\tau_1)}-\bar{A}_1\alpha\gamma e^{-i\omega_c(T_0-\tau_1)}-A_1(1-\gamma)e^{i\omega_c(T_0-\tau_2)}\\
&-\bar{A}_1(1-\gamma)e^{-i\omega_c(T_0-\tau_2)}\\
&+w_c p2i\omega_c\beta x_f(R_1e^{2i\omega_c\tau_0}+R_2e^{-2i\omega_c T_0}),
\end{aligned} \tag{3.22}
$$

$$
\begin{aligned}
&-4J\omega_c^2(R_1e^{2i\omega_c T_0}+R_2e^{-2i\omega_c T_0}\\
&=w_c pri\omega_c(A_2e^{i\omega_c T_0}-\bar{A}_2e^{-i\omega_c T_0})\big(A_1e^{i\omega_c T_0}\\
&+\bar{A}_1e^{-i\omega_c T_0}-A_1\alpha\gamma e^{i\omega c(T_0-\tau_1)}-\bar{A}_1\alpha\gamma e^{-i\omega_c(T_0-\tau_1)}\\
&-A_1(1-\gamma)e^{i\omega_c(T_0-\tau_2)}-A_1(1-\gamma)e^{-i\omega_c(T_0-\tau_2)}\big)\\
&-w_c pr2i\omega_c\beta x_f(R_1e^{2i\omega_c T_0}+R_2e^{-2i\omega_c T_0})
\end{aligned}
$$

$$-K_p(R_1 e^{2i\omega_c T_0} + R_2 e^{-2i\omega_c T_0} + R_3)$$
$$-2i\omega_c K_D(R_1 e^{2i\omega_c T_0} - R_2 e^{-2i\omega_c T_0}). \quad (3.23)$$

For (3.22) and (3.23), equating each of the coefficients of $\exp(2i\omega_c T_0)$, $\exp(-2i\omega_c T_0)$ and $\exp(0)$ on both sides gives

$$P_1 = \Delta_1 A_2 A_1, \quad P_2 = \Delta_2 \bar{A}_2 \bar{A}_1, \quad P_3 = \Delta_{31} A_2 \bar{A}_1 + \Delta_{32} \bar{A}_2 A_1, \quad (3.24)$$

$$R_1 = \Gamma_1 A_2 A_1, \quad R_2 = \Gamma_2 \bar{A}_2 \bar{A}_1, \quad R_3 = \Gamma_{31} A_2 \bar{A}_1 + \Gamma_{32} \bar{A}_2 A_1, \quad (3.25)$$

where

$$\begin{aligned}
\Delta_1 &= \frac{w_c p i\omega_c}{4m\omega_c^2 - 2i\omega_c c - k}(2\beta x_f \Gamma_1 - (1 - \alpha\gamma e^{-i\omega_c \tau_1} - (1-\gamma)e^{-i\omega_c \tau_2})),\\
\Delta_2 &= \frac{w_c p i\omega_c}{-4m\omega_c^2 - 2i\omega_c c + k}((1 - \alpha\gamma e^{-i\omega_c \tau_1} - (1-\gamma)e^{-i\omega_c \tau_2}) + 2\beta x_f \Gamma_2),\\
\Delta_{31} &= \frac{-w_c p i\omega_c}{k}(1 - \alpha\gamma e^{i\omega_c \tau_1} - (1-\gamma)e^{i\omega_c \tau_2}),\\
\Delta_{32} &= \frac{w_c p i\omega_c}{k}(1 - \alpha\gamma e^{-i\omega_c \tau_1} - (1-\gamma)e^{-i\omega_c \tau_2}),
\end{aligned} \quad (3.26)$$

$$\begin{aligned}
\Gamma_1 &= \frac{w_c p r i\omega_c(1 - \alpha\gamma e^{-i\omega_c \tau_1} - (1-\gamma)e^{-i\omega_c \tau_2})}{-4J\omega_c^2 + K_p + 2i\omega_c K_D - 2w_c p r i\omega_c \beta x_f},\\
\Gamma_2 &= \frac{w_c p r i\omega_c(1 - \alpha\gamma e^{i\omega_c \tau_1} - (1-\gamma)e^{i\omega_c \tau_2})}{4J\omega_c^2 - K_p + 2i\omega_c K_D - 2w_c p r i\omega_c \beta x_f},\\
\Gamma_{31} &= \frac{w_c p r i\omega_c}{K_p}(1 - \alpha\gamma e^{i\omega_c \tau_1} - (1-\gamma)e^{i\omega_c \tau_2}),\\
\Gamma_{32} &= \frac{-w_c p r i\omega_c}{K_p}(1 - \alpha\gamma e^{-i\omega_c \tau_1} - (1-\gamma)e^{-i\omega_c \tau_2}).
\end{aligned} \quad (3.27)$$

The solutions (3.20), (3.21) and (3.24) to (3.27) can be substituted into (3.16) and (3.17) and elimination of the terms that lead to secular terms yields

$$\begin{aligned}
(2mi\omega_c + 2c)A_1' &= -w_c p i\omega_c A_2 A_2 \bar{A}_1 \Delta_{31}\gamma\beta - w_c p i\omega_c A_2 \bar{A}_2 A_1 \Delta_{32}\gamma\beta\\
&\quad + w_c p i\omega_c \Delta_1(1 - \alpha\gamma e^{-2i\omega_c \tau_1} - (1-\gamma)e^{-2i\omega_c \tau_2})\bar{A}_2 A_2 A_1\\
&\quad - 2w_c p i\omega_c \Gamma_1(1 - \alpha\gamma e^{i\omega_c \tau_1} - (1-\gamma)e^{i\omega_c \tau_2})A_2 A_1 \bar{A}_1\\
&\quad - 2w_c p\beta x_f A_2' - w_1 p i\omega_c \beta x_f A_2,
\end{aligned} \quad (3.28)$$

$$2i\omega_c J A_2' = w_c p r i\omega_c\big(\alpha\gamma\Delta_{31} A_2 A_2 \bar{A}_1 - \alpha\gamma\Delta_{32} A_2 \bar{A}_2 A_1$$

$$
\begin{aligned}
&- (1-\gamma)\Delta_{31}A_2A_2\bar{A}_1 - (1-\gamma)\Delta_{32}A_2\bar{A}_2A_1 - \Delta_1\bar{A}_2A_2A_1 \\
&+ \alpha\gamma\Delta_1\bar{A}_2A_2A_1 + (1-\gamma)\bar{A}_2A_2A_1\big) \\
&+ w_c pr 2i\omega_c(\Gamma_1A_2A_1\bar{A}_1 - \alpha\gamma\Gamma_1A_2A_1\bar{A}_1e^{i\omega_c\tau_1} \\
&- (1-\gamma)\Gamma_1A_2A_1\bar{A}_1e^{i\omega_c\tau_2}) + 2w_c pr\beta x_f A_2' \\
&- w_1 pri\omega_c\beta x_f A_2 - 2K_D A_2'.
\end{aligned} \tag{3.29}
$$

It will be simpler if we assume $A_2 = hA_1$, where the coefficient h can be estimated by equation (3.12) in a linear case by the Laplace transform method. Taking the Laplace transform of equation (3.12), we can get the transfer function

$$H(s) = (s^2m + sc + k)/(w_c p\beta x_f).$$

Then $h = H(i\omega_c)$ is at the Hopf bifurcation point.

Introducing the polar transformations

$$A_1 = \frac{1}{2}\sigma e^{i\delta}, \quad A_1' = \frac{1}{2}\sigma' e^{i\delta} + i\frac{1}{2}\sigma\delta' e^{i\delta}. \tag{3.30}$$

Substituting (3.30) into (3.19) and separating real and imaginary parts, the following normal form is obtained

$$\sigma' = g_1 w_1\sigma + g_2\sigma^3, \tag{3.31}$$

$$\sigma\delta' = g_3 w_1\sigma + g_4\sigma^3, \tag{3.32}$$

where

$$g_1 = \frac{-m\omega_c^2 ph\beta x_f}{2m^2\omega_c^2 + 2(pw_c\beta x_f - c)^2}, \tag{3.33}$$

$$
\begin{aligned}
g_2 = {}& \frac{m\omega_c pw_c}{8m^2\omega_c^2 + 8(pw_c\beta x_f - c)^2} \cdot \left[\frac{pw_c\beta x_f - c}{m\omega_c}(h\gamma\beta\Delta_{3i} - h\Delta_i + 2h\Gamma_i)\right. \\
&\left. - (h\gamma\beta\Delta_{3r} - h\Delta_r + 2h\Gamma_r)\right],
\end{aligned} \tag{3.34}
$$

$$g_3 = \frac{pw_c\beta x_f + c}{m\omega_c} g_1, \tag{3.35}$$

$$g_4 = -\frac{pw_c h}{8m}(h\gamma\beta\Delta_{3i} - h\Delta_i + 2h\Gamma_i) + \frac{pw_c\beta x_f h + c}{m\omega_c} g_2, \tag{3.36}$$

where

$$
\begin{aligned}
&\Delta = \Delta_r + i\Delta_i = \Delta_1(1 - \alpha\gamma e^{-2i\omega_c\tau_1} - (1-\gamma)e^{-2i\omega_c\tau_2}), \\
&\Delta_3 = \Delta_{31} + \Delta_{32} = \Delta_{3r} + i\Delta_{3i}, \\
&\Gamma = \Gamma_r + i\Gamma_i = \Gamma_1(1 - \alpha\gamma e^{i\omega_c\tau_1} - (1-\gamma)e^{i\omega_c\tau_2}).
\end{aligned} \tag{3.37}
$$

The approximate solutions are expressed by (3.31) and (3.32). The qualitative behavior near the Hopf bifurcation can be determined by the sign of g_2. When $g_2 < 0$, the bifurcation is supercritical and $g_2 > 0$, it is subcritical.

The amplitude of steady-state limit cycles can be obtained from equation (3.31) by setting $\sigma' = 0$. It yields the expression of determining the constant amplitude of the limit cycles

$$\sigma = \sqrt{\frac{g_1}{g_2}(w_c - w)}. \tag{3.38}$$

3.4 Applicability of Lyapunov Method

According to the stability of motion (see [5, 8, 9]), a Lyapunov function is defined in the case of ordinary differential system. For the delay system, in the Lyapunov–Kraovskii method, the Lyapunov functional may provide the requirements instead.

Here a simplified one-degree linear case is discussed. If equation (3.8) with constant $\dot{\theta} = \dot{\theta}_d$ can have a modal transformation then it becomes

$$\begin{aligned} \dot{x} &= y, \\ \dot{y} &= -\frac{c}{m}y - \frac{k}{m}x - \frac{2p\dot{\theta}_d}{m}x + \frac{wp\dot{\theta}_d}{m}\alpha\gamma x(t-\tau_1) \\ &\quad + \frac{wp\dot{\theta}_d}{m}(1-\gamma)x(t-\tau_2) - \frac{wp\dot{\theta}_d}{m}\beta x_f. \end{aligned} \tag{3.39}$$

One could define a functional $V(x(t))$ which satisfies the condition of stability theorem [8]

$$\begin{aligned} V(x(t), y(t)) &= \frac{1}{2}my^2(t) + \frac{1}{2}(k + wp\dot{\theta}_d)x^2(t) \\ &\quad + \frac{1}{2}b\int_{\tau_1}^{0} x^2(s)ds + \frac{1}{2}d\int_{\tau_2}^{0} x^2(s)ds \end{aligned} \tag{3.40}$$

and evaluate $\lim_{\Delta t \to +0}(\Delta V/\Delta t)$ along trajectory of equation (3.8), then

$$\begin{aligned} \frac{d}{dt}V(x(t), y(t)) &= -cy^2(t) + wp\dot{\theta}_d\alpha\gamma y(t)x(t-\tau_1) \\ &\quad + wp\dot{\theta}_d(1-\gamma)y(t)x(t-\tau_2) - wp\dot{\theta}_d\beta x_f + \frac{1}{2}bx^2(t) \\ &\quad - \frac{1}{2}bx^2(t-\tau_1) + \frac{1}{2}dx^2(t) - \frac{1}{2}dx^2(t-\tau_2). \end{aligned} \tag{3.41}$$

When $(d/dt)V \leq 0$, this gives the condition of stability and the uniform asymptotic stability of the trivial solution of equation (3.8). For multi-degree

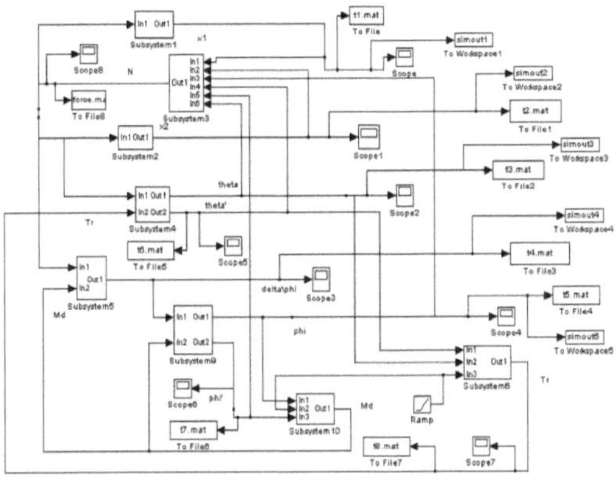

Figure 3. SIMULINK block diagram of the main system.

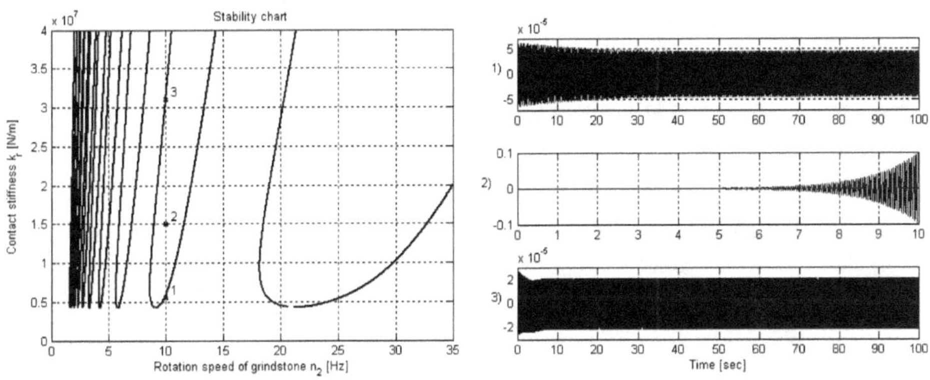

Figure 4. The stability chart and the time domain responses in three reference points.

non-linear equations, one can process similarly as above. This will be our future work.

4. NUMERICAL RESULTS

The MDOF system of (2.7) to (2.10) can be set up in MATLAB/SIMULINK® environment as shown in Figure 3 and the time domain responses can be solved for various sets of parameter values to detect the dynamical behavior of the system. However, here only the SDOF model of (3.6) is considered to illustrate the results of cases (1) and (2) in Section 3. Figure 4 shows the stability chart of (3.6) with the responses of three reference points when only τ_2 is considered. The results are matching very well as one would expect. The chart also provides the stability boundary for the critical contact stiffness (Hopf bifurcation). Figure 5 illustrates the harmonic responses of the (3.6) when the

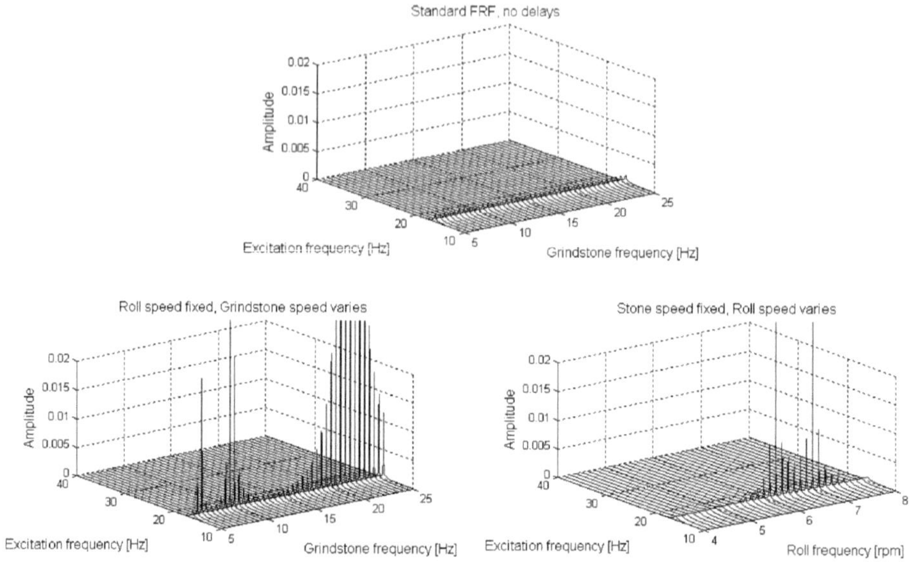

Figure 5. The steady state amplitudes as a function of excitation and rotation frequencies.

contact stiffness is chosen as stable. Important remark is that even if the system is nominally stable the harmonic response amplitudes can become very high if the external excitation is close to the natural frequency of the system. Even if the system would normally have acceptable amplitude levels due to damping, the delay system shows large amplitudes in certain speed regions.

5. CONCLUSIONS

Different stability analysis methods are used to analyze the double time delay grinding system. The classical stability chart provides the stability boundaries for the contact stiffness, which are verified by numerical time domain simulations. The frequency response analysis gives the steady state response amplitudes of the system with the delays. The perturbation analysis provides the approximate solution and the amplitude of limit cycle of the system when the contact stiffness is a non-linear function. Lyapunov–Krasovskii method seems to be applicable to this system and more investigation is required in the future.

REFERENCES

[1] Thompson, R.A., On the Doubly Regenerative Stability of a Grinder: The Theory of Chatter Growth, *Journal of Engineering for Industry*, Vol. 108/75, May 1986.

[2] Moon, F.C., *Dynamics and Chaos in Manufacturing Processes*, John Wiley & Sons, 1998.

[3] Nayfeh, A.H., *Problems in Perturbation*, 1993.

[4] Pratt, J.R. and Nayfeh, A.H., *Chatter Control and Stability Analysis of a Cantilever Boring Bar under Regenerative Cutting Conditions*, The Royal Society, 2000.

[5] Gu, K. and Niculescu, S., Survey on Recent Results in the Stability and Control of Time Delay Systems, *Tansactions of ASME, Journal of Dynamic Systems, Measurement, and Control*, Vol. 125, June 2003.

[6] Altintas, Y., *Manufacturing Automation*, Cambridge University Press, 2000.

[7] Géradin, M. and Rixen, D., *Mechanical Vibrations – Theory and Application to Structural Dynamics*, John Wiley & Sons, 1997.

[8] Driver, R.D., *Ordinary and Delay Differential Equations*, Spinger-Verlag, New York, 1977.

[9] Krasovskii, N.N., *Stability of Motion*, Stanford University Press, 1963.

[10] Landers, R.G., Ulsoy, A.G. and Ma, Y.-H., A Comparison of Model-Based Machining Force Control Approaches, *Int. J. Mach. Tools Manuf.*, Vol. 44, Nos. 7–8, 2004.

[11] Olgac, N. and Huang, C., On the Stability of a Tuned Vibration Absorber for Time Varying Multiple Frequencies, *Journal of Vibration and Control*, Vol. 8, No. 4, 2002.

[12] Stepan, G., Modelling Nonlinear Regenerative Effects in Metal Cutting, *Philos. Trans. Royal Soc. London A*, Vol. 359, 2001.

SUBJECT INDEX

AUTHOR INDEX

LIST OF PARTICIPANTS

Agrawal, Sunil K.
University of Delaware
Dept. of Mechanical Engineering
126 Spence Lab.
Newark, DE 19716-3140
USA
agrawal@mw.udel.edu
tel.: +1/302/831-6536
fax: +1/302/831-3619

Babitsky, Vladimir
Loughborough University
Wolfson School
Loughborough
Leicestershire, LE11 3TU
UK
v.i.babitsky@lboro.ac.uk
tel.: +44/150/9227503
fax: +44/150/9227648

Bachmayer, Mathias
TU München
Lst. f. Steuerungs- u. Regelungstechnik
Arcisstr. 21
80290 München
Germany
bachmayer@lsr.ei.tum.de
tel.: +49/89/289-23417
fax: +49/89/289-28340

Baier, Horst
TU München
Lehrst. f. Leichtbau
Boltzmannstr. 15
85747 Garching
Germany
baier@llb.mw.tum.de
tel.: +49/89/289-16096
fax: +49/89/289-16104

Barrot, François
EPFL-LSRO
MEHO 539
Statcon 9
1015 Lausanne
Switzerland
francois.barrot@epfl.ch

Beadle, Brad
Universität Stuttgart
Institut A f. Mechanik
Allmandring 5b
70550 Stuttgart
Germany
beadle@mecha.uni-stuttgart.de
tel.: +49/711/685-8157
fax: +49/711/685-8169

Bestle, Dieter
Brandenburg University of Technology
Chair of Mechanical Engineering and Vehicle Dynamics
Siemens-Halske-Ring 14
03013 Cottbus
Germany
bestle@tu-cottbus.de
tel.: +49/355/6930-24
fax: +49/355/6930-38

Blajer, Wojciech
TU Radon
Inst. of Applied Mechanics
ul. Krasickiego 54
26-600 Radon
Poland
wblajer@poczta.onet.pl
tel.: +48/48/36171-10
fax: +48/48/36171-32

Bleuler, Hannes
EPFL Lausanne
Lab. de Systemes Robotiques Stat. 9
1015 Lausanne
Switzerland
hannes.bleuler@epfl.ch
tel.: +41/21/693-5927
fax: +41/21/693-3866

Bockstedte, Andreas
TU Hamburg-Harburg
Mech.a.Ocean Engineering
Eissendorfer Str. 42
21073 Hamburg
Germany
bockstedte@tuhh.de
tel.: +49/40/42878-2209
fax: +49/40/42878-2028

Brandenburg, Günter
TU München
Lehrst. f. Feingerätebau und Mikrotechnik
Boltzmannstr. 15
85747 Garching
Germany
brandenb@fgb.mw.tum.de
tel.: +49/89/289-15195
fax: +49/89/289-15192

Buhl, Michael
TU München
Lehrstuhl f. Regelungstechnik
Boltzmannstr. 15
85747 Garching
Germany
buhl@tum.de
tel.: +49/89/289-15670
fax: +49/89/289-15653

Buss, Martin
TU München
Lehrstuhl f. Steuerungs- u. Regelungstechnik
Arcisstr. 21
80290 München
Germany
mb@tum.de
tel.: +49/89/289-28395
fax: +49/89/289-28340

Bremer, Hartmut
Johannes Kepler Universität Linz
Institut für Robotik
Altenberger Str. 69
4040 Linz
Austria
hartmut.bremer@jku.at
tel.: +43/732/2468-9788
fax: +43/732/2468-9792

Chang, Chia-Ou
National Taiwan Univ.
Inst. of Applied Mechanics
Taipei 106
Taiwan
changco@spring.iam.ntu.edu.tw
tel.: +86/2/33665626
fax: +86/2/23639290

Dijk, Johannes
University of Twente
Fac. Engineering Technology
P.O. Box 217
7500 AE Enschede
The Netherlands
vandijk@ctw.utwente.nl
tel.: +31/53/4892601
fax: +31/53/4893631

Elliott, Stephen
University of Southampton
Inst. of Sound and Vibration Research
Highfield
Southampton, SO17 1BJ
UK
sje@isvr.soton.ac.uk
tel.: +44/23/8059-2294
fax: +44/23/8059-3190

Förg, Martin
TU München
Inst. of Applied Mechanics
Boltzmannstr. 15
85747 Garching
Germany
foerg@amm.mw.tum.de
tel.: +49/89/289-15212
fax: +49/89/289-15213

Freymann, Raymond
BMW AG München
Forschung und Technik
Petuelring 130
80788 München
Germany
raymond.freymann@bmw.de
tel.: +49/89/382-45326
fax: +49/89/382-40979

Friedrich, Markus
TU München
Inst. of Applied Mechanics
Boltzmannstr. 15
85747 Garching
Germany
friedrich@amm.mw.tum.de
tel.: +49/89/289-15207
fax: +49/89/289-15213

Gattringer, Hubert
Johannes Kepler Universität Linz
Institut für Robotik
Altenberger Str. 69
4040 Linz
Austria
hubert.gattringer@jku.at
tel.: +43/732/2468-9790
fax: +43/732/2468-9792

Ginzinger, Lucas
TU München
Inst. of Applied Mechanics
Boltzmannstr. 15
85747 Garching
Germany
ginzinger@amm.mw.tum.de
tel.: +49/89/289-15224
fax: +49/89/289-15213

Glasauer, Stefan
LMU München
Lehrstuhl f. Neurologie
Marchioninistr. 15
81377 München
Germany
sglasauer@brain.nefo.med.uni-muenchen.de
tel.: +49/89/7095-4839
fax: +49/89/7095-4801

Günthner, Wolfgang
TU München
Inst. of Applied Mechanics
Boltzmannstr. 15
85747 Garching
Germany
guenthner@amm.mw.tum.de
tel.: +49/89/289-15202
fax: +49/89/289-15213

Guse, Nils
Universität Stuttgart
Institute B of Mechanics
Pfaffenwaldring 9
70550 Stuttgart
Germany
guse@mechb.uni-stuttgart.de
tel.: +49/711/685-6396
fax: +49/711/685-6400

Hagedorn, Peter
TU Darmstadt
Dept. of Applied Mechanics
Hochschulstr. 1
64289 Darmstadt
Germany
hagedorn@mechanik.tu-darmstadt.de
tel.: +49/6151/16-2185
fax: +49/6151/16-4125

Hardeman, Toon
University of Twente
Fac. Engineering Technology
P.O.Box 217
7500 AE Enschede
The Netherlands
hardeman@ctw.utwente.nl
tel.: +31/53/4892567
fax: +31/53/4893631

Herrnberg, Michael
TU München
Lehrstuhl f. Regelungstechnik
Boltzmannstr. 15
85747 Garching
Germany
herrnberg@tum.de
tel.: +49/89/289-15677
fax: +49/89/289-15653

Höfeld, Jürgen
Deutsche Forschungsgemeinschaft
Kennedyallee 40
53175 Bonn-Bad Godesberg
Germany
juergen.hoefeld@dfg.de
tel.: +49/228/885/2321

Horie, Mikio
Tokyo Institute of Technology
Precision and Intelligence Lab
4259-R2-14, Nagatsuta-cho, Midori-ku
Yokohama 226-8503
Japan
mahorie@pi.titech.ac.jp
tel.: +81/45/924-5048
fax: +81/45/924-5977

Huber, Robert
TU München
Inst. of Applied Mechanics
Boltzmannstr. 15
85747 Garching
Germany
huber@amm.mw.tum.de
tel.: +49/89/289-15215
fax: +49/89/289-15213

Huesmann, Alexander
BMW Group Forschung und Technik
Knorrstr. 147
80788 München
Germany
alexander.huesmann@bmw.de
tel.: +49/89/382-21312
fax: +49/89/382-7021312

Kiriazov, Petko
Bulgarian Academy of Sofia
Inst. of Mechanics
Acad. G. Bonchev St. Block 4
1113 Sofia
Bulgaria
kiriazov@bas.bg
tel.: +359/2/979-6442
fax: +359/2/870-7498

Kostadinov, Kostadin
Bulgarian Academy of Sciences
Inst. of Mechanics
Acad. G. Bonchev St. Block 4
1113 Sofia
Bulgaria
kostadinova@imbm.bas.bg
tel.: +359/2/870-6264
fax: +359/2/870-7498

Krajcin, Idriz
University Duisburg-Essen
Chair of Dynamics and Control
Lotharstr. 69
47057 Duisburg
Germany
krajcin@uni-duisburg.de
tel.: +49/203/379-3423
fax: +49/203/379-3027

Krüger, Karin
TU München
Inst. of Applied Mechanics
Boltzmannstr. 15
85747 Garching
Germany
krueger@amm.mw.tum.de
tel.: +49/89/289-15208
fax: +49/89/289-15213

Kugi, Andreas
Univ. Saarland
Lst. f. Systemtheorie u. Regelungstechnik
Gebäude 13
6041 Saarbrücken
Germany
andreas.kugi@lsr.uni-saarland.de
tel.: +49/681/302-64720
fax: +49/681/302-64722

Kühnlenz, Kolja
TU München
Lehrstuhl f. Steuerungs- u. Regelungstechnik
Arcisstr. 21
80290 München
Germany
kolja.kuehnlenz@tum.de
tel.: +49/89/289-23412
fax: +49/89/289-28340

Li, Yanzhi
University of Bath

Dept. Mechanical Engineering
Bath, BA2 7AY
UK
enmyl@bath.ac.uk
tel.: +44/1225/38-4029
fax: +44/1225/38-6928

Lohmann, Boris
TU München
Lehrst. f. Regelungstechnik
Boltzmannstr. 15
85747 Garching
Germany
Lohmann@tum.de
tel.: +49/89/289-15610
fax: +49/89/289-15653

Mizuno, Takeshi
Saitama University
Dept. of Mechanical Engineering
Shimo-Okubo 255, Sakura-ku
Saitama 338-8570
Japan
mizar@mech.saitama-u.ac.jp
tel.: +81/48/858-3455
fax: +81/48/858-3712

Nayfeh, Ali
Virginia Politechnic Inst.
Dept. of Eng. Science & Mechanics
113A Burruss Hall
Blacksburg, VA, 24061
USA
anayfeh@vt.edu
tel.: +1/540/231-5453
fax: +1/540/231-2290

Neubauer, Marcus
Universität Hannover
Inst. f. Dynamik und Schwingungen
Appelstr. 11
30167 Hannover
Germany
neubauer@ifm.uni.hannover.de
tel.: +49/511/762-4168
fax: +49/511/762-4164

Nonami, Kenzo
Chiba University
Dept. of Electronics and Mech. Eng.
1-33, Yayoi-cho, Inage-ku, Chiba-shi
Chiba, 263-8522
Japan
nonami@faculty.chiba-u.jp
tel.: +81/432903195
fax: +81/432903195

Okada, Yohji
Ibaraki University
4-12-1 Nakanarusawa
Hitachi 316-8511
Japan
y.okada@mx.ibaraki.ac.jp
tel.: +81/294/38-5025
fax: +81/294/38-5049

Otter, Martin
DLR
Institut of Robotics and Mechatronics
Wesslinger Str.
82234 Wessling
Germany
Martin.Otter@dlr.de
tel.: +49/8153/28-2473
fax: +49/8153/28-1441

Paschedag, Jörg
TU München
Lehrst. f. Regelungstechnik
Boltzmannstr. 15
85747 Garching
Germany
paschedag@tum.de
tel.: +49/89/289-15664
fax: +49/89/289-15653

Paulitsch, Christoph
Univ. of Southampton
SVR SPCG-Highfield
Southampton, S017 13J
UK
cp@isvr.soton.ac.uk
tel.: +44/02380/59-2689
fax: +44/02380/59-3190

Pfeiffer, Friedrich
TU München
Inst. of Applied Mechanics

Boltzmannstr. 15
85747 Garching
Germany
pfeiffer@amm.mw.tum.de
tel.: +49/89/289-15200
fax: +49/89/289-15213

Ratering, Alexandra
University of Stuttgart
Institut B of Mechanics
Pfaffenwaldring 9
70550 Stuttgart
Germany
ratering@mechb.uni-stuttgart.de
tel.: +49/711/685-6414
fax: +49/711/685-6400

Riebe, Sandor
Volkswagen AG
Brieffach 1710
38436 Wolfsburg
Germany
sandro.riebe@volkswagen.de
tel.: +49/5361/9/49335
fax: +49/5361/9/76373

Sahinkaya, Mehmet N.
University of Bath
Dept. of Mechanical Engineering
Claverton Down
Bath, BA2 7AY
UK
ensmns@bath.ac.uk
tel.: +44/1225/38-4446
fax: +44/1225/38-6928

Schiehlen, Werner
University of Stuttgart
Institute B of Mechanics
Pfaffenwaldring 9
70550 Stuttgart
Germany
schiehlen@mechb.uni-stuttgart.de
tel.: +49/711/685-6388
fax: +49/711/685-6400

Schiller, Frank
TU München
FG Automatisierungstechnik
Boltzmannstr. 15
85747 Garching
Germany
schiller@itm.tum.de
tel.: +49/89/298-16402
fax: +49/89/289-16410

Schlacher, Kurt
University of Linz
Inst. for Automatic Control and Syst. Technology
Altenberger Str. 69
4040 Linz
Austria
kurt.schlacher@jku.at
tel.: +43/732/2468-9730
fax: +43/732/2468-9734

Seto, Kazuto
Nihon University, Col. of Science a. Technology
Dep. of Mechanical Engineering
1-8 Kanda Surugadai
Chiyoda-ku
Tokyo 101
Japan
seto@mech.cst.nihon-u.ac.jp
tel.: +81/3/3259-074
fax: +81/3/3259-074

Skelton, Robert
University of California, San Diego
9500 Dr Gilman
92093 La Jolla
USA
bobskelton@ucsd.edu
tel.: +1/858/822-1054
fax: +1/858/822-3107

Sobolev, Nikolai
Russ. Acad. of Sciences
Inst. f. Problems in Mechanics
pr. Vernadskogo101-1
119526 Moscow
Rus. Feder.
niksobolev@mail.ru
tel.: +7/095/4340207
fax: +7/095/9382048

Sobotka, Marion
TU München
Lehrstuhl f. Steuerungs- u. Regelungstechnik
Arcisstr. 21
80290 München
Germany
marion.sobotka@tum.de
tel.: +49/89/289-23405
fax: +49/89/289-28340

Söffker, Dirk
University Duisburg-Essen
Chair of Dynamics and Control
Lotharstr. 69
47057 Duisburg
Germany
soeffker@uni-duisburg.de
tel.: +49/203/379-3429
fax: +49/203/379-3027

Spaelter, Ulrich
LSRO Lausanne
Lab. de Syst. Robotiques Stat. 9
1015 Lausanne
Switzerland
ulrich.spaelter@epfl.ch
tel.: +41/21/693-3828
fax: +41/21/693-3866

Stépan, Gábor
Budapest Univ. of Technology and Economics
Dep. of Appl. Mechanics
Muegyetem rkt 3
1521 Budapest
Hungary
stepan@mm.bme.hu
tel.: +36/1/463-1369
fax: +36/1/463-3471

Stöbener, Uwe
Halcyonics GmbH
Research & Development
Tuchmacherweg 12
37079 Göttingen
Germany
us@halcyonics.de
tel.: +49/551/999062-20
fax: +49/551/999062-10

Thümmel, Thomas
TU München
Inst. of Applied Mechanics
Boltzmannstr. 15
85747 Garching
Germany
thuemmel@amm.mw.tum.de
tel.: +49/89/289-15205
fax: +49/89/289-15213

Ulbrich, Heinz
TU München
Inst. of Applied Mechanics
Boltzmannstr. 15
85747 Garching
Germany
ulbrich@amm.mw.tum.de
tel.: +49/89/289-15222
fax: +49/89/289-15213

Wisselmann, Dirk
BMW Group Forschung und Technik
Hanauer Str. 46
80992 München
Germany
tel.: +49/89/382-42469
fax: +49/89/382-37117

Wollherr, Dirk
TU München
Lst. f. Steuerungs- u. Regelungstechnik
Arcisstr. 21
80290 München
Germany
dw@tum.de
tel.: +49/89/289-23405
fax: +49/89/289-28340

Wong, Kevin
University of Utah
Dept. of Civil & Environmental Engineering
122S Central Campus Drive
EMRO 210
Salt Lake City, UT 84112
USA
kfwong@civil.utah.edu
tel.: +1/801/585-3379
fax: +1/801/585-5477

Yuan, Lihong
Tampere University of Technology
Korkeakoulunkatu 6
P.O. Box 589
33101 Tampere
Finland
lihong.yuan@tut.fi
tel.: +358/3/3115-4431
fax: +358/3/3115-2307

Zauner, Christoph
TU München
Lst. f. Leichtbau
Boltzmannstr. 15
85747 Garching
Germany
zauner@llb.mw.tum.de
tel.: +49/89/289-16112
fax: +49/89/289-16104

Mechanics

SOLID MECHANICS AND ITS APPLICATIONS

Series Editor: G.M.L. Gladwell

Aims and Scope of the Series

The fundamental questions arising in mechanics are: *Why?*, *How?*, and *How much?* The aim of this series is to provide lucid accounts written by authoritative researchers giving vision and insight in answering these questions on the subject of mechanics as it relates to solids. The scope of the series covers the entire spectrum of solid mechanics. Thus it includes the foundation of mechanics; variational formulations; computational mechanics; statics, kinematics and dynamics of rigid and elastic bodies; vibrations of solids and structures; dynamical systems and chaos; the theories of elasticity, plasticity and viscoelasticity; composite materials; rods, beams, shells and membranes; structural control and stability; soils, rocks and geomechanics; fracture; tribology; experimental mechanics; biomechanics and machine design.

1. R.T. Haftka, Z. Gürdal and M.P. Kamat: *Elements of Structural Optimization*. 2nd rev.ed., 1990 ISBN 0-7923-0608-2
2. J.J. Kalker: *Three-Dimensional Elastic Bodies in Rolling Contact*. 1990 ISBN 0-7923-0712-7
3. P. Karasudhi: *Foundations of Solid Mechanics*. 1991 ISBN 0-7923-0772-0
4. *Not published*
5. *Not published.*
6. J.F. Doyle: *Static and Dynamic Analysis of Structures*. With an Emphasis on Mechanics and Computer Matrix Methods. 1991 ISBN 0-7923-1124-8; Pb 0-7923-1208-2
7. O.O. Ochoa and J.N. Reddy: *Finite Element Analysis of Composite Laminates*. ISBN 0-7923-1125-6
8. M.H. Aliabadi and D.P. Rooke: *Numerical Fracture Mechanics*. ISBN 0-7923-1175-2
9. J. Angeles and C.S. López-Cajún: *Optimization of Cam Mechanisms*. 1991 ISBN 0-7923-1355-0
10. D.E. Grierson, A. Franchi and P. Riva (eds.): *Progress in Structural Engineering*. 1991 ISBN 0-7923-1396-8
11. R.T. Haftka and Z. Gürdal: *Elements of Structural Optimization*. 3rd rev. and exp. ed. 1992 ISBN 0-7923-1504-9; Pb 0-7923-1505-7
12. J.R. Barber: *Elasticity*. 1992 ISBN 0-7923-1609-6; Pb 0-7923-1610-X
13. H.S. Tzou and G.L. Anderson (eds.): *Intelligent Structural Systems*. 1992 ISBN 0-7923-1920-6
14. E.E. Gdoutos: *Fracture Mechanics*. An Introduction. 1993 ISBN 0-7923-1932-X
15. J.P. Ward: *Solid Mechanics*. An Introduction. 1992 ISBN 0-7923-1949-4
16. M. Farshad: *Design and Analysis of Shell Structures*. 1992 ISBN 0-7923-1950-8
17. H.S. Tzou and T. Fukuda (eds.): *Precision Sensors, Actuators and Systems*. 1992 ISBN 0-7923-2015-8
18. J.R. Vinson: *The Behavior of Shells Composed of Isotropic and Composite Materials*. 1993 ISBN 0-7923-2113-8
19. H.S. Tzou: *Piezoelectric Shells*. Distributed Sensing and Control of Continua. 1993 ISBN 0-7923-2186-3
20. W. Schiehlen (ed.): *Advanced Multibody System Dynamics*. Simulation and Software Tools. 1993 ISBN 0-7923-2192-8
21. C.-W. Lee: *Vibration Analysis of Rotors*. 1993 ISBN 0-7923-2300-9
22. D.R. Smith: *An Introduction to Continuum Mechanics*. 1993 ISBN 0-7923-2454-4
23. G.M.L. Gladwell: *Inverse Problems in Scattering*. An Introduction. 1993 ISBN 0-7923-2478-1

Mechanics

SOLID MECHANICS AND ITS APPLICATIONS

Series Editor: G.M.L. Gladwell

24. G. Prathap: *The Finite Element Method in Structural Mechanics.* 1993 ISBN 0-7923-2492-7
25. J. Herskovits (ed.): *Advances in Structural Optimization.* 1995 ISBN 0-7923-2510-9
26. M.A. González-Palacios and J. Angeles: *Cam Synthesis.* 1993 ISBN 0-7923-2536-2
27. W.S. Hall: *The Boundary Element Method.* 1993 ISBN 0-7923-2580-X
28. J. Angeles, G. Hommel and P. Kovács (eds.): *Computational Kinematics.* 1993 ISBN 0-7923-2585-0
29. A. Curnier: *Computational Methods in Solid Mechanics.* 1994 ISBN 0-7923-2761-6
30. D.A. Hills and D. Nowell: *Mechanics of Fretting Fatigue.* 1994 ISBN 0-7923-2866-3
31. B. Tabarrok and F.P.J. Rimrott: *Variational Methods and Complementary Formulations in Dynamics.* 1994 ISBN 0-7923-2923-6
32. E.H. Dowell (ed.), E.F. Crawley, H.C. Curtiss Jr., D.A. Peters, R. H. Scanlan and F. Sisto: *A Modern Course in Aeroelasticity.* Third Revised and Enlarged Edition. 1995 ISBN 0-7923-2788-8; Pb: 0-7923-2789-6
33. A. Preumont: *Random Vibration and Spectral Analysis.* 1994 ISBN 0-7923-3036-6
34. J.N. Reddy (ed.): *Mechanics of Composite Materials.* Selected works of Nicholas J. Pagano. 1994 ISBN 0-7923-3041-2
35. A.P.S. Selvadurai (ed.): *Mechanics of Poroelastic Media.* 1996 ISBN 0-7923-3329-2
36. Z. Mróz, D. Weichert, S. Dorosz (eds.): *Inelastic Behaviour of Structures under Variable Loads.* 1995 ISBN 0-7923-3397-7
37. R. Pyrz (ed.): *IUTAM Symposium on Microstructure-Property Interactions in Composite Materials.* Proceedings of the IUTAM Symposium held in Aalborg, Denmark. 1995 ISBN 0-7923-3427-2
38. M.I. Friswell and J.E. Mottershead: *Finite Element Model Updating in Structural Dynamics.* 1995 ISBN 0-7923-3431-0
39. D.F. Parker and A.H. England (eds.): *IUTAM Symposium on Anisotropy, Inhomogeneity and Nonlinearity in Solid Mechanics.* Proceedings of the IUTAM Symposium held in Nottingham, U.K. 1995 ISBN 0-7923-3594-5
40. J.-P. Merlet and B. Ravani (eds.): *Computational Kinematics '95.* 1995 ISBN 0-7923-3673-9
41. L.P. Lebedev, I.I. Vorovich and G.M.L. Gladwell: *Functional Analysis.* Applications in Mechanics and Inverse Problems. 1996 ISBN 0-7923-3849-9
42. J. Menčik: *Mechanics of Components with Treated or Coated Surfaces.* 1996 ISBN 0-7923-3700-X
43. D. Bestle and W. Schiehlen (eds.): *IUTAM Symposium on Optimization of Mechanical Systems.* Proceedings of the IUTAM Symposium held in Stuttgart, Germany. 1996 ISBN 0-7923-3830-8
44. D.A. Hills, P.A. Kelly, D.N. Dai and A.M. Korsunsky: *Solution of Crack Problems.* The Distributed Dislocation Technique. 1996 ISBN 0-7923-3848-0
45. V.A. Squire, R.J. Hosking, A.D. Kerr and P.J. Langhorne: *Moving Loads on Ice Plates.* 1996 ISBN 0-7923-3953-3
46. A. Pineau and A. Zaoui (eds.): *IUTAM Symposium on Micromechanics of Plasticity and Damage of Multiphase Materials.* Proceedings of the IUTAM Symposium held in Sèvres, Paris, France. 1996 ISBN 0-7923-4188-0
47. A. Naess and S. Krenk (eds.): *IUTAM Symposium on Advances in Nonlinear Stochastic Mechanics.* Proceedings of the IUTAM Symposium held in Trondheim, Norway. 1996 ISBN 0-7923-4193-7
48. D. Ieşan and A. Scalia: *Thermoelastic Deformations.* 1996 ISBN 0-7923-4230-5

Mechanics

SOLID MECHANICS AND ITS APPLICATIONS

Series Editor: G.M.L. Gladwell

49. J.R. Willis (ed.): *IUTAM Symposium on Nonlinear Analysis of Fracture*. Proceedings of the IUTAM Symposium held in Cambridge, U.K. 1997 ISBN 0-7923-4378-6
50. A. Preumont: *Vibration Control of Active Structures*. An Introduction. 1997 ISBN 0-7923-4392-1
51. G.P. Cherepanov: *Methods of Fracture Mechanics: Solid Matter Physics*. 1997 ISBN 0-7923-4408-1
52. D.H. van Campen (ed.): *IUTAM Symposium on Interaction between Dynamics and Control in Advanced Mechanical Systems*. Proceedings of the IUTAM Symposium held in Eindhoven, The Netherlands. 1997 ISBN 0-7923-4429-4
53. N.A. Fleck and A.C.F. Cocks (eds.): *IUTAM Symposium on Mechanics of Granular and Porous Materials*. Proceedings of the IUTAM Symposium held in Cambridge, U.K. 1997 ISBN 0-7923-4553-3
54. J. Roorda and N.K. Srivastava (eds.): *Trends in Structural Mechanics*. Theory, Practice, Education. 1997 ISBN 0-7923-4603-3
55. Yu.A. Mitropolskii and N. Van Dao: *Applied Asymptotic Methods in Nonlinear Oscillations*. 1997 ISBN 0-7923-4605-X
56. C. Guedes Soares (ed.): *Probabilistic Methods for Structural Design*. 1997 ISBN 0-7923-4670-X
57. D. François, A. Pineau and A. Zaoui: *Mechanical Behaviour of Materials*. Volume I: Elasticity and Plasticity. 1998 ISBN 0-7923-4894-X
58. D. François, A. Pineau and A. Zaoui: *Mechanical Behaviour of Materials*. Volume II: Viscoplasticity, Damage, Fracture and Contact Mechanics. 1998 ISBN 0-7923-4895-8
59. L.T. Tenek and J. Argyris: *Finite Element Analysis for Composite Structures*. 1998 ISBN 0-7923-4899-0
60. Y.A. Bahei-El-Din and G.J. Dvorak (eds.): *IUTAM Symposium on Transformation Problems in Composite and Active Materials*. Proceedings of the IUTAM Symposium held in Cairo, Egypt. 1998 ISBN 0-7923-5122-3
61. I.G. Goryacheva: *Contact Mechanics in Tribology*. 1998 ISBN 0-7923-5257-2
62. O.T. Bruhns and E. Stein (eds.): *IUTAM Symposium on Micro- and Macrostructural Aspects of Thermoplasticity*. Proceedings of the IUTAM Symposium held in Bochum, Germany. 1999 ISBN 0-7923-5265-3
63. F.C. Moon: *IUTAM Symposium on New Applications of Nonlinear and Chaotic Dynamics in Mechanics*. Proceedings of the IUTAM Symposium held in Ithaca, NY, USA. 1998 ISBN 0-7923-5276-9
64. R. Wang: *IUTAM Symposium on Rheology of Bodies with Defects*. Proceedings of the IUTAM Symposium held in Beijing, China. 1999 ISBN 0-7923-5297-1
65. Yu.I. Dimitrienko: *Thermomechanics of Composites under High Temperatures*. 1999 ISBN 0-7923-4899-0
66. P. Argoul, M. Frémond and Q.S. Nguyen (eds.): *IUTAM Symposium on Variations of Domains and Free-Boundary Problems in Solid Mechanics*. Proceedings of the IUTAM Symposium held in Paris, France. 1999 ISBN 0-7923-5450-8
67. F.J. Fahy and W.G. Price (eds.): *IUTAM Symposium on Statistical Energy Analysis*. Proceedings of the IUTAM Symposium held in Southampton, U.K. 1999 ISBN 0-7923-5457-5
68. H.A. Mang and F.G. Rammerstorfer (eds.): *IUTAM Symposium on Discretization Methods in Structural Mechanics*. Proceedings of the IUTAM Symposium held in Vienna, Austria. 1999 ISBN 0-7923-5591-1

Mechanics

SOLID MECHANICS AND ITS APPLICATIONS

Series Editor: G.M.L. Gladwell

69. P. Pedersen and M.P. Bendsøe (eds.): *IUTAM Symposium on Synthesis in Bio Solid Mechanics.* Proceedings of the IUTAM Symposium held in Copenhagen, Denmark. 1999 ISBN 0-7923-5615-2
70. S.K. Agrawal and B.C. Fabien: *Optimization of Dynamic Systems.* 1999 ISBN 0-7923-5681-0
71. A. Carpinteri: *Nonlinear Crack Models for Nonmetallic Materials.* 1999 ISBN 0-7923-5750-7
72. F. Pfeifer (ed.): *IUTAM Symposium on Unilateral Multibody Contacts.* Proceedings of the IUTAM Symposium held in Munich, Germany. 1999 ISBN 0-7923-6030-3
73. E. Lavendelis and M. Zakrzhevsky (eds.): *IUTAM/IFToMM Symposium on Synthesis of Nonlinear Dynamical Systems.* Proceedings of the IUTAM/IFToMM Symposium held in Riga, Latvia. 2000 ISBN 0-7923-6106-7
74. J.-P. Merlet: *Parallel Robots.* 2000 ISBN 0-7923-6308-6
75. J.T. Pindera: *Techniques of Tomographic Isodyne Stress Analysis.* 2000 ISBN 0-7923-6388-4
76. G.A. Maugin, R. Drouot and F. Sidoroff (eds.): *Continuum Thermomechanics.* The Art and Science of Modelling Material Behaviour. 2000 ISBN 0-7923-6407-4
77. N. Van Dao and E.J. Kreuzer (eds.): *IUTAM Symposium on Recent Developments in Non-linear Oscillations of Mechanical Systems.* 2000 ISBN 0-7923-6470-8
78. S.D. Akbarov and A.N. Guz: *Mechanics of Curved Composites.* 2000 ISBN 0-7923-6477-5
79. M.B. Rubin: *Cosserat Theories: Shells, Rods and Points.* 2000 ISBN 0-7923-6489-9
80. S. Pellegrino and S.D. Guest (eds.): *IUTAM-IASS Symposium on Deployable Structures: Theory and Applications.* Proceedings of the IUTAM-IASS Symposium held in Cambridge, U.K., 6–9 September 1998. 2000 ISBN 0-7923-6516-X
81. A.D. Rosato and D.L. Blackmore (eds.): *IUTAM Symposium on Segregation in Granular Flows.* Proceedings of the IUTAM Symposium held in Cape May, NJ, U.S.A., June 5–10, 1999. 2000 ISBN 0-7923-6547-X
82. A. Lagarde (ed.): *IUTAM Symposium on Advanced Optical Methods and Applications in Solid Mechanics.* Proceedings of the IUTAM Symposium held in Futuroscope, Poitiers, France, August 31–September 4, 1998. 2000 ISBN 0-7923-6604-2
83. D. Weichert and G. Maier (eds.): *Inelastic Analysis of Structures under Variable Loads.* Theory and Engineering Applications. 2000 ISBN 0-7923-6645-X
84. T.-J. Chuang and J.W. Rudnicki (eds.): *Multiscale Deformation and Fracture in Materials and Structures.* The James R. Rice 60th Anniversary Volume. 2001 ISBN 0-7923-6718-9
85. S. Narayanan and R.N. Iyengar (eds.): *IUTAM Symposium on Nonlinearity and Stochastic Structural Dynamics.* Proceedings of the IUTAM Symposium held in Madras, Chennai, India, 4–8 January 1999 ISBN 0-7923-6733-2
86. S. Murakami and N. Ohno (eds.): *IUTAM Symposium on Creep in Structures.* Proceedings of the IUTAM Symposium held in Nagoya, Japan, 3-7 April 2000. 2001 ISBN 0-7923-6737-5
87. W. Ehlers (ed.): *IUTAM Symposium on Theoretical and Numerical Methods in Continuum Mechanics of Porous Materials.* Proceedings of the IUTAM Symposium held at the University of Stuttgart, Germany, September 5-10, 1999. 2001 ISBN 0-7923-6766-9
88. D. Durban, D. Givoli and J.G. Simmonds (eds.): *Advances in the Mechanis of Plates and Shells The Avinoam Libai Anniversary Volume.* 2001 ISBN 0-7923-6785-5
89. U. Gabbert and H.-S. Tzou (eds.): *IUTAM Symposium on Smart Structures and Structonic Systems.* Proceedings of the IUTAM Symposium held in Magdeburg, Germany, 26–29 September 2000. 2001 ISBN 0-7923-6968-8

Mechanics

SOLID MECHANICS AND ITS APPLICATIONS

Series Editor: G.M.L. Gladwell

90. Y. Ivanov, V. Cheshkov and M. Natova: *Polymer Composite Materials – Interface Phenomena & Processes*. 2001 ISBN 0-7923-7008-2
91. R.C. McPhedran, L.C. Botten and N.A. Nicorovici (eds.): *IUTAM Symposium on Mechanical and Electromagnetic Waves in Structured Media*. Proceedings of the IUTAM Symposium held in Sydney, NSW, Australia, 18-22 Januari 1999. 2001 ISBN 0-7923-7038-4
92. D.A. Sotiropoulos (ed.): *IUTAM Symposium on Mechanical Waves for Composite Structures Characterization*. Proceedings of the IUTAM Symposium held in Chania, Crete, Greece, June 14-17, 2000. 2001 ISBN 0-7923-7164-X
93. V.M. Alexandrov and D.A. Pozharskii: *Three-Dimensional Contact Problems*. 2001 ISBN 0-7923-7165-8
94. J.P. Dempsey and H.H. Shen (eds.): *IUTAM Symposium on Scaling Laws in Ice Mechanics and Ice Dynamics*. Proceedings of the IUTAM Symposium held in Fairbanks, Alaska, U.S.A., 13-16 June 2000. 2001 ISBN 1-4020-0171-1
95. U. Kirsch: *Design-Oriented Analysis of Structures*. A Unified Approach. 2002 ISBN 1-4020-0443-5
96. A. Preumont: *Vibration Control of Active Structures*. An Introduction (2nd Edition). 2002 ISBN 1-4020-0496-6
97. B.L. Karihaloo (ed.): *IUTAM Symposium on Analytical and Computational Fracture Mechanics of Non-Homogeneous Materials*. Proceedings of the IUTAM Symposium held in Cardiff, U.K., 18-22 June 2001. 2002 ISBN 1-4020-0510-5
98. S.M. Han and H. Benaroya: *Nonlinear and Stochastic Dynamics of Compliant Offshore Structures*. 2002 ISBN 1-4020-0573-3
99. A.M. Linkov: *Boundary Integral Equations in Elasticity Theory*. 2002 ISBN 1-4020-0574-1
100. L.P. Lebedev, I.I. Vorovich and G.M.L. Gladwell: *Functional Analysis*. Applications in Mechanics and Inverse Problems (2nd Edition). 2002 ISBN 1-4020-0667-5; Pb: 1-4020-0756-6
101. Q.P. Sun (ed.): *IUTAM Symposium on Mechanics of Martensitic Phase Transformation in Solids*. Proceedings of the IUTAM Symposium held in Hong Kong, China, 11-15 June 2001. 2002 ISBN 1-4020-0741-8
102. M.L. Munjal (ed.): *IUTAM Symposium on Designing for Quietness*. Proceedings of the IUTAM Symposium held in Bangkok, India, 12-14 December 2000. 2002 ISBN 1-4020-0765-5
103. J.A.C. Martins and M.D.P. Monteiro Marques (eds.): *Contact Mechanics*. Proceedings of the 3rd Contact Mechanics International Symposium, Praia da Consolação, Peniche, Portugal, 17-21 June 2001. 2002 ISBN 1-4020-0811-2
104. H.R. Drew and S. Pellegrino (eds.): *New Approaches to Structural Mechanics, Shells and Biological Structures*. 2002 ISBN 1-4020-0862-7
105. J.R. Vinson and R.L. Sierakowski: *The Behavior of Structures Composed of Composite Materials*. Second Edition. 2002 ISBN 1-4020-0904-6
106. Not yet published.
107. J.R. Barber: *Elasticity*. Second Edition. 2002 ISBN Hb 1-4020-0964-X; Pb 1-4020-0966-6
108. C. Miehe (ed.): *IUTAM Symposium on Computational Mechanics of Solid Materials at Large Strains*. Proceedings of the IUTAM Symposium held in Stuttgart, Germany, 20-24 August 2001. 2003 ISBN 1-4020-1170-9

Mechanics

SOLID MECHANICS AND ITS APPLICATIONS

Series Editor: G.M.L. Gladwell

109. P. Ståhle and K.G. Sundin (eds.): *IUTAM Symposium on Field Analyses for Determination of Material Parameters – Experimental and Numerical Aspects.* Proceedings of the IUTAM Symposium held in Abisko National Park, Kiruna, Sweden, July 31 – August 4, 2000. 2003 ISBN 1-4020-1283-7
110. N. Sri Namachchivaya and Y.K. Lin (eds.): *IUTAM Symposium on Nonlinear Stochastic Dynamics.* Proceedings of the IUTAM Symposium held in Monticello, IL, USA, 26 – 30 August, 2000. 2003 ISBN 1-4020-1471-6
111. H. Sobieckzky (ed.): *IUTAM Symposium Transsonicum IV.* Proceedings of the IUTAM Symposium held in Göttingen, Germany, 2–6 September 2002, 2003 ISBN 1-4020-1608-5
112. J.-C. Samin and P. Fisette: *Symbolic Modeling of Multibody Systems.* 2003 ISBN 1-4020-1629-8
113. A.B. Movchan (ed.): *IUTAM Symposium on Asymptotics, Singularities and Homogenisation in Problems of Mechanics.* Proceedings of the IUTAM Symposium held in Liverpool, United Kingdom, 8-11 July 2002. 2003 ISBN 1-4020-1780-4
114. S. Ahzi, M. Cherkaoui, M.A. Khaleel, H.M. Zbib, M.A. Zikry and B. LaMatina (eds.): *IUTAM Symposium on Multiscale Modeling and Characterization of Elastic-Inelastic Behavior of Engineering Materials.* Proceedings of the IUTAM Symposium held in Marrakech, Morocco, 20-25 October 2002. 2004 ISBN 1-4020-1861-4
115. H. Kitagawa and Y. Shibutani (eds.): *IUTAM Symposium on Mesoscopic Dynamics of Fracture Process and Materials Strength.* Proceedings of the IUTAM Symposium held in Osaka, Japan, 6-11 July 2003. Volume in celebration of Professor Kitagawa's retirement. 2004 ISBN 1-4020-2037-6
116. E.H. Dowell, R.L. Clark, D. Cox, H.C. Curtiss, Jr., K.C. Hall, D.A. Peters, R.H. Scanlan, E. Simiu, F. Sisto and D. Tang: *A Modern Course in Aeroelasticity.* 4th Edition, 2004 ISBN 1-4020-2039-2
117. T. Burczyński and A. Osyczka (eds.): *IUTAM Symposium on Evolutionary Methods in Mechanics.* Proceedings of the IUTAM Symposium held in Cracow, Poland, 24-27 September 2002. 2004 ISBN 1-4020-2266-2
118. D. Ieşan: *Thermoelastic Models of Continua.* 2004 ISBN 1-4020-2309-X
119. G.M.L. Gladwell: *Inverse Problems in Vibration.* Second Edition. 2004 ISBN 1-4020-2670-6
120. J.R. Vinson: *Plate and Panel Structures of Isotropic, Composite and Piezoelectric Materials, Including Sandwich Construction.* 2005 ISBN 1-4020-3110-6
121. *Forthcoming*
122. G. Rega and F. Vestroni (eds.): *IUTAM Symposium on Chaotic Dynamics and Control of Systems and Processes in Mechanics.* Proceedings of the IUTAM Symposium held in Rome, Italy, 8–13 June 2003. 2005 ISBN 1-4020-3267-6
123. E.E. Gdoutos: *Fracture Mechanics. An Introduction. 2nd edition.* 2005 ISBN 1-4020-3267-6
124. M.D. Gilchrist (ed.): *IUTAM Symposium on Impact Biomechanics from Fundamental Insights to Applications.* 2005 ISBN 1-4020-3795-3
125. J.M. Hughe, P.A.C. Raats and S. C. Cowin (eds.): *IUTAM Symposium on Physicochemical and Electromechanical Interactions in Porous Media.* 2005 ISBN 1-4020-3864-X
126. H. Ding and W. Chen: *Elasticity of Transversely Isotropic Materials.* 2005ISBN 1-4020-4033-4
127. W. Yang (ed): *IUTAM Symposium on Mechanics and Reliability of Actuating Materials.* Proceedings of the IUTAM Symposium held in Beijing, China, 1–3 September 2004. 2005 ISBN 1-4020-4131-6
128. J.P. Merlet: *Parallel Robots.* 2005 ISBN 1-4020-4132-2

Mechanics

SOLID MECHANICS AND ITS APPLICATIONS

Series Editor: G.M.L. Gladwell

129. G.E.A. Meier and K.R. Sreenivasan (eds.): *IUTAM Symposium on One Hundred Years of Boundary Layer Research.* Proceedings of the IUTAM Symposium held at DLR-Göttingen, Germany, August 12–14, 2004. 2005 ISBN 1-4020-4149-7
130. H. Ulbrich and W. Günthner (eds.): *IUTAM Symposium on Vibration Control of Nonlinear Mechanisms and Structures.* 2005 ISBN 1-4020-4160-8
131. L. Librescu and O. Song: *Thin-Walled Composite Beams.* Theory and Application. 2005 ISBN 1-4020-3457-1

FSC
www.fsc.org
MIX
Papier | Fördert
gute Waldnutzung
FSC® C083411

Kolibri 360 GmbH
Ferdinand-Jühlke-Straße 7
99095 Erfurt, Deutschland
produktsicherheit@kolibri360.de